AF676353

Human Biomonitoring in Health Risk Assessment: Current Practices and Recommendations for the Future

Human Biomonitoring in Health Risk Assessment: Current Practices and Recommendations for the Future

Editor

Christophe Rousselle

MDPI • Basel • Beijing • Wuhan • Barcelona • Belgrade • Manchester • Tokyo • Cluj • Tianjin

Editor
Christophe Rousselle
The French Agency for Food,
Environmental and
Occupational Health & Safety
(ANSES)
France

Editorial Office
MDPI
St. Alban-Anlage 66
4052 Basel, Switzerland

This is a reprint of articles from the Special Issue published online in the open access journal *Toxics* (ISSN 2305-6304) (available at: https://www.mdpi.com/journal/toxics/special_issues/Human_Biomonitoring_Assessment).

For citation purposes, cite each article independently as indicated on the article page online and as indicated below:

LastName, A.A.; LastName, B.B.; LastName, C.C. Article Title. *Journal Name* **Year**, *Volume Number*, Page Range.

ISBN 978-3-0365-7004-4 (Hbk)
ISBN 978-3-0365-7005-1 (PDF)

© 2023 by the authors. Articles in this book are Open Access and distributed under the Creative Commons Attribution (CC BY) license, which allows users to download, copy and build upon published articles, as long as the author and publisher are properly credited, which ensures maximum dissemination and a wider impact of our publications.
The book as a whole is distributed by MDPI under the terms and conditions of the Creative Commons license CC BY-NC-ND.

Contents

toxics

MDPI

Editorial

Editorial for the Special Issue on "Human Biomonitoring in Health Risk Assessment: Current Practices and Recommendations for the Future"

Christophe Rousselle

The French Agency for Food, Environmental and Occupational Health & Safety (ANSES), 14 Rue Pierre et Marie Curie, 94701 Maisons-Alfort, France; christophe.rousselle@anses.fr

In most health risk assessment (HRA) frameworks for chemicals, the default approach for exposure assessment is to estimate the intake from different sources and different routes of exposure. These assessments are often made separately and then added when aggregate exposure scenarios are considered. Various uncertainties are associated with this approach and, depending on the scope of the assessment; it may over- or under-estimate the internal exposure. Having access to data on the internal exposure generated by human biomonitoring gives a complete picture of human exposure and can be used to enhance a chemical risk assessment by providing information on actual human exposure via multiple exposure pathways. An understanding of the contribution that different exposure routes make to the overall exposure (taking into account differences in exposure modifiers, such as age and gender, as well as profession) delivers a starting point for deciding on risk management measures under legislative silos.

Human biomonitoring (HBM) is an important and useful tool for assessing the internal exposure of humans resulting from aggregated exposure to chemicals. HBM can also provide a better estimate of exposure close to the target organ. The inclusion of HBM data could improve HRAs for the general population and workers. Although there are still a number of obstacles that hinder the use of HBM data in HRAs, the growing availability of HBM data offers an opportunity to improve and refine RAs.

This Special Issue intends to illustrate, using case studies, how HBM data could be used to better estimate the internal exposure and resulting risks. Case studies either on exposure from the use of consumer products (cosmetic products, non-food products, etc.) or from exposures via food or water, in the general population or among workers, have contributed to better identify the hurdles that prevent a broader use of HBM data in RAs. New tools such as physiologically based pharmacokinetic (PBPK) models, derivations of health-based guidance values, new approaches for integrating HBM with in vitro/in silico data, and adverse outcome pathways (AOP), through more accurate data on actual internal exposure, could improve HRAs.

The Special Issue "Human Biomonitoring in Health Risk Assessment: Current Practices and Recommendations for the Future"" collected and published 15 contributions focusing on the generation of biomonitoring data on chemicals in different populations, including workers, and on their use in a risk assessment context. The published papers comprise 4 reviews and 11 original articles, among which 8 reported results from the HBM4EU Initiative https://www.hbm4eu.eu/ accessed on 7 January 2023).

Measuring the concentration of a chemical in urine or blood could also be used as a more robust surrogate marker of systemic exposure compared to calculations based on external exposure scenarios. In their review, Willenbockel et al. (2022) [1] quantified the systemic exposure of humans exposed to pesticides used on specific cultures such as tree-grown produce, vines or hops. The analysis revealed that exposure was mainly driven by the application of pesticides and re-entry work, resulting in a higher exposure of

Citation: Rousselle, C. Editorial for the Special Issue on "Human Biomonitoring in Health Risk Assessment: Current Practices and Recommendations for the Future". *Toxics* **2023**, *11*, 168. https://doi.org/10.3390/toxics11020168

Received: 8 February 2023
Accepted: 8 February 2023
Published: 10 February 2023

Copyright: © 2023 by the author. Licensee MDPI, Basel, Switzerland. This article is an open access article distributed under the terms and conditions of the Creative Commons Attribution (CC BY) license (https://creativecommons.org/licenses/by/4.0/).

operators and workers than of residents and bystanders. In nearly all cases, the systemic exposure was below the relevant toxicological reference values.

Blood and urine are the most common matrices investigated in HBM studies as they are easily accessible. However, other biological matrices may be of some interest as they may constitute a better proxy for systemic exposure. In their review, Moriceau et al. (2023) [2] compared the serum/adipose tissue ratio for persistent organic pollutants (POPs) and sought to identify key factors that could explain why for some compounds the ratio differed from 1. All included studies reported high variability in the partition coefficients of POPs and the authors concluded that further research is still needed to better investigate the partitioning of POPs.

In broncho-alveolar lavages (BALs) it is also interesting to analyze inhaled bio-persistent particles, particularly when following occupational exposure. Forest et al. (2022) [3] investigated the relationship between the biomonitoring of nanoparticles in BALs, interstitial lung diseases and occupational exposure to these unintentionally released particles. Their results strengthen the array of presumptions on the contribution of some inhaled particles (from nano- to submicron-sized) to some idiopathic lung diseases.

To interpret the measured concentrations, reference values are necessary. Two types of references values were illustrated in this Special Issue health-based guidance values (HBM-GVs) and a background range in the general population. Meslin et al. (2022) [4] derived HBM-GVs for bisphenol S (BPS) and used results from HBM4EU studies to assess the risk due to the exposure to BPA and BPS in Europe, for workers and the general population. In the same way, Lamkarkach et al. (2022) [5] derived HBM-GVs for dimethylformamide (DMF) for exposed workers. A large database on DMF exposure from studies conducted at workplaces provided dose–response relationships between biomarker concentrations and health effects.

Measuring effect biomarkers and studying their associations with health outcome data could also provide insight into the extent to which chemicals and substitutes have an impact on health. Malondialdehyde (MDA) is an end-product of lipid oxidation in our cells and is present in all biological matrices, including exhaled breath condensate (EBC) and urine. Following a systematic literature review and meta-analysis, Turcu et al. (2022) [6] proposed reference values for MDA in EBC. However, defining the distribution of MDA in EBC measured in reference populations still represents a challenge due to the low number of available studies, different analytical methods, and the questionable methodological quality of many studies. In urine, following a systematic literature review and meta-analysis, Toto et al. (2022) [7] sought to establish urinary MDA concentration ranges for healthy adult populations based on reported values in the available scientific literature. The proposed urinary MDA values should be considered preliminary, as they are based mostly on moderate- to low-quality studies. Bergamaschi et al. (2022) [8] performed a pilot biomonitoring study in workers from a paint production plant exposed to pigment-grade titanium dioxide (TiO_2). They assessed pro-inflammatory cytokines (IL-1β, TNF-α, IL-10, and IL-17), surfactant protein D (SP-D) and Krebs von den Lungen-6 glycoprotein (KL-6) in EBC. Their findings suggest the need for an integrated approach relying on both personal exposure and biomarker assessment to improve the hazard characterization in occupational settings.

HBM data can also be used to perform risk assessments and suggest risk management measures if the internal exposure exceeds the HBM-GVs. Using BP-3 as an example, Rousselle et al. (2022) [9] investigated the benefits and limitations of the use of external versus internal exposure data to explore the usefulness of HBM to support the risk assessment of cosmetic ingredients. The results showed that both approaches did indicate a risk to human health under certain levels of exposure. They also highlighted the need for more robust exposure data on BP-3 and other cosmetic ingredients, and a standardized framework for incorporating HBM data in the risk assessment of cosmetic products. In the same way, Huuskonen et al. (2022) [10] demonstrated, by using the example of ortho-toludine, how HBM data can be used to assess cancer risks for workers and the general population. The urinary mass-balance methodology and generic exposure reconstruction PBPK modelling were both used to estimate the external intake levels corresponding to the observed urinary levels. Domínguez-Morueco (2022) [11] performed a HRA to estimate the risk associated with methylmercury exposure of vulnerable European populations using HBM data. As many data were missing to make a proper assessment, the authors concluded that further HRA refinement is needed with coordinated, widespread HBM data to account for the differences in European exposure and associated risks, so that interventions can be applied to protect vulnerable citizens. Plichta et al. (2022) [12] assessed the risk of dietary exposure to organophosphorus flame retardants (OPFRs) in children using HBM data and estimated how much dietary intake may contribute to the total exposure. The estimated exposure to OPFRs indicates a minimal health risk based on the current knowledge of the available exposure, kinetic and toxicity data.

When no HBM data are available, for example, in the case of prospective assessments, the internal dose of a pesticide in human can be estimated by the expected residue levels in food. Tarrazona et al. (2022) [13] compared consumers' internal exposure to chlorpyrifos based on the urinary marker 3,5,6-trichloro-2-pyridinol (TCPy), using two sources of monitoring data: monitoring of the food chain from the EU program and biomonitoring of European citizens from the HBM4EU initiative, supported by a literature search. Both methods confirmed a drastic reduction in the exposure levels from 2016 onwards. Finally, in Tarvares et al. (2022) [14], a mixture risk assessment based on published HBM data on Cr (VI), Ni and/or PAH occupational co-exposure in Europe was performed and in some situations the sum of risk quotients (SRQ) exceeded 1, which means that risk cannot be excluded, for example, in the welding or waste incineration sectors.

HBM can also be used to assess the impact of management measures taken to reduce external exposures. In their study, Buonaurio et al. (2022) [15] evaluated the effects of traffic on human health comparing biomonitoring data measured during the COVID-19 lockdown, when restrictions led to a 40% reduction in airborne benzene in Rome and a 36% reduction in road traffic, to the same parameters measured in 2021. Their results confirmed a decrease in succinic acid, a product of the Krebs cycle promoting inflammation.

To conclude, a better understanding of population exposure and the exposure of vulnerable groups against health-based human biomonitoring guidance values, provides the basis for effective risk management aimed at reducing the impacts on health. Europe's zero-pollution agenda should depart from an understanding of how European citizens' bodies are polluted with synthetic chemicals and make the reduction in chemical body burden and the associated health impacts a key priority.

We would like to thank all the authors for submitting their original contributions to this Special Issue. We greatly appreciate the support of all the reviewers who spent time evaluating and improving the quality of the manuscripts. We would also like to thank the editors of *Toxics* for their kind invitation, and Selena Li of the *Toxics* Editorial Office for her precious support.

Funding: This research received no external funding.

Institutional Review Board Statement: Not applicable.

Informed Consent Statement: Not applicable.

Data Availability Statement: Not applicable.

Conflicts of Interest: The authors declare no conflict of interest.

References

1. Willenbockel, C.T.; Prinz, J.; Dietrich, S.; Marx-Stoelting, P.; Weikert, C.; Tralau, T.; Niemann, L. A Critical Scoping Review of Pesticide Exposure Biomonitoring Studies in Overhead Cultures. *Toxics* **2022**, *10*, 170. [CrossRef] [PubMed]
2. Moriceau, M.-A.; Cano-Sancho, G.; Kim, M.; Coumoul, X.; Emond, C.; Arrebola, J.-P.; Antignac, J.-P.; Audouze, K.; Rousselle, C. Partitioning of Persistent Organic Pollutants between Adipose Tissue and Serum in Human Studies. *Toxics* **2022**, *11*, 41. [CrossRef] [PubMed]

3. Forest, V.; Pourchez, J.; Pélissier, C.; Durand, S.A.; Vergnon, J.-M.; Fontana, L. Relationship between Occupational Exposure to Airborne Nanoparticles, Nanoparticle Lung Burden and Lung Diseases. *Toxics* **2021**, *9*, 204. [CrossRef] [PubMed]
4. Meslin, M.; Beausoleil, C.; Zeman, F.A.; Antignac, J.-P.; Kolossa-Gehring, M.; Rousselle, C.; Apel, P. Human Biomonitoring Guidance Values (HBM-GVs) for Bisphenol S and Assessment of the Risk Due to the Exposure to Bisphenols A and S, in Europe. *Toxics* **2022**, *10*, 228. [CrossRef] [PubMed]
5. Lamkarkach, F.; Meslin, M.; Kolossa-Gehring, M.; Apel, P.; Garnier, R. Human Biomonitoring Initiative (HBM4EU): Human Biomonitoring Guidance Values Derived for Dimethylformamide. *Toxics* **2022**, *10*, 298. [CrossRef] [PubMed]
6. Turcu, V.; Wild, P.; Hemmendinger, M.; Sauvain, J.-J.; Bergamaschi, E.; Hopf, N.B.; Canu, I.G. Towards Reference Values for Malondialdehyde on Exhaled Breath Condensate: A Systematic Literature Review and Meta-Analysis. *Toxics* **2022**, *10*, 258. [CrossRef] [PubMed]
7. Toto, A.; Wild, P.; Graille, M.; Turcu, V.; Crézé, C.; Hemmendinger, M.; Sauvain, J.-J.; Bergamaschi, E.; Canu, I.G.; Hopf, N.B. Urinary Malondialdehyde (MDA) Concentrations in the General Population—A Systematic Literature Review and Meta-Analysis. *Toxics* **2022**, *10*, 160. [CrossRef] [PubMed]
8. Bergamaschi, E.; Bellisario, V.; Macrì, M.; Buglisi, M.; Garzaro, G.; Squillacioti, G.; Ghelli, F.; Bono, R.; Fenoglio, I.; Barbero, F.; et al. A Biomonitoring Pilot Study in Workers from a Paints Production Plant Exposed to Pigment-Grade Titanium Dioxide (TiO_2). *Toxics* **2022**, *10*, 171. [CrossRef] [PubMed]
9. Rousselle, C.; Meslin, M.; Berman, T.; Woutersen, M.; Bil, W.; Wildeman, J.; Chaudhry, Q. Using Human Biomonitoring Data to Support Risk Assessment of Cosmetic Ingredients—A Case Study of Benzophenone-3. *Toxics* **2022**, *10*, 96. [CrossRef] [PubMed]
10. Huuskonen, P.; Karakitsios, S.; Scholten, B.; Westerhout, J.; Sarigiannis, D.A.; Santonen, T. Health Risk Assessment of Ortho-Toluidine Utilising Human Biomonitoring Data of Workers and the General Population. *Toxics* **2022**, *10*, 217. [CrossRef] [PubMed]
11. Domínguez-Morueco, N.; Pedraza-Díaz, S.; González-Caballero, M.D.C.; Esteban-López, M.; de Alba-González, M.; Katsonouri, A.; Santonen, T.; Cañas-Portilla, A.; Castaño, A. Methylmercury Risk Assessment Based on European Human Biomonitoring Data. *Toxics* **2022**, *10*, 427. [CrossRef] [PubMed]
12. Plichta, V.; Steinwider, J.; Vogel, N.; Weber, T.; Kolossa-Gehring, M.; Murínová, L.P.; Wimmerová, S.; Tratnik, J.S.; Horvat, M.; Koppen, G.; et al. Risk Assessment of Dietary Exposure to Organophosphorus Flame Retardants in Children by Using HBM-Data. *Toxics* **2022**, *10*, 234. [CrossRef] [PubMed]
13. Tarazona, J.V.; González-Caballero, M.D.C.; de Alba-Gonzalez, M.; Pedraza-Diaz, S.; Cañas, A.; Dominguez-Morueco, N.; Esteban-López, M.; Cattaneo, I.; Katsonouri, A.; Makris, K.C.; et al. Improving the Risk Assessment of Pesticides through the Integration of Human Biomonitoring and Food Monitoring Data: A Case Study for Chlorpyrifos. *Toxics* **2022**, *10*, 313. [CrossRef] [PubMed]
14. Tavares, A.M.; Viegas, S.; Louro, H.; Göen, T.; Santonen, T.; Luijten, M.; Kortenkamp, A.; Silva, M.J. Occupational Exposure to Hexavalent Chromium, Nickel and PAHs: A Mixtures Risk Assessment Approach Based on Literature Exposure Data from European Countries. *Toxics* **2022**, *10*, 431. [CrossRef] [PubMed]
15. Buonaurio, F.; Borra, F.; Pigini, D.; Paci, E.; Spagnoli, M.; Astolfi, M.L.; Giampaoli, O.; Sciubba, F.; Miccheli, A.; Canepari, S.; et al. Biomonitoring of Exposure to Urban Pollutants and Oxidative Stress during the COVID-19 Lockdown in Rome Residents. *Toxics* **2022**, *10*, 267. [CrossRef] [PubMed]

Disclaimer/Publisher's Note: The statements, opinions and data contained in all publications are solely those of the individual author(s) and contributor(s) and not of MDPI and/or the editor(s). MDPI and/or the editor(s) disclaim responsibility for any injury to people or property resulting from any ideas, methods, instructions or products referred to in the content.

toxics

MDPI

Article

Occupational Exposure to Hexavalent Chromium, Nickel and PAHs: A Mixtures Risk Assessment Approach Based on Literature Exposure Data from European Countries

Ana Maria Tavares [1,2], Susana Viegas [3,4], Henriqueta Louro [1,2], Thomas Göen [5], Tiina Santonen [6], Mirjam Luijten [7], Andreas Kortenkamp [8] and Maria João Silva [1,2,*]

1 Departamento de Genética Humana, Instituto Nacional de Saúde Dr. Ricardo Jorge (INSA), Avenida Padre Cruz, 1649-016 Lisbon, Portugal; anamptavares@gmail.com (A.M.T.); henriqueta.louro@insa.min-saude.pt (H.L.)
2 ToxOmics–Centre for Toxicogenomics and Human Health, NOVA Medical School, Universidade NOVA de Lisboa, Campo dos Mártires da Pátria, 130, 1169-056 Lisbon, Portugal
3 Public Health Research Centre, NOVA National School of Public Health, Universidade NOVA de Lisboa, 1600-560 Lisbon, Portugal; susana.viegas@ensp.unl.pt
4 Comprehensive Health Research Center (CHRC), 1169-056 Lisbon, Portugal
5 Institute of Occupational, Social and Environmental Medicine (IPASUM), University Erlangen-Nürnberg, Henkestraße 9-11, 91054 Erlangen, Germany; thomas.goeen@fau.de
6 Finnish Institute of Occupational Health, FI-00250 Helsinki, Finland; tiina.santonen@ttl.fi
7 Centre for Health Protection, National Institute for Public Health and the Environment, 3720 BA Bilthoven, The Netherlands; mirjam.luijten@rivm.nl
8 Centre for Pollution Research and Policy, College of Health, Medicine and Life Sciences, Brunel University London, Kingston Lane, Uxbridge, London UB8 3PH, UK; andreas.kortenkamp@brunel.ac.uk
* Correspondence: m.joao.silva@insa.min-saude.pt; Tel.: +351-217519200

Citation: Tavares, A.M.; Viegas, S.; Louro, H.; Göen, T.; Santonen, T.; Luijten, M.; Kortenkamp, A.; Silva, M.J. Occupational Exposure to Hexavalent Chromium, Nickel and PAHs: A Mixtures Risk Assessment Approach Based on Literature Exposure Data from European Countries. *Toxics* **2022**, *10*, 431. https://doi.org/10.3390/toxics10080431

Academic Editor: Giovanna Tranfo

Received: 15 June 2022
Accepted: 25 July 2022
Published: 29 July 2022

Publisher's Note: MDPI stays neutral with regard to jurisdictional claims in published maps and institutional affiliations.

Copyright: © 2022 by the authors. Licensee MDPI, Basel, Switzerland. This article is an open access article distributed under the terms and conditions of the Creative Commons Attribution (CC BY) license (https://creativecommons.org/licenses/by/4.0/).

Abstract: Hexavalent chromium (Cr(VI)), nickel (Ni) and polycyclic aromatic hydrocarbons (PAHs) are genotoxic co-occurring lung carcinogens whose occupational health risk is still understudied. This study, conducted within the European Human Biomonitoring Initiative (HBM4EU), aimed at performing a mixtures risk assessment (MRA) based on published human biomonitoring (HBM) data from Cr(VI), Ni and/or PAHs occupational co-exposure in Europe. After data extraction, Risk Quotient (RQ) and Sum of Risk Quotients (SRQ) were calculated for binary and ternary mixtures to characterise the risk. Most selected articles measured urinary levels of Cr and Ni and a SRQ > 1 was obtained for co-exposure levels in welding activities, showing that there is concern regarding co-exposure to these substances. Similarly, co-exposure to mixtures of Cr(VI), Ni and PAHs in waste incineration settings resulted in SRQ > 1. In some studies, a low risk was estimated based on the single substances' exposure level (RQ < 1), but the mixture was considered of concern (SRQ > 1), highlighting the relevance of considering exposure to the mixture rather than to its single components. Overall, this study points out the need of using a MRA based on HBM data as a more realistic approach to assess and manage the risk at the workplace, in order to protect workers' health.

Keywords: human biomonitoring (HBM); European Human Biomonitoring Initiative (HBM4EU); co-exposure; workplace; mixture risk assessment (MRA)

1. Introduction

Industrial activities such as metal processing, electroplating, chromate production, welding, and thermal cutting make use of chromium and represent the main sources of hexavalent chromium (Cr(VI)) occupational exposure [1,2]. Chromium exposure occurs mainly by inhalation, making the lungs the main target organ for its effects [1]. The occupational exposure to Cr(VI) has been associated with nasal and sinus cancer and with lung, trachea, and bronchus cancers [2]. Cr(VI) compounds have been classified by IARC as carcinogenic to humans (group 1) [3]. The induction of DNA strand breaks, inter-strand cross-links and DNA adducts are deemed to mediate Cr(VI) genotoxicity [1]. This effect is also due to the contribution of oxidative damage originated from reactive oxygen species (ROS) that arise from a series of intracellular Cr(VI) reduction steps. Cytogenetic alterations, such as chromosomal breaks and micronuclei (MN) have also been reported and suggested to be the outcome of double-strand breaks induction [3].

Nickel has been widely used in stainless steel manufacturing, electroplating, foundry applications, printing inks, and prostheses manufacturing, which causes internal levels of Ni significantly to be higher among exposed workers [1]. Exposure by inhalation, ingestion or skin contact can occur in nickel industries [4]. Occupationally inhaled Ni is typically associated with immunological sensitisation, epithelial dysplasia and asthma, and can also cause nasal and lung fibrosis and cancer [1,5,6]. Similarly to Cr(VI), Ni was classified by IARC as carcinogenic to humans [4]. Its genotoxic effects include formation of DNA strand breaks, DNA cross-links and chromosomal aberrations. Both Cr(VI) and Ni can also interfere with DNA repair mechanisms, enhancing the genotoxic effect of other agents [1,7].

Polycyclic aromatic hydrocarbons (PAHs) are widespread environmental contaminants continuously formed during incomplete combustion or pyrolysis of organic material. Their major environmental sources are heating, motor-vehicle exhaust and industrial emissions such as industrial power generation, incinerators, the production of coal tar, coke and asphalt, and petroleum catalytic cracking [8,9]. Recent technology developments have also been using pyrolysis for the production of biofuel from microalgal biomass. However, under inappropriate conditions, this can also result in the emission of PAHs [10,11]. Occupational exposure to PAHs may occur by breathing exhaust fumes, by ingestion of contaminated food and by dermal absorption due to handling of contaminated materials and surfaces. High levels of pollutant mixtures containing PAHs can cause irritation, inflammation, nausea, and vomiting. A long exposure may cause an increased risk of skin and lung cancer. Similarly to Cr(VI) and Ni, PAHs such as benzo(a)pyrene (BaP) can cause ROS-related but also PAH-specific DNA adducts, and chromosomal alterations, reflected in the induction of chromosomal breaks, sister chromatic exchanges (SCEs) and MN [12,13]. Although IARC classifications vary across the existing PAHs, with many of them classified as probably and possibly carcinogenic to humans (group 2A and 2B, respectively) [14], BaP is so far the only one classified as carcinogenic to humans [15].

Occupational co-exposure to Cr(VI) and Ni has been well documented in the literature, due to its frequent occurrence in the same workplace, namely, during welding operations [16,17]. Moreover, a study in Finland demonstrated that Cr, Ni, BaP and other PAHs were amongst the main substances to which welders and flame cutters were co-exposed between 2007 and 2009 [18]. Workers from aluminium production, iron and steel founding industries, waste incineration sites and coke oven plants can be also occupationally co-exposed to PAHs and to metals including Cr and Ni [19,20]. In addition, it has been demonstrated that workers performing painting activities can be highly co-exposed to mixtures of Cr and PAHs as common paint components [19].

Despite the acknowledged evidence of co-exposure, the combined health effects caused by these three substances have been rarely investigated. Current regulatory practices are usually based on single chemical substances [21,22], without integrating the possibility co-exposure or aggregated exposure. The few existing in vitro and in vivo data suggest that combinations of Cr, Ni and PAHs can lead to adverse effect that are higher than those observed for each single substance [23–25]. The multiple chemicals involved in exposure scenarios may lead to myriad interactions with a wide array of underlying mechanisms. This may ultimately result in different health outcomes from those expected from the single substances [26]. In some cases, only one or few components may dominate the overall mixture hazard, and therefore, the identification of the main active components, i.e., the drivers of mixture toxicity, may allow the implementation of more efficient risk reduction measures [27]. Thus, failing to account for the combined effects of co-exposures can lead to an underestimation of the risk. In this sense, the recently published "Chemicals Strategy for Sustainability" claims the need for integrating mixtures risk assessment (MRA) more generally into chemical risk assessment in the scope of different regulatory frameworks [28].

Approaches for a more holistic mixtures risk assessment (MRA) currently remain as research challenges [27]. Knowing the mode of action (MoA) of a substance is essential to perform an accurate MRA. In this regard, two main mathematical models can be used to assess the effects of chemical mixtures: one assumes that individual substances act via a similar MoA and applies dose or concentration addition (CA), while the other model assumes dissimilar MoAs (independent action (IA)) and applies response addition. Synergism and antagonism due to the interaction of substances in a mixture can be defined in relation to this additivity assumption. In this context, the term interaction (synergistic or antagonistic effects) includes all forms of joint actions that deviate from either dose or response addition (Figure 1) [29,30]. If a group of chemicals affect the same toxicological endpoint and/or have the same MoA, as in the case of Cr(VI), Ni and PAHs, it seems plausible to generate a mixtures risk indicator, which requires individual data [31–33]. However, the distinction between similar and dissimilar MoAs in real exposure scenarios with complex chemical mixtures is often unclear [29,34]. There is now a growing consensus that MRAs should start from the default assumption of CA for all mixture components, regardless of the MoAs. If this indicates a significant risk, refined MoA-based assessments may be conducted where the necessary data is available [34,35].

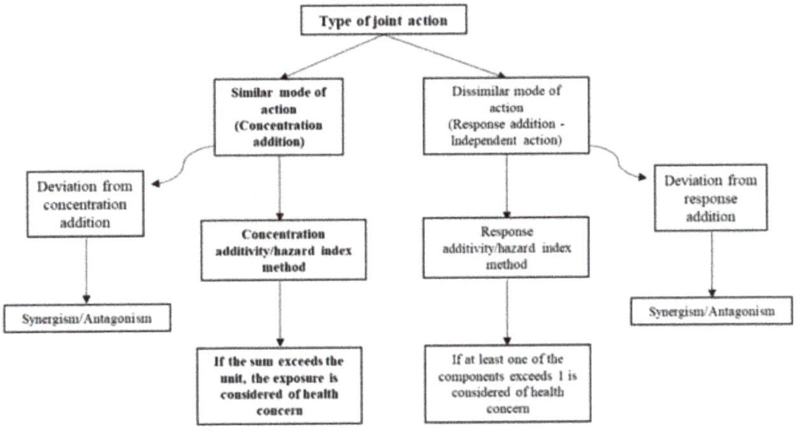

Figure 1. Flowchart representing the types of joint action (similar and dissimilar) and the respective applicable methods in mixtures risk assessment.

The use of human biomonitoring (HBM) data has been gaining momentum in identifying realistic co-exposure patterns in the same environmental or human samples [27]. Total Cr measurements in urine have been used as indicators of occupational exposure [36]; correlation has been also reported between air and urinary nickel levels in several occupational settings [37]; regarding PAHs, the urinary metabolite 1-hydroxypyrene (1-OHP) has been the most frequently analysed biomarker in occupational studies [38]. Besides measuring co-exposure, it is important to detect its early biological effects, and for that purpose, the combination of exposure data with effect biomarkers is essential to assess potential human health effects. Most of these biomarkers do not point to an individual cause, but they may reflect an integrated effect of an aggregated exposure [39]. Micronuclei in human peripheral blood lymphocytes, alkaline comet assay, and oxidative stress biomarkers have been used to assess the genotoxic and carcinogenic effects of several substances [40]. However, few occupational HBM studies present data on effect biomarkers and link them to exposure data.

In the European Human Biomonitoring Initiative (HBM4EU), several substances were identified as research priorities, including Cr(VI), PAHs, and chemical mixtures (https://www.hbm4eu.eu/about-hbm4eu/, accessed on 10 March 2022). Within the HBM4EU scope, a recent multicentric study on the occupational exposure to Cr(VI) has been conducted in a harmonised way in several European countries [40,41], and other field studies are currently underway to provide information on human exposure and co-exposure. This study, developed under the HBM4EU initiative, aimed to perform a MRA for the co-exposure to Cr(VI), Ni and/or PAHs in several occupational settings in Europe, based on external exposure and HBM data gathered from the literature. This work intended to contribute to the development of a consistent and uniform framework applicable to future studies on mixtures exposure and health outcomes.

2. Materials and Methods

2.1. Literature Search and Data Collection

The literature searches were conducted in PubMed® database between May and June 2020, using combinations of key words and/or Medical Subject Headings (MeSH) terms to find articles presenting HBM measurements. Different search strings for the possible combinations (binary and ternary) of Cr(VI), Ni and/or PAHs in occupational settings were used and are detailed in Supplementary Table S1.

Our searches were filtered to retrieve articles published between 2000 and 2020. This time window allowed us to account for the most recent working conditions and technological developments and for new risk management measures, while considering the influence of the regulatory frameworks presently in place. Only peer-reviewed articles which had abstracts and were written in English, Spanish, French or Portuguese were considered. Books or book chapters, comments, editorials, reviews, guidelines, reports, newspaper articles and case studies were excluded from the analysis. Articles meeting the following inclusion criteria were selected: (i) studies presenting biomarkers of exposure for Cr(VI) + Ni, Cr(VI) + PAHs, Ni + PAHs, or Cr(VI) + Ni + PAHs; (ii) HBM data obtained within occupational settings; (iii) studies including data on effect biomarkers (studies with data for a single substance only, but presenting effect biomarkers were still considered); iv) studies performed in EU countries. The following exclusion criteria were also defined: (i) studies conducted outside the EU (UK was still considered); (ii) studies not presenting HBM data for Cr(VI), Ni and/or PAHs, or presenting only graphical data; (iii) studies conducted in non-occupational settings (e.g., environmental studies). Following articles' retrieval, titles and abstracts were first screened, and the selected articles proceeded to full text analysis.

In a second approach, references cited in the retrieved articles were searched, and only those meeting the inclusion criteria and presenting biomarkers of exposure for the three substances were selected. Data on biomarkers of exposure, effect biomarkers (e.g., comet assay, micronucleus assay), occupational settings, and exposure scenarios were extracted from the selected articles, whenever described. All data extracted was presented as arithmetic means, geometric means or medians.

2.2. Determination of the Sum of Risk Quotients

The Sum of Risk Quotients (SRQ), also called Hazard Index (HI) [40], is a regulatory approach to component-based MRA derived from the CA reference model. The SRQ was calculated for each mixture, whenever possible, according to Equation (1):

$$\mathrm{SRQ} = \sum_{i=1}^{n} \frac{\mathrm{EL_i}}{\mathrm{AL_i}} \tag{1}$$

The Risk Quotients (RQ), i.e., the ratio of a substance/metabolite concentration in a biological specimen (EL), and its acceptable level (AL), ($\mathrm{EL_i/AL_i}$), were calculated using data from the selected studies [31,32,42].

Table 1 presents the tolerable risk levels for Cr(VI), Ni and BaP in air used for RQ and SRQ calculations. Because there is no consensus on the acceptable or tolerable risk levels in Europe for occupational activities involving carcinogenic substances, an expert ad hoc group met to debate this topic, and a decision on the use of the tolerable risk level of 4:1000, as defined in the German Technical Rules for Hazardous Substances produced by the Committee on Hazardous Substances from the Federal Institute for Occupational Safety and Health), was reached. An additional advantage was that the "Risk Concept for Carcinogenic Substances" of the AGS, which offers a benchmark for comparison and assessment of exposure to carcinogenic substances, included limit values (in air) for the three components of the mixture under study. It must be stated that these levels still contain a residual risk for cancer [43]. Calculations using Equation (1) were also performed using HBM data obtained from the literature and from documents produced by the Risk Assessment Committee of the European Chemicals Agency (Table 1). For exposure values below the limit of detection (LOD), LOD/2 was used. The Student's t-test was used to compare SRQ values obtained through air and HBM measurements. All data was compiled and analysed using MS Excel 2016.

Table 1. Tolerable risk levels and limit values used for risk assessment calculations.

Substances	Tolerable Levels in Air (μg/m^3) [a]	Limit Values in Urine	
		μg/g Creatinine	μg/L
Chromium (VI)	1.0	1.20 [c]	1.63 [d]
Nickel [b]	6.0	2.21 [d]	3.00 [e]
Benzo(a)pyrene	0.07	1.03 [d] (for 1-OHP)	1.40 [f] (for 1-OHP)

[a] Limit values established in the Technical Rules for Hazardous Substances, AGS, 2021 [43]; [b] Nickel metal, nickel oxide, nickel carbonate, nickel sulphide, sulfidic ores [43]; [c] Based on the regression analysis conducted by Viegas et al. (2022) on U-Cr and inhalable Cr(VI) levels found in welders; equation: y = 0.647 + 0.541x [41]; [d] Conversion from μg/g creatinine to μg/L and from μg/L to μg/g creatinine using the standard creatinine value (~1.36 g/L urine); [e] Value recommended by the SCOEL for a biological guidance value (BGV) for nickel, as referred in [37]; [f] Calculated based on the dose–response relationship between BaP concentration in air and 1-OHP in urine; RAC note from ECHA, 2018 [38].

An RQ or an SRQ < 1 indicates low concern after workers' exposure to the single substance or the mixture, respectively. Conversely, an RQ or an SRQ > 1 indicates that the tolerable risk level was exceeded, and that exposure is likely to have an impact on workers' health [31,32].

3. Results

3.1. Characteristics of the Studies

Overall, 356 articles were retrieved from the PubMed search. After removing duplicates, applying selection criteria, and conducting a second search throughout the chosen articles' references, a total of 22 articles were selected. A flowchart of this selection process is depicted in Figure 2.

Of the 22 articles included in this study, 7 were published before 2010, and only 5 were published in the last five years. Biomarkers of exposure in urine were presented in 19 out of 22 articles, while 5 articles referred biomarkers in blood (blood, plasma and/or erythrocytes) [44–48], 5 articles included biomarkers in exhaled breath condensate (EBC) [49–53], and only 1 article presented measurements in hair and saliva [46]. Besides biomarkers of exposure to Cr, Ni and/or PAHs, most studies also assessed biomarkers of exposure to other substances, such as other heavy metals (Mn, Fe, Al, Hg, As, Be, Pb, Cu, Cd, Co, V), polychlorinated biphenyls (PCBs), and organochlorinated compounds [46–52,54–62]. Table 2 presents urinary levels (median or mean values) retrieved from the selected studies, as well as the description of the respective settings and exposure scenarios. Welding activities were the most frequently assessed, including 9 studies [16,17,44,47,50,52–55], followed by waste incineration assessed in 4 studies in the context of waste management operations [57,59–61].

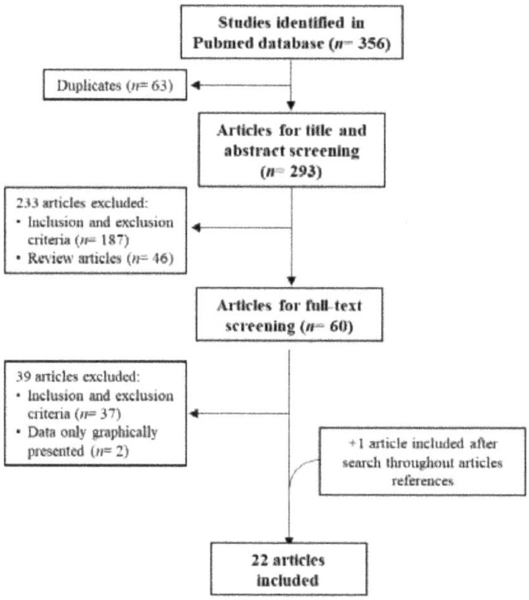

Figure 2. Flowchart of the selection process.

Table 2. Occupational settings, exposure scenarios and median (±SD) exposure values for exposure biomarkers in urine reported in the selected studies (n = 19).

Occupational Setting	Activities	Biomarkers of Exposure in Urine (μg/g Creatinine)			Ref.
		U-Cr	U-Ni	U-1-OHP	
Waste incineration	Incinerator operations, boiler and furnace maintenance, control panel, and waste-gas-washing	0.39 ± 0.24 [a]	3.7 ± 1.9 [a]	<0.04 ± 0.3 [a]	[57]
		0.26 [a]	3.03 [a]	0.1 [a]	[59]
			4.8 ± 3.2 (<4; 11.0) [a]	0.20 [a]	[60]
		0.50 ± 0.58 (exposure 1–3 months) 1.89 ± 4.59 (exposure 3–8 months) 0.26 ± 0.24 [a] (exposure 8–11 months)	9.18 ± 7.47 (exposure 1–3 months) 12.12 ± 8.31 (exposure 3–8 months) 7.69 ± 5.57 [a] (exposure 8–11 months)		[61]
Manufacture of railway vehicles	Welding [e]	0.72 (IQR 0.58–1.20)	1.56 (IQR 1.01–2.48)		[52]
Welding industries		0.32 (IQR 0.97) (pre-shift) 0.54 (IQR 1.39) (post-shift)	1.11 (IQR 1.6) (pre-shift) 1.47 (IQR 2.27) (post-shift)		[50]
		2.3 (0.7–7.3) [b]	1.5 (0.5–3.2) [b]		[55]
	Flux cored arc welding	3.8 (0.48–18.0) (first void) 3.2 (<0.24–30.1) (before work) 3.96 (0.34–40.7) (after work) [a]	1.9 (1.05–5.59) (first void) 2.7 (1.11–4.37) (before work) 2.5 (0.56–5.0) (after work) [a]		[44]

Table 2. *Cont.*

Occupational Setting	Activities	Biomarkers of Exposure in Urine (μg/g Creatinine)			Ref.
		U-Cr	U-Ni	U-1-OHP	
Shipyards and other industries	Flux cored and gas metal arc welding	<1.35 (IQR <0.74–<3.24)	<2.56 (IQR <1.37–<4.39)		[17]
	Gas metal arc welding	0.39	1.57		[54]
	Tungsten inert gas welding	0.71 (0.36–1.27) (before Friday shift) 0.74 (0.41–1.21) (after Friday shift) 0.59 (0.26–1.00) (after weekend break)	0.76 (0.37–1.40) (before Friday shift) 1.11 (0.59–0.79) (after Friday shift) 0.83 (0.31–1.38) (after weekend break)		[53]
		0.85 (IQR 0.53–1.32) (pre-shift) 0.90 (IQR 0.56–1.55) (post-shift)	1.53 (IQR 1.05–3.37) (pre-shift) 1.67 (IQR 0.89–2.97) (post-shift)		[16]
	Gas metal arc welders with massive or flux cored wire of stainless steel; Flux-cored arc welding of mild steel; Tungsten inert gas welding	0.88 [d]	2.07 [d]		[47]
	Stainless steel grinding	1.6 (<0.15–4.6) (first void) 1.40 (<0.14–4.5) (before work) 1.40 (<0.13–5.5) (post shift)	3.79 (0.68–10.6) (first void) 3.39 (0.25–11.1) (before work) 4.56 (<0.53–11.5) (post shift)		[45]
Harbour	Harbour dredgers and lighters	0.22 [d]	0.7 (0.1–7.3)		[56]
Iron and steel industry	Production, polishing and shaving of stainless-steel vessels and other metallurgical processes	0.42 (0.10–19.65)	0.25 (0.12–51.01)		[46]
Dental laboratory	Preparation of prostheses and of metal constructions for dental crowns	0.14 [c,d]	4.35 [c,d]		[62]
Electroplating companies	Processes of electroplating, preparatory work, maintenance activities, polishing of electroplated items, and ancillary tasks	1.10 [d] 1.47 (electroplating workers) [d]	4.26 [d] 4.88 (electroplating workers) [d]		[63]
Workshops specializing in the processing of beryls	Gemstone cutting		≤4 h/week: 1.18 (pre-shift); 0.85 (post-shift) [d] >4 h/week: 0.81 (pre-shift); 0.66 (post-shift) [d]		[58]

U-Cr, Urinary chromium; U-Ni, Urinary nickel; U-1-OHP, Urinary 1-hydroxypyrene; IQR, Interquartile Range; [a] Mean value; [b] Geometric means; [c] Calculated arithmetic mean of pooled urine samples from exposed workers; [d] Reported exposure values converted to μg/g creatinine; [e] Unspecified welding activity.

3.2. Overall Exposure Biomarkers in Urine

Most studies with data on exposure biomarkers in urine reported levels of both urinary chromium (U-Cr) and Ni (U-Ni) ($n = 15$) [16,17,44–47,50,52–56,61–63], as displayed in Table 2.

U-Cr was measured in 17 studies. Five studies reported exposure values exceeding the limits previous calculated (Table 2), among workers performing incineration operations, welding, namely flux cored arc welding, stainless steel grinding and electroplating. The highest value was measured among welders at the end of the workday (3.96 μg/g creatinine) [44]. The lowest exposure value (0.14 μg/g creatinine) was measured among dental technicians who prepared prostheses and metal constructions for dental crowns at a dental laboratory [62]. A low exposure level was also observed among harbour workers who worked on lighters and suction dredgers (0.22 μg/g creatinine) [56].

Similarly, U-Ni was determined in 19 studies. Limit values were exceeded in 8 studies (Table 2), among industry workers performing incineration operations, flux cored arc welding, stainless steel grinding, and electroplating, and also among workers performing prosthesis preparation. The highest level of U-Ni measured (12.12 ± 8.31 μg/g creatinine) was observed in workers performing incineration operations and other related activities for more than 3 months and for less than 8 months, in a hazard waste incinerator [61]. The lowest levels of U-Ni levels (0.25 μg/g creatinine) were detected among workers in the production, polishing and shaving of stainless-steel vessels and other metallurgical processes at an iron and steel industry [46].

Levels of PAH metabolites, namely, 1-OHP in urine (U-1-OHP), were assessed in only 3 studies, all conducted with workers performing operations at waste incinerators. None of the exposure levels reported in the studies exceeded the limit values considered in this study (Table 2).

3.3. Risk Quotients in Welding Activities with Exposure to Cr(VI) and Ni

Table 3 shows the air levels of Cr(VI) and Ni retrieved from seven studies on welding activities. The exposure levels per type of welding activity, as well as the respective RQ and SRQ, were calculated using Equation (1) (see Section 2) and the tolerable risk levels of substances in air. The SRQ values obtained ranged from 0.12 to 252.8, the latter for settings of Gas Metal Arc Welding with massive or flux cored wire of stainless steel (high exposure group). The SRQ exceeded a value of 1 for ten out of fourteen combinations of Cr and Ni exposure levels (Table 3).

All studies conducted on welding activities also reported U-Cr and U-Ni levels, but none of those considered a combined analysis of the 3 substances (i.e., Cr, Ni, PAHs). Table 4 lists the exposure levels in urine per type of welding activity, as well as the respective RQ and SRQ calculated using Equation (1) and limit values presented in Table 1. An SRQ > 1 was obtained in fourteen out of sixteen combinations of U-Cr and U-Ni exposure levels. In ten combinations, the SRQ exceeded a value of one but the respective U-Cr and U-Ni RQ were below the limit values (RQ < 1). The highest SRQ was estimated among "high exposure" workers, who performed gas metal arc welding with massive or flux cored wire of stainless steel (SRQ = 10.9). In another study, the second highest SRQ was also obtained for workers performing flux cored arc welding (SRQ = 4.43) (Table 4).

When comparing the results displayed in Tables 3 and 4 for the same studies, the SRQ obtained using air measurements were significantly higher (about 5-fold higher) than the SRQ determined using HBM data ($p = 0.030$, Student's t-test). The distribution of SRQ from air measurements and HBM data (leaving out the high exposure group) is displayed in Figure 3. A much higher variability across studies was observed when considering the SRQ values obtained from air measurements.

Table 3. External exposure levels of Cr and Ni (air measurements) of workers performing welding activities, and values obtained for RQ and SRQ.

Welding Activity	Cr		Ni		SRQ	Ref.
	Exposure Level (μg/m³) [a]	RQ	Exposure Level (μg/m³)	RQ		
Welding [f]	20.0 [b,c,e]	**20.0**	70.0 [c,e]	**11.7**	**31.7**	[52]
	6.30 [b,g,h]	**6.30**	2.80 [g,h]	0.47	**6.77**	[55]
Flux cored arc welding	11.3 [i]	**11.3**	50.4 [a,i]	**8.40**	**19.7**	[44]
	1.80 [b,c,d]	**1.80**	2.15 [c,d]	0.36	**2.16**	[17]
Gas metal arc welding	7.80 [b,c]	**7.80**	11.0 [c]	**1.83**	**9.63**	[17]
	0.24 [h]	0.24	2.70 [h]	0.45	0.69	[16]
Shielded metal arc welding	0.04 [h]	0.04	0.49 [h]	0.082	0.12	[16]
	17.0 [b,c]	**17.0**	4.00 [c]	0.67	**17.7**	[17]
Tungsten inert gas welding	10.5 [b,c]	**10.5**	2.55 [c,d]	0.43	**10.9**	[17]
	0.23 [h]	0.23	0.67 [h]	0.11	0.34	[16]
	1.00 [d,i]	1.00	1.00 [i]	0.17	**1.17**	[53]
	1.35 [b,d,i]	**1.35**	0.75 [d,i]	0.13	**1.48**	[47]
Gas metal arc welding with massive or flux cored wire of stainless steel (high exposure group)	239.0 [b,i]	**239.0**	82.5 [i]	**13.8**	**252.8**	[47]
Flux-cored arc welding of mild steel	0.60 [b,d,i]	0.60	0.85 [d,i]	0.14	0.74	[47]

Tolerable levels: 1 μg/m³ of Cr(VI) compounds and 6 μg/m³ of nickel compounds [43]; [a] Expressed as median level, except in reference [44] that presents the mean value; [b] Data presented for chromium, without specifying its ionic form. It was assumed that authors were referring to Cr(VI); [c] Inhalable fraction; [d] Values < LOD; LOD/2 was determined; [e] No mean or median values presented; the highest value was chosen; [f] Unspecified welding activity; [g] Geometric mean; [h] Respirable fraction; [i] Unspecified fraction considered as inhalable fraction; RQ > 1 and SRQ > 1 in bold.

Table 4. Exposure levels of Cr and Ni in urine (HBM) of workers performing welding activities, and values obtained for RQ and SRQ.

Welding Activity	U-Cr		U-Ni		SRQ	Ref.
	Exposure Level	RQ	Exposure Level	RQ		
Welding [d]	0.72 μg/g creat	0.60	1.56 μg/g creat [e]	0.71	**1.31**	[52]
	2.30 μg/g creat [b]	**1.92**	3.10 μg/L [b]	**1.03**	**2.95**	[55]
	0.54 μg/g creat (post-shift)	0.45	1.47 μg/g creat [e] (post-shift)	0.67	**1.12**	[50]
Flux cored arc welding	3.96 μg/g creat (after work) [a]	**3.30**	2.50 μg/g creat (after work) [a,e]	**1.13**	**4.43**	[44]
	0.43 μg/g creat [c]	0.36	2.83 μg/L	0.94	**1.30**	[17]
Gas metal arc welding	0.83 μg/g creat [c]	0.69	3.76 μg/L [c]	**1.25**	**1.94**	[17]
	0.39 μg/g creat	0.33	1.57 μg/g creat [e]	0.71	**1.04**	[54]
	0.94 μg/g creat (post-shift)	0.78	2.87 μg/L (post-shift)	0.96	**1.74**	[16]
Shielded metal arc welding	0.74 μg/g creat (post-shift)	0.62	1.20 μg/L (post-shift)	0.40	**1.02**	[16]
	1.17 μg/g creat [c]	0.98	1.98 μg/L	0.66	**1.64**	[17]
Tungsten inert gas welding	0.61 μg/g creat [c]	0.51	0.75 μg/L [c]	0.25	0.76	[17]
	0.78 μg/g creat (post-shift)	0.65	1.31 μg/L (post-shift)	0.44	**1.09**	[16]
	0.74 μg/g creat (after Friday shift)	0.62	1.11 μg/g creat [e] (after Friday shift)	0.50	**1.12**	[53]
	0.50 μg/L [c,e]	0.31	0.75 μg/L [c]	0.25	0.56	[47]
Gas metal arc welders with massive or flux cored wire of stainless steel (high exposure group)	13.53 μg/L [e]	**8.29**	7.92 μg/L	**2.64**	**10.9**	[47]
Flux-cored arc welding of mild steel	0.50 μg/L [c]	0.31	2.70 μg/L	0.90	**1.21**	[47]

Limit values: 1.2 μg/g creatinine for total chromium [41]; 3 μg/L for nickel [37]; U-Cr, Urinary chromium; U-Ni, Urinary nickel; [a] Mean value; [b] Geometric means; [c] Values <LOD; LOD/2 was determined; [d] Unspecified welding activity; [e] Interconversion from μg/g creatinine to μg/L of urine using the standard creatinine value (~1.36 g/L urine), for RQ calculations; RQ > 1 and SRQ > 1 in bold.

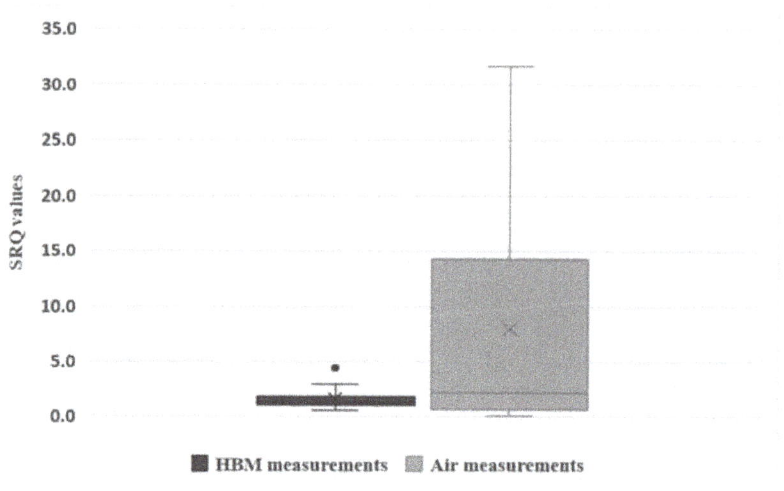

Figure 3. Boxplot showing the SRQ determined using values from air measurements (light grey) or using HBM measurements (dark grey). •—outlier; ×—mean SRQ values (mean (HBM) = 1.62; mean (air) = 7.93).

Among the ten combinations that generated SRQ > 1 based on air Cr and Ni levels, only eight produced SRQ > 1 based on HBM data [17,44,47,52,53,55]. Inversely, in four combinations, SRQ < 1 were derived from external exposure data for gas metal arc welders, shielded metal arc welders, tungsten inert gas welders [16] and flux-cored arc welders of mild steel [47], whereas the corresponding HBM-based SRQ values were 1.74, 1.02, 1.09 and 1.21, respectively, i.e., exceeding the threshold levels of concern.

From another perspective, a more than 17-fold increase in the Cr RQ over Ni RQ was found for the "high exposure group" among gas metal arc welders with massive or flux cored wire of stainless steel [47], suggesting that Cr will be the driver of the mixture toxicity. In line with this observation, a 4-fold increase in Cr RQ over Ni RQ was also noted in another study involving gas metal arc welders [17]. Interestingly, using HBM data, the comparison between the RQ values calculated based on the U-Cr and U-Ni levels showed a 1.8-fold decrease and a 3-fold increase for the "high exposure group" among gas metal arc welders with massive or flux cored wire of stainless steel [47] and for the gas metal arc welders [17], respectively.

3.4. Risk Quotients after Exposure to Cr(VI), Ni and PAHs

Only two studies conducted in hazard waste incinerator settings reported biomarkers of exposure in urine for Cr, Ni, and PAHs, but no air measurements were conducted in these studies [57,59] (Table 2). The levels of urinary Cr, Ni and 1-OHP retrieved and the respective RQ and SRQ values are presented in Table 5. For all exposure scenarios described, SRQ were above one, even among workers performing activities considered of low or non-exposure, such as laboratory and administrative activities, respectively. In both studies, workers were exposed to Cr and PAHs (measured through 1-OHP) at levels below the limit values (RQ < 1); nevertheless, values of SRQ above one were obtained for all activities performed (Table 5).

Table 5. Occupational exposure to Cr, Ni and PAHs in hazard waste incinerators, based on biomarkers of exposure in urine, and respective RQ and SRQ.

Activities	U-Cr		U-Ni		U-1-OHP		SRQ	Ref.
	Exposure Level (μg/g Creat)	RQ	Exposure Level (μg/g Creat)	RQ	Exposure Level (μg/g Creat)	RQ		
Incinerator operations, boiler and furnace maintenance, control panel, and waste-gas-washing	0.39	0.33	3.70 [b]	**1.68**	0.30 [b]	0.29	**2.30**	[57]
	0.26	0.22	3.03 [b]	**1.37**	0.10 [b]	0.10	**1.69**	[59]
Laboratory activities	0.57	0.48	2.30 [b]	**1.04**	0.20 [b]	0.19	**1.71**	[57]
	0.19	0.16	2.55 [b]	**1.16**	0.02 [a,b]	0.019	**1.34**	[59]
Administrative activities	0.22	0.18	3.60 [b]	**1.63**	0.02 [a,b]	0.019	**1.83**	[57]
	0.26	0.22	2.34 [b]	**1.06**	0.02 [a,b]	0.019	**1.30**	[59]

Limit values: 1.2 μg/g creatinine for total chromium [41]; 3 μg/L for nickel [37]; 1.4 μg/L for 1-OHP [38]; [a] Values < LOD (<0.04 μg/g creat); LOD/2 was determined; [b] Interconversion from μg/g creatinine to μg/L of urine using the standard creatinine value (~1.36 g/L urine), for RQ calculations; RQ > 1 and SRQ > 1 in bold.

3.5. Effect Biomarkers

Effect biomarkers were analysed in 6 out of the 22 articles included (Table 6). Three studies assessed chromosomal damage, using the micronucleus assay [56,58,61], and two of them also assessed sister chromatid exchanges (SCE) [56,58]; one study analysed DNA damage by the comet assay [61]. No significant increases in micronuclei and SCE frequencies, or in DNA strand break levels were observed among exposed groups when compared to the control groups. Two studies on welders analysed biomarkers of oxidative stress in EBC [50,53]. A significant increase in H_2O_2/tyrosine, nitrate/tyrosine and cotinine concentrations was shown in welders' EBC compared to controls [50]. However, in tungsten inert gas welders, the biomarkers analysed in EBC only differed significantly between workers' shifts [53]. Another study on welders also measured oxidatively modified guanosines, but no differences were observed across the different welding techniques after creatinine adjustment. Moreover, no comparison was made between exposed and control groups in that study [47].

Table 6. Reported outcomes observed in biomarkers of effect.

Activities [Ref.]	Exposure Biomarkers	Effect Biomarkers	Overall Outcomes		SRQ
Harbour dredgers and lighters [56]	U-Cr (0.22 μg/g creat) and U-Ni (0.7 μg/g creat)	MN	Significantly higher frequency in controls vs. exposed	Controls: 14.2 MNs/1000 cells; Exposed: 11.6 MNs/1000 cells	0.36
		SCE	No significant differences between exposed and controls	Controls: 10.2 SCE rate; Exposed: 10.0 SCE rate	
Incinerator operations, boiler and furnace maintenance, control panel, and waste-gas-washing [61]	U-Cr (short-0.5 ± 0.58; medium-1.89 ± 4.59; long-exposure-0.26 ± 0.24 μg/g creat) U-Ni (short-9.18 ± 7.47; medium-12.12 ± 8.31; long-exposure-7.69 ± 5.7 μg/g creat)	MN	No significant differences between exposed and controls	Controls: 1.3/1000 cells Short-exposure group: 1.6/1000 cells, Medium-exposure group: 1.7/1000 cells Long-exposure group: 1.7/1000 cells	**2.10 (long-exposure)**
		Comet assay		Short exposure group: 6.5 tail factor; Medium exposure group: 6.3 tail factor; Long exposure group: 6.7 tail factor Controls: 7.1 tail factor	

Table 6. *Cont.*

Activities [Ref.]	Exposure Biomarkers	Effect Biomarkers	Overall Outcomes		SRQ
Gemstone cutting [58]	U-Ni ≤4 h/week: 1.18 μg/g creat (pre-shift); 0.85 (post-shift) >4 h/week: 0.81 (pre-shift); 0.66 (post-shift)	SCE	No significant differences between exposed groups	Short exposure group: 9.91 SCE/cell; Long exposure group: 9.70 SCE/ cell	-
Tungsten inert gas welding [53]	U-Cr 0.71 μg/g creat (0.36–1.27) (T1) 0.74 μg/g creat (0.41–1.21) T2) 0.59 μg/g creat (0.26–1.00) (T3) U-Ni 0.76 μg/g creat (0.37–1.40) (T1) 1.11 μg/g creat (0.59–0.79) (T2) 0.83 μg/g creat (0.31–1.38) (T3)	MDA-EBC	Significant decrease at T2	2.79 nM, 2.98 nM and 2.43 nM at T0, T1 and T2, respectively	1.12
		HNE-EBC	Significant increase at T0 than at T2	0.53 nM, 0.48 nM and 0.51 nM at T0, T1 and T2, respectively	
		H_2O_2-EBC	Significant increase at T1 than at T0 and T2	0.18, 0.25, and 0.16 μM at T0, T1 and T2, respectively	
Welding [50]	U-Cr (0.32 μg/g creat (IQR 0.97) (pre-shift) 0.54 μg/g creat (IQR 1.39) (post-shift) U-Ni 1.11 μg/g creat (IQR 1.6) (pre-shift) 1.47 μg/g creat (IQR 2.27) (post-shift)	H_2O_2-EBC	**Significant increase in welders vs. controls**	56 ± 19 pM/μg vs. 6.6 ± 2.2 pM/μg	1.12
		Nitrate/ tyrosine ratio		0.71 ± 0.13 nM/μg vs. 0.22 ± 0.04 nM/μg	
Gas metal arc welders with massive or flux cored wire of stainless steel; Flux-cored arc welding of mild steel; Tungsten inert gas welding in shipyards industries [47]	U-Cr (0.88 μg/g creat) U-Ni (2.07 μg/g creat)	Urinary 8-oxoGuo	No significant differences between welding techniques (adjusted for creatinine)	High-exposure group: 6.59 μg/g creat; Flux-cored arc welding of mild steel: 7.62 μg/g creat; Tungsten inert gas welders: 6.41 μg/g creat	0.56 for TIGW
		Urinary 8-oxodGuo		High-exposure group: 4.28 μg/g creat; Flux-cored arc welding of mild steel: 3.79 μg/g creat; Tungsten inert gas welders: 4.41 μg/g creat	**10.9** for high exposure group [a]
		8-oxodGuo/ 10^6dGuo in white blood cells		High-exposure group: 5.53/10^6dGuo; Flux-cored arc welding of mild steel: 4.79/10^6dGuo; Tungsten inert gas welders: 1.98/10^6dGuo	**1.21** for FCAW of mild steel

8-oxodG: 8-oxo-7,8-dihydro-2′-deoxyguanosine; 8-oxoGuo: 8-oxo-7,8-dihydroguanosine; 10^6dGuo: 10^6 2′-deoxyguanoside; EBC: exhaled breath condensate; FCAW, flux cored arc welding; HNE: 4-hydroxynonenal; MDA: malondialdehyde; MN: micronuclei; SCE: Sister chromatid exchange; TIGW, tungsten inert gas welding; T0, before Friday shift; T1, after Friday shift; T2, before Monday shift; [a] High exposure group corresponding to workers performing gas metal arc welders with massive or flux cored wire of stainless steel.

4. Discussion

The identification of all chemicals present in an occupational setting and the fluctuations in exposures levels during the wide variety of tasks performed in a workday or week can be quite challenging [64]. The knowledge regarding co-exposure to chemicals is quite limited for many workplaces, and the approach commonly followed to identify risk factors is often focused on single chemicals directly resulting from the process or that are used as raw materials.

HBM is an important tool to identify and quantify workers' exposure to emerging or well-known contaminants and can further assist in the evaluation and demonstration of the effectiveness of policy actions or measures already taken at a company level to reduce exposure [39]. In this work, we observed that recent studies conducted in European Union countries and reporting biomarkers of exposure to Cr(VI), Ni and PAHs are scarce. These findings highlight that in most occupational studies and in industrial hygiene routines, the measurement of external exposure remains the common practice to assess workers exposure. However, HBM data is of added value, particularly in the case of substances for which exposure by ingestion, due to hand-to-mouth contact (the case of metals), or by dermal absorption (the case of PAHs) is relevant, as it reflects exposure from all sources and routes of exposure. In this study, we conducted a mixture risk assessment exercise based on both external and internal exposure levels of Cr(VI), Ni and PAHs reported in the literature.

A first general observation was that the SRQ values were significantly higher when using air measurements than using HBM data (urinary biomarkers). Additionally, a higher variability among SRQ values obtained from air measurements was observed, compared with the SRQ values calculated from HBM measurements. The air concentrations of chemicals are influenced by many external factors such as temperature, ventilation, the type of activities being performed, among others, while they do not account, for example, for the effect of individual protection equipment. The lower RQ and SRQ values obtained using HBM data and the lower associated variability, may be mainly related to the use of respiratory protection equipment by workers that will reduce exposure. In addition, uncertainties related to the air–urinary levels correlations, particularly at the low exposure levels, may contribute to this observation. Considering that HBM results give a more accurate estimate of the actual exposure level because they reflect the internal doses of chemicals but also that external exposure levels are usually assessed under occupational health surveillance programs, it seems that this latter approach is able to adequately detect health risks from exposure to single chemicals. Nevertheless, it is important to note that HBM data, while covering all possible routes of workers' exposure, accounting for individual variability (e.g., in chemicals metabolism) and, hence, reflecting the internal doses of chemicals, also provide a more reliable exposure assessment to perform MRA.

Focusing on welding activities where co-exposure to Cr and Ni occurs, an HBM-based SRQ revealed exposure levels of concern in most of the activities, with several exceedances of the limit values, even in the most recent studies. These observations raise concern, since Cr(VI) and Ni are potent human carcinogens [3,4], and cancer risks from welding activities, in particular stainless-steel welding, have been a matter of concern [65]. In some of the analysed studies, welding activities generated RQ < 1 based on U-Cr or U-Ni; however, the associated SRQ was above one. Cr and Ni share genotoxic and non-genotoxic MoAs and both substances may induce a similar health outcome (lung cancer). However, interactive effects are not often considered when performing workplace risk assessment, even though the EU Chemical Agents Directive (98/24/EC) clearly states that in the case of activities involving exposure to several hazardous chemical agents, the risk shall be assessed on the basis of the risk presented by all such chemical agents in combination. Indeed, in the occupational health field, combined effects are never taken into account in individual Observed Effect Levels (OELs) although the employer should assess the risks related to the combined effects of chemicals in the work environment by, e.g., using the hazard index (HI) approach. For example, the EU CAD directive states that "in the case of activities involving exposure to several hazardous chemical agents, the risk shall be assessed on the basis of the risk presented by all such chemical agents in combination". In some countries, such as Finland, the employers have been instructed and advised that they should take possible combined effects into account if the substances have the same MoA. Guidance on HI approach is provided, but unfortunately, this is generally applied mostly in the case of irritant solvents, not in the case of carcinogens. This seems to be the general situation in many European countries, i.e., even though legislation requires consideration of combined effects, this is not very effectively performed in practice. This means that the potential impact of industrial chemicals on workers' health might be overlooked. In light of these findings, regulatory actions put in place at these occupational environments should be re-evaluated in order to reduce workers' exposure.

Although in many of the analysed studies, urinary levels of Cr and Ni were assessed, only two studies focused on hazardous waste incineration reported HBM data for Cr, Ni and

PAHs. However, the co-occurrence of these substances has been recognized in many other industrial activities [66], e.g., in electromechanical repair and car painting centres, sewing work rooms, chemical analysis laboratories [67], coke oven plants, aluminium production, iron and steel founding and welding industries, among others [18–20], showing the paucity of data allowing the assessment of the risks related to those mixtures exposure.

In the context of activities conducted at incinerators, RQ < 1 values were obtained based on U-Cr or U-1-OHP levels that resulted in SRQ > 1. Surprisingly, these results were also observed among administrative workers from the incinerator industries. These findings indicate that if only exposure to single metals is considered, no concern will be raised, reinforcing the importance of considering the mixture for a realistic risk assessment. Furthermore, our results regarding administrative staff highlight the fact that these workers from the same industrial facilities can also present a certain level of exposure possibly due to not wearing protection equipment. Those workers should also be aware of the risks they incur in and should be enrolled in occupational health surveillance schemes or in other occupational health interventions (e.g., dedicated health surveillance programs).

Exposure to chemical mixtures, namely, substances exhibiting similar MoAs, may lead to adverse health effects, even at exposures below the chemical's threshold level. Moreover, from the exposure to multiple chemicals, toxicokinetic and/or toxicodynamic interactions may occur, resulting in new toxic effects usually not observed for single exposures, such as synergism [30,64]. An in vitro study showed that human lung fibroblasts exposed to Cr(VI) and PAHs metabolites resulted in synergistic mutagenic and cell transformation effects [23]. However, an in vivo study demonstrated tissue-specific effects in the liver and gastrointestinal tract, namely, at molecular, gene expression and histopathological level, due to exposure to environmentally relevant Cr(VI) and BaP doses. For instance, BaP reduced mice body weight by more than 10%, but this effect was not modified by co-exposure to Cr(VI). Co-exposure to BaP and Cr(VI) also induced an additive effect in crypt hyperplasia in the small intestine of mice, while in the liver, both substances cause more severe histopathological lesions than those expected from the sum of each single compound (more than additive). Nevertheless, regarding genotoxic effects, co-exposure increased the phosphor-cH2AX-positive cells observed for each single substance, but not in an additive way [25]. These observations in experimental systems have been strengthen by the findings of epidemiological studies contributing to a higher weight of evidence. A study conducted by Wang et al. (2015) on coke oven workers in China detected urinary concentrations of heavy metals, including Cr and Ni, and PAHs metabolites in most of the samples, and reported significantly higher oxidative stress biomarkers among workers exposed to high levels of PAHs and metals, compared to those only exposed to one substance [20]. A study by Campo et al. (2020) observed an additive effect between metals and PAHs on 8-oxodG levels in electric steel foundry workers in Tunisia [68]. Moreover, in a recent study by Zeng et al. (2022), a positive association between co-exposure to PAHs and metals and oxidative stress (dose–response curve of 3-OHPhe or Al with 8-iso-PGF2α) measured among coke oven works was suggestive of a synergistic effect [69].

Effect biomarkers allow us to identify early biological effects from chemicals exposure and contribute to establish exposure–response relationships, explore mechanisms that underlie health outcomes and increase the knowledge on the biological plausibility of epidemiological associations between exposure and adverse health effects [70]. However, in epidemiological studies, effect biomarkers have been used to a lesser extent than exposure biomarkers, thus constituting an understudied field [64]. Among the studies analysed, only six reported data on effect biomarkers [47,50,53,56,58,61]. The most used biomarkers targeted genetic damage: chromosome alterations (MN and SCE in blood lymphocytes), DNA damage (comet assay in blood leukocytes), oxidative DNA adducts formation (8-oxodGuo) or oxidative stress. Comparisons of the values obtained for those biomarkers between exposed and control groups or among workers groups with different exposure levels did not reveal any significant increase associated with exposure, irrespective of the effect biomarker used. Two studies reported statistical comparisons of MN frequencies between incineration workers or harbour dredgers and lighters exposed to Cr and Ni and the respective control groups [56,61]. Both studies failed to find increased MN frequencies in the exposed group. These findings contrast with those from other studies on occupational exposure to Cr(VI) or to Cr(VI) and Ni that reported significantly higher genetic damage in workers comparatively to controls [71–73]. Such differences might also be due to timing issues during samples collection, e.g., the time elapsing between exposure and samples collection, which is determinant to detect DNA damage in blood cells and to other factors related to the study design, e.g., a too small number of participants. More occupational studies comprising, preferably, exposure and effect biomarkers are needed to provide more evidence on their value to link exposure and health effects.

Some limitations in this study must be acknowledged. A considerable degree of heterogeneity was observed amongst studies. Data was stratified by workday, pre- or post-shift sampling, and by duration of exposure, such as hours or months. In terms of statistical treatment of data, some studies presented data per individual measurements; others presented arithmetic means, geometric means, or medians. The units used to express urinary biomarkers also varied from adjustment to µg/g of creatinine, to µg/L, µmol/mol of creatinine, and nmol/L. These aspects possibly prevented us from making a more accurate comparison of the data retrieved. In the HBM4EU initiative, the importance of harmonizing methodologies and designs in HBM studies has been highlighted in recent publications [40,74], reinforcing the importance of joint efforts to produce comparable data that can be further used to regulatory actions in the context of different frameworks. Contextual information was also lacking in some studies, such as information on possible exposure routes. However, since we analysed HBM data, all exposure routes were considered in our RQ and SRQ calculations. In the analysed studies, there were also exposure biomarkers in blood cells, plasma or exhaled breath condensate, although we only considered urinary exposure biomarkers. However, these are the most commonly used in HBM studies due to its simplicity of sampling, non-invasive nature and the existing validated methods for measurement.

Another important limitation were the uncertainties related to the conversion of urinary levels from air levels to be used as limit values for our MRA. For Cr, we have used a regression equation that moderately correlates new exposure data from welders [41]; for Ni, regression equations are also very unsure, especially at low exposure levels and in the case of poorly soluble Ni compounds present in welding [37]; for 1-OHP, some uncertainties must be also considered due to the fact that for PAHs we are often dealing with mixtures with variable pyrene vs. BaP ratios. Furthermore, regarding the attempt to identify the potential driver of the mixtures' toxicity, a simplified approach, i.e., the comparison of the RQ values calculated for each single component, was used. It is important to note that the toxicity of each component in a mixture can be substantially different, and that the effect of the mixture thereby depends not only on each chemical's concentration but also on its relative contribution to the toxicity observed. This means that a small quantity of a very potent carcinogen (e.g., benzo[a]pyrene) in the mixture can have a much larger biological effect than a large amount of a less potent chemical (e.g., Ni). Nevertheless, the RQ value accounts for the toxicity of the chemical because it allows the comparison of exposure doses or concentrations with available reference values (e.g., no observed effect or lowest effect concentrations) derived from toxicity data.

A recent survey on field experts on the use of HBM data in chemical risk assessment revealed that the lack of HBM-based guidance values for the majority of substances has greatly limited the use of HBM data for risk assessment [75]. There is no agreement on the acceptable risk levels in Europe for occupational activities involving carcinogenic substances. Given that this was a crucial point for this work, because this can significantly impact the RQ and SRQ results, the topic was debated with an expert group in occupational health and risk assessment. Those experts recognized that the tolerable risk level of 1:4000, as defined in the German Technical Rules for Hazardous Substances (TRGS), produced by the Committee on Hazardous Substances (AGS) of the Federal Institute for Occupational

Safety and Health), that describe the state-of-the-art knowledge on occupational medicine and industrial hygiene for activities with hazardous substances, could be followed. According to TRGS, above the defined cross-substance tolerable risk limit of 4:1000, there is a high risk for activities involving carcinogenic substances that is classified as not tolerable. Likewise, the "Risk Concept for Carcinogenic Substances" of the AGS offers a benchmark for comparison and assessment of exposure to carcinogenic substances, including limit values (in air) for the three components of the mixture under study, which was considered an advantage, assuming that those values' derivation followed a similar rationale, decreasing the possible uncertainty of selecting limit values derived by different approaches or for different population groups.

A linear no-threshold dose–response relationship has been generally assumed for carcinogenic chemicals, regarding cancer risk assessment. This relationship applies to Cr(VI) that acts as a direct genotoxic carcinogen (Cr-ternary DNA adduct is considered the critical lesion) for which no threshold can be assumed and linear extrapolation is commonly applied [36]. Likewise, many PAHs, including BaP, are carcinogenic and share the same direct genotoxic mechanism of action (metabolic activation to electrophilic intermediates and formation of DNA adducts leading to mutation), indicating a non-threshold dose-effect relationship [38]. However, nickel compounds act through indirect and secondary (via inflammation) genotoxicity mechanisms, and a threshold MoA for their carcinogenic effects [37] is considered. For these compounds, a "hockey stick"-shaped dose–response curve has been assumed, which likely contains a breakpoint at 0.005 mg/m^3 or 0.006 mg/m^3 [37]. Given that Ni does not produce effects at exposures below the indicated limit value in air, only the effect of Cr(VI) (exposure to the binary mixture) or of the mixture of PAHs and Cr(VI) (ternary mixture) should be considered at such exposure levels. This does not affect the overall RQ and SRQ values calculated from air measurements data (Table 3), in which a single study described air values below the limit value but contributing to a SRQ above 1 [53].

Several authors have addressed some potential limitations of the CA mathematical model that may hamper its application to environmentally realistic assessment cases, where chemicals with different modes of action or showing different dose–response relationships can co-occur. This may be the case of mixtures of genotoxicants, for which the application of the CA model is not consensual. Kortemkamp et al. (2014) [76] assessed the applicability of either CA or IA to several mixtures of aneugenic and clastogenic chemicals by using the micronucleus assay in CHO-K1 cells. The authors concluded that grouping the chemicals into common assessment groups and using hybrid CA/IA prediction models could provide a better prediction for the effect of mixtures consisting of chemicals with different mechanisms of genotoxic action than using either the CA or the IA model. In such cases, the application of CA did not offer a good description of the observed effect and neither did the IA that led to an underestimation of the joint effect. The uncertainties associated with estimating low-level effects also precluded firm conclusions regarding the use of the IA at such low levels. In addition, it is worth noting that the use of the IA requires knowledge on the precise effect magnitude that each component would provoke if present individually at the concentration found in the mixture. This information is only accessible through comprehensive dose–response analysis of each mixture component, which is difficult to find [30]. With this perspective, the use of the CA model to the mixture herein assessed under realistic exposure scenarios seemed to be the best option, even though Ni exposure will probably not contribute to the joint effect at exposure levels below the dose–response breakpoint.

As the substances in the mixture under study are carcinogenic to humans, it is recommended that exposure should be reduced to values close to zero among workers exposed to these substances. It is also important to keep in mind that within occupational settings, substitution of such carcinogens by other substances may not be possible in a short timeframe, and therefore, occupational exposure to non-threshold carcinogens should be limited to levels as low as reasonably achievable (ALARA principle) to protect workers' health.

Overall, this study highlights that risk assessment frameworks only focused on single substances are insufficient in light of the increasing evidence underlining the importance of considering cumulative exposure to multiple chemicals in order to more adequately estimate the risks they pose [27]. Understanding exposure to realistic concentrations of chemical mixtures and their associated health effects will require increased collaboration across different scientific fields, namely, toxicology, epidemiology, exposure science, risk assessment and statistics, for a proper integration of data from each discipline [26]. More MRA studies based on HBM data obtained in real occupational exposure scenarios are needed for other industrial settings. Aiming to contribute to a reliable assessment and management of chemical risks, the HBM4EU initiative is currently building capacities to establish a European Human Biomonitoring Platform in order to harmonize HBM activities in EU countries. This will deliver comparable data on human exposure to chemicals and mixtures of chemicals as a basis for policy making to improve chemical safety (https://www.hbm4eu.eu/about-hbm4eu/, accessed on 10 March 2022). An occupational study aiming to assess external and internal exposure, as well as effect biomarkers, in workers from an aircraft maintenance company exposed to Cr(VI), Ni and PAHs is already ongoing. The results are expected to give a better insight into the co-exposure and early health effects of this mixture in exposed workers.

Supplementary Materials: The following supporting information can be downloaded at: https://www.mdpi.com/article/10.3390/toxics10080431/s1, Table S1: Database searches.

Author Contributions: Conceptualization, A.M.T., S.V., H.L., T.G., T.S., M.L., A.K. and M.J.S.; methodology, A.M.T., S.V., H.L. and M.J.S.; formal analysis, A.M.T. and H.L.; investigation, A.M.T., H.L. and M.J.S.; resources, S.V., T.G. and T.S.; data curation, A.M.T.; writing—original draft preparation, A.M.T., H.L. and M.J.S.; writing—review and editing, A.M.T., S.V., H.L., T.G., T.S., M.L., A.K. and M.J.S.; visualization, A.M.T.; supervision, M.J.S.; project administration, M.L. and A.K. All authors have read and agreed to the published version of the manuscript.

Funding: This research was funded by European Unions' Horizon 2020 research and innovation Programme under grant agreement No 733032 HBM4EU and co-funded by UIDP/00009/2020; UIDB/00009/2020 (Centre for Toxicogenomics and Human Health—ToxOmics, FCT).

Institutional Review Board Statement: Not applicable.

Informed Consent Statement: Not applicable.

Data Availability Statement: Not applicable.

Conflicts of Interest: The authors declare no conflict of interest. The funders had no role in the design of the study, in the collection, analyses, or interpretation of data, in the writing of the manuscript, or in the decision to publish the results.

References

1. Annangi, B.; Bonassi, S.; Marcos, R.; Hernández, A. Biomonitoring of humans exposed to arsenic, chromium, nickel, vanadium, and complex mixtures of metals by using the micronucleus test in lymphocytes. *Mutat. Res.-Rev. Mutat. Res.* **2016**, *770*, 140–161. [CrossRef] [PubMed]
2. CDC/NIOSH. *Criteria for a Recommended Standard. Occupational Exposure to Hexavalent Chromium*; DHHS (NIOSH) Publication No. 2013–128; Centres for Desease Control and Prevention: Atlanta, GA, USA, 2013.
3. IARC. Chromium (VI) Compounds. In *IARC Monographs on the Evaluation of Carcinogenic Risks to Humans*; IARC: Lyon, France, 2018; pp. 147–167.
4. IARC. Arsenic, Metals, Fibres and Dusts: Nickel and Nickel Compounds. In *IARC Monographs on the Evaluation of Carcinogenic Risks to Humans*; IARC: Lyon, France, 2012; pp. 169–218.
5. Barchowsky, A.; O'Hara, K.A. Metal-induced cell signaling and gene activation in lung diseases. *Free Radic. Biol. Med.* **2003**, *34*, 1130–1135. [CrossRef]
6. Grimsrud, T.K.; Berge, S.R.; Haldorsen, T.; Andersen, A. Exposure to different forms of nickel and risk of lung cancer. *Am. J. Epidemiol.* **2002**, *156*, 1123–1132. [CrossRef] [PubMed]
7. Das, K.K.; Reddy, R.C.; Bagoji, I.B.; Das, S.; Bagali, S.; Mullur, L.; Khodnapur, J.P.; Biradar, M.S. Primary concept of nickel toxicity—An overview. *J. Basic Clin. Physiol. Pharmacol.* **2019**, *30*, 141–152. [CrossRef]

8. Hartwig, A.; Arand, M.; Epe, B.; Guth, S.; Jahnke, G.; Lampen, A. *Mode of Action-Based Risk Assessment of Genotoxic Carcinogens*; Springer: Berlin/Heidelberg, Germany, 2020; pp. 1787–1877. ISBN 0123456789.
9. Šimko, P. Polycyclic aromatic hydrocarbons. *Saf. Anal. Foods Anim. Orig.* **2016**, *3*, 441–460. [CrossRef]
10. Li, G.; Hu, R.; Wang, N.; Yang, T.; Xu, F.; Li, J.; Wu, J.; Huang, Z.; Pan, M.; Lyu, T. Cultivation of microalgae in adjusted wastewater to enhance biofuel production and reduce environmental impact: Pyrolysis performances and life cycle assessment. *J. Clean. Prod.* **2022**, *355*, 131768. [CrossRef]
11. Huang, Z.; Zhang, J.; Pan, M.; Hao, Y.; Hu, R.; Xiao, W.; Li, G.; Lyu, T. Valorisation of microalgae residues after lipid extraction: Pyrolysis characteristics for biofuel production. *Biochem. Eng. J.* **2022**, *179*, 108330. [CrossRef]
12. Kim, K.H.; Jahan, S.A.; Kabir, E.; Brown, R.J.C. A review of airborne polycyclic aromatic hydrocarbons (PAHs) and their human health effects. *Environ. Int.* **2013**, *60*, 71–80. [CrossRef]
13. European Commission. *SCOEL/REC/404 Polycyclic Aromatic Hydrocarbon Mixtures Containing Benzo[a]pyrene (PAH). Recommendation from the Scientific Committee on Occupational Exposure Limits*; Publications Office of the European Union: Luxembourg, 2016. [CrossRef]
14. Jameson, C.W. Polycyclic aromatic hydrocarbons and associated occupational exposures. In *Tumour Site Concordance and Mechanisms of Carcinogenesis*; Baan, R., Stewart, B., Straif, K., Eds.; International Agency for Research on Cancer (IARC): Lyon, France, 2019; Chapter 7; pp. 59–64.
15. IARC. Benzo(a)pyrene. In *Encyclopedia of Toxicology*, 3rd ed.; Elsevier: Amsterdam, The Netherlands, 2014; pp. 423–428. ISBN 9780123864543.
16. Pesch, B.; Lehnert, M.; Weiss, T.; Kendzia, B.; Menne, E.; Lotz, A.; Heinze, E.; Behrens, T.; Gabriel, S.; Schneider, W.; et al. Exposure to hexavalent chromium in welders: Results of the WELDOX II field study. *Ann. Work Expo. Health* **2018**, *62*, 351–361. [CrossRef]
17. Weiss, T.; Pesch, B.; Lotz, A.; Gutwinski, E.; Van Gelder, R.; Punkenburg, E.; Kendzia, B.; Gawrych, K.; Lehnert, M.; Heinze, E.; et al. Levels and predictors of airborne and internal exposure to chromium and nickel among welders-Results of the WELDOX study. *Int. J. Hyg. Environ. Health* **2013**, *216*, 175–183. [CrossRef]
18. Lißner, L.; Kuhl, T.; Kauppinen, T.; Uuksulainen, S. *Exposure to Carcinogens and Work-Related Cancer: A Review of Assessment Methods European Risk Observatory—European Risk Observatory Report*; European Agency for Safety and Health at Work; Publications Office of the European Union: Luxembourg, 2014. [CrossRef]
19. IARC. Chemical agents and related occupations. *IARC Monogr. Eval. Carcinog. Risks Hum.* **2012**, *100 Pt F*, 9–562.
20. Wang, T.; Feng, W.; Kuang, D.; Deng, Q.; Zhang, W.; Wang, S.; He, M.; Zhang, X.; Wu, T.; Guo, H. The effects of heavy metals and their interactions with polycyclic aromatic hydrocarbons on the oxidative stress among coke-oven workers. *Environ. Res.* **2015**, *140*, 405–413. [CrossRef]
21. Kortenkamp, A.; Faust, M. Regulate to reduce chemical mixture risk. *Science* **2018**, *361*, 224–226. [CrossRef]
22. Rotter, S.; Beronius, A.; Boobis, A.R.; Hanberg, A.; van Klaveren, J.; Luijten, M.; Machera, K.; Nikolopoulou, D.; van der Voet, H.; Zilliacus, J.; et al. Overview on legislation and scientific approaches for risk assessment of combined exposure to multiple chemicals: The potential EuroMix contribution. *Crit. Rev. Toxicol.* **2018**, *48*, 796–814. [CrossRef]
23. Feng, Z.; Hu, W.; Rom, W.N.; Costa, M.; Tang, M.S. Chromium(VI) exposure enhances polycyclic aromatic hydrocarbon-DNA binding at the *p53* gene in human lung cells. *Carcinogenesis* **2003**, *24*, 771–778. [CrossRef]
24. Peng, C.; Muthusamy, S.; Xia, Q.; Lal, V.; Denison, M.S.; Ng, J.C. Micronucleus formation by single and mixed heavy metals/loids and PAH compounds in HepG2 cells. *Mutagenesis* **2015**, *30*, 593–602. [CrossRef]
25. Sánchez-Martín, F.J.; Fan, Y.; Carreira, V.; Ovesen, J.L.; Vonhandorf, A.; Xia, Y.; Puga, A. Long-term coexposure to hexavalent chromium and B[a]P causes tissue-specific differential biological effects in liver and gastrointestinal tract of mice. *Toxicol. Sci.* **2015**, *146*, 52–64. [CrossRef]
26. Hernández, A.F.; Tsatsakis, A.M. Human exposure to chemical mixtures: Challenges for the integration of toxicology with epidemiology data in risk assessment. *Food Chem. Toxicol.* **2017**, *103*, 188–193. [CrossRef]
27. Drakvik, E.; Altenburger, R.; Aoki, Y.; Backhaus, T.; Bahadori, T.; Barouki, R.; Brack, W.; Cronin, M.T.D.; Demeneix, B.; Hougaard Bennekou, S.; et al. Statement on advancing the assessment of chemical mixtures and their risks for human health and the environment. *Environ. Int.* **2020**, *134*, 105267. [CrossRef]
28. EU. Commission Chemicals Strategy for Sustainability Towards a Toxic-Free Environment. *J. Chem. Inf. Model.* **2013**, *53*, 1689–1699.
29. Kienzler, A.; Bopp, S.K.; van der Linden, S.; Berggren, E.; Worth, A. Regulatory assessment of chemical mixtures: Requirements, current approaches and future perspectives. *Regul. Toxicol. Pharmacol.* **2016**, *80*, 321–334. [CrossRef] [PubMed]
30. EFSA Scientific Committee; More, S.J.; Bampidis, V.; Benford, D.; Bennekou, S.H.; Bragard, C.; Halldorsson, T.I.; Hernández-Jerez, A.F.; Koutsoumanis, K.; Naegeli, H.; et al. Guidance on harmonised methodologies for human health, animal health and ecological risk assessment of combined exposure to multiple chemicals. *EFSA J.* **2019**, *17*, 77. [CrossRef] [PubMed]
31. Greim, H.; Snyder, R. (Eds.) *Toxicology and Risk Assessment*; John Wiley & Sons, Ltd.: Chichester, UK, 2008; ISBN 9780470868959.
32. Lebret, E. Information Needs from Policy Makers Translated in Terms of Functionality of HBM Mixture Information Deliverable Report Deadline: November 2018. 2020. Available online: https://www.hbm4eu.eu/result/deliverables/ (accessed on 14 July 2021).
33. Sarigiannis, D.A.; Hansen, U. Considering the cumulative risk of mixtures of chemicals—A challenge for policy makers. *Environ. Health A Glob. Access Sci. Source* **2012**, *11* (Suppl. S1), S18. [CrossRef] [PubMed]

34. Kortenkamp, A.; Mengelers, M.; Vinggaard, A.M.; Silva, M.J.; Slama, R.; Vermeulen, R.; Vlaanderen, J.; Ottenbros, I.; Viegas, S.; Gomes, B.; et al. Deliverable Report AD 15.1 Plan for Development of Case Studies. 2019. Available online: https://core.ac.uk/download/pdf/196530897.pdf (accessed on 4 January 2021).
35. Beronius, A.; Zilliacus, J.; Hanberg, A.; Luijten, M.; van der Voet, H.; van Klaveren, J. Methodology for health risk assessment of combined exposures to multiple chemicals. *Food Chem. Toxicol.* **2020**, *143*, 111520. [CrossRef]
36. European Commission. *SCOEL/REC/386 Chromiun VI Compounds. Recommendation from the Scientific Committee on Occupational Exposure Limits*; European Commission: Luxembourg, 2017.
37. European Chemicals Agency (ECHA). *Committee for Risk Assessment—RAC. Opinion on Scientific Evaluation of Occupational Exposure Limits for Nickel and Its Compounds*; European Chemicals Agency (ECHA): Helsinki, Finland, 2018.
38. European Chemicals Agency (ECHA). RAC—Committee for Risk Assessment. In *Note on Reference Dose-Response Relationship for the Carcinogenicity of Pitch, Coal Tar, High Temperature and on PBT and vPvB Properties*; European Chemicals Agency (ECHA): Helsinki, Finland, 2018.
39. Viegas, S.; Jeddi, M.Z.; Hopf, N.B.; Bessems, J.; Palmen, N.; Galea, K.S.; Jones, K.; Kujath, P.; Duca, R.C.; Verhagen, H.; et al. Biomonitoring as an underused exposure assessment tool in occupational safety and health context—Challenges and way forward. *Int. J. Environ. Res. Public Health* **2020**, *17*, 5884. [CrossRef]
40. Santonen, T.; Alimonti, A.; Bocca, B.; Duca, R.C.; Galea, K.S.; Godderis, L.; Göen, T.; Gomes, B.; Hanser, O.; Iavicoli, I.; et al. Setting up a collaborative European human biological monitoring study on occupational exposure to hexavalent chromium. *Environ. Res.* **2019**, *177*, 108583. [CrossRef]
41. Viegas, S.; Martins, C.; Bocca, B.; Bousoumah, R.; Duca, R.C.; Galea, K.S.; Godderis, L.; Iavicoli, I.; Janasik, B.; Jones, K.; et al. HBM4EU Chromates Study: Determinants of Exposure to Hexavalent Chromium in Plating, Welding and Other Occupational Settings. *Int. J. Environ. Res. Public Health* **2022**, *19*, 3683. [CrossRef]
42. Teuschler, L.K.; Hertzberg, R.C. Current and future risk assessment guidelines, policy, and methods development for chemical mixtures. *Toxicology* **1995**, *105*, 137–144. [CrossRef]
43. Ausschuss für Gefahrstoffe (AGS). *Technische Regeln für Gefahrstoffe. Risikobezogenes Maßnahmenkonzept für Tätigkeiten mit Krebser—Zeugenden Gefahrstoffen*; Committee on Hazardous Substances, 2021; Volume 165. Available online: https://www.baua.de/DE/Angebote/Rechtstexte-und-Technische-Regeln/Regelwerk/TRGS/TRGS-910.html (accessed on 7 April 2021).
44. Stridsklev, I.C.; Schaller, K.H.; Langård, S. Monitoring of chromium and nickel in biological fluids of stainless steel welders using the flux-cored-wire (FCW) welding method. *Int. Arch. Occup. Environ. Health* **2004**, *77*, 587–591. [CrossRef]
45. Stridsklev, I.C.; Schaller, K.H.; Langård, S. Monitoring of chromium and nickel in biological fluids of grinders grinding stainless steel. *Int. Arch. Occup. Environ. Health* **2007**, *80*, 450–454. [CrossRef]
46. Gil, F.; Hernández, A.F.; Márquez, C.; Femia, P.; Olmedo, P.; López-Guarnido, O.; Pla, A. Biomonitorization of cadmium, chromium, manganese, nickel and lead in whole blood, urine, axillary hair and saliva in an occupationally exposed population. *Sci. Total Environ.* **2011**, *409*, 1172–1180. [CrossRef]
47. Pesch, B.; Lotz, A.; Koch, H.M.; Marczynski, B.; Casjens, S.; Käfferlein, H.U.; Welge, P.; Lehnert, M.; Heinze, E.; Van Gelder, R.; et al. Oxidatively damaged guanosine in white blood cells and in urine of welders: Associations with exposure to welding fumes and body iron stores. *Arch. Toxicol.* **2015**, *89*, 1257–1269. [CrossRef]
48. Wultsch, G.; Nersesyan, A.; Kundi, M.; Mišík, M.; Setayesh, T.; Waldherr, M.; Vodicka, P.; Vodickova, L.; Knasmüller, S. Genotoxic and Cytotoxic Effects in Exfoliated Buccal and Nasal Cells of Chromium and Cobalt Exposed Electroplaters. *J. Toxicol. Environ. Health-Part A Curr. Issues* **2017**, *80*, 651–660. [CrossRef]
49. Hoffmeyer, F.; Raulf-Heimsoth, M.; Weiss, T.; Lehnert, M.; Gawrych, K.; Kendzia, B.; Harth, V.; Henry, J.; Pesch, B.; Brüning, T. Relation between biomarkers in exhaled breath condensate and internal exposure to metals from gas metal arc welding. *J. Breath Res.* **2012**, *6*, 027105. [CrossRef]
50. Gube, M.; Ebel, J.; Brand, P.; Göen, T.; Holzinger, K.; Reisgen, U.; Kraus, T. Biological effect markers in exhaled breath condensate and biomonitoring in welders: Impact of smoking and protection equipment. *Int. Arch. Occup. Environ. Health* **2010**, *83*, 803–811. [CrossRef]
51. Hoffmeyer, F.; Weiß, T.; Lehnert, M.; Pesch, B.; Berresheim, H.; Henry, J.; Raulf-Heimsoth, M.; Broding, H.C.; Bünger, J.; Harth, V.; et al. Increased metal concentrations in exhaled breath condensate of industrial welders. *J. Environ. Monit.* **2011**, *13*, 212–218. [CrossRef]
52. Hulo, S.; Chérot-Kornobis, N.; Howsam, M.; Crucq, S.; de Broucker, V.; Sobaszek, A.; Edme, J.L. Manganese in exhaled breath condensate: A new marker of exposure to welding fumes. *Toxicol. Lett.* **2014**, *226*, 63–69. [CrossRef]
53. Riccelli, M.G.; Goldoni, M.; Andreoli, R.; Mozzoni, P.; Pinelli, S.; Alinovi, R.; Selis, L.; Mutti, A.; Corradi, M. Biomarkers of exposure to stainless steel tungsten inert gas welding fumes and the effect of exposure on exhaled breath condensate. *Toxicol. Lett.* **2018**, *292*, 108–114. [CrossRef]
54. Persoons, R.; Arnoux, D.; Monssu, T.; Culié, O.; Roche, G.; Duffaud, B.; Chalaye, D.; Maitre, A. Determinants of occupational exposure to metals by gas metal arc welding and risk management measures: A biomonitoring study. *Toxicol Lett.* **2014**, *231*, 135–141. [CrossRef]
55. Lehnert, M.; Weiss, T.; Pesch, B.; Lotz, A.; Zilch-Schöneweis, S.; Heinze, E.; Van Gelder, R.; Hahn, J.U.; Brüning, T. Reduction in welding fume and metal exposure of stainless steel welders: An example from the WELDOX study. *Int. Arch. Occup. Environ. Health* **2014**, *87*, 483–492. [CrossRef]

56. Wegner, R.; Radon, K.; Heinrich-Ramm, R.; Seemann, R.; Riess, A.; Koops, F.; Poschadel, B.; Szadkowski, D. Biomonitoring results and cytogenetic markers among harbour workers with potential exposure to river silt aerosols. *Occup. Environ. Med.* **2004**, *61*, 247–253. [CrossRef] [PubMed]
57. Schuhmacher, M.; Domingo, J.L.; Agramunt, M.C.; Bocio, A.; Müller, L. Biological monitoring of metals and organic substances in hazardous-waste incineration workers. *Int. Arch. Occup. Environ. Health* **2002**, *75*, 500–506. [CrossRef] [PubMed]
58. Wegner, R.; Heinrich-Ramm, R.; Nowak, D.; Olma, K.; Poschadel, B.; Szadkowski, D. Lung function, biological monitoring, and biological effect monitoring of gemstone cutters exposed to beryls. *Occup. Environ. Med.* **2000**, *57*, 133–139. [CrossRef] [PubMed]
59. Agramunt, M.C.; Domingo, A.; Domingo, J.L.; Corbella, J. Monitoring internal exposure to metals and organic substances in workers at a hazardous waste incinerator after 3 years of operation. *Toxicol. Lett.* **2003**, *15*, 83–91. [CrossRef]
60. Mari, M.; Schuhmacher, M.; Domingo, J.L. Levels of metals and organic substances in workers at a hazardous waste incinerator: A follow-up study. *Int. Arch. Occup. Environ. Health* **2009**, *82*, 519–528. [CrossRef]
61. Wultsch, G.; Mišík, M.; Nersesyan, A.; Knasmueller, S. Genotoxic effects of occupational exposure measured in lymphocytes of waste-incinerator workers. *Mutat. Res.-Genet. Toxicol. Environ. Mutagen.* **2011**, *28*, 3–7. [CrossRef]
62. Kettelarij, J.; Nilsson, S.; Midander, K.; Lidén, C.; Julander, A. Snapshot of cobalt, chromium and nickel exposure in dental technicians. *Contact Dermat.* **2016**, *75*, 370–376. [CrossRef]
63. Beattie, H.; Keen, C.; Coldwell, M.; Tan, E.; Morton, J.; McAlinden, J.; Smith, P. The use of bio-monitoring to assess exposure in the electroplating industry. *J. Expo. Sci. Environ. Epidemiol.* **2017**, *27*, 47–55. [CrossRef]
64. Zare Jeddi, M.; Hopf, N.B.; Viegas, S.; Price, A.B.; Paini, A.; van Thriel, C.; Benfenati, E.; Ndaw, S.; Bessems, J.; Behnisch, P.A.; et al. Towards a systematic use of effect biomarkers in population and occupational biomonitoring. *Environ. Int.* **2021**, *146*, 106257. [CrossRef]
65. Cherrie, J.W.; Levy, L. Managing Occupational Exposure to Welding Fume: New Evidence Suggests a More Precautionary Approach is Needed. *Ann. Work Expo. Health* **2019**, *64*, 1–4. [CrossRef]
66. Maître, A.; Collot-Fertey, D.; Anzivino, L.; Marques, M.; Hours, M.; Stoklov, M. Municipal waste incinerators: Air and biological monitoring of workers for exposure to particles, metals, and organic compounds. *Occup. Environ. Med.* **2003**, *60*, 563–569. [CrossRef] [PubMed]
67. Colman Lerner, J.E.; Elordi, M.L.; Orte, M.A.; Giuliani, D.; de los Angeles Gutierrez, M.; Sanchez, E.Y.; Sambeth, J.E.; Porta, A.A. Exposure and risk analysis to particulate matter, metals, and polycyclic aromatic hydrocarbon at different workplaces in Argentina. *Environ. Sci. Pollut. Res.* **2018**, *25*, 8487–8496. [CrossRef] [PubMed]
68. Campo, L.; Hanchi, M.; Sucato, S.; Consonni, D.; Polledri, E.; Olgiati, L.; Saidane-Mosbahi, D.; Fustinoni, S. Biological Monitoring of Occupational Exposure to Metals in Electric Steel Foundry Workers and Its Contribution to 8-Oxo-7,8-Dihydro-2′-Deoxyguanosine Levels. *Int. J. Environ. Res. Public Health* **2020**, *17*, 1811. [CrossRef] [PubMed]
69. Zeng, H.; Fang, B.; Hao, K.; Wang, H.; Zhang, L.; Wang, M.; Hao, Y.; Wang, X.; Wang, Q.; Yang, W.; et al. Combined effects of exposure to polycyclic aromatic hydrocarbons and metals on oxidative stress among healthy adults in Caofeidian, China. *Ecotoxicol. Environ. Saf.* **2022**, *230*, 113168. [CrossRef] [PubMed]
70. Olea, N.; Fernández, M.F.; Mustieles, V.; David, A.; Barouki, R.; Fini, J.; Demeneix, B.; Lambrechts, N.; Schoeters, G.; Vingaard, A.M.; et al. Criteria for Prioritization of Biomarkers of Effect Deliverable Report WP 14 Effect Biomarkers. 2017. Available online: https://www.hbm4eu.eu/wp-content/uploads/2017/03/HBM4EU_D14.1_Criteria-Prioritization-Biomarkers-of-Effect.pdf (accessed on 12 January 2021).
71. Sudha, S.; Prathyumnan, S.; Keyan, K.S.; Joseph, S.; Vasudevan, B.S.G.; Sasikala, K. Evaluation of DNA damage induction and repair inhibition in welders exposed to hexavalent chromium. *Asian Pac. J. Cancer Prev.* **2010**, *11*, 95–100.
72. Balachandar, V.; Arun, M.; Mohana Devi, S.; Velmurugan, P.; Manikantan, P.; Karthick Kumar, A.; Sasikala, K.; Venkatesan, C. Evaluation of the genetic alterations in direct and indirect exposures of hexavalent chromium [Cr(VI)] in leather tanning industry workers North Arcot District, South India. *Int. Arch. Occup. Environ. Health* **2010**, *83*, 791–801. [CrossRef] [PubMed]
73. El Safty, A.M.K.; Samir, A.M.; Mekkawy, M.K.; Fouad, M.M. Genotoxic Effects Due to Exposure to Chromium and Nickel Among Electroplating Workers. *Int. J. Toxicol.* **2018**, *37*, 234–240. [CrossRef]
74. Galea, K.S.; Porras, S.P.; Viegas, S.; Bocca, B.; Bousoumah, R.; Duca, R.C.; Godderis, L.; Iavicoli, I.; Janasik, B.; Jones, K.; et al. HBM4EU chromates study—Reflection and lessons learnt from designing and undertaking a collaborative European biomonitoring study on occupational exposure to hexavalent chromium. *Int. J. Hyg. Environ. Health* **2021**, *234*, 113725. [CrossRef]
75. Louro, H.; Heinälä, M.; Bessems, J.; Buekers, J.; Vermeire, T.; Woutersen, M.; van Engelen, J.; Borges, T.; Rousselle, C.; Ougier, E.; et al. Human biomonitoring in health risk assessment in Europe: Current practices and recommendations for the future. *Int. J. Hyg. Environ. Health* **2019**, *222*, 727–737. [CrossRef]
76. Ermler, S.; Scholze, M.; Kortenkamp, A. Genotoxic mixtures and dissimilar action: Concepts for prediction and assessment. *Arch. Toxicol.* **2014**, *88*, 799–814. [CrossRef]

toxics MDPI

Article

Improving the Risk Assessment of Pesticides through the Integration of Human Biomonitoring and Food Monitoring Data: A Case Study for Chlorpyrifos

Jose V. Tarazona [1,2,*], Maria del Carmen González-Caballero [1], Mercedes de Alba-Gonzalez [1], Susana Pedraza-Diaz [1], Ana Cañas [1], Noelia Dominguez-Morueco [1], Marta Esteban-López [1], Irene Cattaneo [2], Andromachi Katsonouri [3], Konstantinos C. Makris [4], Thorhallur I. Halldorsson [5,6], Kristin Olafsdottir [7], Jan-Paul Zock [8], Jonatan Dias [9], Annelies De Decker [10], Bert Morrens [11], Tamar Berman [12], Zohar Barnett-Itzhaki [12,13], Christian Lindh [14], Liese Gilles [15], Eva Govarts [15], Greet Schoeters [15,16], Till Weber [17], Marike Kolossa-Gehring [17], Tiina Santonen [18] and Argelia Castaño [1,*]

1 National Centre for Environmental Health, Instituto de Salud Carlos III, 28220 Madrid, Spain; mcgonzalez@isciii.es (M.d.C.G.-C.); malba@isciii.es (M.d.A.-G.); spedraza@isciii.es (S.P.-D.); acanas@isciii.es (A.C.); ndominguez@isciii.es (N.D.-M.); m.esteban@isciii.es (M.E.-L.)
2 European Food Safety Authority (EFSA), I-43126 Parma, Italy; irene.cattaneo@ext.efsa.europa.eu
3 Cyprus State General Laboratory, Ministry of Health, Nicosia 1451, Cyprus; akatsonouri@sgl.moh.gov.cy
4 Cyprus International Institute for Environmental and Public Health, Cyprus University of Technology, Limassol 3036, Cyprus; konstantinos.makris@cut.ac.cy
5 Faculty of Food Science and Nutrition, School of Health Sciences, University of Iceland, 102 Reykjavik, Iceland; tih@hi.is
6 Department of Epidemiology Research, Statens Serum Institut, DK-2300 Copenhagen, Denmark
7 Department of Pharmacology and Toxicology, University of Iceland, 107 Reykjavik, Iceland; stinaola@hi.is
8 National Institute for Public Health and the Environment (RIVM), Bilthoven, 3720 BA De Bilt, The Netherlands; jan-paul.zock@rivm.nl
9 Wageningen Food Safety Research (WFSR), 6700 AE Wageningen, The Netherlands; jonatan.dias@wur.nl
10 APB Provinciaal Instituut voor Hygiëne, 2000 Antwerpen, Belgium; annelies.dedecker@provincieantwerpen.be
11 Department of Sociology, University of Antwerp, 2020 Antwerpen, Belgium; bert.morrens@uantwerpen.be
12 Ministry of Health, Jerusalem 9446724, Israel; tamar.berman@moh.gov.il (T.B.); zoharba@ruppin.ac.il (Z.B.-I.)
13 Ruppin Research Group in Environmental and Social Sustainability, Ruppin Academic Center, Emek Hefer 4025000, Israel
14 Division of Occupational and Environmental Medicine, Institute of Laboratory Medicine, Lund University, 22363 Lund, Sweden; christian.lindh@med.lu.se
15 VITO Health, Flemish Institute for Technological Research (VITO), 2400 Mol, Belgium; liese.gilles@vito.be (L.G.); eva.govarts@vito.be (E.G.); greet.schoeters@vito.be (G.S.)
16 Department of Biomedical Sciences, University of Antwerp, 2020 Antwerp, Belgium
17 German Environment Agency (UBA), 14195 Berlin, Germany; till.weber@uba.de (T.W.); marike.kolossa@uba.de (M.K.-G.)
18 Finnish Institute of Occupational Health, P.O. Box 40 Helsinki, Finland; tiina.santonen@ttl.fi
* Correspondence: jose.tarazona@efsa.europa.eu (J.V.T.); castano@isciii.es (A.C.)

Citation: Tarazona, J.V.; González-Caballero, M.d.C.; Alba-Gonzalez, M.d.; Pedraza-Diaz, S.; Cañas, A.; Dominguez-Morueco, N.; Esteban-López, M.; Cattaneo, I.; Katsonouri, A.; Makris, K.C.; et al. Improving the Risk Assessment of Pesticides through the Integration of Human Biomonitoring and Food Monitoring Data: A Case Study for Chlorpyrifos. *Toxics* **2022**, *10*, 313. https://doi.org/10.3390/toxics10060313

Academic Editor: Christophe Rousselle

Received: 5 May 2022
Accepted: 30 May 2022
Published: 9 June 2022

Publisher's Note: MDPI stays neutral with regard to jurisdictional claims in published maps and institutional affiliations.

Copyright: © 2022 by the authors. Licensee MDPI, Basel, Switzerland. This article is an open access article distributed under the terms and conditions of the Creative Commons Attribution (CC BY) license (https://creativecommons.org/licenses/by/4.0/).

Abstract: The risk assessment of pesticide residues in food is a key priority in the area of food safety. Most jurisdictions have implemented pre-marketing authorization processes, which are supported by prospective risk assessments. These prospective assessments estimate the expected residue levels in food combining results from residue trials, resembling the pesticide use patterns, with food consumption patterns, according to internationally agreed procedures. In addition, jurisdictions such as the European Union (EU) have implemented large monitoring programs, measuring actual pesticide residue levels in food, and are supporting large-scale human biomonitoring programs for confirming the actual exposure levels and potential risk for consumers. The organophosphate insecticide chlorpyrifos offers an interesting case study, as in the last decade, its acceptable daily intake (ADI) has been reduced several times following risk assessments by the European Food Safety Authority (EFSA). This process has been linked to significant reductions in the use authorized in the EU, reducing consumers' exposure progressively, until the final ban in 2020, accompanied by setting all EU maximum residue levels (MRL) in food at the default value of 0.01 mg/kg. We present a comparison of estimates of the consumer's internal exposure to chlorpyrifos based on the urinary marker 3,5,6-trichloro-2-pyridinol (TCPy), using two sources of monitoring data: monitoring of the food chain from the EU program and biomonitoring of European citizens from the HB4EU project, supported by a literature search. Both methods confirmed a drastic reduction in exposure levels from 2016 onwards. The margin of exposure approach is then used for conducting retrospective risk assessments at different time points, considering the evolution of our understanding of chlorpyrifos toxicity, as well as of exposure levels in EU consumers following the regulatory decisions. Concerns are presented using a color code, and have been identified for almost all studies, particularly for the highest exposed group, but at different levels, reaching the maximum level, red code, for children in Cyprus and Israel. The assessment uncertainties are highlighted and integrated in the identification of levels of concern.

Keywords: human biomonitoring; chlorpyrifos; pesticide exposure; pesticide risk assessment; HBM4EU

1. Introduction

In most jurisdictions around the globe, pesticides are subjected to specific authorization requirements [1]. The regulatory decisions are supported by scientifically-based pre-marketing assessments and re-assessments, complemented by enforcement actions including monitoring programs. The authorization of use patterns, also named good agricultural practice (GAP), is supported by setting maximum residue levels in food commodities, established by the Codex Alimentarius at international level as well as by specific legislation within each jurisdiction, such as Regulation (EC) No 396/2005 in the EU.

The risk assessment process follows the standard paradigm for chemical risk assessment. The hazard assessment identifies possible hazards and establishes health-based guidance values (HBGV): the acceptable daily intake (ADI) for chronic life-long exposures, the acute reference dose (ARfD) for exposures within a single day or single meal, and the acceptable operator exposure level (AOEL) for non-oral exposures. The dietary exposure assessment combines the expected or measured pesticide residue levels in food commodities with the consumption of each commodity according to standard diets. While efforts for developing methodologies for combining the risk of concurrent exposure to different pesticides are ongoing, most current regulatory decisions are still based on a substance-by-substance approach.

While these prospective risk assessments are essential for preventing consumer exposure to levels of concern, they are based on agreed assumptions and not on actual consumer exposure. Human biomonitoring (HBM) of relevant markers complements the information with retrospective assessments, and is emerging as a key tool for calibrating and validating the regulatory risk assessment models. The European Human Biomonitoring Initiative (HBM4EU, www.hbm4eu.eu, accessed on 4 May 2022) is a European Joint Program, which aims to harmonize and use biomonitoring to understand human exposure to chemicals in the environment, in occupational settings or in non-occupational settings through the use of consumer products, and the related health risks, in order to improve chemical risk management and to support policy making [2]. In the frame of HBM4EU, aligned biomonitoring studies from across Europe [3,4] provided harmonized internal exposure data to specific chemical substances in different age groups of the general population. Studies included in HBM4EU-aligned studies met a set of inclusion criteria, such as minimal number of participants, minimal set of variables, specific time period for data collection, laboratory QA/QC, as well as specific conditions for reporting. These criteria are not always explicitly considered in published monitoring studies. In order to further explore the comparison of both monitoring approaches, this work includes a second set of human biomonitoring studies, retrieved from a scientific literature search. As chlorpyrifos has been extensively studied, it was decided to focus on a single EU country, not included in the selected aligned studies, and with a number of available monitoring studies covering at least a decade.

It is particularly interesting to compare the risk estimates based on the occurrence of the substance in food with those based on biomonitoring in humans, as both represent real exposure levels; and the organophosphate insecticide chlorpyrifos offers a stimulating case study. In the last decade, the initial acceptable daily intake (ADI), established in 2006 as 0.01 mg/kg body weight per day, has been reduced several times following EFSA assessments, triggering a reduction in the authorized use and maximum residue levels (MRLs) and, therefore, a progressive reduction in the estimated exposure of European citizens. In 2012, a review was required due to new toxicological studies, and two years later the EFSA proposed a reduction of the ADI to 0.001 mg/kg body weight per day [5]. The EFSA assessment triggered a re-assessment of the MRLs in 2015 [6], which were incorporated into the EU legislation in 2016. In 2019, the EFSA concluded that toxicological reference values for this substance could not be established [7]. Similar conclusions were obtained for chlorpyrifos-methyl [8]. Subsequently, in 2020, the approval of the active substance chlorpyrifos was not renewed and existing authorizations for plant protection products containing chlorpyrifos in the EU Member States were revoked; and all MRL were set at the default value of 0.01 mg/kg as indicated in the EU legislation.

Chlorpyrifos exposure can be monitored in humans through two urine biomarkers, 3,5,6-trichloro-2-pyridinol (TCPy) and the group of alkyl phosphates (AP). Biological reference values for both biomarkers have been proposed [9]. TCPy is a common metabolite with the closely related pesticide chlorpyrifos-methyl and has been selected for the work presented herein. The previously proposed reference values should be updated considering the recently raised concerns on genotoxicity potential and developmental neurotoxicity. Following the EFSA assessment that an ADI can no longer be proposed, the risk assessment presented here is based on the alternative approach, quantifying the margins of exposure for a set of selected points of departure (PoD).

2. Materials and Methods

2.1. Data and Information Sources

Toxicity data were retrieved from EFSA conclusions and related documents published in the EFSA Journal https://efsa.onlinelibrary.wiley.com/journal/18314732 (accessed on 25 March 2022) or publicly available through the Open-EFSA web https://open.efsa.europa.eu/ (accessed on 25 March 2022). A literature search using Web Of Science and SCOPUS was conducted for retrieving toxicokinetic information focusing on studies with human volunteers.

Aggregated (percentiles for the full dataset) HBM data (TCPy) were obtained from HBM4EU-aligned studies measuring this metabolite of chlorpyrifos. These included six studies on adults (PT INSEF, Study CH, IL RAVMABAT, FR ESTEBAN, IS Diet_HBM, DE ESB), with an age range of 20–39 years, from France, Germany, Iceland, Israel, Portugal and Switzerland, conducted in the period 2014–2021; and six studies on children (FR Esteban, SI SLOCRP, NL SPECIMEN, BE 3xG, CY Organiko, IL RAVMABAT), with an age range of 6–11 years, from Belgium, Cyprus, France, Israel, Slovenia and the Netherlands, conducted in the period 2014–2020.

A literature search, using Web Of Science and SCOPUS, was conducted for retrieving published monitoring data. Following a pre-screening, Spain was selected as a relevant country for covering published human biomonitoring studies from projects other than HBM4EU, providing a complementary assessment for children and adult exposure in Southern Europe. The search focused on data from Spain and the list of studies was completed with a secondary search covering the references and citations of the identified studies.

2.2. Estimation of HBM Values

The HBMPoD represents the human biomonitoring PoD estimated through the adaptation of the general equation proposed under HBM4EU [10] for deriving human biomonitoring guidance values (HBM-GV). The HBGV is replaced by the PoD and combined with the molar urinary excretion fraction of the metabolite TCPy (Fue(TCPy)), corrected by the molecular weight, MW, differences. The adapted equation is:

$$\mathrm{HBM} - \mathrm{PoD(endpoint)} = \frac{\mathrm{PoD(endpoint)} \times \left[\frac{\mathrm{MW(TCPy)} \times \mathrm{Fue\ (TCPy)}}{\mathrm{MW(chlorpyrifos)}}\right]}{\text{Daily urinary excretion adjusted to the bw}} \quad (1)$$

The hazard assessment is based on the EFSA conclusions published in 2011 [11], 2014 [5] and 2019 [7]. The reasoning for the evolution of the proposed ADI since 2011 and the most recent conclusion, are analyzed. Several points of departure—PoD (also named toxicological reference points—RP) have been selected to cover the most relevant observed effects. These PoDs represent the no-observed-adverse-effect level (NOAEL), lowest-observed-adverse-effect level (LOAEL) or benchmark dose (BMD) for specific effects extracted from the dose–response curve of the relevant studies. The uncertainties mentioned in the EFSA assessment [6] are considered for the interpretation of the margins of exposure (MoEs).

The toxicokinetic information was retrieved from a literature search (SCOPUS and Web of Science) focusing on human data.

2.3. *Estimation of Urinary TCPy Levels from EU Food Monitoring Data*

Urinary TCPy levels (µg/L) were calculated from the reported dietary risk data for different European diets (expressed in highest calculated exposure in % of ADI) retrieved from the EFSA Pesticide Residues Intake Model (PRIMo) [12] included in the European Union Annual Reports on pesticide residues in food from 2012 to 2019 [13–20]. Considering that TCPy is a common metabolite for chlorpyrifos and chlorpyrifos methyl, the external dose in mg/kg bw per day was estimated from the reported risk and the selected ADI value of both compounds. The obtained values were converted into molar units (mol/kg bw/day) and the molar contribution of TCPy from chlorpyrifos and chlorpyrifos methyl was summed up for each relevant diet. The total urinary TCPy concentration (µg/L) was then calculated, applying the TCPy molar excretion fraction of 0.7 [21] and assuming a urinary output of 24 mL/kg bw/day [22]. It should be noted that this estimation does not cover possible exposure from non-dietary sources.

2.4. *Risk Characterization*

Both prospective and retrospective risk assessments are based on the MoE approach; applied to each selected HBM-PoD endpoint to express the risk as the MoE(endpoint) according to the following equation:

$$MoE(endpoint) = \frac{HBM - PoD(endpoint)}{Exposure\ level} \quad (2)$$

Prospective biomarker exposure estimations were calculated by applying toxicokinetic considerations to chlorpyrifos and chlorpyrifos-methyl exposure estimations, as published in EFSA Annual Reports on monitored levels in food.

Retrospective biomarker exposure estimations included data provided by the HBM4EU project and a literature search focused on published biomonitoring data for Spain (SCOPUS and Web of Science).

The risk characterisation was based on individualised comparisons of the observed MoEs with the relevant uncertainty factors, covering both generic (i.e., interspecies and intraspecies differences, non-threshold genotoxicity mechanisms) and specific uncertainties connected to the hazard, toxicokinetic and exposure data.

3. Results and Discussion

3.1. Evolution of Chlorpyrifos Hazard Characterisation in the EU

Chlorpyrifos was included in Annex I to Directive 91/414/EEC in 2006 by CD 2005/72/EC and the following reference doses were established: ADI and AOEL: 0.01 mg/kg bw per day, and ARfD: 0.1 mg/kg bw per day based on brain cholinesterase (AChE) inhibition and neurotoxic findings, respectively, as critical endpoints.

In 2013, after revision of new toxicological studies and scientific papers, the EFSA peer-review expert meeting agreed on the use of the red blood cell (RBC) AChE inhibition, instead of brain AChE inhibition, to derive the reference values, which were established as 0.001 mg/kg bw per day for both ADI and AOEL, and 0.005 mg/kg bw for ARfD [5].

After the process of renewal of approval, initiated in 2017, and the PPR 01 Experts' meeting in 2019, experts agreed that the point of departure (PoD) for chlorpyrifos should be the developmental neurotoxicity (DNT) LOAEL of 0.3 mg/kg per day [7]. Regarding the general assessment, it was concluded that there were several uncertainties: the genotoxicity potential remains unclarified and the effects recorded in the DNT rat study indicate a concern that was supported by the epidemiological evidence related to developmental neurological outcomes in children for chlorpyrifos. In any case, overall, no reference values could be set because of the unclear genotoxicity potential of chlorpyrifos. Moreover, the approval of the active substance chlorpyrifos was not renewed and all existing authorizations for plant protection products containing chlorpyrifos in EU Member States were revoked by February 2020. The pesticide residues and MRLs (mg/kg) for chlorpyrifos were set at the default lowest limits of analytical determination (LODs) of 0.01 mg/kg in all products. The regulation was in force as of 13 November 2020.

In line with the more recent EFSA assessment [7], the following PoDs for chlorpyrifos were selected for this risk assessment:

- Overall PoD based on the DNT study on rats, as adverse effects were observed at the lowest tested dose the PoD was the LOAEL of 0.3 mg/kg bw per day
- Relevant long-term NOAEL 0.1 mg/kg bw per day, also applicable to parental toxicity and maternal NOAEL
- Short-term NOAEL for red blood cells AChE inhibition 0.1 mg/kg bw per day, same value as above but related to short-term exposures
- Relevant offspring NOAEL 1 mg/kg bw per day
- Relevant reproductive NOAEL 5 mg/kg bw per day
- Relevant carcinogenicity NOAEL 10 mg/kg bw per day (highest dose tested)

The specific toxicity of the metabolite TCPy, which is part of the residue definition in food, was also assessed and EFSA proposed an ADI of 0.06 mg/kg bw per day [5–7].

The EFSA assessment for chlorpyrifos-methyl [8] proposes the same overall PoD (based on chlorpyrifos study) and also for long-term toxicity. No value is proposed for short-term AChE, and the other PoDs are slightly higher: 3, 10 and 40 mg/kg bw per day for offspring, reproductive and carcinogenicity, respectively [8].

3.2. Proposed HBM-PoDs

The metabolism of chlorpyrifos has been extensively studied in both animals and humans [7,23]. Basically, the fraction of chlorpyrifos absorbed in the intestine is nearly completely converted to equimolar amounts of TCPy and alkyl phosphate metabolites; and both have been used as potential human biomarkers [24]. Alkyl phosphates cover a broad group of pesticides, while TCPy is only common in the closely related pesticide chlorpyrifos-methyl; thus TCPy was selected for this assessment. In a pharmacokinetic study with six human volunteers [21], the percentage of the administered oral dose recovered in urine as TCPy ranged between 49–81% with an average of 70%. The predicted percentage of absorbed dose in the same volunteers ranged between 52–84%, with an average of 72%. The good correlation between absorption and molar TCPy excretion confirms the capacity of TCPy as a biomarker with equimolar conversion of the absorbed dose. Other studies with human volunteers [25] and animals [5,7] have identified higher oral absorption rates, up to

93%. Based on these findings, the value of 70% was selected as the central estimate within a 50–93% range, corresponding to a molar urinary excretion fraction, Fue, of 0.7 (0.5–0.93 range). The molar ratio of TCPy to chlorpyrifos is 0.566. The 24 h daily urinary excretion selected by HBM4EU is 0.02 L/kg bw for adults and 0.03 L/kg bw for children [26].

Following this selection and the equation above, Table 1 presents the proposed HBM-PoDs to be used in the risk characterization.

Table 1. Proposed human biomonitoring points of departure (HBM-PoD) for chlorpyrifos following the EFSA [6] hazard characterization.

Endpoint	EFSA PoD Value mg/kg bw day	HBM-PoD Adults mg/L	HBM-PoD Children mg/L
Overall (based on DNT LOAEL)	0.3 [1]	5.94	3.96
Long-term and maternal toxicity NOAEL	0.1	1.98	1.32
Short-term NOAEL for red blood cells AChE	0.1	1.98	1.32
Offspring NOAEL	1	19.81	13.21
Reproductive NOAEL	5	99.05	66.03
Carcinogenic NOAEL	10	198.10	132.07

[1] This value is based on an LOAEL, as effects were observed at the lowest tested dose.

In line with the EFSA assessment [8], the proposed HMB-PoDs are also relevant for exposure to chlorpyrifos-methyl and or a combination of both pesticides; being slightly conservative for the last three endpoints in the cases where chlorpyrifos-methyl is the main contributor.

3.3. Prospective Exposure Assessment Based on Monitored Levels in Food

The predicted TCPy urinary levels in the EU population, estimated from monitoring of chlorpyrifos in foodstuffs, are presented in Figure 1, using data from the European Union Annual Reports on pesticide residues in food from 2012 to 2019 [13–20]. Each EU report includes estimations for several diets; some diets cover generic clusters, while others are provided by the EU Member States for generic or specific national population groups, including both adults and children. Since 2012, the EFSA has updated PRIMo several times and the updates have included modifications in the diets in each version. In order to use exclusively publicly available data, the estimations were conducted for the diets reported for each year. The summary statistics are presented in Figure 1 as box and whisker plots presenting the range (vertical lines), upper and lower quartiles (box), if needed outliers covering the maximum and the minimum estimations (single dots), the average value (x) and the 50th percentile (horizontal line in the box).

The results showed a drastic reduction in exposure levels between 2015 and 2016, reflecting the regulatory measures adopted in the EU, as well as a general tendency towards exposure reduction between 2012 and 2015 and between 2016 and 2019. Individual estimates for each diet are reported in Table S1 at the Supplementary Material section.

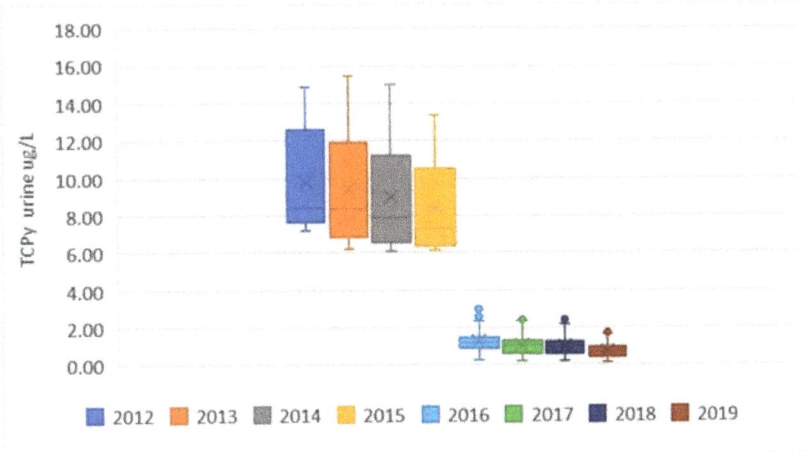

Figure 1. Predicted TCPy levels in EU consumers, estimated from chlorpyrifos levels in food extracted from the EU annual monitoring programs between 2012 and 2019. Data presented as box and whisker plots (see text for details).

3.4. *Retrospective Exposure Assessment Based on Human Biomonitoring*

3.4.1. HBM4EU Data

Aggregated TCPy HBM data from HBM4EU-aligned studies measuring this biomarker for chlorpyrifos exposure in the general European population are presented in Table 2. These studies cover children from 6–11 years of age, in the period 2014–2020 and young adults 20–39 years of age, in the period 2014–2021 from different European countries and Israel. The 50th and 95th percentiles were selected for representing the exposure levels of the average population and the highest exposed group, respectively. In addition, the upper level of the 95th confidence interval of the 95th percentile has been selected for assessing the uncertainty in the exposure levels of the highest exposed group.

Table 2. Selected aggregated TCPy HBM data from HBM4EU-aligned studies. For France, some values are not reported (n.r.) due to high percentage of samples below the level of detection.

Population Group	Country	P50 µg/L	P95 µg/L	Upper 95 CI µg/L
Children	Belgium	1.22	3.24	5.05
	Cyprus	6.52	13.82	15.74
	France	n.r.	n.r.	n.r.
	Israel	2.80	18.38	28.84
	Slovenia	0.61	3.08	4.92
	The Netherlands	1.13	3.49	5.55
Adults	France	n.r.	n.r	0.06
	Germany	0.82	2.87	3.87
	Iceland	0.61	2.07	3.30
	Israel	2.75	11.22	55.22
	Portugal	1.86	7.35	8.37
	Switzerland	0.97	3.64	4.72

3.4.2. Spanish Human Biomonitoring Data

The available biomonitoring studies covering the Spanish population have been recently reviewed Yusa et al. [27]. No other studies were identified in the literature search. The studies cover children and several adult groups, and are distributed between 2003 and 2019. Setting comparisons is a challenge due to differences in both the study design and the reported information. Roca et al. [28] only reported creatinine adjusted values, the data were corrected to unadjusted values using the correlation derived from creatinine adjusted vs. non-adjusted data from Fernandez et al. [29]. Figure 2 summarizes the comparison of TCPy urinary levels estimated from food with those from human monitoring for the general population, including data for pregnant and lactating women while excluding farm workers. As the data for Spanish adults only covers the 2016–2019 period, data from the other Iberian country, Portugal, have been added to support the comparison.

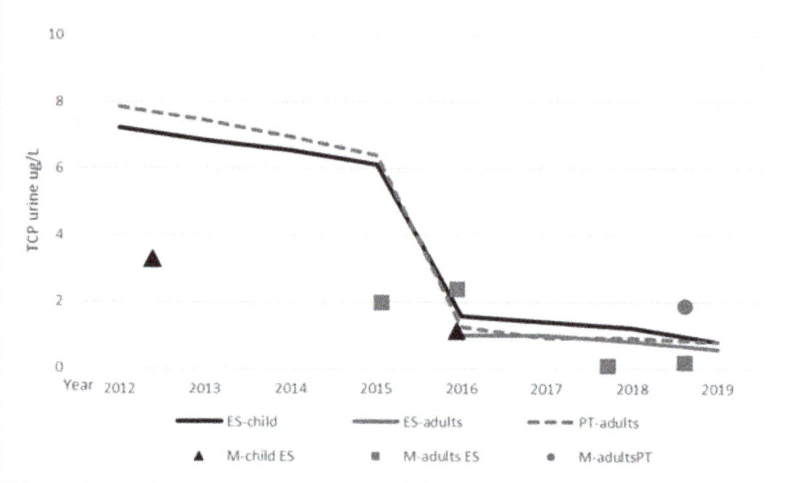

Figure 2. TCPy levels estimated from food (lines) vs. human monitoring (M-) for Spain and Portugal.

Although the data available for the comparison are limited, the results suggest a general agreement between median measured levels and those estimated from food monitoring and the diets for Spain and Portugal. For five databases, information on both 50th and 95th percentiles was provided, the 95/50 ratio was around four for three databases and around 10 for the other two, confirming high individual variability.

3.5. *Estimated MoEs for the Prospective Assessment*

Figures 3–5 present the time evolution (from 2012 to 2019) of the estimated MoEs for the different EU diets and endpoints. European Union annual reports on pesticide residues in food from 2012 to 2019 included modifications in the diets used in different years. The comparison was based on the minimum, average and maximum estimations for the children and adult/general diets used each year. The proposed thresholds (see Section 4, Risk Characterization) have been included in the figures to facilitate the results' interpretation.

In line with the exposure reduction, a clear temporal trend is observed for all PoDs, with a significant point of inflection in 2016, coinciding with the modification of the MRLs in this same year, following the 2014 EFSA conclusion and 2015 EFSA MRL review report. Further modifications in the MRLs and also regarding actual use contributed to additional temporal trends.

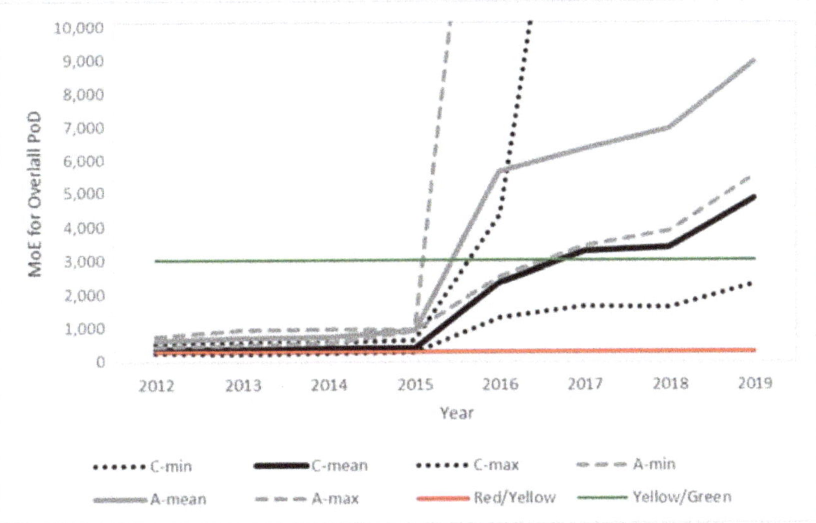

Figure 3. Time evolution of the margins of exposure for children (black) and adults (grey) for the overall PoD predicted from the food monitoring data. Solid lines represent the mean value and dotted lines the maximum and minimum values for children (C) and adults (A), respectively. Color lines indicate the thresholds between the risk levels (see Section 4 for details).

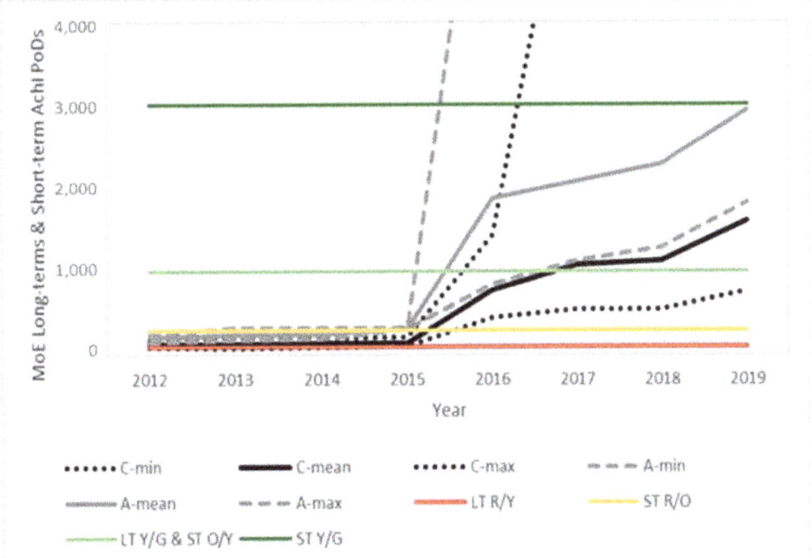

Figure 4. Time evolution of the margins of exposure for children (black) and adults (grey) for the long-term and short-term AChE PoDs predicted from the food monitoring data. Solid lines represent the mean value and dotted lines the maximum and minimum values for children (C) and adults (A), respectively. Color lines indicate the thresholds between the risk levels (see Section 4 for details).

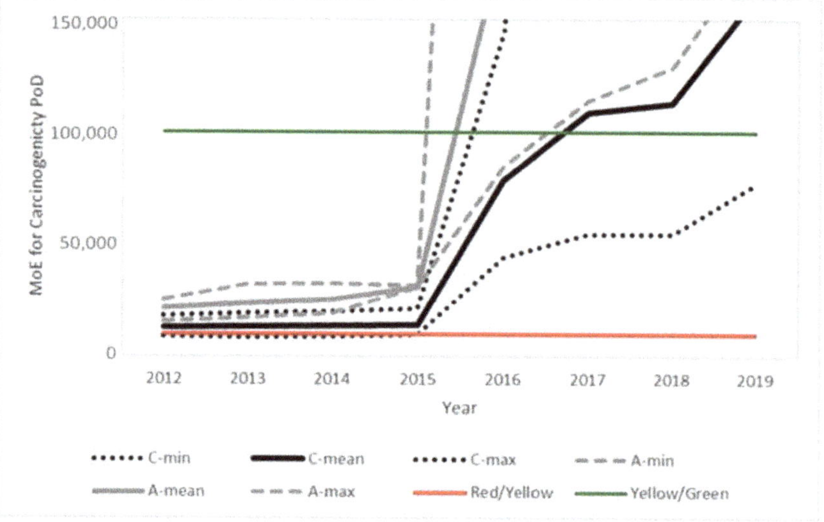

Figure 5. Time evolution of the margins of exposure for children (black) and adults (grey) for the carcinogenicity PoD predicted from the food monitoring data. Solid lines represent the mean value and dotted lines the maximum and minimum values for children (C) and adults (A), respectively. Color lines indicate the thresholds between the risk levels (see Section 4 for details).

3.6. Estimated MoEs for the Retrospective Assessment

3.6.1. HBM4EU Data

The estimated MoE ranges for the HBM4EU-aligned studies reporting TCPy urinary levels are summarized in Table 3. The MoEs have been estimated for the percentiles 50th; 95th and its upper 95th confidence interval, in order to characterize the risk for the average population, the high exposed group, and the highest exposed individuals, respectively. Individual estimations for each country and population group are reported in Table S2 in the Supplementary Material section.

Table 3. Margins of exposure ranges for the HBM4EU-aligned studies and different endpoints. P50 represents the average population, P95 the high exposed group and upper CI P95 the most exposed individuals.

Population Group	Endpoint	MoE Range P50	MoE Range P95	MoE Range Upper CI P95
Children 6–11 2014–2020	Overall LOAEL	607–6462	215–1287	137–804
	Long-term	203–2154	72–429	46–268
	Short-term AChE	203–2154	72–429	46–268
	Carcinogenicity	47,180–215,519	7184–42,926	4579–26,824
Adults 20–39 2014–2021	Overall LOAEL	2159–7244	529–2870	108–1800
	Long-term	720–2415	176–957	36–600
	Short-term AChE	720–2415	176–957	36–600
	Carcinogenicity	72,010–241,585	17,648–95,701	3587–60,030

3.6.2. Spanish and Portuguese Populations

The estimated MoE ranges for the available Spanish studies reporting TCPy urinary levels and for the HBM4EU dataset from Portugal are summarized in Table 4. MoEs have been estimated for the 50th percentile (or the geomean if not available); 95th (or the 75th if not available), and the maximum value (or upper 95th confidence interval), in order to characterize the risk for the average population, the high exposed group, and the highest exposed individuals, respectively.

Table 4. Margins of exposure for the different endpoints estimated for the studies covering the Spanish population. HBM4EU data for Portugal are included for comparison.

Study	Endpoint	MoE P50	MoE P95	MoE Max
Roca et al. [28] Valencia Children 6–11 N = 125; 2010	Overall LOAEL Long-term Short-term AChE Carcinogenicity	1201 400 400 4007	320 107 107 1066	33 11 11 112
Fernandez et al. [29] Valencia Children 5–12 N = 568; 2016	Overall LOAEL Long-term Short-term AChE Carcinogenicity	3504 1168 1168 116,876	357 119 119 11,920	39 13 13 1292
Suarez et al. [30] Andalusia Adolescents 15–17 N = 117; 2017–2019	Overall LOAEL Long-term Short-term AChE Carcinogenicity	247,500 82,500 82,500 5,282,675	58,929 19,643 19,643 1,257,780	4091 1364 1364 87,317
Llop et al. [31] Valencia Pregnant women N = 573; 2003–2006	Overall LOAEL Long-term Short-term AChE Carcinogenicity	12,122 4041 4041 404,186	1800 600 600 60,030	51 17 17 1689
Fernandez et al. [32] Valencia Lactating women N = 116; 2015	Overall LOAEL Long-term Short-term AChE Carcinogenicity	2970 990 990 99,050	752 251 251 25,076	354 118 118 11,792
Gari et al., [33] Catalonia Adults N = 80; year not reported	Overall LOAEL Long-term Short-term AChE Carcinogenicity	2475 825 825 82,542		675 225 225 22,511
Gari et al., [33] Catalonia Farm workers N = 45; year not reported	Overall LOAEL Long-term Short-term AChE Carcinogenicity	1414 471 471 47,167		297 99 99 9905
HBM4EU Portugal Adults N = 296; 2019–2020	Overall LOAEL Long-term Short-term AChE Carcinogenicity	3193 1064 1064 106,477	808 269 269 26,952	710 237 237 23,680

4. Risk Characterization

The interpretation of the MoEs required specific consideration, based on the selected PoD and the associated health concerns. Some were based on the extrapolation factors proposed for the derivation of HBGV [34–36], while others were specifically associated with the assessment of substances with potential genotoxicity concerns [37].

As a first step, any MoE below 100 for a PoD based on animal studies should be considered as a clear concern, considering the minimum factor of 100 required to cover intra- and interspecies extrapolation when setting HBGVs.

The overall PoD is based on a LOAEL instead of a NOAEL, requiring an additional factor. According to the ECHA guidance, a minimum factor of three is generally applicable, while a factor of up to 10 may be needed on some occasions. Consequently, any MoE below 300 represented a concern, and the concerns may be extended to MoEs between 300 and 1000, as the EFSA assessment [7] identified a number of relevant limitations for this study, including reduced exposure duration that would also trigger the need for an additional factor.

When the MoE was based on a short-term PoD, additional considerations were needed for assessing the monitoring data reflecting chronic exposure. An uncertainty factor (UF) of two is proposed by the EFSA and ECHA for the extrapolation of subchronic to chronic studies, while larger factors are needed for acute to subacute and subacute to subchronic extrapolations. The available study included acute (single dose) and subacute (repeated exposure from postnatal day 11 to 22), and a factor of five was identified between the acute and subacute NOAELs. For the subacute to subchronic extrapolation, ECHA recommends a factor of three, the extrapolation from subacute to chronic is not recommended by EFSA. Consequently a factor of three could be considered, however, it should be noted that according to the EFSA assessment [7], for chlorpyrifos the same NOAEL for RBC AChE activity inhibition of 0.1 mg/kg bw per day is applicable for both short-term and long-term exposure, thus the use of an additional factor of three is conservative in this specific case.

For the assessment of carcinogenicity, EFSA [35] proposed that an MoE of 10,000 or higher based on the benchmark dose level ($BMLDL_{10}$) from an animal carcinogenicity study, and taking into account all uncertainties, could be considered of low concern for public health. Although a $BMLDL_{10}$ was not available, the EFSA assessment concluded that for chlorpyrifos, carcinogenicity was of low concern and proposed a PoD based on the NOAEL that could be used as surrogate. This approach has been developed for impurities and other substances not intentionally added to food. As chlorpyrifos is no longer authorized in the EU, the approach is applicable for an assessment based on the current situation, although the monitoring data represented residues in line with the MRLs established at the time of monitoring.

One additional element is the consideration of the severity and nature of the observed effect. Specifically for pesticides the legislation indicates that an additional factor may be considered and applied for some effects including development neurotoxicity, and factors up to 10 have been applied in some pesticide risk assessments.

Based on these considerations, the MoEs can be grouped into four risk categories:

- RED: Confirmed concern: the MoE is lower than the requirements for uncertainty factors in standard assessments, i.e., lower than 100 for the NOAELs, lower than 300 for the LOAELs, and lower than 10,000 for the carcinogenicity of genotoxic substances
- ORANGE: Possible concern: the MoE is lower than the requirements for uncertainty factors in standard assessments plus the upper range regarding additional considerations for the extrapolation (factor of 10 for NOAEL to LOAEL, and factor of 3 for subacute to subchronic extrapolation)
- YELLOW: Concerns cannot be excluded. MoEs higher than those above but not offering an additional margin of 10.
- GREEN: Risk cannot be excluded due to the concerns on genotoxicity but it is expected to be very low: MOEs providing an additional margin of at least 10 from those of possible or confirmed concern.

According with this proposal, a summary of the risk characterization is graphically presented in Table 5.

Table 5. Graphic representation of the identified risk levels. Color code: RED: confirmed concern; ORANGE: possible concern (only applicable to OA and ST); YELLOW: concerns cannot be excluded; GREEN: risk cannot be excluded due to the concerns on genotoxicity but is expected to be very low. The letters refer to the PoD(s) reaching the reported concern level: AO—overall PoD; ST—short term AChE, LT—long term NOAEL, C—carcinogenicity, All—all PoDs at this level.

Study	Population Group	Country/Region	Average Population	High Exposed Group	Highest Individual
Roca et al. [28]	Children	ES-Valencia	C	C	All
Fernandez [29]	Children	ES-Valencia	ST	ST	All
Suarez et al. [30]	adolescents	ES-Andalusia			ST/C

Table 5. *Cont.*

Study	Population Group	Country/Region	Average Population	High Exposed Group	Highest Individual
HBM4EU-aligned studies	Children	*Belgium*	OA/ST	All	ST
	Children	*Cyprus*	ST	All	All
	Children	*Israel*	All	All	All
	Children	*Netherlands*	OA/ST	All	ST
	Children	*Slovenia*	OA/ST	All	ST
Llop et al. [31]	Pregnant women	ES-Valencia		All	All
Fernandez [32]	lactating women	ES-Valencia	All	OA/ST	OA/ST
Gari et al. [33]	Adults	ES-Catalonia	All	No data	OA/ST
Gari et al. [33]	farm workers	ES-Catalonia	All	No data	OA/ST
HBM4EU-aligned studies	Adults	*Germany*	OA/ST	All	All
	Adults	*Iceland*	OA/ST	All	All
	Adults	*Israel*	All	OA/ST	All
	Adults	*Portugal*	OA/LT/ST	OA/ST	OA/ST
	Adults	*Switzerland*	OA/LT/ST	All	All

The comparison of the published biomonitoring data for Spanish population groups with the HBM4EU-aligned studies indicated some common elements as well as relevant differences. Commonalities were observed regarding the diversity of the results but with a higher risk for children than for adults, and the high relevance of the overall and short-term PoD endpoints. The consistency is particularly relevant when comparing the Spanish data with the aligned HBM4EU data for Portugal, further supporting the similarities observed in the levels estimated from the food monitoring data for the two Iberian countries (Figures 3–5). The highest risks were identified in a study conducted by Roca et al. [28] in 2014, before the 2016 modification of MRLs in the EU, and for Israel, not covered by the EU MRL regulation. Some Spanish datasets focused on a specific population groups. An interesting finding, to be further investigated, was the low risk observed for Spanish pregnant women, observed by Llop et al. [31] for samples taken in 2003–2006, when high MRLs were allowed in the EU, and confirmed by Bravo et al. [38] for samples collected in 2016–2017. The measured TCPy levels were similar or lower than those reported in other areas of the world [39–41], thus the risk assessment conclusions can be extrapolated to other regions.

The variability within the sampled population is evidenced by the change in color, and was large for some, but not all, databases.

5. Conclusions

Human biomonitoring data provide information on actual exposure levels, and in combination with guidance values, on risk levels including intrapopulation variability. The main limitation is the availability of proper urinary biomarkers; for chlorpyrifos TCPy also covers exposure to the closely related pesticide chlorpyrifos-methyl, as well as direct exposure to the metabolite itself, as TCPy is part of the residue definition in several food commodities. The proposal presented in this study demonstrates that this approach can be extended to those cases when an HBGV cannot be established. The combination of several PoDs, covering endpoints with different levels of concern for public health, and different exposure values, provide informative risk characterizations to support decision making. For chlorpyrifos, the monitoring data have confirmed the need for action, providing support to the regulatory decisions adopted in the EU. Promising results have been obtained regarding the comparison of the prospective assessment using monitoring data in food, and retrospective assessments using human biomonitoring; however additional studies are needed to generalize this opportunity; and this approach should be further explored using other pesticides in order to improve current predictive models through a calibration exercise. These improved models could be applied for pesticides with no suitable biomarkers for human biomonitoring, as well as for updating the premarketing risk assessments supporting the authorization and setting of maximum pesticide residue levels.

Supplementary Materials: The following supporting information can be downloaded at: https://www.mdpi.com/article/10.3390/toxics10060313/s1, Table S1: Individual TPCy estimations for each PRIMo diet; Table S2: Individual MOE estimations for each HBM4EU country and population group.

Author Contributions: Conceptualization, J.V.T.; formal analysis, J.V.T. and I.C.; resources, A.K., K.C.M., T.I.H., K.O., J.-P.Z., J.D., A.D.D., B.M., T.B., Z.B.-I., C.L., L.G., E.G., G.S., T.W., M.K.-G., writing—original draft preparation, J.V.T., M.d.C.G.-C., M.d.A.-G., S.P.-D. and I.C.; writing—review and editing, A.C. (Ana Cañasare), N.D.-M., M.E.-L., A.K., K.C.M., T.I.H., K.O., J.-P.Z., J.D., A.D.D., B.M., T.B., Z.B.-I., C.L., L.G., E.G., G.S., T.W., M.K.-G., T.S. and A.C. (Argelia Castaño); visualization, J.V.T.; supervision, J.V.T. and A.C. (Argelia Castaño); project administration, T.S. All authors have read and agreed to the published version of the manuscript.

Funding: This study was part of the HBM4EU project receiving funding from the European Union's Horizon 2020 research and innovation programme under grant agreement No. 733032.

Institutional Review Board Statement: Not applicable.

Informed Consent Statement: Not applicable as only aggregated human monitoring data are reported. For details see HBM4EU project https://www.hbm4eu.eu/ (accessed on 4 May 2022).

Data Availability Statement: Summary of EFSA toxicity assessments, PRIMo model, and monitored levels in food are available through the documents published in the EFSA Journal https://efsa.onlinelibrary.wiley.com/journal/18314732, (accessed on 25 March 2022) at Open-EFSA web https://open.efsa.europa.eu/, (accessed on 25 March 2022), or Zenodo https://zenodo.org/record/6322020#.YkSAbChBzD4, (accessed on 25 March 2022).

Acknowledgments: This study was part of the HBM4EU project receiving funding from the European Union's Horizon 2020 research and innovation programme under grant agreement No 733032. The authors want to thank Lars Niemann for his relevant comments and suggestions.

Conflicts of Interest: The authors declare no conflict of interest. Jose Tarazona is on leave on personal grounds from his position as Research Professor at the Spanish National Centre for Environmental Health and employed as staff scientist by the European Food Safety Authority (EFSA). Irene Cattaneo is a trainee at EFSA. The views expressed in this publication are those of the authors and should not be interpreted as representing the official position of the European Food Safety Authority (EFSA). EFSA assumes no responsibility or liability for any errors or inaccuracies that may appear.

References

1. OECD. *Report on the OECD WORKSHOP on Sustainable Pest Management in Practice: Anticipating and Adapting to Changes in the Pesticides Regulatory Landscapes*; OECD: Paris, France, 2018.
2. Ganzleben, C.; Antignac, J.-P.; Barouki, R.; Castaño, A.; Fiddicke, U.; Klanova, J.; Lebret, E.; Olea, N.; Sarigiannis, D.; Schoeters, G.; et al. Human biomonitoring as a tool to support chemicals regulation in the European Union. *Int. J. Hyg. Environ. Health* **2017**, *220 Pt 2A*, 94–97. [CrossRef]
3. Gilles, L.; Govarts, E.; Rambaud, L.; Vogel, N.; Castaño, A.; López, M.E.; Martin, L.R.; Koppen, G.; Remy, S.; Vrijheid, M.; et al. HBM4EU combines and harmonises human biomonitoring data across the EU, building on existing capacity—The HBM4EU survey. *Int. J. Hyg. Environ. Health* **2021**, *237*, 113809. [CrossRef] [PubMed]
4. López, M.E.; Göen, T.; Mol, H.; Nübler, S.; Haji-Abbas-Zarrabi, K.; Koch, H.M.; Kasper-Sonnenberg, M.; Dvorakova, D.; Hajslova, J.; Antignac, J.P.; et al. The European human biomonitoring platform-Design and implementación of a laboratory quality assurance/quality control (QA/QC) programme for selected priority chemicals. *Int. J. Hyg. Environ. Health* **2021**, *234*, 113740. [CrossRef] [PubMed]
5. EFSA (European Food Safety Authority). Conclusion on the peer review of the pesticide human health risk assessment of the active substance chlorpyrifos. *EFSA J.* **2014**, *12*, 3640. [CrossRef]
6. EFSA (European Food Safety Authority). Reasoned opinion on the refined risk assessment regarding certain maximum residue levels (MRLs) of concern for the active substance chlorpyrifos. *EFSA J.* **2015**, *13*, 4142. [CrossRef]

7. EFSA (European Food Safety Authority). Statement on the available outcomes of the human health assessment in the context of the pesticides peer review of the active substance chlorpyrifos. *EFSA J.* **2019**, *17*, e05809. [CrossRef]
8. EFSA (European Food Safety Authority). Updated statement on the available outcomes of the human health assessment in the context of the pesticides peer review of the active substance chlorpyrifos-methyl. *EFSA J.* **2019**, *17*, e05908. [CrossRef]
9. Bouchard, M.; Carrier, G.; Brunet, R.C.; Bonvalot, Y.; Gosselin, N.H. Determination of Biological Reference Values for Chlorpyrifos Metabolites in Human Urine Using a Toxicokinetic Approach. *J. Occup. Environ. Hyg.* **2005**, *2*, 155–168. [CrossRef]
10. Apel, P.; Rousselle, C.; Lange, R.; Sissoko, F.; Kolossa-Gehring, M.; Ougier, E. Human biomonitoring initiative (HBM4EU)—Strategy to derive human biomonitoring guidance values (HBM-GVs) for health risk assessment. *Int. J. Hyg. Environ. Health* **2020**, *230*, 113622. [CrossRef]
11. European Food Safety Authority. Conclusion on the peer review of the pesticide risk assessment of the active substance chlorpyrifos. *EFSA J.* **2011**, *9*, 1961. [CrossRef]
12. EFSA (European Food Safety Authority); Anastassiadou, M.; Brancato, A.; Cabrera, L.C.; Ferreira, L.; Greco, L.; Jarrah, S.; Kazocina, A.; Leuschner, R.; Magrans, J.O.; et al. Pesticide Residue Intake Model-EFSA PRIMo revision 3.1. *EFSA Support. Publ.* **2019**, *16*, 1605E. [CrossRef]
13. European Food Safety Authority. The 2012 European Union Report on pesticide residues in food. *EFSA J.* **2014**, *12*, 3942. [CrossRef]
14. European Food Safety Authority. The 2013 European Union report on pesticide residues in food. *EFSA J.* **2015**, *13*, 4038. [CrossRef]
15. EFSA (European Food Safety Authority). The 2014 European Union report on pesticide residues in food. *EFSA J.* **2016**, *14*, e04611. [CrossRef]
16. EFSA (European Food Safety Authority). The 2015 European Union report on pesticide residues in food. *EFSA J.* **2017**, *15*, e04791. [CrossRef]
17. EFSA (European Food Safety Authority). The 2016 European Union report on pesticide residues in food. *EFSA J.* **2018**, *16*, e05348. [CrossRef]
18. EFSA (European Food Safety Authority). Scientific report on the 2017 European Union report on pesticide residues in food. *EFSA J.* **2019**, *17*, e05743. [CrossRef]
19. EFSA (European Food Safety Authority); Medina-Pastor, P.; Triacchini, G. The 2018 European Union report on pesticide residues in food. *EFSA J.* **2020**, *18*, e06057. [CrossRef]
20. EFSA (European Food Safety Authority); Cabrera, L.C.; Pastor, P.M. The 2019 European Union report on pesticide residues in food. *EFSA J.* **2021**, *19*, e06491. [CrossRef]
21. Nolan, R.; Rick, D.; Freshour, N.; Saunders, J. Chlorpyrifos: Pharmacokinetics in human volunteers. *Toxicol. Appl. Pharmacol.* **1984**, *73*, 8–15. [CrossRef]
22. Aylward, L.L.; Irwin, K.; St-Amand, A.; Nong, A.; Hays, S.M. Screening-level Biomonitoring Equivalents for tiered interpretation of urinary 3-phenoxybenzoic acid (3-PBA) in a risk assessment context. *Regul. Toxicol. Pharmacol.* **2018**, *92*, 29–38. [CrossRef] [PubMed]
23. Eaton, D.L.; Daroff, R.B.; Autrup, H.; Bridges, J.; Buffler, P.; Costa, L.G.; Coyle, J.; McKhann, G.; Mobley, W.C.; Nadel, L.; et al. Review of the Toxicology of Chlorpyrifos With an Emphasis on Human Exposure and Neurodevelopment. *Crit. Rev. Toxicol.* **2008**, *38* (Suppl. 2), 1–125. [CrossRef] [PubMed]
24. Arnold, S.M.; Morriss, A.; Velovitch, J.; Juberg, D.; Burns, C.J.; Bartels, M.; Aggarwal, M.; Poet, T.; Hays, S.; Price, P. Derivation of human Biomonitoring Guidance Values for chlorpyrifos using a physiologically based pharmacokinetic and pharmacodynamic model of cholinesterase inhibition. *Regul. Toxicol. Pharmacol.* **2015**, *71*, 235–243. [CrossRef] [PubMed]
25. Griffin, P.; Mason, H.; Heywood, K.; Cocker, J. Oral and dermal absorption of chlorpyrifos: A human volunteer study. *Occup. Environ. Med.* **1999**, *56*, 10–13. [CrossRef] [PubMed]
26. Lange, R.; Apel, P.; Rousselle, C.; Charles, S.; Sissoko, F.; Kolossa-Gehring, M.; Ougier, E. The European Human Biomonitoring Initiative (HBM4EU): Human biomonitoring guidance values for selected phthalates and a substitute plasticizer. *Int. J. Hyg. Environ. Health* **2021**, *234*, 113722. [CrossRef]
27. Yusà, V.; Fernández, S.F.; Dualde, P.; López, A.; Lacomba, I.; Coscollà, C. Exposure to non-persistent pesticides in the Spanish population using biomonitoring: A review. *Environ. Res.* **2022**, *205*, 112437. [CrossRef]
28. Roca, M.; Miralles-Marco, A.; Ferré, J.; Pérez, R.; Yusà, V. Biomonitoring exposure assessment to contemporary pesticides in a school children population of Spain. *Environ. Res.* **2014**, *131*, 77–85. [CrossRef]
29. Fernández, S.F.; Pardo, O.; Corpas-Burgos, F.; Yusà, V. Exposure and cumulative risk assessment to non-persistent pesticides in Spanish children using biomonitoring. *Sci. Total Environ.* **2020**, *746*, 140983. [CrossRef]
30. Suárez, B.; Vela-Soria, F.; Castiello, F.; Olivas-Martinez, A.; Acuña-Castroviejo, D.; Gómez-Vida, J.; Olea, N.; Fernández, M.F.; Freire, C. Organophosphate pesticide exposure, hormone levels, and interaction with PON1 polymorphisms in male adolescents. *Sci. Total Environ.* **2021**, *769*, 144563. [CrossRef]
31. Llop, S.; Murcia, M.; Iñiguez, C.; Roca, M.; González, L.; Yusà, V.; Rebagliato, M.; Ballester, F. Distributions and determinants of urinary biomarkers of organophosphate pesticide exposure in a prospective Spanish birth cohort study. *Environ. Health* **2017**, *16*, 46. [CrossRef]

32. Fernández, S.F.; Pardo, O.; Adam-Cervera, I.; Montesinos, L.; Corpas-Burgos, F.; Roca, M.; Pastor, A.; Vento, M.; Cernada, M.; Yusà, V. Biomonitoring of non-persistent pesticides in urine from lactating mothers: Exposure and risk assessment. *Sci. Total Environ.* **2020**, *699*, 134385. [CrossRef] [PubMed]
33. Garí, M.; González-Quinteiro, Y.; Bravo, N.; Grimalt, J.O. Analysis of metabolites of organophosphate and pyrethroid pesticides in human urine from urban and agricultural populations (Catalonia and Galicia). *Sci. Total Environ.* **2018**, *622–623*, 526–533. [CrossRef] [PubMed]
34. EFSA Scientific Committee. Guidance on selected default values to be used by the EFSA Scientific Committee, Scientific Panels and Units in the absence of actual measured data. *EFSA J.* **2012**, *10*, 2579. [CrossRef]
35. EFSA Scientific Committee; More, S.; Bampidis, V.; Benford, D.; Bragard, C.; Halldorsson, T.; Bennekou, S.H.; Koutsoumanis, K.; Machera, K.; Naegeli, H.; et al. Statement on the derivation of Health-Based Guidance Values (HBGVs) for regulated products that are also nutrients. *EFSA J.* **2021**, *19*, e06479. [CrossRef] [PubMed]
36. ECHA. *Guidance on Information Requirements and Chemical Safety Assessment. Chapter, R.8: Characterisation of Dose [Concentration]-Response for Human Health*; ECHA: Helsinki, Finland, 2012.
37. EFSA Scientific Committee. Scientific Opinion on the applicability of the Margin of Exposure approach for the safety assessment of impurities which are both genotoxic and carcinogenic in substances added to food/feed. *EFSA J.* **2012**, *10*, 2578. [CrossRef]
38. Bravo, N.; Peralta, S.; Grimalt, J.O.; Martínez, M.; Rovira, J.; Schuhmacher, M. Organophosphate metabolite concentrations in maternal urine during pregnancy. *Environ. Res.* **2020**, *182*, 109003. [CrossRef]
39. Li, Y.; Wang, X.; McKenzie, J.F.; Mannetje, A.T.; Cheng, S.; He, C.; Leathem, J.; Pearce, N.; Sunyer, J.; Eskenazi, B.; et al. Pesticide exposure in New Zealand school-aged children: Urinary concentrations of biomarkers and assessment of determinants. *Environ. Int.* **2022**, *163*, 107206. [CrossRef]
40. Dalmolin, S.P.; Dreon, D.B.; Thiesen, F.V.; Dallegrave, E. Biomarkers of occupational exposure to pesticides: Systematic review of insecticides. *Environ. Toxicol. Pharmacol.* **2020**, *75*, 103304. [CrossRef]
41. Trunnelle, K.J.; Bennett, D.H.; Tulve, N.S.; Clifton, M.S.; Davis, M.D.; Calafat, A.M.; Moran, R.; Tancredi, D.J.; Hertz-Picciotto, I. Urinary pyrethroid and chlorpyrifos metabolite concentrations in Northern California families and their relationship to indoor residential insecticide levels, part of the Study of Use of Products and Exposure Related Behavior (SUPERB). *Environ. Sci. Technol.* **2014**, *48*, 1931–1939. [CrossRef]

toxics

Article

Methylmercury Risk Assessment Based on European Human Biomonitoring Data

Noelia Domínguez-Morueco [1], Susana Pedraza-Díaz [1,*], María del Carmen González-Caballero [1], Marta Esteban-López [1], Mercedes de Alba-González [1], Andromachi Katsonouri [2], Tiina Santonen [3], Ana Cañas-Portilla [1,*] and Argelia Castaño [1]

1 National Centre for Environmental Health, Instituto de Salud Carlos III, 28220 Madrid, Spain; ndominguez@isciii.es (N.D.-M.); mcgonzalez@isciii.es (M.d.C.G.-C.); m.esteban@isciii.es (M.E.-L.); malba@isciii.es (M.d.A.-G.); castano@isciii.es (A.C.)
2 Cyprus State General Laboratory, Ministry of Health, P.O. Box 28648, Nicosia 2081, Cyprus; akatsonouri@sgl.moh.gov.cy
3 Finnish Institute of Occupational Health, P.O. Box 40, 00032 Työterveyslaitos, Finland; tiina.santonen@ttl.fi
* Correspondence: spedraza@isciii.es (S.P.-D.); acanas@isciii.es (A.C.-P.)

Citation: Domínguez-Morueco, N.; Pedraza-Díaz, S.; González-Caballero, M.d.C.; Esteban-López, M.; de Alba-González, M.; Katsonouri, A.; Santonen, T.; Cañas-Portilla, A.; Castaño, A. Methylmercury Risk Assessment Based on European Human Biomonitoring Data. *Toxics* **2022**, *10*, 427. https://doi.org/10.3390/toxics10080427

Academic Editor: Cristina M. L. Carvalho

Received: 17 June 2022
Accepted: 27 July 2022
Published: 28 July 2022

Publisher's Note: MDPI stays neutral with regard to jurisdictional claims in published maps and institutional affiliations.

Copyright: © 2022 by the authors. Licensee MDPI, Basel, Switzerland. This article is an open access article distributed under the terms and conditions of the Creative Commons Attribution (CC BY) license (https://creativecommons.org/licenses/by/4.0/).

Abstract: A risk assessment (RA) was conducted to estimate the risk associated with methylmercury (MeHg) exposure of vulnerable European populations, using Human Biomonitoring (HBM) data. This RA was performed integrating published data from European HBM surveys and earlier EFSA approaches (EFSA 2012). Children/adolescents (3 to 17 years old) and women of childbearing age (18 to 50 years old) were selected as relevant study population groups for this RA. Two types of HBM datasets were selected: HBM studies (n = 18) with mercury (Hg) levels (blood and hair, total Hg and/or MeHg) in the general population in different EU countries and the DEMOCOPHES harmonized study in child–mother pairs (hair, total Hg) in 17 EU countries as a reference. Two approaches were included in the RA strategy: the first one was based on estimations of the fraction of children/adolescents and women of childbearing age, respectively, from the EU general population exceeding the HBM-I value established by the German Human Biomonitoring Commission, measured as Hazard Quotients (HQ); and the second approach was based on estimations of the fraction of the two population groups exceeding the Tolerable Weekly Intake (TWI) (or their equivalent to Tolerable Daily Intake (TDI)) defined by EFSA in 2012. The HQ approach showed that for both groups, the risk varies across EU countries and that some EU areas are close to or exceeding the exposure guidance values. This is the case of Spain and Portugal, which showed the highest HQ (GM and/or P95), probably due to their higher fish consumption. Results from the EFSA approach show that hair values of children/adolescents and women of childbearing age (both in selected HBM studies and in DEMOCOPHES study) are below the TDI of 1.9 µg/g; therefore, in general, the European population does not exceed the daily average/intake dose for MeHg and/or Hg. A possible risk underestimation was identified in our assessment since for many studies no data on P95 were available, causing loss of relevant information for risk characterization on the upper bound. In addition, data from other European countries also with high seafood consumption, such as France, Greece or Iceland, were not available. For this reason, further RA refinement is needed with harmonized and more widespread HBM data to account for differences in European exposure and associated risks, so that interventions to protect vulnerable citizens, can be applied.

Keywords: HBM4EU; human biomonitoring; mercury; methylmercury; risk assessment

1. Introduction

Mercury (Hg) is a highly toxic heavy metal that poses a significant global threat to human beings and the environment. Together with its various compounds, it can cause severe impacts on human health, including irreversible damage to the central nervous system [1].

Elemental Hg is released into the atmosphere from many natural and anthropogenic sources. Once released, Hg can move around the globe and can remain in circulation for thousands of years. Mercury in water bodies presents the greatest risk to humans because it is converted by microorganisms into methylmercury (MeHg), which is the most toxic form of this metal, easily absorbed by animals and bioaccumulated throughout the food chain [2]. Among the MeHg effects are irreversible damage to the central nervous system, even at very low levels depending on the age exposure window. Fetuses, newborns and children are amongst the most vulnerable and sensitive to the adverse neurodevelopmental effects caused by environmental MeHg exposure. Other relevant population groups are adolescents, as they are completing their developmental period, and pregnant women or women of childbearing age, due to the potential transfer to the fetuses (MeHg can cross the placenta and act on the developing nervous system) [2–4].

In Europe, the most significant source of human exposure to MeHg is the diet, especially in populations with high seafood consumption [3], such as in coastal regions—the Mediterranean region of Europe or Arctic region. Exposure levels are also influenced by the type of fish consumed (eating high predatory fish entails a higher risk), as well as their capture areas.

The U.S. Environmental Protection Agency (EPA), the World Health Organization (WHO) and the European Food Safety Authority (EFSA) [4–7], have compiled the adverse effects of MeHg in humans, including neurocognitive, motoric, auditory and visual electrophysiological responses, as well as neurobehavioral effects. Pivotal studies for dose-response analysis were derived from the cohorts in the Faroe Islands where whale meat is predominant in the diet [8,9] and in Seychelles with a high fish diet [10]. Different guidance values have been established in order to protect human health, having as point of departure (PoD) the findings observed in the aforementioned cohorts. EFSA set a tolerable weekly intake (TWI) for MeHg of 1.3 μg/kg of body weight [4]. The Human Biomonitoring Commission of the German Environment Agency derived Human Biomonitoring (HBM) values for total Hg in blood for children and adults of 5 μg/L(HBM-I) and 15 μg/L (HBM-II), respectively [11].

In the Minamata Convention on Mercury [12], measures were adopted to reduce global Hg emissions. Due to the long-range Hg transport and its bioaccumulation and biomagnification in the environment, these reduction measures will not have an immediate result and, therefore, the evaluation of the current exposure to MeHg in the general population and especially in vulnerable populations continues to be important [13]. This can be performed by estimating the external exposure to Hg, through the determination of the concentration of Hg in specific matrices such as food, water and air [14]; however, this ambient monitoring approach does not provide data about the real uptake by the human body [15].

In this context, human biomonitoring (HBM) is a very useful tool because it provides information on the real internal exposure to Hg. Different biological matrices are used to measure Hg, blood and hair being the most relevant specifically for MeHg, whereas urinary Hg is mainly reflecting exposure to inorganic mercury. Blood Hg levels reflect more recent exposure. The measurement of total Hg in blood reflects exposure to all forms, organic, inorganic and elemental mercury, with MeHg accounting for around 60–91% [16]. Hair is the preferred choice for many studies because it provides a simple, integrative and non-invasive sample for estimating long-term average exposure to MeHg since 89–91% of the total Hg is estimated to be in this organic form [17,18]. Once incorporated in the hair, Hg does not return to the blood; therefore, it provides a good long-term marker of exposure to MeHg [19]. In addition, Hg levels in hair show a direct relationship with blood total Hg levels through the traditional conversion ratio 250:1 established by WHO and recently reviewed and updated to 280:1 by [20].

Although HBM studies are currently recognized as an appropriate tool for exposure and risk assessment [21], their use is still limited in the European risk assessment schemes [22]. In this context, the European Human Biomonitoring Initiative (HBM4EU) has established a European-wide HBM program to generate knowledge on human internal exposure to chemical pollutants and their potential health impacts in Europe, to support policy makers' efforts to regulate chemical safety and improve public health in Europe (https://www.hbm4eu.eu/ accessed on 17 June 2022) [23].

The aim of this study is to present how European HBM data on Hg can be used in different risk assessment schemes for MeHg in order to estimate the risk for vulnerable European populations and to identify potential challenges.

2. Materials and Methods

2.1. Study Population and HBM Dataset

Children/adolescents from 3 to 17 years old and women of childbearing age in the range of 18 to 50 years old were selected as relevant population groups for this RA.

A systematic search of available Human Biomonitoring data from EU countries on methylmercury and total mercury was conducted through IPCHEM (The Information Platform for Chemical Monitoring), which is the European Commission's reference access point for searching, accessing and retrieving chemical occurrence data collected and managed in Europe, and based on the HBM4EU Scoping Document [24], for the two population groups described above. In addition, SCOPUS, PubMed and Web of Science were consulted (accessed on 9 March 2022). The criteria for selecting specific HBM datasets were the following: 1) studies with available data of Hg/MeHg levels in hair or blood expressed as geometric mean (GM) and/or 95th percentile (P95), and 2) with participants from at least one of the two population groups of interest, as defined previously. The bibliographic search strategy is detailed in the Supplementary Materials.

2.2. RA Based on Integrating Data from European HBM Surveys

The methodology used in the current MeHg RA is based on the risk-based approach for non-cancer effects published by St-Amand et al. [25] and recently verified by [26]. For non-cancer endpoints, hazard quotients (HQ) are calculated as the ratio of the biomarker concentration to the chemical specific Biomonitoring Equivalents (BE) or Human Biomonitoring Guidance Values (HBM-GVs) (Equation (1)):

$$HQ = \frac{[\text{Biomarker}]}{\text{BE or HBM} - \text{GV}} \tag{1}$$

where biomarker concentration, [biomarker], is the geometric mean (GM) or upper bound (P95 in this case, reflecting the reasonable worst-case estimate). BEs are based on reference dose (RfD) or tolerable daily intake (TDI) values. As there are no BEs for Hg and, thus, for MeHg, the HBM-GVs derived for Hg by the German HBM Commission [11] were used for HQ calculation. In this case, the HBM-I value for total Hg in blood, 5 μg/L, was selected for evaluation of the risk, since it is the most restrictive HBM-GV and since blood can be used to document recent exposure to MeHg. Those studies in which Hg data (GM and/or P95) were only available in hair, were converted to blood levels using the hair to blood conversion ratio of 280:1 [20].

2.3. RA Based on EFSA (2012) Approach

The EFSA Panel on Contaminants in the Food Chain (CONTAM Panel) delivered a scientific opinion on the risks to human health related to the presence of inorganic mercury and MeHg in food in 2012 [4], establishing a tolerable weekly intake (TWI) for MeHg of 1.3 μg/kg body weight (b.w.), expressed as Hg. In order to evaluate the chronic exposure to MeHg, we compared this value to the Hg concentrations found in the selected HBM surveys. For this, a conversion of TWI to hair content according to the WHO [19] was carried out. It is estimated that a daily MeHg average intake of 0.1 μg per kg body weight per day (0.1 μg/kg b.w./d) by an adult woman results in hair Hg concentrations of about 1 μg/g and this relationship is generally directly proportional [19]. Thereby, a TWI of 1.3 μg/kg b.w for MeHg, equivalent to a TDI of 0.19 μg/kg b.w./d, corresponds to hair levels of 1.9 µg/g. This derived value of 1.9 µg/g was compared to exposure levels (total Hg in hair) in the two study population groups, children/adolescents and women of childbearing age from the EU general population.

3. Results and Discussion

3.1. HBM Dataset

A total of thirty-five European HBM studies were identified fulfilling the above-indicated criteria (available GM and/or P95 hair or blood Hg/MeHg levels in children/adolescents 3–17 years old and/or women of childbearing age 18–50 years old) (Table 1). These included 17 individual DEMOCOPHES studies and 18 additional European HBM studies. Figure 1 shows the geographical distribution of all the studies included in this RA.

Table 1. Human biomonitoring (HBM) studies with Hg and MeHg data in blood and hair in the selected populations.

Country	Study	Year	Population Age (*N*)	Hg				MeHg				Refs.
				Blood (µg/L)		Hair (µg/g)		Blood (µg/L)		Hair (µg/g)		
				GM	P95	GM	P95	GM	P95	GM	P95	
Belgium	FLESH	2007–2011	Adolescents 14–15 y (206)	–	–	0.19	–	–	–	0.12	–	[27]
			Mothers 20–40 y (242)	–	–	0.35	–	–	–	0.26	–	
	DEMOCOPHES-BE	2010–2012	Children 6–11 y (127)	–	–	0.20	–	–	–	–	–	[28]
			Mothers <45 y (127)	–	–	0.37	–	–	–	–	–	
Cyprus	DEMOCOPHES-CY	2010–2012	Children 6–11 y (60)	–	–	0.33	–	–	–	–	–	[28]
			Mothers <45 y (60)	–	–	0.46	–	–	–	–	–	
Czech Republic	CZ-HBM	2001–2003	Children 8–10 y (333)	0.43	1.44	–	–	–	–	–	–	[29]
	CZ-HBM	1996–2008	Children 8–10 y (344 [a]/347) *		1.47		0.52	–	–	–	–	[30]
	DEMOCOPHES-CZ	2010–2012	Children 6–11 y (120)	–	–	0.10	–	–	–	–	–	[28]
			Mothers <45 y (120)	–	–	0.16	–	–	–	–	–	
	CZ-HBM	2015	Women 18–50 y (n.a.)	0.80	0.90	–	–	–	–	–	–	[31]
	CZ-HBM	2016	Children 5 & 9 y (419)	0.32	1.03	–	–	–	–	–	–	[32]
Denmark	DEMOCOPHES-DK	2010–2012	Children 6–11 y (144)	–	–	0.25	–	–	–	–	–	[28]
			Mothers <45 y (144)	–	–	0.39	–	–	–	–	–	

Table 1. *Cont.*

Country	Study	Year	Population Age (N)	Hg Blood (μg/L) GM	Hg Blood (μg/L) P95	Hg Hair (μg/g) GM	Hg Hair (μg/g) P95	MeHg Blood (μg/L) GM	MeHg Blood (μg/L) P95	MeHg Hair (μg/g) GM	MeHg Hair (μg/g) P95	Refs.
Germany	GerES II	1990–1992	Children 6–17 y (712)	0.33	1.40	–	–	–	–	–	–	[33]
Germany	GerES IV	2003–2006	Children 3–14 y (1240)	0.23	1.00	–	–	–	–	–	–	[33]
Germany	DEMOCOPHES-DE	2010–2012	Children 6–11 y (120)	–	–	0.06	–	–	–	–	–	[28]
			Mothers <45 y (120)	–	–	0.11	–	–	–	–	–	
Hungary	DEMOCOPHES-HU	2010–2012	Children 6–11 y (119)	–	–	0.03	–	–	–	–	–	[28]
			Mothers <45 y (119)	–	–	0.04	–	–	–	–	–	
Ireland	DEMOCOPHES-IE	2010–2012	Children 6–11 y (120)	–	–	0.10	–	–	–	–	–	[28]
			Mothers <45 y (120)	–	–	0.16	–	–	–	–	–	
Italy	PROBE	2008–2010	Adolescents 13–15 y (252)	0.94	3.55	–	–	–	–	–	–	[34]
Italy	–	2007–2009	Pregnant women n.a. (606 [a]/ 604 [b]/236 [c]/ 220 [d])	3.14	–	1.06	–	4.46	–	1.67	–	[35]
Luxembourg	DEMOCOPHES -LU	2010–2012	Children 6–11 y (56)	–	–	0.18	–	–	–	–	–	[28]
			Mothers <45 y (56)	–	–	0.39	–	–	–	–	–	
Poland	DEMOCOPHES-PL	2010–2012	Children 6–11 y (120)	–	–	0.07	–	–	–	–	–	[28]
			Mothers <45 y (120)	–	–	0.14	–	–	–	–	–	
Portugal	DEMOCOPHES-PT	2010–2012	Children 6–11 y (120)	–	–	1.03	–	–	–	–	–	[28]
			Mothers <45 y (120)	–	–	1.20	–	–	–	–	–	
Romania	DEMOCOPHES-RO	2010–2012	Children 6–11 y (120)	–	–	0.09	–	–	–	–	–	[28]
			Mothers <45 y (120)	–	–	0.10	–	–	–	–	–	
Slovakia	DEMOCOPHES-SK	2010–2012	Children 6–11 y (129)	–	–	0.09	–	–	–	–	–	[28]
			Mothers <45 y (129)	–	–	0.13	–	–	–	–	–	

Table 1. *Cont.*

Country	Study	Year	Population Age (N)	Hg Blood (μg/L) GM	Hg Blood (μg/L) P95	Hg Hair (μg/g) GM	Hg Hair (μg/g) P95	MeHg Blood (μg/L) GM	MeHg Blood (μg/L) P95	MeHg Hair (μg/g) GM	MeHg Hair (μg/g) P95	Refs.
Slovenia	SLO-HBM	2008–2009, 2011–2014	Women 19–39 y (535 [a]/503 [b])	1.1	4.06	0.27	0.99	–	–	–	–	[36]
	PHIME project	2011–2014	Children 6–11 y (174)	0.77	–	0.18	–	–	–	–	–	[37]
			Women 20–35 y (127)	1.04	–	0.24	–	–	–	–	–	
	DEMOCOPHES-SI	2010–2012	Children 6–11 y (120)	–	–	0.17	–	–	–	–	–	[28]
			Mothers <45 y (120)	–	–	0.23	–	–	–	–	–	
Spain	–	1996	Children 6–16 y (233)	–	–	0.77	–	–	–	–	–	[38]
	BIOAMBIENT.ES	2009–2010	Women >18 y (918 [a]/327 [b])	6.27	16.90	1.87	4.6	–	–	–	–	[39]
	INMA Project	2008–2012	Children 4–5 y (1252)	–	–	0.98	–	–	–	–	–	[40]
	DEMOCOPHES-ES	2010–2012	Children 6–11 y (120)	–	–	0.88	–	–	–	–	–	[28]
			Mothers <45 y (120)	–	–	1.49	–	–	–	–	–	
	BIOVAL programme	2016	Children 6–11 y (611)	–	–	0.79	3.25	–	–	–	–	[41]
	BETTERMILK Project	2017	Breastfeeding mothers 20–45 y (120)	–	–	1.22	–	–	–	–	–	[42]
	HEALS-EXHES	2016–2017	Children cord blood (53)	2.87	7.91	–	–	–	–	–	–	[43,44]
			Mothers GM 34 y (53)	2.05	6.98	–	–	–	–	–	–	
Sweden	DEMOCOPHES-SE	2010–2012	Children 6–11 y (100)	–	–	0.18	–	–	–	–	–	[28]
			Mothers <45 y (100)	–	–	0.25	–	–	–	–	–	
	–	2016–2017	Children/adolescents 12, 15 & 18 y (1099)	0.66	2.10	–	–	–	–	–	–	[45]
Switzerland	DEMOCOPHES-CH	2010–2012	Children 6–11 y (120)	–	–	0.08	–	–	–	–	–	[28]
			Mothers <45 y (120)	–	–	0.15	–	–	–	–	–	

Table 1. *Cont.*

Country	Study	Year	Population Age (N)	Hg				MeHg				Refs.
				Blood (μg/L)		Hair (μg/g)		Blood (μg/L)		Hair (μg/g)		
				GM	P95	GM	P95	GM	P95	GM	P95	
United Kingdom	DEMOCOPHES-UK	2010–2012	Children 6–11 y (21)	–	–	0.19	–	–	–	–	–	[28]
			Mothers <45 y (21)	–	–	0.15	–	–	–	–	–	
17 EU countries	DEMOCOPHES-17	2010–2012	Children 6–11 y (1836)	–	–	0.14	1.29	–	–	–	–	[28]
			Mothers <45 y (1839)	–	–	0.23	1.89	–	–	–	–	

[a] N for total Hg in blood. [b] N for total Hg in hair. [c] N for MeHg in blood. [d] N for MeHg in hair. * Missing years: Hg in blood 1997, 2000, 2002, 2003, 2004, 2005, 2007; Hg in hair 2004, 2005, 2007. n.a. not available.

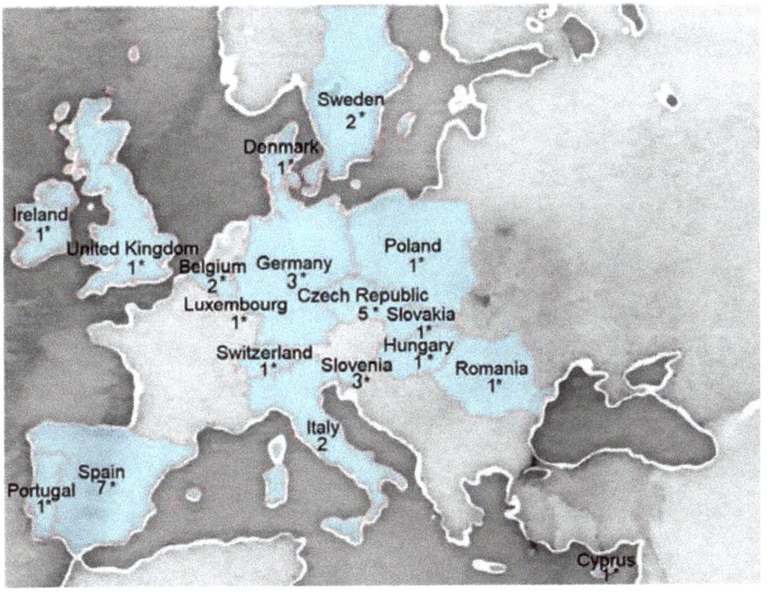

Figure 1. Distribution of European HBM studies included in the MeHg risk assessment. * Indicates countries included in the DEMOCOPHES study. Numbers in the figure indicate the number of studies in each country.

The DEMOCOPHES study was used as reference, since it is a well-defined cross-sectional survey of exposure to environmental chemicals, including Hg, performed in 17 European countries using a harmonized European protocol (developed by its twin project, COPHES) [28] and applying strict Quality Control/Quality Assurance (QA/QC) measures in the chemical analysis of the hair samples [46]. In DEMOCOPHES, a total of 1875 child/mother pairs were recruited from urban and rural areas in the participating countries, while excluding exposure hotspots. The exposure biomarkers included the hair Hg concentration (expressed as GM for the individual studies and as GM and P95 for the whole study (DEMOCOPHES-17)) of samples collected from September 2011 to February 2012 from children aged 6–11 years old and their mothers (<45 years old).

Among the international studies retrieved, only two reported levels of MeHg in children and/or women, one in hair [27], and the other one in blood and hair [35], while the remaining studies reported only total Hg levels.

To assess the relevant results obtained across Europe for children/adolescents and women of childbearing age from the 18 European HBM studies recorded in Table 1 (excluding DEMOCOPHES data) and to compare them to the DEMOCOPHES data, a range of GM and a range of P95 of MeHg and/or Hg blood concentrations was compiled (Table 2). As mentioned earlier, for those studies with data for only MeHg and/or Hg concentrations in hair, the hair-to-blood conversion ratio of 280:1 recently proposed by Esteban et al. [20] was used. Previous studies [20,47,48] have shown that there is a high interindividual variation in the calculated hair-to-blood ratios in different populations and even at different ages. In the study by Esteban et al. [20], focused on European populations, the geometric mean hair to blood ratio of 280 predicted blood values equally well to the regression equation of hair Hg vs. blood Hg, while when the ratio of 250 recommended by WHO was used, blood Hg estimations were significantly less accurate from those predicted by the regression equation or by using the 280 ratio.

Table 2. Range of geometric means (GM) and/or 95th percentiles (P95) of MeHg and/or Hg concentrations in blood (µg/L) (hair values are expressed as blood concentration using hair to blood ratio of 280:1) from European HBM studies and GM and P95 values for the whole DEMOCOPHES-17 reference study.

	DEMOCOPHES		Other HBM Studies	
	Children (17 Studies)	Mothers (17 Studies)	Children/Adolescents (13 Studies)	Women (9 Studies)
Range of GM	0.09–3.69	0.14–5.31	0.23–3.50	0.80–6.27
Range of P95	–	–	1.03–11.6	0.90–16.9
$GM_{DEMOCOPHES\text{-}17}$	0.51	0.82	–	–
$P95_{DEMOCOPHES\text{-}17}$	4.60	6.75	–	–

When conducting the HBM data search, we observed a high degree of heterogeneity in the data available for the different European studies not only related to the biomarkers/matrices measured, but also to the descriptive statistics reported. This limited the analysis, which was mostly performed using GM values, losing risk characterization on the upper bound.

Data from 18 studies, in addition to the 17 included in DEMOCOPHES, were retrieved; this made information available for 18 European countries, although the degree of representativeness of the data collected is limited. In addition, we observed that data from some other European countries with high seafood consumption, such as France, Greece or Iceland, and which could be relevant for the MeHg risk assessment performed here, were not available.

Most of the studies did not perform Hg speciation analysis (Table 1); therefore, we used total Hg concentrations as surrogate because MeHg is the main contributor to the levels found both in blood and hair as reported in the literature [16–18]. This was also observed from the studies which had total Hg and MeHg measured [27,35]. However, in the case of the Belgian FLESH study [27], 63% and 74% of the total Hg in hair was in the form of MeHg in adolescents and women, respectively. This could be associated with different factors relative to the region or lifestyle, such as seafood consumption patterns (lower consumption) in these groups. In the study carried out by Valent et al. [35], the percentages seem to be higher although they cannot be calculated as the number of samples in which speciation was done differs from the number in which total Hg was measured.

3.2. RA Based on Integrating Data from European HBM Surveys

The HQ were calculated as the ratio between the GM or the P95 of the MeHg and/or Hg blood concentrations retrieved in the selected HBM datasets (Table 2) and the HBM-I value for Hg in blood of 5 μg/L (Equations (2) and (3)), following a conservative approach as mentioned earlier:

$$\mathrm{HQ\ (GM)} = \frac{\text{GM of HBM results distribution}}{\text{Selected internal threshold (HBM} - \text{GV of 5 μg/L)}} \quad (2)$$

$$\mathrm{HQ\ (P95)} = \frac{\text{P95 of HBM results distribution}}{\text{Selected internal threshold (HBM} - \text{GV of 5 μg/L)}} \quad (3)$$

HQs near or exceeding 1 indicate that exposure levels are near or exceeding selected exposure guidance value.

Table 3 shows the HQ calculations for the GM and/or P95 of Hg levels in blood for the range of GM and P95 of all the studies together for the selected populations. Supplementary Table S1 shows the HQs calculated for each study included in this RA.

Table 3. HQ calculations for the GM and/or P95 of Hg levels in blood for all HBM studies considered for the selected populations.

	Children/Adolescents		Women of Childbearing Age	
	European HBM Studies	**DEMOCOPHES**	**European HBM Studies**	**DEMOCOPHES**
$\mathbf{HQ_{range\ GM}}$	0.05 to 0.70	0.02 to 0.74	0.16 to 1.25	0.03 to 1.06
$\mathbf{HQ_{range\ P95}}$	0.21 to 2.32	–	0.18 to 3.38	–

In the case of children/adolescents, the HQs calculated for the range of GMs for the HBM studies included in this work (excluding DEMOCOPHES) do not exceed the value of 1 (Figure 2). However, the P95 for some Spanish studies are over 1 ($HQ_{P95\ HEALS\text{-}EXHES}$ = 1.58 and $HQ_{P95\ BIOVAL}$ = 2.32, Supplementary Table S1).

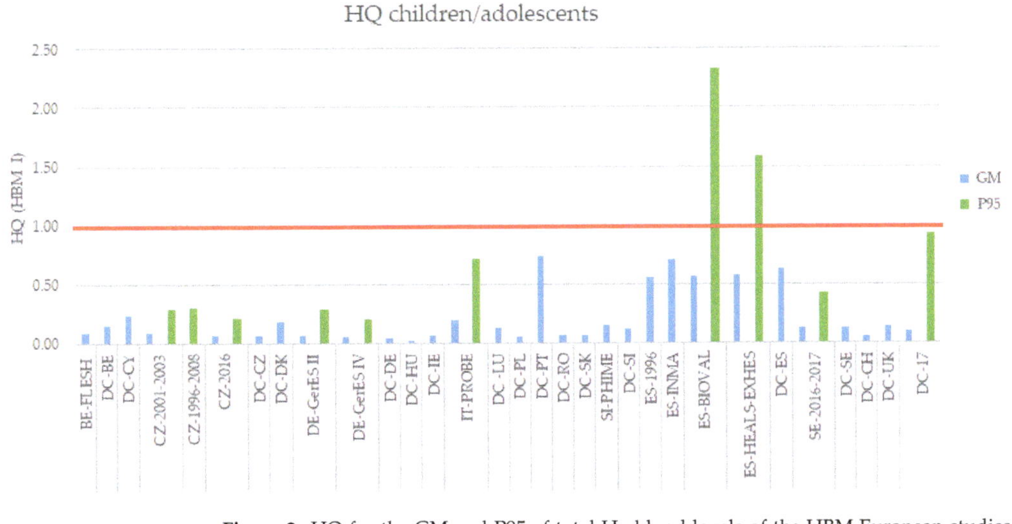

Figure 2. HQ for the GM and P95 of total Hg blood levels of the HBM European studies in children/adolescents (3–17 years old). DEMOCOPHES Study (DC). Note: MeHg levels in the case of FLESH study.

The HQs for DEMOCOPHES study, both for the range of GMs of the 17 individual studies, as well as for the GM of the study as a whole, do not exceed the value of 1(Figure 2). However, the P95 of DEMOCOPHES-17 is close to the value of 1 ($HQ_{P95\ DEMOCOPHES\text{-}17}$ = 0.92). The GM of other individual countries, for example Portugal, are close ($HQ_{GM\ DEMOCOPHES\text{-}PT}$ = 0.74).

For women of childbearing age, the HQ calculated for the range of GMs and P95 in the selected European HBM studies (excluding DEMOCOPHES) exceed the value of 1 only in two studies from Spain ($HQ_{GM\ BIOAMBIENT.ES\text{-}SP}$ = 1.25 and $HQ_{P95\ BIOAMBIENT.ES\text{-}SP}$ = 3.38; $HQ_{P95\ HEALS}$ = 1.40) and one study also from Spain, was close to 1 ($HQ_{GM\ BETTERMILK}$ = 0.87) (Figure 3).

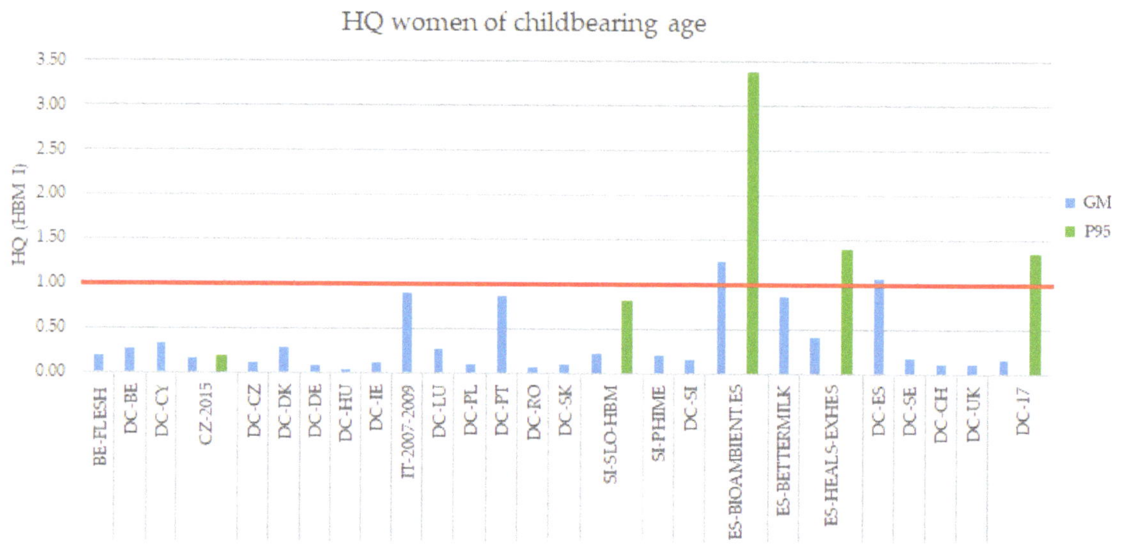

Figure 3. HQ for the GM and P95 of Hg blood levels of the HBM European studies in women of childbearing age (18–50 years old). DEMOCOPHES Study (DC). Note: MeHg levels in the case of FLESH and Valent et al., 2013 [35] studies.

The HQ for DEMOCOPHES showed that, in contrast to the results observed for the children/adolescents, both the upper part of the range of GMs of the 17 individual studies in DEMOCOPHES, as well as the P95 of the whole study are close or exceeding the value of 1 ($HQ_{P95\ DEMOCOPHES}$ = 1.35 and $HQ_{range\ GM\ DEMOCOPHES}$ = 0.03–1.06). Spain and Portugal are among the countries with GM exceeding or close to the guidance values ($HQ_{GM\ DEMOCOPHES\text{-}ES}$ = 1.06; $HQ_{GM\ DEMOCOPHES\text{-}PT}$ = 0.86) (Figure 3), the latter being consistent with the findings in the case of children/adolescents.

Overall, risk assessment based on HQ using the HBM-I guidance value showed that the risk varies in the different EU countries and some EU areas are close to or exceeding this exposure guidance value, both for children/adolescents and women of childbearing age. This is the case of Spain and Portugal, which were the countries with highest HQ, probably due to their higher seafood consumption frequency or species consumed, as was already described in DEMOCOPHES [28]. The guidance value selected here was the HBM-I, being a most conservative HBM-GV for evaluating the risk associated to MeHg and/or Hg exposure. At exposure levels exceeding the HBM-I value, but falling below the HBM-II value, adverse health effects cannot be excluded with sufficient certainty and a follow-up examination should be performed to determine whether there is a continued elevated exposure [19]. Since the main MeHg exposure source is seafood consumption, the recommendation to minimize risk would be to develop harmonized dietary advice. This would allow benefitting from the seafood consumption while protecting the health of vulnerable populations, such as pregnant women and children. Recommendations on how to reduce the uptake of MeHg were published by EFSA in 2015 [5]. However, due to the variety of fish species consumed across Europe, EFSA's Scientific Committee suggested that it is not possible to make general recommendations and that the situation should be evaluated in each country. Most European countries, however, do not have official recommendations on the fish consumption. The effectiveness of the proper implementation of dietary advice to pregnant women to lower mercury intake has been demonstrated in a recent study in Denmark [49]. More recently, it has also been addressed in the HBM4EU-MOM study, where a harmonized intervention focusing on dietary advice to reduce prenatal exposure to mercury in 5 European coastal countries has been carried out [50]. Proper and harmonized communication of these findings to the European population, and particularly to vulnerable groups, is paramount.

In the DEMOCOPHES study, we have observed that the ranges of GMs and P95 were smaller in comparison to those obtained in the rest of European datasets, with a similar geographical distribution. This could be due to the fact that DEMOCOPHES used a harmonized protocol for all participating countries and the rest of European studies have followed different protocols. However, it is necessary to point out that for the individual DEMOCOPHES studies and some other HBM studies in this RA, no P95 values were available, and the overall risk characterization had to be performed mostly using GM.

Furthermore, the publicly available HBM data do not include information from several European countries with high seafood consumption, such as other Mediterranean countries, which could probably affect significantly the results obtained.

3.3. RA Based on EFSA (2012) Approach

Exposure levels (Hg in hair, GM and P95) in children/adolescents and women of child-bearing age from the EU general population were compared with the available exposure guideline value for Hg in hair of 1.9 μg/g derived here, obtained after transformation of the TWI for MeHg of 1.3 μg/kg b.w established by EFSA (detailed in Section 2.3).

In the case of the children/adolescents, only one study exceeded the 1.9 μg/g of Hg in hair, namely the BIOVAL program, conducted in Spain in 2016, with a P95 value of 3.25 μg/g (Figure 4). However, when evaluating GM values, no study exceeded this value; therefore, in terms of daily average or intake dose, the recommended values by EFSA are not being exceeded.

In the case of women (Figure 5), only one study from Spain, BIOAMBIENT.ES, had Hg hair levels close to or over the 1.9 μg/g reference value with a GM of 1.87 μg/g and a P95 value of 4.6 μg/g, respectively. Amongst the other studies, those with GM closer to the reference value included a study carried out in pregnant women in Italy with a hair MeHg GM value of 1.67 μg/g [35], further two studies in Spain, DEMOCOPHES-ES (GM of 1.49 μg/g) and BETTERMILK project (GM of 1.22 μg/g), and DEMOCOPHES-PT in Portugal (GM of 1.20 μg/g). As indicated earlier, there were no data on P95 values for any of these studies, which hampers the risk assessment as these are all countries with high seafood consumption, and with higher GM values, and could potentially exceed the HQ of 1 if the risk assessment is based on P95 levels representing reasonable worst-case scenario. This research confirms previous findings for five European countries (Belgium, Ireland, Italy, Portugal and Spain), that showed that larger subgroups of the population in the southern European countries (up to 11% in Portugal) were potentially at risk for a MeHg exposure above the TWI value, while this risk was much lower in Ireland and Belgium [51]. Nevertheless, in general terms, the European population, as represented by the studies included in this RA, does not exceed the daily average/intake dose for MeHg and/or Hg.

Most European HBM studies included in this RA measured total Hg in blood and/or in hair, and only two studies analyzing the different Hg species, were identified. Speciation analysis is necessary to differentiate between inorganic/elemental Hg and MeHg exposure. However, speciation of Hg requires complex and lengthy analytical procedures and expensive reagents and equipment, which are not routinely available in analytical laboratories. Furthermore, in the case of hair, it is accepted that total Hg analysis can be used as surrogate of MeHg analysis, due to the high percentage (89–91%) of this form in the total Hg concentration found in hair; therefore, it is common that HBM studies measure total Hg in hair [20]. In addition, as mentioned earlier, it is also recognized that MeHg accounts for a high proportion (60–91%) of the total Hg in blood in general population without, for example, occupational exposure to inorganic Hg [16]. The development and implementation of harmonized methodologies would allow an adequate characterization of the exposure to each type of Hg in vulnerable populations in Europe. Therefore, the appropriate selection of the matrices and biomarkers of exposure is crucial to reduce the potential uncertainties associated to the conversion rates in the risk assessment [20].

Figure 4. Exposure levels (Hg in hair, GM and P95) in children/adolescents (3–17 years old) from the EU general population compared to the internal exposure guideline value for Hg in hair derived from EFSA (1.9 µg/g, in red). DEMOCOPHES Study (DC). Note: MeHg levels in the case of FLESH study.

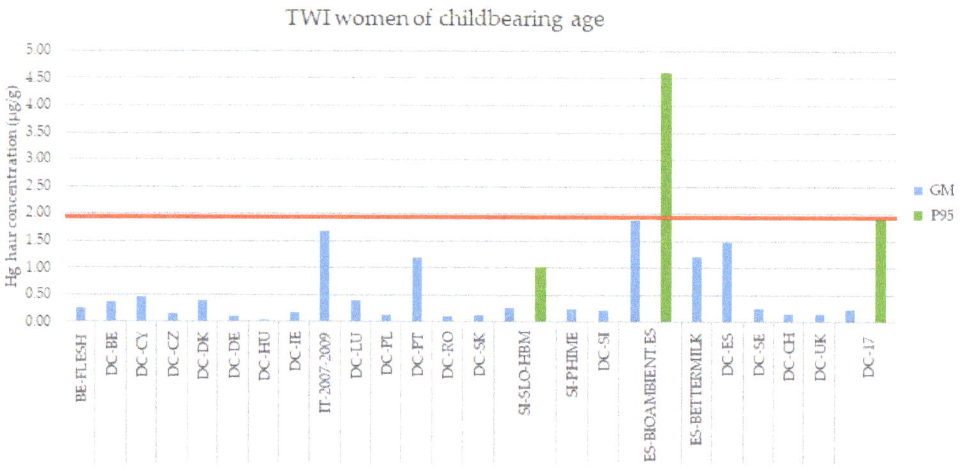

Figure 5. Exposure levels (Hg in hair, GM and P95) in women of childbearing age (18–50 years old) from the EU population compared to the internal exposure guideline value for Hg in hair derived from EFSA (1.9 µg/g in red). DEMOCOPHES Study (DC). Note: MeHg levels in the case of FLESH and Valent et al., 2013 [35] studies.

4. Conclusions

The followed HQ approach showed that for both children/adolescents and women of childbearing age, there is a high variability in the risk across EU countries. EU areas with high seafood consumption (Spain and Portugal) showed the highest HQ, close to or exceeding the exposure guidance value. Using the approach based on the TDI derived here in both studied populations, hair values were below TDI of 1.9 μg/g, suggesting that the European population does not generally exceed the daily average/intake dose for MeHg and/or Hg. However, since some countries may be at higher risk due to their diet and that climate change can increase the MeHg levels in fish [52], efforts to develop dietary recommendations focusing on the vulnerable populations should be planned and implemented. This would be especially important in Mediterranean countries with fish consumption patterns that, as seen here, can lead to HQ values above 1.

The HBM approach provides a direct way to assess the factors contributing to the risk and to take actions for risk communication and management in a more appropriate and personalized way, as needed.

The work reported here highlights the need for further harmonization in HBM studies. Although a significant effort has been made to study the real exposure to MeHg of vulnerable populations throughout Europe, the lack of harmonization in the HBM studies, both in terms of the biomarkers used and the descriptive statistics reported, has limited the risk assessment in this case. The availability of data on relevant determinants, e.g., seafood consumption, and, ideally, on health outcomes would be of importance for a more comprehensive analysis, in order to assess health risk resulting from Hg and/or MeHg exposure

In addition, a possible data underestimation was identified as a limitation, since for many studies the P95 data were not available and, thereby, losing an important information of the upper bound for risk characterization. In addition, HBM data from other European countries with high fish consumption, such as France, Greece or Iceland, were not available. For this reason, as a recommendation, further RA refinement is needed with more widespread harmonized HBM data to account for differences in European exposure.

Supplementary Materials: The following supporting information can be downloaded at: https://www.mdpi.com/article/10.3390/toxics10080427/s1, Search strategy on the bibliographic databases and Table S1: Hazard Quotients calculated for each study included in this risk assessment.

Author Contributions: Formal analysis, N.D.-M. and M.d.C.G.-C.; writing—original draft preparation, N.D.-M., S.P-D. and M.d.C.G.-C.; writing—review and editing: A.C.-P., M.E.-L., M.d.A.-G., A.K., T.S. and A.C.; supervision, S.P.-D., A.C.-P. and A.C. All authors have read and agreed to the published version of the manuscript.

Funding: This study was part of the HBM4EU project receiving funding from the European Union's Horizon 2020 research and innovation program under grant agreement No. 733032.

Institutional Review Board Statement: Not applicable.

Informed Consent Statement: Not applicable as only aggregate human monitoring data are reported. For details see HBM4EU project https://www.hbm4eu.eu/, accessed on 17 June 2022.

Data Availability Statement: Aggregate data included in this study were publicly available from IPCHEM (https://ipchem.jrc.ec.europa.eu/, accessed on 9 March 2022), SCOPUS (https://www.scopus.com, accessed on 9 March 2022), PubMed (https://pubmed.ncbi.nlm.nih.gov, accessed on 9 March 2022) and Web of Science (institutional access through the Fundación Española para la Ciencia y Tecnología, Spanish Ministry of Science and Innovation). These sites were last accessed on 9 March 2022.

Acknowledgments: This study was part of the HBM4EU project receiving funding from the European Union's Horizon 2020 research and innovation program under grant agreement No 733032. The authors want to thank Miguel Juliá Molina for his contribution in Figure 1.

Conflicts of Interest: The authors declare no conflict of interest.

References

1. Agency for Toxic Substances and Disease Registry (ATSDR). *Toxicological Profile for Mercury*; U.S. Department of Health and Human Services, Public Health Service: Atlanta, GA, USA, 1999.
2. Dórea, J.G. Persistent, bioaccumulative and toxic substances in fish: Human health considerations. *Sci. Total Environ.* **2008**, *400*, 93–114. [CrossRef] [PubMed]
3. Marín, S.; Pardo, O.; Baguena, R.; Font, G.; Yusà, V. Dietary exposure to trace elements and health risk assessment in the region of Valencia, Spain: A total diet study. *Food Addit. Contam. Part A Chem. Anal. Control Expo. Risk Assess.* **2017**, *34*, 228–240. [CrossRef] [PubMed]
4. European Food Safety Authority (EFSA). Scientific Opinion on the risk for public health related to the presence of mercury and methylmercury in food. *EFSA J.* **2012**, *10*, 2985.
5. European Food Safety Authority (EFSA). Statement on the benefits of fish/seafood consumption compared to the risks of methylmercury in fish/seafood. *EFSA J.* **2015**, *13*, 3982. [CrossRef]
6. World Health Organization (WHO). *The World Health Report—Primary Health Care: Now More Than Ever*; WHO: Geneva, Switzerland, 2008.
7. National Research Council (US) Committee on the Toxicological Effects of Methylmercury. *Toxicological Effects of Methylmercury*; National Academies Press (US): Washington, DC, USA, 2000.
8. Weihe, P.; Grandjean, P. Cohort studies of Faroese children concerning potential adverse health effects after the mothers' exposure to marine contaminants during pregnancy. *Acta Vet. Scand.* **2012**, *54*, 1. [CrossRef]
9. Grandjean, P.; Weihe, P.; White, R.F.; Debes, F.; Araki, S.; Yokoyama, K.; Murata, K.; Sørensen, N.; Dahl, R.; Jørgensen, P.J. Cognitive Deficit in 7-Year-Old Children with Prenatal Exposure to Methylmercury. *Neurotoxicol. Teratol.* **1997**, *19*, 417–428. [CrossRef]
10. Van Wijngaarden, E.; Thurston, S.W.; Myers, G.J.; Harrington, D.; Cory-Slechta, D.A.; Strain, J.; Watson, G.; Zareba, G.; Love, T.; Henderson, J.; et al. Methyl mercury exposure and neurodevelopmental outcomes in the Seychelles Child Development Study Main cohort at age 22 and 24years. *Neurotoxicol. Teratol.* **2016**, *59*, 35–42. [CrossRef]
11. Schulz, C.; Angerer, J.; Ewers, U.; Kolossa-Gehring, M. The German Human Biomonitoring Commission. *Int. J. Hyg. Environ. Health* **2007**, *210*, 373–382. [CrossRef]
12. United Nations. Minamata Convention on Mercury. 2013. Available online: https://www.unep.org/resources/report/minamata-convention-mercury (accessed on 16 June 2022).
13. Castaño, A.; Cutanda, F.; Esteban, M.; Pärt, P.; Navarro, C.; Gómez, S.; Rosado, M.; López, A.; López, E.; Exley, K.; et al. Fish consumption patterns and hair mercury levels in children and their mothers in 17 EU countries. *Environ. Res.* **2015**, *141*, 58–68. [CrossRef]
14. U.S. Environmental Protection Agency. *Water Quality Criterion for the Protection of Human Health: Methylmercury*; EPA-823-R-01e001; U.S. Environmental Protection Agency: Washington, DC, USA, 2001.
15. Angerer, J.; Ewers, U.; Wilhelm, M. Human biomonitoring: State of the art. *Int. J. Hyg. Environ. Health* **2007**, *210*, 201–228. [CrossRef]
16. Kim, B.G.; Jo, E.M.; Kim, G.Y.; Kim, D.S.; Kim, Y.M.; Kim, R.B.; Suh, B.S.; Hong, Y.S. Analysis of methylmercury concentration in the blood of Koreans by using cold vapor atomic fluorescence spectrophotometry. *Ann. Lab. Med.* **2012**, *32*, 31–37. [CrossRef]
17. Cernichiari, E.; Toribara, T.Y.; Liang, L.; Marsh, D.O.; Berlin, M.W.; Myers, G.J.; Cox, C.; Shamlaye, C.F.; Choisy, O.; Davidson, P. The biological monitoring of mercury in the Seychelles study. *NeuroToxicology* **1995**, *16*, 613–628.
18. Miklavčič, A.; Cuderman, P.; Mazej, D.; Tratnik, J.S.; Krsnik, M.; Planinšek, P.; Osredkar, J.; Horvat, M. Biomarkers of low-level mercury exposure through fish consumption in pregnant and lactating Slovenian women. *Environ. Res.* **2011**, *111*, 1201–1207. [CrossRef]
19. WHO (World Health Organization). *Guidance for Identifying Populations at Risk from Mercury Exposure*; UNEP DTIE Chemicals Branch and WHO Department of Food Safety, Zoonoses and Foodborne Diseases: Geneva, Switzerland, 2008; p. 176.
20. Esteban-López, M.; Arrebola, J.P.; Juliá, M.; Pärt, P.; Soto, E.; Cañas, A.; Pedraza-Díaz, S.; González-Rubio, J.; Castaño, A. Selecting the best non-invasive matrix to measure mercury exposure in human biomonitoring surveys. *Environ. Res.* **2022**, *204 Pt D*, 112394. [CrossRef]
21. Casteleyn, L.; Biot, P.; Viso, A.C. From Human biomarkers to human biomonitoring in environmental health in Europe—Highlights of the Conference held in Paris on 4–5 November 2008. *Bull. Epidemiol. Hebd.* **2009**, *16*, 2–5.
22. Louro, H.; Heinälä, M.; Bessems, J.; Buekers, J.; Vermeire, T.; Woutersen, M.; van Engelen, J.; Borges, T.; Rousselle, C.; Ougier, E.; et al. Human biomonitoring in health risk assessment in Europe: Current practices and recommendations for the future. *Int. J. Hyg. Environ. Health* **2019**, *222*, 727–737. [CrossRef]
23. Ganzleben, C.; Antignac, J.P.; Barouki, R.; Castaño, A.; Fiddicke, U.; Klánová, J.; Lebret, E.; Olea, N.; Sarigiannis, D.; Schoeters, G.R.; et al. Human biomonitoring as a tool to support chemicals regulation in the European Union. *Int. J. Hyg. Environ. Health* **2017**, *220*, 94–97. [CrossRef]
24. Scoping Document on Mercury and Its Organic Compounds. Available online: https://www.hbm4eu.eu/hbm4eu-substances/mercury/ (accessed on 9 March 2022).
25. St-Amand, A.; Werry, K.; Aylward, L.L.; Hays, S.M.; Nong, A. Screening of population level biomonitoring data from the Canadian Health Measures Survey in a risk-based context. *Toxicol. Lett.* **2014**, *231*, 126–134. [CrossRef]

26. Coscollà, C.; Sánchez, A.; Corpas-Burgos, F.; López, A.; Pérez, R.; Kuligowski, J.; Vento, M.; Yusà, V. Exposure and Risk Assessment of Hg, Cd, As, Tl, Se, and Mo in Women of Reproductive Age Using Urinary Biomonitoring. *Environ. Toxicol. Chem.* **2021**, *40*, 1477–1490. [CrossRef]
27. Schoeters, G.; Colles, A.; Hond, E.D.; Croes, K.; Vrijens, J.; Baeyens, W.; Nelen, V.; Mieroop, E.V.D.; Covaci, A.; Bruckers, L.; et al. Chapter 2F the flemish environment and health study (FLEHS)—Second survey (2007–2011): Establishing reference values for biomarkers of exposure in the flemish population. In *Biomarkers and Human Biomonitoring*; The Royal Society of Chemistry: London, UK, 2012; Volume 1, pp. 135–165.
28. Hond, E.D.; Govarts, E.; Willems, H.; Smolders, R.; Casteleyn, L.; Kolossa-Gehring, M.; Schwedler, G.; Seiwert, M.; Fiddicke, U.; Castaño, A.; et al. First Steps toward Harmonized Human Biomonitoring in Europe: Demonstration Project to Perform Human Biomonitoring on a European Scale. *Environ. Health Perspect.* **2015**, *123*, 255–263. [CrossRef]
29. Batáriová, A.; Spěváčková, V.; Beneš, B.; Čejchanová, M.; Šmíd, J.; Černá, M. Blood and urine levels of Pb, Cd and Hg in the general population of the Czech Republic and proposed reference values. *Int. J. Hyg. Environ. Health* **2006**, *209*, 359–366. [CrossRef] [PubMed]
30. Puklová, V.; Krsková, A.; Černá, M.; Čejchanová, M.; Řehůřková, I.; Ruprich, J.; Kratzer, K.; Kubínová, R.; Zimová, M. The mercury burden of the Czech population: An integrated approach. *Int. J. Hyg. Environ. Health* **2010**, *213*, 243–251. [CrossRef] [PubMed]
31. Czech Republic National Institute of Public Health. *Environmental Health Monitoring System in the Czech Republic—Summary Report, 2015*, 1st ed.; TIGIS: Prague, Czech, 2016; ISBN 978-80-7071-352-5.
32. Czech Republic National Institute of Public Health. *Environmental Health Monitoring System in the Czech Republic—Summary Report, 2016*, 1st ed.; TIGIS: Prague, Czech, 2017; ISBN 978-80-7071-365-5.
33. Schulz, C.; Conrad, A.; Becker, K.; Kolossa-Gehring, M.; Seiwert, M.; Seifert, B. Twenty years of the German Environmental Survey (GerES): Human biomonitoring—Temporal and spatial (West Germany/East Germany) differences in population exposure. *Int. J. Hyg. Environ. Health* **2007**, *210*, 271–297. [CrossRef] [PubMed]
34. Pino, A.; Amato, A.; Alimonti, A.; Mattei, D.; Bocca, B. Human biomonitoring for metals in Italian urban adolescents: Data from Latium Region. *Int. J. Hyg. Environ. Health* **2012**, *215*, 185–190. [CrossRef]
35. Valent, F.; Mariuz, M.; Bin, M.; Little, D.; Mazej, D.; Tognin, V.; Tratnik, J.; McAfee, A.J.; Mulhern, M.S.; Parpinel, M.; et al. Associations of Prenatal Mercury Exposure From Maternal Fish Consumption and Polyunsaturated Fatty Acids With Child Neurodevelopment: A Prospective Cohort Study in Italy. *J. Epidemiol.* **2013**, *23*, 360–370. [CrossRef]
36. Tratnik, J.S.; Falnoga, I.; Mazej, D.; Kocman, D.; Fajon, V.; Jagodic, M.; Stajnko, A.; Trdin, A.; Šlejkovec, Z.; Jeran, Z.; et al. Results of the first national human biomonitoring in Slovenia: Trace elements in men and lactating women, predictors of exposure and reference values. *Int. J. Hyg. Environ. Health* **2019**, *222*, 563–582. [CrossRef]
37. Trdin, A.; Falnoga, I.; Fajon, V.; Živković, I.; Tratnik, J.S.; Prpić, I.; Špirić, Z.; Horvat, M. Mercury speciation in meconium and associated factors. *Environ. Res.* **2019**, *179*, 108724. [CrossRef]
38. Batista, J.; Schuhmacher, M.; Domingo, J.; Corbella, J. Mercury in hair for a child population from Tarragona Province, Spain. *Sci. Total Environ.* **1996**, *193*, 143–148. [CrossRef]
39. Castaño, A.; Pedraza-Díaz, S.; Cañas, A.; Pérez-Gómez, B.; Ramos, J.; Bartolomé, M.; Pärt, P.; Soto, E.; Motas, M.; Navarro, C.; et al. Mercury levels in blood, urine and hair in a nation-wide sample of Spanish adults. *Sci. Total Environ.* **2019**, *670*, 262–270. [CrossRef]
40. Llop, S.; Murcia, M.; Amorós, R.; Julvez, J.; Santa-Marina, L.; Soler-Blasco, R.; Rebagliato, M.; Iñiguez, C.; Aguinagalde, X.; Iriarte, G.; et al. Postnatal exposure to mercury and neuropsychological development among preschooler children. *Eur. J. Epidemiol.* **2020**, *35*, 259–271. [CrossRef]
41. Pérez, R.; Suelves, T.; Molina, Y.; Corpas-Burgos, F.; Yusà, V.; on behalf of the BIOVAL task force. Biomonitoring of mercury in hair of children living in the Valencian Region (Spain). Exposure and risk assessment. *Chemosphere* **2018**, *217*, 558–566. [CrossRef]
42. Yusà, V.; Pérez, R.; Suelves, T.; Corpas-Burgos, F.; Gormáz, M.; Dualde, P.; Coscolla, C.; Quiles, J.; Roca, M.; Vento, M. Biomonitoring of mercury in hair of breastfeeding mothers living in the Valencian Region (Spain). Levels and predictors of exposure. *Chemosphere* **2017**, *187*, 106–113. [CrossRef]
43. Bocca, B.; Ruggieri, F.; Pino, A.; Rovira, J.; Calamandrei, G.; Mirabella, F.; Martínez, M.; Domingo, J.L.; Alimonti, A.; Schuhmacher, M. Human biomonitoring to evaluate exposure to toxic and essential trace elements during pregnancy. Part B: Predictors of exposure. *Environ. Res.* **2020**, *182*, 109108. [CrossRef]
44. Bocca, B.; Ruggieri, F.; Pino, A.; Rovira, J.; Calamandrei, G.; Martínez, M.Á.; Domingo, J.L.; Alimonti, A.; Schuhmacher, M. Human biomonitoring to evaluate exposure to toxic and essential trace elements during pregnancy. Part A. concentrations in maternal blood, urine and cord blood. *Environ. Res.* **2019**, *177*, 108599. [CrossRef]
45. Almerud, P.; Zamaratskaia, G.; Lindroos, A.K.; Bjermo, H.; Andersson, E.M.; Lundh, T.; Ankarberg, E.H.; Lignell, S. Cadmium, total mercury, and lead in blood and associations with diet, sociodemographic factors, and smoking in Swedish adolescents. *Environ. Res.* **2021**, *197*, 110991. [CrossRef]
46. Esteban, M.; Schindler, B.K.; Jiménez, J.A.; Koch, H.M.; Angerer, J.; Rosado, M.; Gómez, S.; Casteleyn, L.; Kolossa-Gehring, M.; Becker, K.; et al. Mercury analysis in hair: Comparability and quality assessment within the transnational COPHES/DEMOCOPHES project. *Environ. Res.* **2015**, *141*, 24–30. [CrossRef]

47. Budtz-Jørgensen, E.; Grandjean, P.; Jørgensen, P.J.; Weihe, P.; Keiding, N. Association between mercury concentrations in blood and hair in methylmercury-exposed subjects at different ages. *Environ. Res.* **2004**, *95*, 385–393. [CrossRef]
48. Packull-McCormick, S.; Ratelle, M.; Lam, C.; Napenas, J.; Bouchard, M.; Swanson, H.; Laird, B.D. Hair to blood mercury concentration ratios and a retrospective hair segmental mercury analysis in the Northwest Territories, Canada. *Environ. Res.* **2021**, *203*, 111800. [CrossRef]
49. Kirk, L.E.; Jørgensen, J.S.; Nielsen, F.; Grandjean, P. Public health benefits of hair-mercury analysis and dietary advice in lowering methylmercury exposure in pregnant women. *Scand. J. Public Health* **2017**, *45*, 444–451. [CrossRef]
50. Namorado, S.; Assunção, R.; Santiago, S.; Esteban-López, M.; Domínguez-Morueco, N. HBM4EU-MOM: Intervene to raise awareness to specific dietary recommendations and reduce prenatal exposure to mercury. In Proceedings of the 33rd Annual Conference of the International Society for Environmental Epidemiology, Online, 23–26 August 2021. E-poster No. P-473.
51. Jacobs, S.; Sioen, I.; Jacxsens, L.; Domingo, J.L.; Sloth, J.J.; Marques, A.; Verbeke, W. Risk assessment of methylmercury in five European countries considering the national seafood consumption patterns. *Food Chem. Toxicol.* **2017**, *104*, 26–34. [CrossRef]
52. Schartup, A.; Thackray, C.P.; Qureshi, A.; Dassuncao, C.; Gillespie, K.; Hanke, A.; Sunderland, E.M. Climate change and overfishing increase neurotoxicant in marine predators. *Nature* **2019**, *572*, 648–650. [CrossRef]

toxics MDPI

Article

Human Biomonitoring Initiative (HBM4EU): Human Biomonitoring Guidance Values Derived for Dimethylformamide

Farida Lamkarkach [1,*], Matthieu Meslin [1], Marike Kolossa-Gehring [2], Petra Apel [2] and Robert Garnier [3]

1 ANSES 14 Rue Pierre et Marie Curie, 94701 Maisons-Alfort, France; matthieu.meslin@anses.fr
2 German Environment Agency (UBA), Corrensplatz 1, 14195 Berlin, Germany; marike.kolossa@uba.de (M.K.-G.); petra.apel@uba.de (P.A.)
3 Paris Poison Centre, Toxicology Department (FeTox), APHP, Lariboisière-Fernand-Widal Hospital, 200 Rue du Faubourg Saint-Denis, 75010 Paris, France; robert.garnier@aphp.fr
* Correspondence: farida.lamkarkach@anses.fr

Abstract: Within the European Joint Program on Human Biomonitoring HBM4EU, human biomonitoring guidance values (HBM-GVs) for the general population (HBM-GV$_{GenPop}$) or for occupationally exposed adults (HBM-GV$_{Worker}$) are derived for prioritized substances including dimethylformamide (DMF). The methodology to derive these values that was agreed upon within the HBM4EU project was applied. A large database on DMF exposure from studies conducted at workplaces provided dose–response relationships between biomarker concentrations and health effects. The hepatotoxicity of DMF has been identified as having the most sensitive effect, with increased liver enzyme concentrations serving as biomarkers of the effect. Out of the available biomarkers of DMF exposure studied in this paper, the following were selected to derive HBM-GV$_{Worker}$: total N-methylformamide (tNMF) (sum of N-hydroxymethyl-N-methylformamide and NMF) and N-acetyl-S-(N-methylcarbamoyl)cysteine (AMCC) in urine. The proposed HBM-GV$_{Worker}$ is 10 mg·L^{-1} or 10 mg·g^{-1} creatinine for both biomarkers. Due to their different half-lives, tNMF (representative of the exposure of the day) and AMCC (representative of the preceding days' exposure) are complementary for the biological monitoring of workers exposed to DMF. The levels of confidence for these HBM-GV$_{Worker}$ are set to "high" for tNMF and "medium-low" for AMCC. Therefore, further investigations are required for the consolidation of the health-based HBM-GV for AMCC in urine.

Keywords: HBM4EU; dimethylformamide; DMF; HBM-GV; guidance value; biomarker; human biomonitoring; toxicokinetics; health effects; liver; carcinogenicity; reprotoxic effects

Citation: Lamkarkach, F.; Meslin, M.; Kolossa-Gehring, M.; Apel, P.; Garnier, R. Human Biomonitoring Initiative (HBM4EU): Human Biomonitoring Guidance Values Derived for Dimethylformamide. *Toxics* **2022**, *10*, 298. https://doi.org/10.3390/toxics10060298

Academic Editors: Giovanna Tranfo and Luis Alberto Henríquez-Hernández

Received: 22 April 2022
Accepted: 26 May 2022
Published: 31 May 2022

Publisher's Note: MDPI stays neutral with regard to jurisdictional claims in published maps and institutional affiliations.

Copyright: © 2022 by the authors. Licensee MDPI, Basel, Switzerland. This article is an open access article distributed under the terms and conditions of the Creative Commons Attribution (CC BY) license (https://creativecommons.org/licenses/by/4.0/).

1. Introduction

The European Joint Program on Human Biomonitoring (HBM4EU) is a joint effort of 30 countries and the European Environment Agency, co-funded by the European Commission, within the framework of Horizon 2020 [1]. With a project duration from 2017 to 2022, HBM4EU aims to harmonize and advance human biomonitoring in Europe by studying the internal exposure of European citizens to chemicals and its impact on health according to jointly agreed-upon harmonized procedures. The project generates scientific knowledge to answer concrete policy-relevant questions for Europe and thus builds bridges between science and policy, benefiting society by improving public health.

HBM-GVs are derived according to the methodology set within HBM4EU and detailed in Apel et al. (2020). Values are specifically derived for the general population (HBM-GV$_{GenPop}$) and for workers (HBM-GV$_{Worker}$). They indicate the concentration of a compound or its metabolite(s) in a biological matrix (e.g., blood, urine) at and below which no health risk is anticipated (according to current knowledge) [2].

The HBM4EU Consortium identified eighteen substances or substance groups of high priority to answer open policy-relevant questions via targeted research. HBM-GVs have previously been established for several substances (e.g., cadmium [3], BPA [4], phthalates and DINCH [5], and pyrrolidones [6]). The selected substances included DMF as part of the aprotic solvent group.

As DMF is a widely used chemical in numerous industrial sectors, many workers are expected to be occupationally exposed to it. DMF is readily biodegradable and not known for its bioaccumulation potential; thus, long-term environmental exposure is unlikely [7]. Therefore, HBM-GVs are proposed only for workers (HBM-GV_{Worker}) in the present work.

DMF is predominantly used as an industrial solvent in the synthesis of fine chemicals (e.g., active pharmaceutical ingredients and crop protection ingredients) and for the production of polyurethane-coated textiles (e.g., artificial leather, rain protection clothing, footwear, medical mattress covers, surgical incise films, etc.). DMF is also used as a solvent in the production of synthetic fibers and for the formulation of mixtures, as a gas stabilizer in acetone cylinders, as a cleaning solvent, as a laboratory chemical, etc. [8]. In 2019, the Risk Assessment Committee (RAC) and the Committee for Socio-economic Analysis (SEAC) of ECHA emitted an opinion on an Annex XV dossier proposing restrictions on DMF [7]. RAC concluded that "manufacturers, importers and downstream users of the substance on its own (regardless of whether DMF is a (main) constituent, an impurity or a stabilizer) or in mixtures in a concentration equal or greater than 0.3% shall use in their chemical safety assessment and safety data sheets a worker based harmonized Derived No Effect Level (DNEL) value for long-term inhalation exposure of 6 $mg{\cdot}m^{-3}$ and a worker based harmonized DNEL for long-term dermal exposure of 1.1 $mg{\cdot}kg^{-1}$ $bw{\cdot}d^{-1}$." [7].

According to the harmonized classification, labelling, and packaging (CLP) regulation (EC N° 1272/2008), DMF may damage an unborn child, is harmful in contact with skin, can cause serious eye irritation, and is harmful if inhaled. Until 2018, DMF was not classified as to its carcinogenicity to humans by the International Agency for Research on Cancer (IARC) due to inadequate evidence in humans and evidence suggesting lack of carcinogenicity in animal studies [9]. In 2018, IARC re-evaluated DMF as probably carcinogenic for humans (Group 2A) based on limited evidence in humans and sufficient evidence in experimental animal studies [10]. DMF is not categorized for its carcinogenic properties in the European Union.

For the present work, a review of health effects associated with DMF exposure was conducted to provide an HBM-GV_{Worker} for occupationally exposed people, with respect to the current methodology set in the HBM4EU project [2]. To achieve this, an assessment of the extensive database on DMF (including toxicokinetic and toxicodynamic data) was performed.

2. Materials and Methods

2.1. General Methodology to Derive HBM-GVs in the Framework of the HBM4EU Project

An HBM-GV corresponds to a biomarker concentration in a biological matrix and represents a value at and below which adverse human health effects generated by the substance exposure are not to be expected, according to current knowledge. HBM-GVs are derived within the HBM4EU project according to a systematic and transparent methodology [2] and by taking into account the feedback provided by competent experts from the 30 HBM4EU participating countries, which was thereby mutually agreed upon within the HBM4EU consortium.

The methodological approach was developed on the basis of the procedure described in the German Human Biomonitoring Commission's position paper [11,12], by the team from the Summit Toxicology consulting firm [13,14], and in the guidance document elaborated by the French Agency for Food, Environmental, and Occupational Health and Safety (ANSES) [15].

The HBM4EU method for deriving an HBM-GV can be divided into the following steps:

- Selection of the relevant biomarker(s): a biomarker is defined as any substance, structure, or process that can be measured in the body or its degradation product(s) which influences or predicts the incidence of outcome or disease. Biomarkers can be classified into biomarkers of exposure (BME), biomarkers of effects, or biomarkers of susceptibility [2]. This first step consists of the data collection on the substance and its metabolites (i.e., toxicokinetic and toxicodynamic data). Based on these data, biomarkers of exposure and/or effect are identified and then chosen according to defined criteria: specificity, sensitivity, half-life, sampling conditions, invasiveness, background level, and analytical methods [15].
- The derivation of HBM-GVs for the selected biomarkers can then be conducted through three possible options (decision tree described in Figure 1). When the corresponding data are available, the preferred option is to base HBM-GV(s) identification on the relationship between internal concentrations of the selected biomarker(s) and the occurrence of adverse effects. The second possible option is to derive HBM-GVs from external limit values (i.e., Occupational Exposure Levels [OEL] or Toxicity Reference Values [TRV]) proposed by relevant European or non-European bodies. The last option consists of the derivation of HBM-GVs on the basis of critical effects observed in animal toxicological studies. These options are described in more detail in Apel et al. (2020) [2].

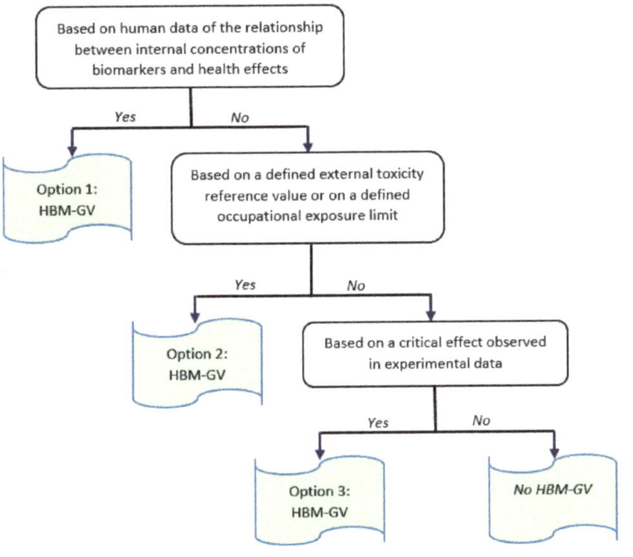

Figure 1. Decision tree for determining HBM-GVs.

If possible, option 1 is preferred for deriving an HBM-GV (Figure 1). For this, the following steps are required:

i. Choice of the critical effect which is considered to be the most sensitive among all adverse effects that may arise from exposure to the substance (e.g., changes in morphology, physiology, growth, development, reproduction, or life span resulting in an impairment of functional capacity, an impairment of the capacity to offset additional stress, or an increase in sensitivity).
ii. Selection of the key study and identification of a point of departure (POD) with the most informative studies, i.e., well-conducted human studies adequately reporting measured internal concentration levels of a substance, sampling times, analytical methods used, and the relationships between concentrations of a substance or its metabolites in human biological media and the occurrence of adverse effects. If

relevant and qualitatively acceptable human studies are available, a key human study together with a Point of Departure (POD) is selected.

iii. Application of assessment factors (AFs), when necessary, to obtain the HBM-GVs. These can be divided into an AF_H for the intraspecies variability or possible other AFs to compensate for the potential remaining uncertainties in the derived HBM-GV, especially regarding the possible deficiencies or data gaps in the available data sets [2].

2.2. Methodology Used for Deriving HBM-GV$_{Worker}$ for DMF

For the present work, the general methodology to derive HBM-GVs in the framework of the HBM4EU project was applied for DMF using the data issued from:

- the reports by the American Conference of Governmental Industrial Hygienists (ACGIH) [16,17], the German Research Foundation or Deutsche Forschungsgemeinschaft (DFG) [18,19], the European Chemicals Agency (ECHA) [7], the International Agency for Research on Cancer (IARC) ([10]), and the Scientific committee for occupational exposure limits (SCOEL) [20];
- for more recent and specific publications, a bibliographical research, which was conducted in Medline and Scopus until 2021 with the following keywords: Dimethylformamide, DMF, guidance value, toxicity reference value (TRV), biomarker of exposure, biomonitoring, toxicokinetic, health effects, liver, carcinogenicity, and reprotoxic effects.

Additionally, a global level of confidence (i.e., high, medium, or low) is attributed to each derived HBM-GV$_{Worker}$ to reflect the uncertainties related to its derivation. This level of confidence is mainly based on the quality of the available data [2].

3. Results

3.1. Identification of Possible Biomarkers of Exposure

DMF absorption has been well-studied in humans (workers and volunteers); absorption via inhalation is high, with 60–90% of the inhaled dose retained in the respiratory tract [21,22]. Dermal absorption in humans from direct contact is also very high (up to 40%, depending on temperature and humidity) [23–25]. In the study by Nomiyama et al. (2001), in which the authors evaluated the difference between the absorption of DMF vapor by dermal and inhalation routes, it was estimated that skin and lung absorption contributed for 40.4% and 59.6% of total absorption, respectively [26]. In a study conducted in exposed workers, Lauwerys et al. (1980) reported that in workers without gloves, the amount of DMF absorbed through the skin may be more than twice that absorbed by inhalation [27].

DMF and its metabolites are distributed throughout the organism, and quite uniformly in the different tissues [7,16]. In rodents, they freely cross the placenta [28,29]; the corresponding data are not available for humans.

DMF is rapidly metabolized in the liver. N-hydroxymethyl-N-methylformamide (HMMF) arises from DMF mainly through enzymatic oxidation by the cytochrome P450 enzyme system (CYP2E1) [30]. Then, demethylation leads to the formation of N-methylformamide (NMF). The concentrations of HMMF and NMF in urine are grouped as total NMF (tNMF), as it is difficult to analyze these metabolites separatly due to the thermal decomposition of the hydroxyl derivative in the injection port of the gas chromatograph [31,32]. NMF can be further oxidized to N-(hydroxymethyl)formamide (HMF) and formamide. N-acetyl-S-(N-methylcarbamoyl)cysteine (AMCC) is another metabolite formed after exposure to DMF [33]. AMCC is the end product of the enzymatic breakdown of S-(N-methylcarbamoyl)glutathione. The latter is formed by the reaction between glutathione and the probable reactive metabolic intermediate, methyl isocyanate (MIC) [22,33]. Unlike in rodents, AMCC is a major metabolite of DMF in humans [34] (Figure 2).

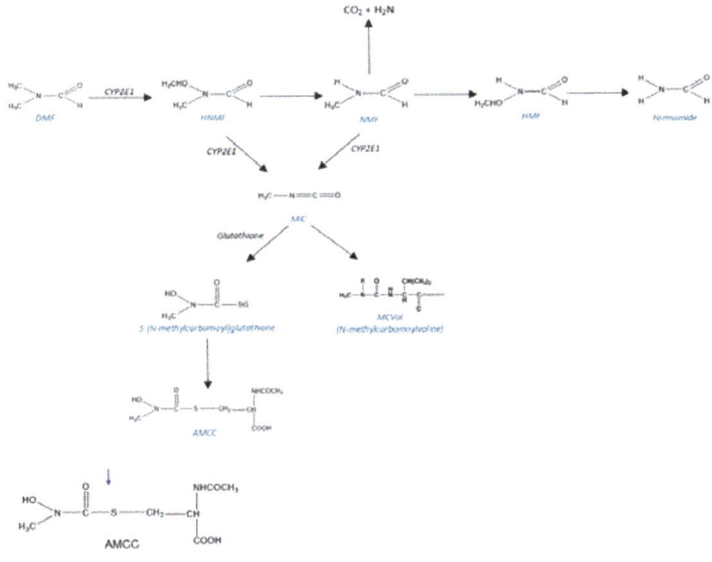

Figure 2. A scheme for the metabolic pathways of DMF (as described in the text).

As MIC is a reactive substance, it also produces adducts to proteins; N-methylcarbamoy lvaline (MCVal) adducts at the N-terminal position of the globin chains of hemoglobin are also identified as possible biomarkers of DMF exposure [35,36]. Nε-(N-methylcarbamoyl) lysine adducts to globin were also identified in humans occupationally exposed to DMF [37]; however, data on this possible biological indicator of exposure are very scarce.

DMF metabolism has been reported to be saturable; it is able to inhibit its own metabolism [7,38–40]. It is competitively inhibited by simultaneous alcohol exposure (and potentially by the exposure to any other potential substrate of CYP2E1) [41]. DMF reciprocally inhibits ethanol metabolism; it is a strong inhibitor of aldehyde-dehydrogenase, and this effect is responsible for alcohol intolerance reactions (resulting from the accumulation of acetaldehyde) [42]. Usual alcohol consumption induces CYP2E1 activity and consequently DMF metabolism into HMMF (and probably also into MIC) [7].

The hepatic biotransformation of DMF and the urinary excretion of its metabolites (HMMF, NMF, formamide, and AMCC) are the main elimination pathways of DMF. In a study by Mraz and Nohova (1992), the authors analyzed DMF metabolite excretion in 10 volunteers during and after an 8-h exposure to DMF. After exposure to 30 mg·m^{-3} or 60 mg·m^{-3}, the yield of compounds measured in urine was similar: 0.3% DMF, 22.3% tNMF (HMMF + NMF), 13.2% tHMF (HMF + formamide), and 13.4% AMCC. The half-lives of tNMF, AMCC, tHMF, and DMF were 4, 23, 7 and 2 h, respectively, after 30 mg·m^{-3} exposure [22]. A slower elimination, without significant accumulation over the workweek, was reported after skin exposure of volunteers to liquid DMF with an average half-life of 7–8 h for tNMF [23,24].

Saillenfait et al. (1997), who treated lactating rats with a single oral dose of 100 mg/kg bw [^{14}C]-DMF on lactation day 14, measured DMF, HMMF, and NMF in the rats' milk at concentrations equal to those in the plasma. In addition, they noted that 60–70% of the absorbed dose of [^{14}C]-DMF was excreted in the urine and 3–4% in the feces [29].

There is no study evaluating the pulmonary elimination, possibly due to strong adsorption of DMF on the walls of the measuring device [22].

A recent overview of key studies on toxicokinetic data in humans and animals is presented in ECHA's opinion on DMF restriction [7].

Based on all these data, several metabolites can be identified as potential BME for biological monitoring of DMF at the workplace:

- Unchanged DMF in urine;
- total NMF or tNMF (which is the sum of HMMF and NMF) in urine;
- AMCC in urine;
- MCVal in blood; and
- formamide in urine.

Advantages and limits of each BME are detailed in Section 3.4.

3.2. Identification and Characterization of the Dangers Associated with DMF Exposure

As the HBM4EU strategy for deriving an HBM-GV recommends giving priority to human data for the characterization of the relationship between the risk of adverse effects and internal concentrations of the selected biomarker(s), only human data were initially considered. This initial retrieval provided a sufficiently extensive database on the effects on humans, allowing the identification of HBM-GVs.

In published human studies of health effects, DMF exposure was mostly assessed via airborne DMF measurements and/or via biomonitoring (mainly tNMF and AMCC in urine, and less often MCVal in blood).

The following effects have been studied: acute toxicity; skin, eye, and respiratory tract irritation; chronic toxicity (effects on liver, gastro-intestinal effects, and alcohol intolerance); genotoxicity; carcinogenicity; and reproductive and developmental effects.

A few publications have reported on DMF irritative effects at the workplace; effects such as chemical burns of the skin and eyes were observed after direct contact with liquid DMF [43]. Workers reported eye and upper airway irritation after exposure to 10 ppm or more [17].

It appears that the main systemic target for toxicity after acute or chronic exposure to DMF is the liver, both in humans and laboratory animals. Experimentally, the NOAEL and LOAEL for hepatotoxicity in the most sensitive species are 12 $mg \cdot kg^{-1}$ $bw \cdot d^{-1}$ and 60 $mg \cdot kg^{-1}$ $bw \cdot d^{-1}$, respectively, for the ingestion route, and 25 and 100 ppm, respectively, for the inhalation route, the rat being the most sensitive species in both cases [18]. In humans, acute high-dose exposure to DMF can cause damage to the liver and even lead to death, although in most reported cases, the damage was reversible [44]. He et al. (2010) observed that DMF could produce acute toxic hepatitis as well as chronic hepatic damage such as hepatic cirrhosis [45]. More recently, Wu et al. (2017) reported that liver damages in DMF-exposed workers include hepatitis, liver fibrosis, and cirrhosis [46]. Many authors investigated liver function via serum liver enzyme concentrations (e.g., alanine aminotransferase (ALT); aspartate aminotransferase (AST); and gamma glutamyltranspeptidase (γGT)) together with subjective symptoms and clinical signs in workers exposed to DMF. According to the MAK commission, an increase of serum AST and ALT is the most suitable parameter for detecting DMF-induced effects on the liver [18].

Table 1 presents those studies where biological evaluation of exposure and a search for hepatic damage was performed. In some of them, DMF exposure was also evaluated through air measurements. However, DMF airborne concentrations are not always consistently associated with the occurrence of health effects, as DMF systemic contamination may also result from direct skin contact and vapor–skin absorption [45].

Some authors considered confounding factors, e.g., alcohol consumption (based on questionnaires, in almost all cases). Workers who did not consume alcohol tolerated much higher concentrations of DMF without changes in liver functions [47,48].

Several studies have reported DMF-related disorders of the digestive system; the symptoms included abdominal pain, anorexia, nausea, vomiting, and diarrhea [28]. In some studies, these effects on the digestive tract were reported in workers exposed to low DMF air concentrations (below 10 ppm or even 5 ppm), but with probable direct skin contact [49–51].

Table 1. Summary of occupational studies reporting health effects of DMF exposure with biomonitoring data.

Reference	Subjects	Exposure DMF in the Air (DMFa) Metabolites *	Results/Observations
Lyle et al., 1979 [52] England	Workers (DMF used as solvent) N = 102) 3-year follow-up	DMFa Range: <10 to 200 ppm (30–600 mg·m^{-3}) tNMF (tNMFu) Range: <10 to 77 µL/L *(probable error in the unit)*	Alcohol intolerance reactions Facial flushing and other symptoms in 19 workers 26 of the 34 episodes occurred after the workers had consumed alcoholic drinks Liver function not investigated
Yonemoto and Suzuki, 1980 [53] Japan	Workers (synthetic leather factory) N = 11 (biomonitoring data for 9 of them)	DMFa Range: 0–5 ppm (0–15 mg·m^{-3}) (TWA) Post-Shift (PS) tNMFu Range: 0.4–19.56 mg·d^{-1}	No effect on serum biochemistry (liver enzymes) Alcohol intolerance: 6/11 workers said to be less tolerant than before
Lauwerys et al., 1980 [27] Belgium	Workers in an acrylic fiber factory N = 22 (+28 controls)	DMFa Mean: 13 (1.3–46.6) mg·m^{-3} (4.5 (0.4–15.3) ppm) Stationary sampling tNMFu <40–50 mg·g^{-1} cr (PS)	No effects on serum biochemistry (liver enzymes not elevated) Signs of alcohol intolerance in some workers
Catenacci et al., 1984 [54] Italy Quoted by SCOEL (2006) [20]	N = 54 (employed > 5 y) acrylic fiber plant 2 groups exposed and 54 controls)	Group 1 (N = 28) DMFa Mean (range): 6 (4–8) ppm (18 (12–24) mg·m^{-3}) tNMFu 22.3 mg·L^{-1} Group 2 (N = 26) DMFa Mean (range): 1 (0.6–1.6) ppm (3 (1.8–4.8) mg·m^{-3}) tNMFu: 7 mg·L^{-1}	No significant effects on liver enzymes in the 2 groups
Sakai et al., 1995 [55] Japan	Workers (N = 10) Polyurethane production 2.5-year follow-up	DMFa Geometric mean (GM): 2.5–10.4 ppm (7.5–31.2 mg·m^{-3}) PS tNMFu Mean: 24.7 mg·g^{-1} cr AMCCu Mean: 22.0 mg·g^{-1} cr	No effects on liver enzymes

Table 1. *Cont.*

Reference	Subjects	Exposure DMF in the Air (DMFa) Metabolites *	Results/Observations
Fiorito et al., 1997 [50] Italy	N = 75 (employed) synthetic leather production and 75 controls (unexposed workers)	DMFa Group 1 (Washing) N = 10 GM: 21.5 $mg \cdot m^{-3}$ (7.2 ppm) Range: 5–40 $mg \cdot m^{-3}$ Group 2 (Production) N = 12 GM: 18.7 $mg \cdot m^{-3}$ (6.2 ppm) Range: 5–40 $mg \cdot m^{-3}$ tNMFu (N = 22): GM: 13.6 $mg \cdot L^{-1}$ or 13.4 $mg \cdot g^{-1}$ cr PS	Elevation of liver enzymes (12/75) [$p < 0.01$] Alcohol intolerance in 50% of exposed workers and facial flushing (38%), palpitations (30%), headache (22%), body flushing (15%), and tremors (14%) Gastrointestinal symptoms (stomach pain, nausea, loss of appetite) in 50% of exposed workers
Wrbitzky and Angerer, (1998) [47]; Wrbitzky, (1999) [48] Germany	Polyacrylic fiber production N = 126 (total of exposed workers)	DMFa Mean (SD): 4.1 ± 7.4 (<0.1–37.9) ppm (12.3 ± 22.2 $mg \cdot m^{-3}$) tNMFu Mean (SD): 14.9 ± 18.7 (0.9–100) $mg \cdot L^{-1}$ 9.1 ± 11.4 (0.5–62.3) $mg \cdot g^{-1}$ cr	Effects on liver enzymes Synergetic effect of alcohol consumption on liver enzymes activity
	Finishing N = 55	DMFa Mean (SD): 14.2 ± 2.2 (>0.1–13.7) ppm (42.6 ± 6.6 $mg \cdot m^{-3}$) tNMFu Mean (SD): 4.5 ± 4.3 $mg \cdot g^{-1}$ cr	Effects on liver enzymes in alcohol consumers
	Dyeing N = 12	DMFa Mean (SD): 2.5 ± 3.1 (0.1–9.8) ppm (7.5 ± 9.3 $mg \cdot m^{-3}$) tNMFu Mean (SD): 6.7 ± 5.4 (0.8–17.2) $mg \cdot g^{-1}$ cr	No effects on liver enzymes in workers not drinking alcohol Reduced alcohol consumption in workers drinking alcohol
	Dry spinning N = 28	DMFa Mean (SD): 6.4–9.6 (0.8–36.9) ppm (19.2 ± 28.8 $mg \cdot m^{-3}$) tNMFu Mean (SD): 11.6 ± 13.1 (0.9–62.3 $mg \cdot g^{-1}$ cr)	
	Wet spinning N = 30	DMFa Mean (SD): 7.3 ± 10.2 (0.3–37.9) ppm (21.9 ± 30.6 $mg \cdot m^{-3}$) tNMFu Mean (SD): 16.0 ± 15.9 (0.4–54.0) $mg \cdot g^{-1}$ cr	

Table 1. *Cont.*

Reference	Subjects	Exposure DMF in the Air (DMFa) Metabolites *	Results/Observations
He et al., 2010 [45] China	Synthetic leather and other resins production N = 79 (58 men and 21 women)	Group 1 (N = 33): Low exposure DMFa Min-Max: Not detected- <4.55 mg·m^{-3} (1.6 ppm) DMFu GM: 0.26 mg·g^{-1} creatinine tNMFu GM: 1.80 mg·g^{-1} creatinine AMCCu GM: 4.25 mg·g^{-1} creatinine Group 2 (N = 24): Medium exposure DMFa (Mean): 9 mg·m^{-3} (3 ppm) DMFu (GM): 0.53 mg·g^{-1} creatinine tNMFu (GM): 9.6 mg·g^{-1} creatinine AMCCu (GM): 25.4 mg·g^{-1} creatinine Group 3 (N = 22): High exposure DMFa (Mean): 36 mg·m^{-3} (12 ppm) DMFu (GM): 1.78 mg·g^{-1} creatinine tNMFu (GM): 26.5 mg·g^{-1} creatinine AMCCu (GM): 45.5 mg·g^{-1} creatinine	About 60% of subjects with urine AMCC concentration above 40 mg·g^{-1} cr had raised liver enzyme activities Statistically more workers with raised liver enzymes in group 3 (high exposure group) than in group 1 (administrative staff of the factory); $p < 0.05$
Kilo et al., 2016 [51] Germany	Synthetic fiber production N = 220 workers and 175 Controls	Mean ± SD DMFa: 6.2 ± 7.6 mg·m^{-3}; 2.1 ± 2.5 ppm tNMFu: 7.75 (±8.82) mg·L^{-1} AMCCu: 9.42 (±10.42) mg·g^{-1} cr McVal: 83.3 (±83.1) nmol·g^{-1} globin	None of the tested liver enzyme activities showed a positive association with any of the three exposure markers Alcohol intolerance reactions (not influencing alcohol consumption behavior)

Table 1. *Cont.*

Reference	Subjects	Exposure DMF in the Air (DMFa) Metabolites *	Results/Observations
Wu et al., 2017 [46] China	Synthetic leather production N = 698 And 188 controls	3 exposure groups: Median (range) Low exposure group tNMFu (N = 228): 0.0025 mg·L^{-1} (ND-0.11) AMCCu (N = 227): 2.18 mg·L^{-1} (ND-16,95) MCVal (N = 232): 15.19 nmol·mol^{-1} globin (ND-29.37) Moderate exposure groups tNMFu (N = 227): 1.78 mg·L^{-1} (0.11–3.88) AMCCu (N = 228): 44.9 mg·L^{-1} (16.95–86.62) MCVal (N = 234): 46.00 (29.37–63.95) nmol·mol^{-1} globin High exposure groups tNMFu (N = 227): 9.59 mg·L^{-1} (>3.88) AMCCu (N = 227): 148.01 mg·L^{-1} (>86.62) MCVal (N = 232): 87.01 (63.95–) nmol/mol globin	Liver injury assessed by measurement of liver enzyme levels and compared to reference value ranges (AST and ALT: 0–45, γGT: 8–58U/L) Statistically more workers with raised liver enzymes only in high-exposure group for tNMF, in both moderate- and high-exposure groups for AMCCu and MCVal ($p < 0.05$)

* DMFu: DMF in urine; tNMFu: Total NMF in urine; AMCCu: AMCC in urine; MCVal: MCVal in blood; SD: Standard deviation.

Exposure to DMF in combination with subsequent alcohol consumption can induce an alcohol intolerance reaction (with flushing of the facial skin, neck, and arms). This alcohol intolerance results from the accumulation of acetaldehyde due to the inhibition of aldehyde dehydrogenase [18,52]. Alcohol intolerance has been observed after DMF exposure in both rats and humans [51,52,56]. Kilo et al. (2017) observed signs of alcohol intolerance (flushing reaction) in almost half of the workers at 4.43 mg·L^{-1}, 4.84 mg·L^{-1}, and 60.5 nmol·g^{-1} globin for urine tNMF, urine AMCC, and MCVal adducts to hemoglobin, respectively. Alcohol intolerance can be observed for very low DMF exposure, but with a large interindividual variability due to the corresponding interindividual variability of aldehyde dehydrogenase activity [51]. Wolff (1972) reported that 83% of East Asian subjects (Japanese, Taiwanese, and Koreans) responded with a marked visible facial flushing after drinking small amounts of alcohol. In contrast, after similar doses, only 3% of Caucasian subjects showed visible flushing [57]. Thus, Antabuse effects reported in occupational studies must be interpreted in light of the differences in alcohol sensitivity (based on genetic polymorphisms of the enzymes alcohol dehydrogenase and aldehyde dehydrogenase) [58].

IARC (2018) [10] reviewed a large body of data to assess the genotoxicity of DMF [9,59–66]. Overall, there is no proof of the genotoxicity of DMF; experimental studies both in vitro, in prokaryotes and mammalian cells, and in vivo, in rodents, generally gave negative results. The genotoxic effects reported in a few field studies cannot constitute evidence of genotoxic effects of DMF due to confounding factors, in particular from co-exposures.

According to IARC [10], there is sufficient evidence for the carcinogenicity of DMF in laboratory animals. The carcinogenicity of DMF has been demonstrated in two inhalation (whole body) studies in rats and mice and in one inhalation (whole body) and ingestion (in drinking water) study in male rats. In an inhalation study in mice by Senoh et al. (2004), exposure to DMF increased the incidence of hepatocellular adenoma, hepatocellular carcinoma, and the aggregate of hepatocellular adenoma and carcinoma and hepatoblastoma; this was observed both in male and female mice with a dose relationship effect [67]. A second study, conducted in mice by Malley et al. (1994), provided negative results [68]. A dose-dependent increase in the incidence of hepatocellular adenomas, hepatocellular carcinomas, and aggregated adenoma and hepatocellular carcinomas was observed in rats of both sexes in an inhalation study [67]. The second study conducted by inhalation in rats was negative [68]. In the combined respiratory and oral study by Ohbayashi et al. (2009) in male rats, exposure to DMF increased the incidence of hepatocellular adenomas and aggregated hepatocellular adenoma and carcinoma. An increased incidence of hepatocellular carcinoma was observed in the group treated orally only [69].

The IARC (2018) determined that there is limited evidence of DMF carcinogenicity in humans based on three publications. The first reported a cluster of 3 cases of testicular cancer in 153 mechanics in a US military aviation repair shop [70]. The second study was motivated by the first, and also reported a cluster of three cases of testicular cancer in workers exposed to DMF in a tannery [71]. The third study was a retrospective study conducted in an acrylic fiber factory; it did not identify an increased risk of testicular cancer associated with DMF exposure. However, it showed an excess risk of oropharyngeal cancers [72]. This retrospective study was followed by a case-control study conducted in this factory and three other similar plants. The latter study identified 11 cases of testicular cancer, but did not show an excess risk of this cancer associated with exposure to DMF (OR: 0.99; 95% CI: 0.22–4.44) [73]. More recently and after the last IARC evaluation, the results of a Korean cohort study were published. The cohort was constituted by 11,953 workers with one or more urine tNMF measurements between 2001 and 2004 for DMF occupational exposure biomonitoring. Their mortality was matched with the mortality data of the Korean National Statistical Office and followed up for cancer mortality between 2000 and 2011. DMF exposure was estimated as low, medium, or high, according to tNMF concentrations in urine measuring <7.5 mg·L^{-1}, 7.5–15 mg·L^{-1}, or >15 mg·L^{-1}, respectively. Overall cancer mortality was significantly elevated in medium and high exposure groups with adjusted hazard ratios (HRadj) of 2.72 (95% CI: 1.09–6.81) and ≥2.41 (95% CI 1.03–5.66), respectively.

HRadj were also significantly elevated for lung cancer in the medium exposure group (HRadj 14.36, 95% CI 1.41–146.86) and for hepatocellular carcinoma in the high exposure group (HRadj 3.73, 95% CI 1.05–13.24). As exposure evaluation was only transversal and during a brief period, as the cumulated duration of exposure of each worker is unknown, and as major confounding factors including alcohol consumption and smoking status were not taken into account, these results should be interpreted with caution [74].

The data considered by the experts of IARC as constituting limited evidence of the carcinogenicity of DMF in humans and of an increased risk of testicular cancer associated with exposure to this substance would justify more cautious conclusions. Indeed, this evaluation is based on only two clusters of cases, with a negative case-control study. Moreover, the neoplastic lesions observed in the animal studies were localized only in the liver and not in the testis, and could result from DMF hepatotoxicity and capacity for inducing oxidative stress [10]. In addition, there is no evidence that DMF is genotoxic.

DMF is recognized as toxic for reproduction and classified as such according to the CLP regulation. In humans, the study by Chang et al. (2004) on 12 workers in a synthetic leather factory exposed to DMF showed that these workers had reduced sperm mobility when compared to 8 controls. This decrease was proportional to urinary tNMF concentration but not to DMF air concentration. However, considering the small size of the group and the fact that the number of subjects exposed to DMF and having consumed alcohol (8/12; 66.7%) was significantly higher than that of the controls (3/8; 37.5%), these results should be considered with caution [75].

In animals, numerous studies in rodents have demonstrated the effects of DMF on female fertility and on development. In most studies, embryo/foetotoxic effects included a reduction in the body weight of the offspring and a reduction in number and size of litters. Teratogenic effects included various skeletal malformations (especially craniofacial and sternal malformations). In rats, embryo/foetotoxicity generally occurred at maternally toxic doses or concentrations, and teratogenicity was also not reported in the absence of maternal toxicity. However, in mice and rabbits, embryo/foetotoxicity and/or signs of teratogenicity were observed at doses which did not generate maternal toxicity. Exposure to DMF is consistently reported to result in umbilical hernia in rabbit developmental toxicity studies, whereas gallbladder agenesis and sternal malformations were only observed in the two most reliable studies (after dermal and inhalation exposure). The lowest concentration level causing malformations in rabbits was 150 ppm (NOAEC: 50 ppm) [76].

Based on these results, the RAC concluded that DMF was responsible for skeletal developmental disorders in all three species (rats, mice and rabbits), the rabbit being the most sensitive species to developmental toxicity of DMF [7]. In oral studies the NOAEL for effects on fertility was 219 mg/kg bw/d in mice [77] and the NOAEL for effects on fertility and development was 166 mg/kg bw/d in rats [76]. In respiratory toxicity studies, the LOAEC for developmental effects was 150 ppm in rabbits, with a 50 ppm NOAEC [7,76]. In dermal exposure studies, it was not possible to identify a NOAEL, the LOAEL was 94 mg/kg bw/d in rats [76]. According to the RAC, the transposability to humans of the effects observed in animals is plausible [7]. Orally, the NOAEL in rabbits was 44.1 $mg{\cdot}kg^{-1}$ bw/d^{-1}.

3.3. Choice of the Critical Effect

In exposed workers, the critical systemic effects of DMF (i.e., those occurring for the lowest exposure levels) are the hepatotoxic effects.

The results of studies in animals showed hepatotoxic, reproductive, and carcinogenic effects as the most relevant. Therefore, for these endpoints, lower points of departure found in animals are detailed in Table 2.

Table 2. Lower points of departure for relevant adverse effects reported in animal studies.

Route	Effects on Liver	Reproductive Effects	Carcinogenic Effects on Liver
Inhalation	NOAEL: 25 ppm LOAEL: 100 ppm (Rats and mice) [68]	NOAEL: 25 ppm LOAEC: 150 ppm (Rabbits) [76]	LOAEC: 200 ppm (mice) LOAEC: 400 ppm (rats) [67]
Oral	NOAEL = 238 mg/kg bw/d LOAEL = 475 mg/kg bw/d (Rats) (BASF (1977) unpublished data, quoted by ECHA [7])	NOAEL = 166 mg/kg bw/d LOAEL = 503 mg/kg bw/d (Rats) [76] NOAEL: 44.1 mg/kg bw/d (Rabbits) [78]	LOAEL = 800 ppm (Rats) [69]
Dermal	-	LOAEL: 94 mg/kg/d (Rats) 100 mg/kg/d (Rabbits) [76]	-

DMF is responsible for reprotoxic effects. As shown in Table 2, the LOAELs for these effects in the most sensitive species are slightly higher than the corresponding values for hepatotoxic effects, by inhalation (which is the relevant route of exposure for workers).

DMF produced carcinogenic effects in rats and mice. However, DMF is probably not a genotoxic substance; the tumors induced in laboratory animals were always hepatic and occurred after repeated exposure to hepatotoxic doses. From these observations, it may be inferred that DMF is probably a threshold carcinogen, and as highlighted in Table 2, protection against hepatotoxic effects would also protect against carcinogenic effects.

DMF can induce alcohol intolerance in some individuals at lower levels than those responsible for hepatotoxicity. This is documented by numerous case reports and epidemiological studies. However, the great interindividual variability of alcohol tolerance and the indirect character of this adverse effect (which needs alcohol intake to occur) makes it unsuitable as a critical effect, for the fixation of reference exposure limits applicability to all workers.

Considering that hepatotoxic effects are the critical effects of DMF exposure, hepatic enzyme activity (AST and ALT) is a suitable parameter for detecting DMF toxicity in exposed workers.

3.4. Choice of Relevant Biomarkers

The possible biomarkers for DMF exposure surveillance were previously identified (see above Section 3.1). The available data on the associations between these biomarkers and health effects in exposed workers are presented in Table 1.

In this section, advantages and limits of each possible BME are described in order to select one or more reference BME(s) on the basis of their specificity, sensitivity, and adequacy to occupational exposure biomonitoring, or the availability of analytical methods.

3.4.1. Unchanged DMF in Urine

Unchanged DMF can be detected in the urine of occupationally exposed people. It is a specific BME to DMF. However, unchanged DMF is excreted only at low levels; there are very few data on urine DMF in exposed workers or volunteers, and due to its very short half-life, this BME is not adapted to usual occupational exposure.

3.4.2. Total NMF in Urine

The excretion half-life of tNMF in urine is short (4 h). Therefore, the concentration of urinary tNMF at end of the shift represents the exposure during a working day, and urine samples for biomonitoring can be collected at the end of any shift or exposure period. The suitability of this biomarker for the derivation of an $HBM\text{-}GV_{Worker}$ has been confirmed by the observed positive association between tNMF concentration in urine and health effects in workers, reported by occupational studies over several decades. These studies provide data on tNMF levels in urine associated with liver effects [27,45,46,48,50,51,54,55]. According to Wu et al. (2017), tNMF is the best biomarker for predicting liver injury, based on the statistical significance of differences between workers exposed to DMF with and without liver injury (p = 0.001, 0.054, and 0.043, for tNMF, AMCC in urine, and MCVal in blood, respectively) [46]. A good correlation between tNMF in urine and individual airborne DMF concentration has been reported by some authors [27,32,79–82]. Urinary tNMF could not be detected using gas chromatography, with a limit of detection of 0.1 mg·L^{-1} in people who were not exposed to DMF according to Will et al. (1997) [83]. That makes urinary tNMF a good and specific biomarker of DMF occupational exposure, especially as non-occupational exposure is rare. However, it should be noted that tNMF elimination may be delayed after skin absorption, as shown by Mràz and Nohovà (1992) [22], and in the case of concomitant alcohol consumption [41]. Moreover, some analytical issues are raised by Kawai et al. (1992), who reported that results can be affected by the temperature of the gas chromatography injection port (gas chromatographic methods are commonly used for the analysis of tNMF in urine samples of exposed workers) [32]. Therefore, to obtain complete degradation of HMMF into NMF, the recommended temperature is 250 °C or above [16].

3.4.3. AMCC in Urine

The excretion half-life of AMCC in urine is long (23 h). This results in the progressive elevation of urinary concentration of AMCC over consecutive days of exposure. Therefore, urine sampling for biomonitoring of occupational DMF exposure should be performed at the end of the shift and at the end of the workweek (or at least after several previous shifts) to reflect the cumulative DMF load of the preceding working days.

AMCC urinary level is associated with health effects, especially with hepatotoxicity; AMCC in urine reflects the production of MIC, the presumed reactive intermediate of DMF. AMCC elimination in urine is not delayed by skin exposure [16]. There are limited data showing that AMCC urinary levels are reduced after alcohol consumption due to the inhibition of DMF metabolism [16]. Occupational studies showed positive associations between levels of urinary AMCC and abnormal liver function [45,46,51,55] and between airborne DMF and urinary AMCC levels [55,79,81,82]. According to He et al. (2010), AMCC is an ideal biomarker of exposure in health risk assessments following exposure to DMF, as it is highly correlated with the production of its hepatotoxic metabolite(s) [45].

However, AMCC can be found in urine at low levels in the general population, as reported by Käfferlein and Angerer, Schettgen et al. (2008), and Kenwood et al. (2021), especially in active or passive smokers, as MIC is present in cigarette smoke [84–86].

3.4.4. MCVal Adducts to Globin in Blood

MCVal has emerged as another possible biomarker of exposure and possibly as a biomarker of effect for DMF. Since adducts accumulate over the lifetime of the erythrocyte (120 days), MCVal can reflect cumulated DMF exposure over the preceding 4 months. However, due to its kinetics, MCVal requires approximately 100 days to reach a steady state. It is therefore a long-term exposure indicator, for which sampling should only be conducted after several months of exposure.

Two recent studies reported a positive relationship between MCVal and liver effects [46,51]. Seitz et al. (2019) showed a correlation between levels of MCVal in workers and airborne DMF concentration. As for AMCC, the formation of MCVal is directly linked to the presumed reactive intermediate, MIC [82].

For some authors, despite practical and technical difficulties linked to this BME, MCVal is the best biomarker reflecting both cumulated DMF exposure of the last months and hepatotoxic risk. The stability of MCVal (compared to tNMF and AMCC) is very important in some scenarios for risk assessment (episodic or irregular exposure, residents living around factories, or workers recently losing or leaving their jobs). Hepatic damage may persist several days or weeks after an overexposure, which has become undetectable through urine tNMF or even AMCC measurements. For example, Wu et al. (2017) observed five workers with abnormal liver enzyme activities who had low levels of tNMF but high concentration of MCVal [46].

3.4.5. Formamide in Urine

There are limited data on formamide as a biomarker of DMF exposure. Moreover, urine formamide is not a specific biomarker of DMF exposure, as it may also reflect NMF or formamide exposure as shown by Mràz and Nohovà (1992), who detected formamide in urines of control workers [24].

3.4.6. Conclusion on BME Selection

Table 3 summarizes the advantages and limits of each potential BME.

Table 3. Advantages and limits of the relevant BME.

Analyte	Biological Matrix	Advantages	Limits
Total NMF	Urine	- Short half-life: concentration at the end of shift is a good estimate of the exposure of the same day - Good specificity: not found in the general population - Strong association with health (hepatic) effects - Good correlation with airborne DMF	- Delayed excretion after skin absorption - Influenced by alcohol consumption - Analytical methods should be adapted to the measure of tNMF (NMF + HMMF)
AMCC	Urine	- Long half-life: concentration at the end of shift and at the end of the week is a good estimate of the exposure during the workweek - Elimination not delayed by skin exposure - Directly linked to MIC formation and hepatotoxic effects - Good correlation with mean airborne DMF concentration of the preceding days	- Might be found in general population, especially in active or passive smokers
MCVal	Blood	- Very stable, good indicator of the cumulated exposure of the last months - Directly linked to MIC formation - Dose response association with health (hepatic) effects - Acceptable correlation with airborne DMF	- Limited data on the associations with external exposure and health effects - Invasive sampling - High technical requirements and cost of the measurement
DMF	Urine	- Specific	- Limited data on the associations with external exposure and health effects - Very short half-life (2 h) - Only low levels excreted at high absorbed doses.
Formamide	Urine	None	- No data on the associations with external exposure and health effects - Not specific, can be found in the absence of DMF exposure.

There are not enough data on the associations between unchanged DMF or formamide in urine and health effects or external exposure to derive HBM-GV_{Worker} for these BMEs.

Despite the theoretical advantages presented by MCVal, the paucity of data on the associations of this potential biomarker with external exposure and health effects, together with the invasive character of the associated sampling and the high technical demands, do not allow the selection of this BME for deriving HBM-GVs.

Among the possible BMEs of DMF exposure, total NMF and AMCC in urine are the most relevant and rather complementary. Notably, both have been recommended by many countries for biological monitoring of occupational exposure to DMF.

Thus, tNMF and AMCC in urine were selected as the BME for deriving HBM-GV_{Worker} for DMF biomonitoring.

3.4.7. Analytical Methods

The analytical protocols and methods for HBM measurement of DMF in HBM4EU participants' member states are detailed in the "Prioritised list of biomarkers, matrices and analytical methods for the 2nd prioritization round of substances" [87]. NMF is usually measured in urine samples of exposed workers without prior derivatization using GC-MS (gas chromatography-mass spectrometry), GC-NPD (Nitrogen-phosphorus detectors), or GC-FPD (Flame Photometric Detector). Limits of detection are between 0.2 (GC-NPD) and 0.5 mg·L^{-1} (GC-MS). Kawai et al. (1992) reported that results can be affected by the temperature of the GC injection port [32]. Therefore, a temperature of 250 °C or above is recommended to obtain complete degradation of HMMF into NMF to measure tNMF in urine. AMCC in urine samples has been analyzed by a combination of liquid chromatography with mass spectrometry and LC-MS/MS after solid phase extraction. LODs of 5 and 5.5 µg L^{-1} were achieved [82,88].

3.5. Published Limit Values for Urine tNMF and AMCC in Occupational Setting

Recognized national and international agencies or organizations provide limit values for tNMF and/or AMCC in urine. Table 4 compiles these proposed limit values, indicating how they were produced.

Table 4. Existing limit values for urine tNMF and AMCC in an occupational setting.

Agency	Reference Value for Airborne DMF (Key Studies and Critical Effect)	Biomarker	Approach/ Endpoint	Key Study	Internal TRV and Sampling Time
SCOEL, 2006 [20]	8h-TWA = 5 ppm (Liver damage in rats and mice, exposed by inhalation, whole body) [68]	tNMF in urine	Correlation based on the OEL of 5 ppm	Studies in workers [32,48,55,79–81,89,90]	BLV = 15 mg·L^{-1} Post-shift
ACGIH, 2017 [16]	TLV-TWA = 5 ppm Liver damage in rats and mice and irritation in humans (eyes and upper respiratory tract) [49,68,91,92]	tNMF in urine	Relation between BME levels and effects on liver	Studies in workers [27,45,48,55]	BEI = 30 mg·L^{-1} End of shift
		AMCC in urine	Relation between BME levels and effects on liver	Studies in workers [45,55]	BEI = 30 mg·L^{-1} End of shift and end of workweek
DFG, 2019 [19]	MAK value = 5 ppm (Liver damage in rats and mice, exposed by inhalation, whole body) [68]	tNMF in urine	Correlation based on the MAK value of 5 ppm	Studies in workers [82]	BAT = 20 mg·L^{-1} End of exposure or end of shift
		AMCC in urine			BAT = 25 mg·g^{-1}·cr End of exposure or end of shift; Long-term exposure indicator: sampling at the end of a shift after several previous shifts

BLV: Biological limit value; BEI: Biological Exposure Index; BAT: Biologischer Arbeitsstoff Toleranzwert; 8h-TWA: Time weighted average on 8 h; TLV-TWA: threshold limit value—time weighted average; MAK value: Maximale Arbeitsplatz-Konzentration.

As shown in Table 4, available data allowed the derivation of limit values for tNMF and AMCC in urine via two approaches. These approaches correspond to the first and second options described in the HBM4EU project (Figure 1). Option 1 was retained by ACGIH experts who derived limit values for both tNMF and AMCC in urine from workplace studies showing relationships between concentrations of these biomarkers and health effects. Option 2 was used by SCOEL and DFG, who produced limit values for tNMF (SCOEL and DFG) and AMCC (DFG) using correlations between external exposure and the concentration of biomarkers and based on the occupational exposure limit (OEL) for DMF. As shown in Figure 1, option 2 should be used for the determination of HBM-GVs only in those cases where option 1 is inapplicable. Moreover, concerning DMF, option 2 is not the best choice, as this substance is readily absorbed through the skin. This is abundantly documented by experimental data and in the workplace [27,47,80]. The absorption of DMF vapor through the skin and by inhalation was evaluated in a study from Nomiyama et al. (2001) [26]. In this study, it was estimated that skin and lung absorption contributed to 40% and 60% of vapor absorption, respectively. DMF absorption through the skin is even higher when direct contact with liquid DMF is possible (see Section 3.1). For these reasons, the limit values for tNMF and AMCC proposed by SCOEL and DFG are not the first choice. Thus, option 1 is preferred for deriving an HBM-GV (Figure 1). This is the option which was retained by ACGIH experts [16]. Their proposal of a 30 mg·L^{-1} (end of shift) limit value was based on the results of the studies by Lauwerys et al. (1980), Sakai et al. (1995), Fiorito et al. (1997), Wrbitzky (1999), and He et al. (2010) [27,45,48,50,55]. It did not take into account the more recent publications by Kilo et al. (2016) and Wu et al. (2017) [46,51]. In the same way, the ACGIH proposal of a 30 mg·L^{-1} limit value for AMCC in urine was based on only two publications [45,55], and did not take into account the more recent studies by Kilo et al. (2016) and Wu et al. (2017) [46,51].

Finally, none of the limit values previously published for urine tNMF and AMCC in occupational settings can be retained, either because of the inadequacy of the method applied for their production (SCOEL, DFG) or because several essential publications were not taken into account for their elaboration (ACGIH). Consequently, the identification of a POD for the production of HBM-GV$_{Worker}$ for tNMF and AMCC in urine must be performed, using the option 1 method of the HBM4EU decision tree (Figure 1).

3.6. Choice of Key Studies and Identification of a POD for tNMF in Urine

Four of the occupational studies reporting the association of urine tNMF concentrations with hepatic damage or its absence (Table 1) cannot be retained because of methodological flaws:

- the study by Lyle et al. [52], because it reports imprecise results with evident errors in measurements and/or units of t-NMF concentrations in urine;
- the study by Yonemoto et al. [53], because the unit used for tNMF urinary excretion (mg·d^{-1}) is inadequate for deriving HBM-GV$_{Worker}$; and
- the papers by Catenacci et al. (1984) [54] and Fiorito et al. (1997) [50], because the analytic method used by these authors for tNMF measurements gave underestimated results [16,18,20].

Considering the other six studies testing for hepatic damage associated with tNMF concentration in urine, three studies were conducted in European countries and the other three in Asian countries.

Despite the interest and relevance of the studies conducted by Lauwerys et al. (1980) [27], Wrbitzky and Angerer (1998) [47], and Wrbitzky (1999) [48], they cannot be retained to derive an HBM-GV$_{Worker}$ because of the flaws detailed below:

- Lauwerys et al. (1980) [27] observed no effect on liver enzymes up to 40–50 mg·g^{-1} creatinine (cr) tNMF in urine of 22 workers exposed to DMF during five consecutive days. The authors underlined that in the factory, the selection criteria (not disclosed) at the beginning of employment were rather severe and could have led to recruitment bias so that the results obtained may not reflect responses in any worker [16];

- The two publications by Wrbitzky and Angerer (1998) and Wrbitzky (1999) reporting on the same study conducted in a cohort of 126 workers showed that liver damage was significantly more frequent in the exposed group than in controls. Mean tNMF concentration in the exposed group was 9.1 $mg{\cdot}g^{-1}$ creatinine (14.9 $mg{\cdot}L^{-1}$). However, considering the working areas, it was observed that liver damage was unexpectedly associated with the lowest exposure group (mean urine tNMF: 4.5 $mg{\cdot}g^{-1}$ creatinine) and could be explained by a higher alcohol consumption. In the other three areas, no excess of liver damage was observed for mean urine tNMF concentrations of 6.7, 11.6, and 16 $mg{\cdot}g^{-1}$ creatinine [47,48].

Finally, four studies can be selected to derive an $HBM\text{-}GV_{Worker}$:

- In a recent European study by Kilo et al. (2017), no excess risk of liver damage was observed in a cohort of 220 workers exposed to DMF with a mean concentration of 7.75 $mg{\cdot}L^{-1}$ tNMF in urine, compared with 175 controls [51].
- The other three studies were conducted in Asian people:
- Sakai et al. (1995) reported no effects on liver enzymes of DMF exposure in 10 workers during à 2.5-year-follow-up. The mean tNMF concentration in the urine of these workers was 24.47 $mg{\cdot}g^{-1}{\cdot}cr$ [55];
- He et al. (2010) also reported that, when their cohort of 79 workers was divided into three groups, a significantly elevated risk of liver damage (liver enzyme elevation) according to DMF exposure was observed only in the group with the highest exposure (mean tNMF concentration: 26.5 $mg{\cdot}g^{-1}{\cdot}cr$) [45]; and
- Wu et al. (2017) measured liver enzyme activity in a cohort of 698 workers exposed to DMF and in 188 controls. They also measured tNMF urine concentration in exposed workers. A significantly elevated risk of liver damage was observed only for the third tertile of tNMF distribution (median tNMF concentration: 9.59 $mg{\cdot}L^{-1}$). The lower limit for the benchmark dose with a benchmark response of 10% above the adverse response rate of liver injury seen in the control group ($BMDL_{10}$) was 14 $mg{\cdot}L^{-1}$ (tNMF) [46].

From the above data, it appears that liver damage was observed in groups of workers with mean tNMF concentrations in urine of:

- 26.5 $mg{\cdot}g^{-1}{\cdot}cr$ [45] or
- a median concentration of 9.59 $mg{\cdot}L^{-1}$ and a $BMDL_{10}$ value of 14 $mg{\cdot}L^{-1}$ [46].

No liver damage was observed when the mean tNMF concentration in urine was:

- 7.75 $mg{\cdot}L^{-1}$ [51] or
- 9.6 $mg{\cdot}g^{-1}{\cdot}cr$ [45] or
- 24.47 $mg{\cdot}g^{-1}{\cdot}cr$ [55].

Based on the overall results, it seems appropriate to identify a value of 10 $mg{\cdot}L^{-1}$ or 10 $mg{\cdot}g^{-1}$ cr as an $HBM\text{-}GV_{Worker}$ for tNMF in urine to prevent health effects of DMF exposure.

3.7. Choice of the Key Study and POD for AMCC in Urine

As described above, the biomonitoring of the urinary AMCC is relevant to evaluate DMF cumulative exposure of the previous days. Moreover, AMCC results from the formation of MIC, the reactive intermediate probably responsible for DMF hepatotoxicity. There are few studies providing data on the association of urinary AMCC concentrations with liver damage. However, the same four studies which were used for the elaboration of an $HBM\text{-}GV_{Worker}$ for tNMF can be used to derive an $HBM\text{-}GV_{Worker}$ for AMCC [45,46,51,55]:

- In the German study by Kilo et al. (2016), no effects on liver enzymes were observed in a cohort of 220 workers with a mean AMCC urine concentration of 9.42 $mg{\cdot}g^{-1}$ cr when they were compared to 175 controls; however, the range of the measured AMCC urine concentrations was very large (standard deviation: 10.42 $mg{\cdot}g^{-1}$ cr) [51];

- in the Japanese study by Sakaï et al. (1995), no effects on liver enzymes were observed in 10 workers exposed to DMF during à 2.5-year follow-up. The mean AMCC concentration in the urine of these workers was 22 mg·g^{-1}·cr [55];
- in the study from China, He et al. (2010) also reported that, when their cohort of 79 workers was divided into three groups, a significantly elevated risk of liver damage (liver enzyme elevation) according to DMF exposure was observed only in the group with the highest exposure (mean AMCC concentration in urine: 45.5 mg·g^{-1} cr). Geometric mean values for the concentration of AMCC in urine of workers from the low and medium exposure groups were 4.25 mg·g^{-1}·cr and 25.4 mg·g^{-1}·cr, respectively [45];
- a second Chinese study by Wu et al. (2017) measured liver enzyme activity in a cohort of 698 workers exposed to DMF and in 188 controls. They also measured AMCC urine concentration in exposed workers. A significantly elevated risk of liver damage was observed only for the second and the third tertiles of the AMCC distribution (median AMCC concentrations: 44.09 mg·L^{-1} and 148.01 mg·L^{-1}, respectively). The median and maximal AMCC concentrations in the low exposure group (1st tertile), with no detectable liver damage excess, were 2.18 mg·L^{-1} and 16.95 mg·L^{-1}, respectively. The lower limit for the benchmark dose with a benchmark response of 10% above the adverse response rate of liver injury seen in the control group ($BMDL_{10}$) was 155 mg·L^{-1} (AMCC) [46].

Based on these four studies, NOAEL values for AMCC in urine are between 9.42 mg·g^{-1}·cr and 25.4 mg·g^{-1}·cr with LOAEL of 44 mg·L^{-1} or 45.5 mg·g^{-1}·cr. Due to the large interval of NOAEL values and the large margin between NOAEL and LOAEL values, an $HBM\text{-}GV_{Worker}$ of 10 mg·g^{-1}·cr is recommended for AMCC in urine. This proposal is conservative.

4. Discussion

Literature on DMF biomonitoring offers a large database on robust relationships between health effects and urine levels of tNMF, and to a lesser extent of AMCC, the critical effect being hepatic damage (characterized by an elevation of serum hepatic enzyme activity). Limit values have previously been proposed by SCOEL (2006), ACGIH (2017), and DFG (2019) for tNMF and AMCC in urine of workers exposed to DMF [16,19,20]. However, a recent study by Wu et al. (2017) identified a lower threshold value for hepatotoxic effects than those previously identified by SCOEL, ACGIH, or DFG for tNMF concentration in urine (15 mg·L^{-1} to 30 mg·L^{-1}: see Table 4) [46].

Moreover, in its opinion on a restriction dossier for DMF [7], the RAC preferred deriving the $DNEL_{inhalation}$ value for workers from the Kilo et al. (2016) study [51], which was also not taken into account for the evaluations by SCOEL (2006) [20] and ACGIH (2017) [16].

This background justified a new evaluation of the published data for the derivation of guidance values for tNMF and AMCC concentrations in urine of workers.

The present evaluation identified a pool of studies for the derivation of an $HBM\text{-}GV_{Worker}$ for tNMF in urine. From the results provided by these studies, an $HBM\text{-}GV_{Worker}$ of 10 mg·L^{-1} (or 10 mg·g^{-1}·cr) was agreed on and is hereby recommended. Considering tNMF half-life time, end-of-shift urine sampling at any day of the workweek can be performed for this biomarker.

In addition to this recommendation, an $HBM\text{-}GV_{Worker}$ has been derived for AMCC which is an indicator of cumulative exposure after several days. Due to the dispersion of NOAEL values and the large margin between NOAEL and LOAEL values, a conservative limit value of 10 mg·g^{-1}·cr was agreed on and is provisionally recommended. Considering AMCC half-life time, end-of-shift urine sampling at any day of the workweek can be performed for this biomarker.

In the current state of knowledge, tNMF should be considered as the most reliable biomonitoring indicator for DMF exposure. As tNMF and AMCC are not alternative, but rather complementary biomarkers of DMF exposure, further studies on the association of AMCC urine level with health effects in workers exposed to DMF should be encouraged. They will probably allow the identification of a higher $HBM\text{-}GV_{Worker}$ for AMCC.

Acute and chronic alcohol consumption interferes with DMF metabolism and tNMF and AMCC elimination kinetics; thus, information on alcohol habits and alcohol consumption on the sampling day should be collected to allow interpretation of the measurement results.

As DMF LOAEL and NOAEL for developmental effects in the most sensitive species are higher than the corresponding values for hepatotoxic effect, and as DMF carcinogenic effects in animals follow hepatic damage, the HBM-GV_{Worker} for tNMF and AMCC which were established to protect against hepatotoxic effects are expected to also protect from developmental toxicity and carcinogenic effects. Alcohol intolerance can be observed at lower exposure levels in some individuals. Consequently, workers exposed to DMF should be informed of the risk of alcohol consumption in the exposure periods and at least a week after their end.

As mentioned in Apel et al. (2020), a level of confidence (low, medium, or high) could be attributed to each calculated HBM-GV. The level of confidence should reflect the uncertainties identified during the derivation of the value and could constitute a good incentive to later revise values with an estimated 'lower' level of confidence [2].

An attribution of a global level of confidence is suggested for the proposed HBM-GV_{Worker} (for both BME), considering the assessment of the various uncertainties, and detailed in the table below (Table 5).

Table 5. Level of Confidence (LoC) regarding HBM-GV_{Worker} recommended for urinary tNMF and AMCC.

	Urinary tNMF	Urinary AMCC
Regarding the nature and quality of the toxicological data	The database on DMF is based on a large number of both human and animal studies, and data on tNMF are robust and consistent. LoC: High	The database on DMF is based on a large number of both human and animal studies, but available studies reporting results for AMCC are limited. LoC: Medium
Regarding the critical endpoint and mode of action	The confidence in the evidence of effects on the liver function is high. The effects on the liver after DMF exposure are well-studied in humans (workplace) and animals. LoC: High	The confidence in the evidence of effects on the liver function is high. The effects on the liver after DMF exposure are well-studied (in humans and animals). LoC: High
Regarding the selected key studies for identification of the POD and their results	The database gives several robust occupational studies with many subjects and consistent results for tNMF. The approach consists of the selection of a pool of studies (from 1980 to 2017) carried out on Asians and Caucasian people (to consider the genetic variability due to ethnicity). LoC: High	The HBM-GV_{Worker} is based on a the same four studies used for the derivation of tNMF HBM-GV_{Worker}. However, the results of these studies indicate a large interval of NOAEL values together with a large margin between NOAEL and LOAEL values. LoC: Low
Global LoC	High	Low-medium

It should be specified that the levels of confidence proposed in this document are not determined according to an established recognized methodology, but rather rely on expert judgment regarding the reliability of the data and the calculation method used to derive the HBM-GVs.

5. Conclusions

The present work provides an HBM-GV_{Worker} which can be used to assess and limit occupational DMF exposure. In the current state of knowledge, tNMF should be considered as the most reliable biomonitoring indicator for DMF exposure. As tNMF and AMCC are not alternative, but rather complementary biomarkers of DMF exposure, further studies on the association of AMCC urine levels with health effects in workers exposed to DMF should be encouraged. They will probably allow the derivation of a higher HBM-GV_{Worker} for AMCC.

Moreover, it is important to underline that a population living near industries using/producing DMF may also be significantly exposed [93]. Studies for the characterization of DMF environmental exposure and its effects should be encouraged before considering the elaboration of an HBM-GV_{GenPop}.

Author Contributions: Conceptualization, F.L. and R.G.; methodology, F.L. and R.G.; validation, P.A. and M.K.-G.; formal analysis, M.M., F.L. and R.G.; writing—original draft preparation, F.L. and R.G.; writing—review and editing, M.M., P.A., M.K.-G., F.L. and R.G.; supervision, F.L. and P.A.; project administration, M.K.-G. All authors have read and agreed to the published version of the manuscript.

Funding: The HBM4EU project has received funding from the European Union's Horizon 2020 research and innovation program under grant agreement No 733032 and received co-funding from the author's organizations.

Institutional Review Board Statement: Not applicable.

Informed Consent Statement: Not applicable.

Data Availability Statement: Not applicable.

Acknowledgments: The authors are grateful to Greet Schoeters and Christophe Rousselle, for their support and discussions and to the National Experts from National Hubs who commented (HBM4EU project).

Conflicts of Interest: The authors declare no conflict of interest. The funders had no role in the design of the study; in the collection, analyses, or interpretation of data; in the writing of the manuscript, or in the decision to publish the results.

References

1. Ganzleben, C.; Antignac, J.-P.; Barouki, R.; Castaño, A.; Fiddicke, U.; Klánová, J.; Lebret, E.; Olea, N.; Sarigiannis, D.; Schoeters, G.R.; et al. Human Biomonitoring as a Tool to Support Chemicals Regulation in the European Union. *Int. J. Hyg. Environ. Health* **2017**, *220*, 94–97. [CrossRef] [PubMed]
2. Apel, P.; Rousselle, C.; Lange, R.; Sissoko, F.; Kolossa-Gehring, M.; Ougier, E. Human Biomonitoring Initiative (HBM4EU)—Strategy to Derive Human Biomonitoring Guidance Values (HBM-GVs) for Health Risk Assessment. *Int. J. Hyg. Environ. Health* **2020**, *230*, 113622. [CrossRef] [PubMed]
3. Lamkarkach, F.; Ougier, E.; Garnier, R.; Viau, C.; Kolossa-Gehring, M.; Lange, R.; Apel, P. Human Biomonitoring Initiative (HBM4EU): Human Biomonitoring Guidance Values (HBM-GVs) Derived for Cadmium and Its Compounds. *Environ. Int.* **2021**, *147*, 106337. [CrossRef] [PubMed]
4. Ougier, E.; Zeman, F.; Antignac, J.-P.; Rousselle, C.; Lange, R.; Kolossa-Gehring, M.; Apel, P. Human Biomonitoring Initiative (HBM4EU): Human Biomonitoring Guidance Values (HBM-GVs) Derived for Bisphenol A. *Environ. Int.* **2021**, *154*, 106563. [CrossRef]
5. Lange, R.; Apel, P.; Rousselle, C.; Charles, S.; Sissoko, F.; Kolossa-Gehring, M.; Ougier, E. The European Human Biomonitoring Initiative (HBM4EU): Human Biomonitoring Guidance Values for Selected Phthalates and a Substitute Plasticizer. *Int. J. Hyg. Environ. Health* **2021**, *234*, 113722. [CrossRef]
6. David, M.; Gerofke, A.; Lange, R.; Kolossa-Gehring, M.; Apel, P. The European Human Biomonitoring Initiative (HBM4EU): Human Biomonitoring Guidance Values (HBM-GVs) for the Aprotic Solvents N-Methyl-2-Pyrrolidone (NMP) and N-Ethyl-2-Pyrrolidone (NEP). *Int. J. Hyg. Environ. Health* **2021**, *238*, 113856. [CrossRef]
7. European Chemical Agency (ECHA). *Opinion of the Committee for Risk Assessment and Opinion of the Committee for Socio-Economic Analysis on an Annex XV Dossier Proposing Restrictions of the Manufacture, Placing on the Market or Use of a Substance within the EU*; European Chemical Agency (ECHA): Helsinki, Finland, 2019.

8. European Chemical Agency (ECHA). *Background Document for N,N-Dimethylformamide (DMF)*; Document Developed in the Context of ECHA's Fifth Recommendation for the Inclusion of Substances in Annex XIV; European Chemical Agency (ECHA): Helsinki, Finland, 2014.
9. International Agency for Research on Cancer (IARC). *IARC Monographs on the Evaluation of Carcinogenic Risks to Humans*; World Health Organization: Lyon, France, 1999; Volume 71, Part Two.
10. International Agency for Research on Cancer (IARC). Some Chemicals. In *IARC Monographs on the Evaluation of Carcinogenic Risks to Humans*; World Health Organization: Lyon, France, 2018; Volume 115.
11. Kommission HBM des Umweltbundesamtes. Grundsatzpapier Zur Ableitung von HBM-Werten. Bundesgesundheitsbl Gesundheitsforsch Gesundheitsschutz. 2014, pp. 138–147. Available online: https://www.umweltbundesamt.de/sites/default/files/medien/377/dokumente/art_10.1007_s00103-013-1867-2-2.pdf (accessed on 20 April 2022).
12. Apel, P.; Angerer, J.; Wilhelm, M.; Kolossa-Gehring, M. New HBM Values for Emerging Substances, Inventory of Reference and HBM Values in Force, and Working Principles of the German Human Biomonitoring Commission. *Int. J. Hyg. Environ. Health* **2017**, *220*, 152–166. [CrossRef]
13. Hays, S.M.; Becker, R.A.; Leung, H.W.; Aylward, L.L.; Pyatt, D.W. Biomonitoring Equivalents: A Screening Approach for Interpreting Biomonitoring Results from a Public Health Risk Perspective. *Regul. Toxicol. Pharmacol.* **2007**, *47*, 96–109. [CrossRef]
14. Hays, S.M.; Aylward, L.L.; LaKind, J.S.; Bartels, M.J.; Barton, H.A.; Boogaard, P.J.; Brunk, C.; DiZio, S.; Dourson, M.; Goldstein, D.A.; et al. Guidelines for the Derivation of Biomonitoring Equivalents: Report from the Biomonitoring Equivalents Expert Workshop. *Regul. Toxicol. Pharmacol.* **2008**, *51*, S4–S15. [CrossRef]
15. ANSES. *Document de Référence Pour l'élaboration de Valeurs Limites d'exposition à Des Agents Chimiques En Milieu Professionnel*; Agence Nationale de Sécurité Sanitaire Pour l'Alimentation, l'Environnement et Le Travail: Paris, France, 2017; 142p.
16. American Conference of Governmental Industrial Hygienists (ACGIH). *Biological Exposure Index Documentation for N,N-Dimethylformamide*; American Conference of Governmental Industrial Hygienists: Washington, DC, USA, 2017.
17. American Conference of Governmental Industrial Hygienists (ACGIH). *N,N-Dimethylformamide*; American Conference of Governmental Industrial Hygienists: Washington, DC, USA, 2018.
18. Käfferlein, H.U. Addendum Zu N,N-Dimethylformamid. In *Biologische Arbeitsstoff-Toleranz-Werte (BAT-Werte) und Expositionsäquivalente für Krebs-Erzeugende Arbeitsstoffe (EKA) und Biologische Leitwerte (BLW)*; Drexler, H., Greim, H., Eds.; DFG MAK Commission: Germany, 2006.
19. Göen, T.; Drexler, H.; Hartwig, A. *Addendum to N,N-Dimethylformamide [N,N-Dimethylformamid, Addendum] BAT Value Documentation in German*; DFG MAK Commission: Germany, 2019.
20. *SCOEL/SUM/121*; Recommendation from the Scientific Committee for Occupational Exposure Limits on N,N-Dimethylformamide. The Scientific Committee on Occupational Exposure Limits (SCOEL), 2006.
21. Brugnone, F.; Perbellini, L.; Gaffuri, E.; Apostoli, P. Biomonitoring of Industrial Solvent Exposures in Workers' Alveolar Air. *Int. Arch. Occup. Environ. Health* **1980**, *47*, 245–261. [CrossRef]
22. Mráz, J.; Nohová, H. Absorption, Metabolism and Elimination of N,N-Dimethylformamide in Humans. *Int. Arch. Occup. Environ. Health* **1992**, *64*, 85–92. [CrossRef] [PubMed]
23. Maxfield, M.E.; Barnes, J.R.; Azar, A.; Trochimowicz, H.T. Urinary Excretion of Metabolite Following Experimental Human Exposures to DMF or to DMAC. *J. Occup. Med.* **1975**, *17*, 506–511. [PubMed]
24. Mráz, J.; Nohová, H. Percutaneous Absorption of N,N-Dimethylformamide in Humans. *Int. Arch. Occup. Environ. Health* **1992**, *64*, 79–83. [CrossRef]
25. Tsuda, Y.; Miyauchi, H.; Minozoe, A.; Tanaka, S.; Arito, H.; Tsukahara, T.; Nomiyama, T. Seasonal Difference in Percutaneous Absorption of N,N-Dimethylformamide as Determined Using Two Urinary Metabolites. *J. Occup. Health* **2014**, *56*, 252–259. [CrossRef] [PubMed]
26. Nomiyama, T.; Nakashima, H.; Chen, L.L.; Tanaka, S.; Miyauchi, H.; Yamauchi, T.; Sakurai, H.; Omae, K. N,N-Dimethylformamide: Significance of Dermal Absorption and Adjustment Method for Urinary N-Methylformamide Concentration as a Biological Exposure Item. *Int. Arch. Occup. Environ. Health* **2001**, *74*, 224–228. [CrossRef] [PubMed]
27. Lauwerys, R.R.; Kivits, A.; Lhoir, M.; Rigolet, P.; Houbeau, D.; Buchet, J.P.; Roels, H.A. Biological Surveillance of Workers Exposed to Dimethylformamide and the Influence of Skin Protection on Its Percutaneous Absorption. *Int. Arch. Occup. Environ. Health* **1980**, *45*, 189–203. [CrossRef]
28. World Health Organization. *N,N-Dimethylformamide*; World Health Organization (WHO): Geneva, Switzerland, 2001.
29. Saillenfait, A.M.; Payan, J.P.; Beydon, D.; Fabry, J.P.; Langonne, I.; Sabate, J.P.; Gallissot, F. Assessment of the Developmental Toxicity, Metabolism, and Placental Transfer of N,N-Dimethylformamide Administered to Pregnant Rats. *Fundam. Appl. Toxicol.* **1997**, *39*, 33–43. [CrossRef]
30. Mráz, J.; Jheeta, P.; Gescher, A.; Hyland, R.; Thummel, K.; Threadgill, M.D. Investigation of the Mechanistic Basis of N,N-Dimethylformamide Toxicity. Metabolism of N,N-Dimethylformamide and Its Deuterated Isotopomers by Cytochrome P450 2E1. *Chem. Res. Toxicol.* **1993**, *6*, 197–207. [CrossRef]
31. Scailteur, V.; de Hoffmann, E.; Buchet, J.P.; Lauwerys, R. Study on in Vivo and in Vitro Metabolism of Dimethylformamide in Male and Female Rats. *Toxicology* **1984**, *29*, 221–234. [CrossRef]

32. Kawai, T.; Yasugi, T.; Mizunuma, K.; Watanabe, T.; Cai, S.X.; Huang, M.Y.; Xi, L.Q.; Qu, J.B.; Yao, B.Z.; Ikeda, M. Occupational Dimethylformamide Exposure. 2. Monomethylformamide Excretion in Urine after Occupational Dimethylformamide Exposure. *Int. Arch. Occup. Environ. Health* **1992**, *63*, 455–460. [CrossRef]
33. Mráz, J.; Turecek, F. Identification of N-Acetyl-S-(N-Methylcarbamoyl)Cysteine, a Human Metabolite of N,N-Dimethylformamide and N-Methylformamide. *J. Chromatogr.* **1987**, *414*, 399–404. [CrossRef]
34. Mráz, J.; Cross, H.; Gescher, A.; Threadgill, M.D.; Flek, J. Differences between Rodents and Humans in the Metabolic Toxification of N,N-Dimethylformamide. *Toxicol. Appl. Pharmacol.* **1989**, *98*, 507–516. [CrossRef]
35. Käfferlein, H.U.; Angerer, J. N-Methylcarbamoylated Valine of Hemoglobin in Humans after Exposure to N,N-Dimethylformamide: Evidence for the Formation of Methyl Isocyanate? *Chem. Res. Toxicol.* **2001**, *14*, 833–840. [CrossRef] [PubMed]
36. Angerer, J.; Göen, T.; Krämer, A.; Käfferlein, H.U. N-Methylcarbamoyl Adducts at the N-Terminal Valine of Globin in Workers Exposed to N,N-Dimethylformamide. *Arch. Toxicol.* **1998**, *72*, 309–313. [CrossRef] [PubMed]
37. Mráz, J.; Cimlová, J.; Stránský, V.; Nohová, H.; Kicová, R.; Simek, P. N-Methylcarbamoyl-Lysine Adduct in Globin: A New Metabolic Product and Potential Biomarker of N, N-Dimethylformamide in Humans. *Toxicol. Lett.* **2006**, *162*, 211–218. [CrossRef]
38. Hundley, S.G.; Lieder, P.H.; Valentine, R.; Malley, L.A.; Kennedy, G.L. Dimethylformamide Pharmacokinetics Following Inhalation Exposures to Rats and Mice. *Drug Chem. Toxicol.* **1993**, *16*, 21–52. [CrossRef]
39. Hundley, S.G.; McCooey, K.T.; Lieder, P.H.; Hurtt, M.E.; Kennedy, G.L. Dimethylformamide Pharmacokinetics Following Inhalation Exposure in Monkeys. *Drug Chem. Toxicol.* **1993**, *16*, 53–79. [CrossRef]
40. Henschler, D. *Occupational Toxicants: Critical Data Evaluation for MAK Values and Classification of Carcinogens*; Greim, H., Ed.; Wiley-VCH: Weinheim, Germany, 1998; Volume 8.
41. Eben, A.; Kimmerle, G. Metabolism Studies of N, N-Dimethylformamide. III. Studies about the Influence of Ethanol in Persons and Laboratory Animals. *Int. Arch. Occup. Environ. Health* **1976**, *34*, 109–126. [CrossRef]
42. Kim, K.-W.; Chung, Y.H. Hepatotoxicity in Rats Treated with Dimethylformamide or Toluene or Both. *Toxicol. Res.* **2013**, *29*, 187–193. [CrossRef]
43. Garnier, R.; Chataigner, D.; Perez-Trigalou, B.; Efthymiou, M.L. Intoxications professionnelles par le diméthylformamide. *Arch. Mal. Prof.* **1992**, *53*, 111–120.
44. Li, M.-J.; Zeng, T. The Deleterious Effects of N,N-Dimethylformamide on Liver: A Mini-Review. *Chem. Biol. Interact.* **2019**, *298*, 129–136. [CrossRef] [PubMed]
45. He, J.; Wang, P.; Zhu, J.; Wu, G.; Ji, J.; Xue, Y. Role of Urinary Biomarkers of N,N-Dimethylformamide in the Early Detection of Hepatic Injury among Occupational Exposed Workers. *Int. Arch. Occup. Environ. Health* **2010**, *83*, 399–406. [CrossRef] [PubMed]
46. Wu, Z.; Liu, Q.; Wang, C.; Xu, B.; Guan, M.; Ye, M.; Jiang, H.; Zheng, M.; Zhang, M.; Zhao, W.; et al. A Comparative Benchmark Dose Study for N, N-Dimethylformamide Induced Liver Injury in a Chinese Occupational Cohort. *Toxicol. Sci.* **2017**, *158*, 140–150. [CrossRef]
47. Wrbitzky, R.; Angerer, J. N,N-Dimethylformamide–Influence of Working Conditions and Skin Penetration on the Internal Exposure of Workers in Synthetic Textile Production. *Int. Arch. Occup. Environ. Health* **1998**, *71*, 309–316. [CrossRef]
48. Wrbitzky, R. Liver Function in Workers Exposed to N,N-Dimethylformamide during the Production of Synthetic Textiles. *Int. Arch. Occup. Environ. Health* **1999**, *72*, 19–25. [CrossRef]
49. Cai, S.X.; Huang, M.Y.; Xi, L.Q.; Li, Y.L.; Qu, J.B.; Kawai, T.; Yasugi, T.; Mizunuma, K.; Watanabe, T.; Ikeda, M. Occupational Dimethylformamide Exposure. 3. Health Effects of Dimethylformamide after Occupational Exposure at Low Concentrations. *Int. Arch. Occup. Environ. Health* **1992**, *63*, 461–468. [CrossRef] [PubMed]
50. Fiorito, A.; Larese, F.; Molinari, S.; Zanin, T. Liver Function Alterations in Synthetic Leather Workers Exposed to Dimethylformamide. *Am. J. Ind. Med.* **1997**, *32*, 255–260. [CrossRef]
51. Kilo, S.; Göen, T.; Drexler, H. Cross-Sectional Study on N,N-Dimethylformamide (DMF); Effects on Liver and Alcohol Intolerance. *Int. Arch. Occup. Environ. Health* **2016**, *89*, 1309–1320. [CrossRef]
52. Lyle, W.H.; Spence, T.W.; McKinneley, W.M.; Duckers, K. Dimethylformamide and Alcohol Intolerance. *Br. J. Ind. Med.* **1979**, *36*, 63–66. [CrossRef]
53. Yonemoto, J.; Suzuki, S. Relation of Exposure to Dimethylformamide Vapor and the Metabolite, Methylformamide, in Urine of Workers. *Int. Arch. Occup. Environ. Health* **1980**, *46*, 159–165. [CrossRef]
54. Catenacci, G.; Grampella, D.; Terzi, R.; Sala, A.; Pollini, G. Hepatic Function in Subjects Exposed to Environmental Concentrations of DMF Lower than the Actually Proposed TLV. *G. Ital. Med. Lav.* **1984**, *6*, 157–158. [PubMed]
55. Sakai, T.; Kageyama, H.; Araki, T.; Yosida, T.; Kuribayashi, T.; Masuyama, Y. Biological Monitoring of Workers Exposed to N,N-Dimethylformamide by Determination of the Urinary Metabolites, N-Methylformamide and N-Acetyl-S-(N-Methylcarbamoyl) Cysteine. *Int. Arch. Occup. Environ. Health* **1995**, *67*, 125–129. [CrossRef] [PubMed]
56. Hanasono, G.K.; Fuller, R.W.; Broddle, W.D.; Gibson, W.R. Studies on the Effects on N,N′-Dimethylformamide on Ethanol Disposition and on Monoamine Oxidase Activity in Rats. *Toxicol. Appl. Pharmacol.* **1977**, *39*, 461–472. [CrossRef]
57. Wolff, P.H. Ethnic Differences in Alcohol Sensitivity. *Science* **1972**, *175*, 449–450. [CrossRef] [PubMed]
58. Chan, A.W. Racial Differences in Alcohol Sensitivity. *Alcohol Alcohol.* **1986**, *21*, 93–104.
59. Hennebrüder, K.; Angerer, J. Determination of DMF Modified DNA Base N4-Methylcarbamoylcytosine in Human Urine Using off-Line Sample Clean-up, Two-Dimensional LC and ESI-MS/MS Detection. *J. Chromatogr. B Analyt. Technol. Biomed. Life Sci.* **2005**, *822*, 124–132. [CrossRef]

60. Chen, Y. Induction of DNA strand breakage of human peripheral blood cells by N,N-dimethylformamide. *Chin. J. Health Lab. Technol.* **2014**, *14*, 166–168.
61. Shieh, D.-B.; Chen, C.-C.; Shih, T.-S.; Tai, H.-M.; Wei, Y.-H.; Chang, H.-Y. Mitochondrial DNA Alterations in Blood of the Humans Exposed to N,N-Dimethylformamide. *Chem. Biol. Interact.* **2007**, *165*, 211–219. [CrossRef]
62. Major, J.; Hudák, A.; Kiss, G.; Jakab, M.G.; Szaniszló, J.; Náray, M.; Nagy, I.; Tompa, A. Follow-up Biological and Genotoxicological Monitoring of Acrylonitrile- and Dimethylformamide-Exposed Viscose Rayon Plant Workers. *Environ. Mol. Mutagenesis* **1998**, *31*, 301–310. [CrossRef]
63. Cheng, T.J.; Hwang, S.J.; Kuo, H.W.; Luo, J.C.; Chang, M.J. Exposure to Epichlorohydrin and Dimethylformamide, Glutathione S-Transferases and Sister Chromatid Exchange Frequencies in Peripheral Lymphocytes. *Arch. Toxicol.* **1999**, *73*, 282–287. [CrossRef]
64. McQueen, C.A.; Way, B.M.; Williams, G.M. Genotoxicity of Carcinogens in Human Hepatocytes: Application in Hazard Assessment. *Toxicol. Appl. Pharmacol.* **1988**, *96*, 360–366. [CrossRef]
65. Antoine, J.L.; Arany, J.; Léonard, A.; Henrotte, J.; Jenar-Dubuisson, G.; Decat, G. Lack of Mutagenic Activity of Dimethylformamide. *Toxicology* **1983**, *26*, 207–212. [CrossRef]
66. Koudela, K.; Spazier, K. Results of Cytogenetic Examination of Persons Working in an Environment of Increased Concentration of Dimethylformamide Vapors in the Atmosphere. *Prac. Lèk.* **1981**, *33*, 121–123.
67. Senoh, H.; Aiso, S.; Arito, H.; Nishizawa, T.; Nagano, K.; Yamamoto, S.; Matsushima, T. Carcinogenicity and Chronic Toxicity after Inhalation Exposure of Rats and Mice to N,N-Dimethylformamide. *J. Occup. Health* **2004**, *46*, 429–439. [CrossRef]
68. Malley, L.A.; Slone, T.W.; Van Pelt, C.; Elliott, G.S.; Ross, P.E.; Stadler, J.C.; Kennedy, G.L. Chronic Toxicity/Oncogenicity of Dimethylformamide in Rats and Mice Following Inhalation Exposure. *Fundam. Appl. Toxicol.* **1994**, *23*, 268–279. [CrossRef] [PubMed]
69. Ohbayashi, H.; Umeda, Y.; Senoh, H.; Kasai, T.; Kano, H.; Nagano, K.; Arito, H.; Fukushima, S. Enhanced Hepatocarcinogenicity by Combined Inhalation and Oral Exposures to N,N-Dimethylformamide in Male Rats. *J. Toxicol. Sci.* **2009**, *34*, 53–63. [CrossRef]
70. Ducatman, A.M.; Conwill, D.E.; Crawl, J. Germ Cell Tumors of the Testicle among Aircraft Repairmen. *J. Urol* **1986**, *136*, 834–836. [CrossRef]
71. Levin, S.M.; Baker, D.B.; Landrigan, P.J.; Monaghan, S.V.; Frumin, E.; Braithwaite, M.; Towne, W. Testicular Cancer in Leather Tanners Exposed to Dimethylformamide. *Lancet* **1987**, *2*, 1153. [CrossRef]
72. Chen, J.L.; Fayerweather, W.E.; Pell, S. Cancer Incidence of Workers Exposed to Dimethylformamide and/or Acrylonitrile. *J. Occup. Med.* **1988**, *30*, 813–818. [CrossRef]
73. Walrath, J.; Fayerweather, W.E.; Gilby, P.G.; Pell, S. A Case-Control Study of Cancer among Du Pont Employees with Potential for Exposure to Dimethylformamide. *J. Occup. Med.* **1989**, *31*, 432–438.
74. Yoon, J.-H.; Yoo, C.-I.; Ahn, Y.-S. N,N-Dimethylformamide: Evidence of Carcinogenicity from National Representative Cohort Study in South Korea. *Scand. J. Work Environ. Health* **2019**, *45*, 396–401. [CrossRef] [PubMed]
75. Chang, H.-Y.; Tsai, C.-Y.; Lin, Y.-Q.; Shih, T.-S.; Lin, Y.-C. Urinary Biomarkers of Occupational N,N-Dimethylformamide (DMF) Exposure Attributed to the Dermal Exposure. *J. Expo. Anal. Environ. Epidemiol.* **2004**, *14*, 214–221. [CrossRef] [PubMed]
76. Hellwig, J.; Merkle, J.; Klimisch, H.J.; Jäckh, R. Studies on the Prenatal Toxicity of N,N-Dimethylformamide in Mice, Rats and Rabbits. *Food Chem. Toxicol.* **1991**, *29*, 193–201. [CrossRef]
77. Fail, P.A.; George, J.D.; Grizzle, T.B.; Heindel, J.J. Formamide and Dimethylformamide: Reproductive Assessment by Continuous Breeding in Mice. *Reprod. Toxicol.* **1998**, *12*, 317–332. [CrossRef]
78. Merkle, J.; Zeller, H. [Studies on acetamides and formamides for embryotoxic and teratogenic activities in the rabbit (author's transl)]. *Arzneimittelforschung* **1980**, *30*, 1557–1562.
79. Käfferlein, H.U.; Göen, T.; Müller, J.; Wrbitzky, R.; Angerer, J. Biological Monitoring of Workers Exposed to N,N-Dimethylformamide in the Synthetic Fibre Industry. *Int. Arch. Occup. Environ. Health* **2000**, *73*, 113–120. [CrossRef]
80. Yang, J.S.; Kim, E.A.; Lee, M.Y.; Park, I.J.; Kang, S.K. Biological Monitoring of Occupational Exposure to N,N-Dimethylformamide–the Effects of Co-Exposure to Toluene or Dermal Exposure. *Int. Arch. Occup. Environ. Health* **2000**, *73*, 463–470. [CrossRef]
81. Imbriani, M.; Maestri, L.; Marraccini, P.; Saretto, G.; Alessio, A.; Negri, S.; Ghittori, S. Urinary Determination of N-Acetyl-S-(N-Methylcarbamoyl)Cysteine and N-Methylformamide in Workers Exposed to N, N-Dimethylformamide. *Int. Arch. Occup. Environ. Health* **2002**, *75*, 445–452. [CrossRef]
82. Seitz, M.; Kilo, S.; Eckert, E.; Müller, J.; Drexler, H.; Göen, T. Validity of Different Biomonitoring Parameters for the Assessment of Occupational Exposure to N,N-Dimethylformamide (DMF). *Arch. Toxicol.* **2018**, *92*, 2183–2193. [CrossRef]
83. Will, W.; Schulz, G. N,N-Dimethylformamide (DMF) [Biomonitoring Methods, 1996]. In *The MAK-Collection for Occupational Health and Safety*; John Wiley & Sons, Ltd: Hoboken, NJ, USA, 2012; pp. 97–108. ISBN 978-3-527-60041-0.
84. Käfferlein, H.U.; Angerer, J. Determination of N-Acetyl-S-(N-Methylcarbamoyl)Cysteine (AMCC) in the General Population Using Gas Chromatography-Mass Spectrometry. *J. Environ. Monit.* **1999**, *1*, 465–469. [CrossRef]
85. Schettgen, T.; Musiol, A.; Kraus, T. Simultaneous Determination of Mercapturic Acids Derived from Ethylene Oxide (HEMA), Propylene Oxide (2-HPMA), Acrolein (3-HPMA), Acrylamide (AAMA) and N,N-Dimethylformamide (AMCC) in Human Urine Using Liquid Chromatography/Tandem Mass Spectrometry. *Rapid. Commun. Mass Spectrom.* **2008**, *22*, 2629–2638. [CrossRef] [PubMed]

86. Kenwood, B.M.; Bagchi, P.; Zhang, L.; Zhu, W.; Chambers, D.M.; Blount, B.C.; De Jesús, V.R. Characterization of US Population Levels of Urinary Methylcarbamoyl Mercapturic Acid, a Metabolite of N,N-Dimethylformamide and Methyl Isocyanate, in the National Health and Nutrition Examination Survey (NHANES) 2005–2006 and 2011–2016. *Environ. Sci. Pollut. Res.* **2021**, *28*, 16781–16791. [CrossRef] [PubMed]
87. HBM4EU. *Deliverable 9.5 Prioritised List of Biomarkers, Matrices and Analytical Methods for the 2nd Prioritisation Round of Substances*; HBM4EU, 2022.
88. Alwis, K.U.; Blount, B.C.; Britt, A.S.; Patel, D.; Ashley, D.L. Simultaneous Analysis of 28 Urinary VOC Metabolites Using Ultra High Performance Liquid Chromatography Coupled with Electrospray Ionization Tandem Mass Spectrometry (UPLC-ESI/MSMS). *Anal. Chim. Acta* **2012**, *750*, 152–160. [CrossRef] [PubMed]
89. Wang, V.-S.; Shih, T.-S.; Cheng, C.-C.; Chang, H.-Y.; Lai, J.-S.; Lin, C.-C. Evaluation of Current Biological Exposure Index for Occupational N, N-Dimethylformamide Exposure from Synthetic Leather Workers. *J. Occup. Environ. Med.* **2004**, *46*, 729–736. [CrossRef]
90. Kim, H.-A.; Kim, K.; Heo, Y.; Lee, S.-H.; Choi, H.-C. Biological Monitoring of Workers Exposed to N, N-Dimethylformamide in Synthetic Leather Manufacturing Factories in Korea. *Int. Arch. Occup. Environ. Health* **2004**, *77*, 108–112. [CrossRef]
91. Tomasini, M.; Todaro, A.; Piazzoni, M.; Peruzzo, G.F. Exposure to dimethylformamide: Study of 14 cases. *Med. Lav.* **1983**, *74*, 217–220.
92. Cirla, A.M.; Pisati, G.; Invernizzi, E.; Torricelli, P. Epidemiological Study on Workers Exposed to Low Dimethylformamide Concentrations. *G. Ital. Med. Lav.* **1984**, *6*, 149–156.
93. Wang, C.; Huang, C.; Wei, Y.; Zhu, Q.; Tian, W.; Zhang, Q. Short-Term Exposure to Dimethylformamide and the Impact on Digestive System Disease: An Outdoor Study for Volatile Organic Compound. *Environ. Pollut.* **2014**, *190*, 133–138. [CrossRef]

toxics

MDPI

Article

Biomonitoring of Exposure to Urban Pollutants and Oxidative Stress during the COVID-19 Lockdown in Rome Residents

Flavia Buonaurio [1], Francesca Borra [1], Daniela Pigini [2], Enrico Paci [2], Mariangela Spagnoli [2], Maria Luisa Astolfi [1], Ottavia Giampaoli [3,4], Fabio Sciubba [3,4], Alfredo Miccheli [3,4], Silvia Canepari [3], Carla Ancona [5] and Giovanna Tranfo [2,*]

[1] Department of Chemistry, Sapienza University of Rome, 00185 Rome, Italy; flavia.buonaurio@uniroma1.it (F.B.); francesca.borra@libero.it (F.B.); marialuisa.astolfi@uniroma1.it (M.L.A.)
[2] Department of Occupational and Environmental Medicine, Epidemiology and Hygiene, INAIL, Monte Porzio Catone, 00144 Rome, Italy; d.pigini@inail.it (D.P.); e.paci@inail.it (E.P.); m.spagnoli@inail.it (M.S.)
[3] Department of Environmental Biology, Sapienza University of Rome, 00185 Rome, Italy; ottavia.giampaoli@uniroma1.it (O.G.); fabio.sciubba@uniroma1.it (F.S.); alfredo.miccheli@uniroma1.it (A.M.); silvia.canepari@uniroma1.it (S.C.)
[4] NMR-Based Metabolomics Laboratory (NMLab), Sapienza University of Rome, 00185 Rome, Italy
[5] Department of Epidemiology, Lazio Regional Health Service, 00154 Rome, Italy; c.ancona@deplazio.it
* Correspondence: g.tranfo@inail.it; Tel.: +39-0694181436

Abstract: Background: The objective of this study is to evaluate the effects of traffic on human health comparing biomonitoring data measured during the COVID-19 lockdown, when restrictions led to a 40% reduction in airborne benzene in Rome and a 36% reduction in road traffic, to the same parameters measured in 2021. Methods: Biomonitoring was performed on 49 volunteers, determining the urinary metabolites of the most abundant traffic pollutants, such as benzene and PAHs, and oxidative stress biomarkers by HPLC/MS-MS, 28 elements by ICP/MS and metabolic phenotypes by NMR. Results: Means of s-phenylmercaptutric acid (SPMA), metabolites of naphthalene and nitropyrene in 2020 are 20% lower than in 2021, while 1-OH-pyrene was 30% lower. A reduction of 40% for 8-oxo-7,8-dihydroguanosine (8-oxoGuo) and 8-oxo-7,8-dihydro-2-deoxyguanosine (8-oxodGuo) and 60% for 8-oxo-7,8-dihydroguanine (8-oxoGua) were found in 2020 compared to 2021. The concentrations of B, Co, Cu and Sb in 2021 are significantly higher than in the 2020. NMR untargeted metabolomic analysis identified 35 urinary metabolites. Results show in 2021 a decrease in succinic acid, a product of the Krebs cycle promoting inflammation. Conclusions: Urban pollution due to traffic is partly responsible for oxidative stress of nucleic acids, but other factors also have a role, enhancing the importance of communication about a healthy lifestyle in the prevention of cancer diseases.

Keywords: benzene; COVID-19 pandemic; lockdown; Rome; elements; urban traffic; oxidative stress; metabolomics

Citation: Buonaurio, F.; Borra, F.; Pigini, D.; Paci, E.; Spagnoli, M.; Astolfi, M.L.; Giampaoli, O.; Sciubba, F.; Miccheli, A.; Canepari, S.; et al. Biomonitoring of Exposure to Urban Pollutants and Oxidative Stress during the COVID-19 Lockdown in Rome Residents. *Toxics* **2022**, *10*, 267. https://doi.org/10.3390/toxics10050267

Academic Editor: Christophe Rousselle

Received: 26 April 2022
Accepted: 20 May 2022
Published: 21 May 2022

Publisher's Note: MDPI stays neutral with regard to jurisdictional claims in published maps and institutional affiliations.

Copyright: © 2022 by the authors. Licensee MDPI, Basel, Switzerland. This article is an open access article distributed under the terms and conditions of the Creative Commons Attribution (CC BY) license (https://creativecommons.org/licenses/by/4.0/).

1. Introduction

In the year 2020, there was a viral pandemic known as COVID-19 (Coronavirus Disease-2019), generated by the SARS-CoV-2 virus (Severe Acute Respiratory Syndrome-Coronavirus 2) still ongoing, to varying degrees in almost all countries [1].

According to WHO, there were 4,251,460 confirmed cases and 84,997 deaths worldwide on December 28th 2020, of which 102,418 cases and 3358 deaths were registered in Italy on the same date [2].

The first two Italian cases were confirmed on 30 January 2020 [3]; the subsequent outbreaks detected in northern and central Italy led to a series of restrictions, such as local quarantines, but with the increase in infections, new stricter restriction measures were gradually introduced and extended to the entire national territory [4].

On 9 March 2020, a national lockdown started, during which face-to-face teaching in schools and universities was suspended and working from home was promoted wherever possible; travel for unnecessary reasons, sports activities, demonstrations and events were prohibited, and museums and theatres were closed; non-essential retail commercial activities, catering services, religious celebrations, and gatherings of people in public places were suspended by law [4].

The restrictive measures were extended several times until 3 May 2020, until they were almost completely removed, starting from 15 June 2020 [5].

In addition to the various social, economic and political consequences [6,7], the effects of the lockdown were also evident at the environmental level, especially in terms of a drastic decrease in vehicular traffic, whose effects were visible above all in large urban centers, such as Rome [8] and Milan [9]. A reduction in emissions was also measured in China [10].

The lockdown scenario was a unique condition, which rendered possible the direct measurement of the effects of reduced exposure to traffic pollution on the same group of subjects for a long period [11]. Therefore, this study is the first one that aims to investigate the reduction in exposure biomarkers related to traffic pollution in Rome, due to the restrictions imposed by the Government to contain the COVID-19 pandemic.

Traffic-related air pollution is a global concern due to the adverse health effects on humans [12]. Both short- and long-term exposure to air pollutants increase the risk of respiratory and cardiovascular diseases in the population [13]. Outdoor air pollution is an established risk factor for lung cancer incidence and mortality [14]. Among traffic-related air pollutants, there are benzene, polycyclic aromatic hydrocarbons (PAHs) and toxic metals [15–17]. Benzene and PAHs can penetrate biological membranes and generate reactive oxygen species (ROS), which can affect inflammation associated with immune responses [18,19]. Benzene is a known carcinogen that causes hematotoxicity even at exposure levels below 1 ppm [20,21] and has been associated with nucleic acid oxidation biomarkers [22]. It has an influence on the central nervous system and it is involved in metabolic function, such as insulin resistance [23]. Furthermore, benzene is reported to be the cause of myeloid leukemia, myelodysplastic syndrome and a probable cause of non-Hodgkin lymphoma [24]. For this reason, the Directive 2008/50/EC on ambient air quality and cleaner air for Europe has set a limit value at 5 $\mu g/m^3$ of airborne benzene as the annual average since January 2010 [25]. Polycyclic aromatic hydrocarbons (PAHs) are also a class of carcinogenic agents [26]. Their origin is the incomplete combustion and pyrolysis of organic materials, especially wood, coal, oil and waste. PAHs containing four rings or less typically remain in a gaseous form when released into the atmosphere, particularly benzo[a]pyrene, which is classified by IARC (International Agency for Research on Cancer, Lyon, France) as a class 1 carcinogen. The PAH biomarkers 1-hydroxypyrene (1-OHPy), 6-hydroxynitropyrene (6-OHNPy) and 3-hydroxybenzo[a]pyrene (3-OHBaPy) have been demonstrated to correlate well with oxidative stress biomarkers [27]. Short-term exposure to PAHs has been reported to cause impaired lung function with asthmatics and thrombotic effects in people with coronary heart disease. Furthermore, exposure to high levels of PAHs can generate acute health effects, such as eye irritation, nausea, vomiting, diarrhea and confusion. In addition, skin, lung, bladder and gastrointestinal cancers have been reported as major long-term effects of PAH exposure [28].

Both essential and non-essential elements, including heavy metals, have also been detected in urban air particulate matter (PM) [16]. The composition and content of heavy metals vary greatly among atmospheric particles of different sizes and origins [16]. Vehicular traffic is one of the major sources of heavy metals in urban areas [29]. In road dust, contributions are due to worn out tires, friction and brake discs and to the wear of the road surface [30]. The most widespread and documented elements include cadmium (Cd), cobalt (Co), chromium (Cr), copper (Cu), mercury (Hg), nickel (Ni), lead (Pb), antimony (Sb) and zinc (Zn) [17,31–34]. Toxic elements in road dust enter the human body mainly via three routes: direct inhalation, ingestion and dermal contact [35–38]. In addition, road dust can be resuspended by people walking or cycling and by the vehicles themselves, resulting in greater exposure levels around the road [32]. Toxic elements can affect human health, causing DNA damage, with induces mutagenic, teratogenic and carcinogenic effects [39–41]. Therefore, human exposure to road dust may increase health risks, including carcinogenic contamination [42,43]. Urinary metals were positively associated with oxidative stress biomarkers at exposure levels relevant to the general population [44].

The objective of this study is to evaluate the effects of traffic on human health comparing biomonitoring data measured during the COVID-19 lockdown, when restrictions led to a drastic reduction in road traffic in Rome, with the same parameters measured in 2021. Therefore, the exposure to benzene, PAHs and 28 elements was assessed by the biological monitoring of urine samples collected during the lockdown months (March–April 2020) and in the same period of the following year, by a group of 47 Rome resident volunteers, who worked from home during 2020. The exposure biomarkers were correlated to effect biomarkers, namely the urinary oxidative stress biomarkers produced by the oxidation of guanine residues in DNA and RNA [45].

1-hydroxynaphtalene (1-OHNAP) and 2-hydroxynaphtalene (2-OHNAP) are the metabolites of naphthalene, the most abundant PAHs in human urine. Urinary 1-OHPy is the biomarker of occupational exposure to polycyclic aromatic hydrocarbons mixtures. The major urinary metabolite of benzo[a]pyrene, 3-OHBaPy, has been considered closely related to genotoxicity [46]. 1-nitropyrene has been used as a molecular marker for diesel exhaust gases, which significantly contributes to particulate-associated toxicity [47]. Urinary metabolites of nitropyrene have been evaluated for their usefulness as markers of short-term exposure to diesel exhaust [48].

Cotinine is the main metabolite of nicotine and a biomarker for smoking [49]. Cigarette smoking is a major source of benzene exposure in active smokers and can affect the levels of biological markers of benzene exposure in non-smokers exposed to second-hand smoke [41,50].

Oxidatively generated damage to DNA and RNA plays an important role in cancer development, cardiovascular and neurodegenerative diseases, diabetes, pulmonary fibrosis, and much more [51,52]. Guanine is the DNA base most susceptible to oxidation because of its low redox potential, leading to the formation of 8-oxo-7,8-dihydroguanine (8-oxoGua), the most common lesion, and of 8-Oxo-7,8-dihydro-20-deoxyguanosine (8-oxodGuo). 8-oxoGua and 8-oxodGuo found in human urine originate from DNA repair mechanisms and possibly also from the turnover of oxidatively damaged DNA. The oxidative modifications of RNA guanine can lead to the formation of both 8-oxo-7,8-dihydroguanine (8-oxoGua) and 8-oxo-7,8-dihydroguanosine (8-oxoGuo) [53,54]. Attacks of ROS on DNA and RNA, therefore, lead to the urinary elimination of 8-oxoGua, 8-oxodGuo and 8-oxoGuo, which are considered biomarkers of oxidatively generated damage on DNA and RNA: they are always detectable in the general population as the result of the exposure to oxidative stress agents, which can have different origins [45].

The same urine samples were also subjected to NMR untargeted metabolomic analysis: air pollutant exposures have been associated with metabolic pathways primarily related to oxidative stress and inflammation, as assessed through untargeted metabolomics in 23 studies [55].

2. Materials and Methods

2.1. Enrollment and Sampling Campaigns

The studied group consisted of 47 volunteers, 20 males and 27 females, aged 47.51 $\pm$ 17.33 years, mainly researchers and their families, working from home during the lockdown; only 2 were smokers.

Participants received and signed an informed consent and a brief questionnaire about physical characteristics (sex, age and work); a numeric code was assigned to each subject to protect their privacy. Urine samples were collected by the volunteers during the lockdown period (March–April 2020) and in the same period of the following year, and immediately frozen until being transferred to the laboratories for the analyses. All the urine samples were analyzed by liquid chromatography/tandem mass spectrometry (HPLC/MS-MS), inductively coupled plasma mass spectrometer (ICP-MS), cold vapor generation atomic fluorescence spectrometry (CV-AFS) and by proton nuclear magnetic resonance (^{1}H-NMR). The urinary metabolites of benzene S-phenylmercapturic acid (SPMA), five PAHs, three biomarkers of oxidative stress to nucleic acids and 28 elements were determined, following the same pattern of previous studies on groups of general population volunteers [27,56].

The study was a non-interventional/observational study based on the definitions of the European Clinical Trials Directive 2001/20/EC, for which the approval of an ethics committee was not requested [57]; it was conducted according to the Declaration of Helsinki and followed the International Code of Ethics for Occupational Health Professionals of the International Committee of Occupational Health [58].

2.2. Targeted Monitoring

2.2.1. Instrumentation and Chemical Supplies

The HPLC/MS-MS is an API 4000 triple-quadrupole mass spectrometry detector equipped with a Turbo Ion Spray (TIS) probe (AB Sciex, Framingham, MA, USA) coupled to a Series 200 LC quaternary pump (PerkinElmer, Norwalk, CT, USA). Detection was carried out in the multiple reaction monitoring mode (MRM) and parameters were optimized for the analytes by the automated infusion quantitative optimization procedure and then refined by flow injection analysis (FIA) using pure standards. Urine samples were stored at −25 °C and analyzed immediately after thawing. The 1.5 version of Analyst® software (AB Sciex, Framingham, MA, USA) was used for instrument control. The analytical results are expressed in µg/g of creatinine. Urinary creatinine was determined by the Jaffè method using the alkaline picrate test with UV/Vis detection at 490 nm [59].

The analytical reference standards of 1- and 2-OHNaP, 1-OHPy, 6-OHNPy 3-OHBaPy, SPMA, Cotinine, 8-oxoGua, 8-oxodGuo and 8-oxoGuo were purchased by Spectra 2000 s.r.l (Rome, Italy). The isotope labelled internal standards Cotinine d_3, 2-OHNAP d_7, 1-OHPy d_9, 3-OHBaPy d_{11}, DL-SPMA-3,3 d_2, ($^{13}C^{15}N_2$) 8-oxodGuo and ($^{13}C^{15}N_2$) 8-oxoGuo were obtained from CDN Isotopes Inc. (Pointe-Claire, QC, Canada). ($^{13}C^{15}N_2$) 8-oxoGua (98%) was obtained from Cambridge Isotope Laboratories Inc. (Tewksbury, MA. USA); 3-NO_2Tyr was purchased by Cayman Chemical Company (USA) and 3NO_2Tyr d3 from TRC (Toronto, Canada). Multi-elemental standard solutions for ICP-MS analysis and Hg standard solution for CV-AFS determination were obtained from VWR International (Milan, Italy) and SCP Science (Baie D'Urfé, Quebec, Canada), respectively. For the preparation of ICP-MS internal standards, single standard solutions of In, Rh, Sc and Th (Merck KGaA, Darmstadt, Germany) and Y (Panreac Química, Barcelona, Spain) were used. 5-methylCytidine (5-MeCyt), glacial acetic acid 30%, NH_3, dimethyl sulfoxide, sodium hydroxide solution (50–52% in water), sodium borohydride and CHROMASOLV® gradient grade 99.9% methanol and acetonitrile for HPLC/MS 99.9%, carbon disulfide low benzene content, and n-exane 99.9% were obtained from Sigma Aldrich (Milan, Italy). Sodium hydroxide (98%, anhydrous pellets, RPE for analysis, ACS-ISO), hydrochloric acid (30% suprapure) and nitric acid (67% suprapure) were purchased by Carlo Erba Reagents (Milan, Italy). Purified water was obtained from a Milli-Q Plus system (Millipore Milford, MA, USA). The SPE cartridges, Sep-Pak Plus C18 (10 mL, 500 mg), were supplied by Waters (Waters S.p.A. Milan, Italy). Anotop 10LC syringe filter device (0.2 m pore size, 10 mm diameter) and syringe filter with cellulose nitrate membranes (0.45 mm pore size) were purchased from Whatman Inc. (Maidstone, UK) and GVS Filter Technology (Indianapolis, IN, USA), respectively. A Sinergi LUNA C8 column (250 × 4.6 mm, 4 µm), a Kinetex® 2.6 µm Polar C18 100 Å (150 × 4.6 mm) (Torrance, CA, USA) and Discovery C18 (150 × 4.6 mm, 5 µm) provided by Merck KGaA, (Darmstadt, Germany) were used for the study.

2.2.2. SPMA and Cotinine Analysis

DL S-phenylmercapturic acid (DL-SPMA), cotinine and the deuterium-labeled internal standards DL-SPMA-3,3-d_2 and cotinine-d_3 were determined with the following method [21]. Briefly: A total of 25 μL of internal standard (deuterated SPMA solution in methanol 1 mg/L) and 50 μL of 6N HCl were added to 5 mL of urine and, subsequently, subjected to solid-phase extraction (SPE) on Sep-pak Plus C18, 500 mg, 6 mL; 20 μL of the eluate was then analyzed using a Discovery C18, 150 × 4.6 mm reverse phase column, with a gradient of acetonitrile/methanol 90/10 *v/v* and acetic acid 0.5% *v/v* in water. The precursor→product ionic transitions monitored were in the negative ion mode. 238.1→109.1 for SPMA, and 240.1→109.1 for $SPMAd_2$ and in the positive ion mode 177.3→80.10 for cotinine and 180.3→80.10 for cotinine-d_3.

2.2.3. PAH Analysis

PAH metabolites were determined following a modified version of a published method [60]: A total of 25 μL of internal standard mixture (deuterated PAH standard solution in methanol 1 mg/L), 50 μL of β-glucuronidase-arilsulfatase enzyme from Elix Pomatia and 0.5 mL of acetate buffer 0.1 M at pH 5 were added to 5 mL of urine and, subsequently, incubated in a thermostatic bath at 38 °C for 16 h to hydrolyze the PAH metabolite conjugates. After enzymatic hydrolysis, the samples were then extracted with n-exane and 20 μL injected in the HPLC/MS/MS for the analysis; a reverse phase Luna C8 250 × 4.6 (Phenomenex, Torrance, CA, USA) column was used with a gradient of methanol and water. In the negative ion mode, the precursor→product ionic transitions monitored were 217.1→189.1 for 1-OHPy and 226.0→198.1 for 1-OHPy-d_9, 262.0→231.9 for 6-OHNPy, 267.1→239 for 3-OHBaPy and 278.0→250.0 for 3-OHBaPy-d_{11}, 142.9→115.2 for 1 and 2-OHNAP, and 149.8→122 for 2-OHNAP-d7.

2.2.4. Determination of Oxidative Stress Biomarkers

The urinary concentrations of 8-oxoGua, 8-oxoGuo, 8-oxodGuo and 3-NO_2Tyr were determined according to a previously described method [61] with modifications in the sample thawing, dilution solvents, chromatographic column and mobile phases. The samples were thawed in lukewarm water at around 37 °C, vortexed and centrifuged at 10,000× *g* for 5 min; the urine supernatant was added with a mixture of mixture internal standard, ((13C15N2) 8-oxoGua, (13C15N2) 8-oxoGuo, (13C15N2) 8-oxodGuo and 3-NO_2Tyr d3) and injected into the HPLC-MS/MS system. The reference standards of the analytes were firstly dissolved in DMSO, then in methanol and finally diluted with water. The chromatographic column was a Kinetex® 2.6 μm Polar C18 100 Å (150 × 4.6 mm) and the mobile phase consisted of a gradient of a mixture of acetonitrile/methanol 90/10 *v/v* and 0.5% acetic acid in water. The same method was also used for 5-methylCytidine (5-MeCyt) and cotinine, but after diluting the sample 1:100, and after adding the internal standard (Cotinine d3), using a different chromatographic column, Discovery C18 (150 × 4.6 mm, 5 μm). The precursor/product ionic transitions monitored (positive ion mode) were 168.0→140.0 and 171.0→143.0 for 8-oxoGua and its internal standard, 284.3→168.0 and 287.13→171.1 for 8-oxodGuo and its internal standard, 300.24→168.2 and 303.24→171.0 for 8-oxoGuo and its internal standard, 226.99→181.0 and 229.99→184.0 for 3-NO_2Tyr and its internal standard; 257.95→126.100; 180.3→80.10 was the transition monitored for 5-MeCyt, 177.3→80.10 for cotinine and 180.3→80.10 for cotinine-d3 were used as internal standard for both 5-MeCyt and cotinine.

2.2.5. Determination of Urinary Elements

All elements except Hg were analyzed with a quadruple ICP-MS (820-MS; Bruker, Bremen, Germany) equipped with a concentric glass nebulizer (0.4 mL min^{-1}; MicroMistTM; Analytik Jena AG, Jena, Germany) and a cyclonic spray chamber. Analyses were performed in standard mode and collision–reaction interface (CRI) mode. For As, Ca, Cr, Fe, Mn, Se and V, the assays were run in the CRI mode with He and H_2 (99.9995% purity; SOL Spa, Monza, Italy) as cell gases. For B, Ba, Be, Bi, Cd, Co, Cs, Cu, K, Li, Mg, Mo, Na, Ni, Pb, Rb, Sb, Sn, Sr, Te, Tl and Zn, the assays were run in standard mode. A CV-AFS (AFS 8220 Titan, FullTech Instruments, Rome, Italy) was used for Hg determination. The preparations of the calibration standards and the detailed ICP-MS and CV-AFS operating conditions were described in previous papers [62–64]. The limits of determination (LoDs) and quantification (LoQs) were calculated, respectively, as three and ten times σ/b, where σ is the standard deviation of blank determination (ten replicates) and b is the slope of the calibration curve. Results for LoDs and LoQs were shown in Supplementary Table S1. Urine samples (1 mL) were diluted 1:5 with 3% HCl or 1:10 with 2% HNO_3 and filtered using a syringe filter with cellulose nitrate membranes prior to CV-AFS or ICP-MS analysis, respectively.

2.3. *Analytical Determination of Urinary Metabolic Profiles*

2.3.1. Sample Preparation for NMR Analysis

Untargeted biomonitoring was performed on 34 pairs of matched samples. An amount of 1200 µL of urine was centrifuged at 11,000× *g* for 15 min at 4 °C to remove the cellular debris. A total of 100 µL of a trimethylsilylpropionic-2,2,3,3-d4 acid (TSP) in D_2O solution (2 mM final concentration) as internal standard were added to 1 mL of centrifuged samples and the pH was measured and adjusted at pH = 7 by adding small amounts of NaOH or HCl. Finally, 700 µL of sample were transferred in precision tubes for NMR experiments.

2.3.2. ^{1}H-NMR Spectroscopy

^{1}H-NMR spectra were acquired at 298 K using a JEOL JNM-ECZR spectrometer (JEOL Ltd., Tokyo, Japan) equipped with a magnet operating at 14.09 Tesla and 600.17 MHz for 1H frequency. All the spectra were recorded with 64k points and 64 scans, setting spectral width to 9.03 KHz (15 ppm), with a pre-saturation pulse length of 2.00 s and a relaxation delay of 5.72 s, for an acquisition time of 5.81 s. The identification step was achieved by two-dimensional experiments (^{1}H-^{1}H Homonuclear Total Correlation Spectroscopy (TOCSY), ^{1}H-^{13}C Heteronuclear Single Quantum Correlation (HSQC) and Heteronuclear Multiple Bond Correlation (HMBC)) on selected samples and confirmed by literature comparison. TOCSY experiments were recorded at 298 K with a spectral width of 15 ppm in both dimensions, using 8 k × 256 data points matrix, repetition time of 3.00 s and 80 scans, with a mixing time of 80.00 ms. HSQC experiments were acquired with a spectral width of 9.03 KHz (15 ppm) in proton dimension and 30 KHz (200 ppm) in the carbon dimension, using 8 k × 256 data point matrix for the proton and the carbon dimensions, respectively, with a repetition delay of 2 s and 96 scans. One-dimensional NMR spectra were processed and quantified by using the ACD Lab 1D-NMR Manager ver. 12.0 software (Advanced Chemistry Development, Inc., Toronto, ON, Canada); 2D-NMR spectra were processed using JEOL Delta v5.3.1 software (JEOL Ltd., Tokyo, Japan). All the NMR spectra were manually phased, baseline corrected and referenced to the chemical shift of the TSP methyl resonance at $\delta = 0.00$. The quantification of metabolites was obtained by comparing the integrals of their diagnostic resonances with the internal standard TSP integral and normalized for their number of protons. Metabolite levels are expressed as µmol mmol^{-1} of creatinine, referred at its $\delta = 4.05$ ppm resonance.

2.4. *Data Analysis and Statistics*

The descriptive statistics of the results are presented in tables (mean and standard deviation, median and 5/95th percentiles, and minimum and maximum values). Data below LOD were substituted with LOD/2. Data were analyzed with the *t*-test on log-transformed data to determine differences between the time series. Statistical analysis was performed using IBM SPSS Statistics 25 (IBM, Armonk, NY, USA), with the significance level set at $p \leq 0.05$. The results of the HPLC/MS/MS analysis are presented in Table 1, while the results of ICP/MS are shown in Supplementary Materials Table S1.

Table 1. Results of targeted metabolomics expressed in µg/g of creatinine on 47 pairs of matched samples. Asterisks indicate statistical significance ($p < 0.01$).

Analyte	8-oxoGua	8-oxoGuo	8-oxodGuo	3-NO_2Tyr	5-MeCyt	SPMA	1-OHPy	6-OHNPy	3-OHBaPy	1-OHNAP	2-OHNAP
						YEAR 2020					
mean	15.99 *	6.73 *	2.26 *	33.91	11.33	0.31 *	0.05 *	>LOD	0.04	0.43	4.65
SD	14.54	4.31	1.25	27.60	6.99	0.20	0.04	>LOD	0.04	0.76	6.12
median	9.72	5.36	2.21	21.74	9.89	0.23	0.04	>LOD	0.02	0.27	2.89
5th perc	4.02	2.14	0.36	12.04	5.03	0.13	0.02	>LOD	0.01	0.01	1.03
95th perc	47.85	15.46	4.38	103.52	21.85	0.73	0.13	0.01	0.14	1.63	12.85
min	2.07	1.67	0.09	7.96	4.33	0.09	0.01	>LOD	0.01	0.01	0.52
max	64.46	19.54	5.31	126.20	43.13	0.94	0.18	>LOD	0.17	4.62	39.22
						YEAR 2021					
mean	46.50	10.60	3.64	32.95	12.72	0.39	0.07	>LOD	0.03	0.46	6.08
SD	43.98	5.61	1.84	24.81	8.05	0.23	0.03	>LOD	0.03	1.15	6.80
median	28.80	8.66	3.14	23.11	9.89	0.31	0.06	>LOD	0.02	0.12	3.66
5th perc	6.89	4.52	1.42	14.08	5.12	0.14	0.03	>LOD	0.01	0.01	0.76
95th perc	118.19	22.39	5.98	69.41	30.86	0.83	0.13	0.01	0.07	1.82	19.48
min	4.43	3.89	1.05	3.17	3.68	0.14	0.02	>LOD	0.01	0.01	0.01
max	232.30	24.50	10.74	142.61	38.48	1.07	0.16	0.02	0.18	7.43	37.32

Statistical multivariate analysis for untargeted metabolomics was applied to the data matrix combining NMR and HPLC-MS/MS metabolites. Unsupervised PCA and supervised PLS-DA were applied to the entire log-transformed dataset after centering and auto-scaling. The multivariate analysis was carried out using Unscrambler 10.5 software (CAMO, Oslo, Norway). In order to identify a statistically significant variation of the single variables between the two years considered, a Shapiro–Wilk test was applied to define the normality of distribution and then a paired Student's *t*-test was applied, using Sigmaplot 12.0 software (Systat Software, Inc., San Jose, CA, USA). A *p*-value of 0.05 was considered a threshold for statistical significance.

3. Results and Discussions

In Italy, a lockdown due to the COVID-19 pandemic began on 9 March 2020. During this period, the schools were closed and working from home was promoted where possible; travel for unnecessary reasons, sports activities, demonstrations and events was prohibited; non-essential retail commercial activities, catering services, religious celebrations and gatherings of people in public places were suspended.

In addition to the various social, economic and political consequences, the lockdown effects were also evident at the environmental level, especially in terms of a drastic decrease in vehicular traffic, visible above all in large urban centers, such as Rome.

Figure 1 shows the IMR (detected mobility index) recorded by ANAS S.p.A. (the national road agency), which compares the average monthly traffic data in Italy in March 2019, February and March 2020, and March 2021 [65,66].

The restrictive lockdown measures reduced the IMR in the Lazio region (where Rome is located) from February to March 2020 of 36%. The data concerning the month of March 2021 show how the restoration of normal activities brought the mobility index back to a value higher of 48% compared to March 2020.

In order to evaluate the decrease in the airborne concentrations of benzene and PM2.5, the data from the air quality monitoring network of ARPA (the regional agency for environmental protection) of Lazio were used, and in particular those provided in the online weekly bulletins [67–69].

The weekly values of benzene and PM2.5 airborne concentrations measured by monitoring devices located in five different areas of Rome were averaged. Figure 2 shows the weekly trend of the concentration of airborne benzene in the years 2019, 2020 and 2021. A seasonal decreasing trend can be observed for all the three years, from week 10 to week 30, but the values recorded for the weeks of the lockdown (week 11 to 21, marked in red in the figure) show a reduction in benzene emissions of about 25% compared to the previous year (2019) and of 40% compared to the next (2021), while for PM2.5, there were no significant changes (data not shown).

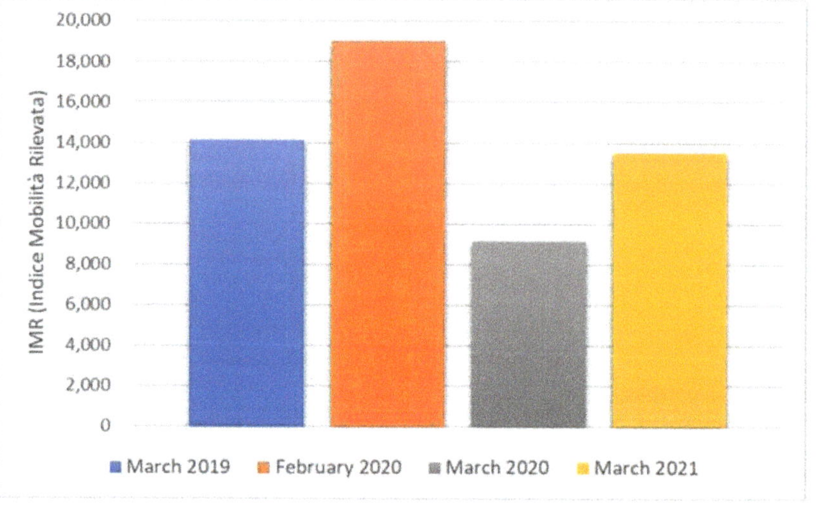

Figure 1. Average monthly traffic data in Italy in March 2019, February and March 2020, and March 2021. IMR (Indice Mobilità Rilevata) is a mobility index expressed as the mean number of vehicles/day, calculated by ANAS S.p.A.

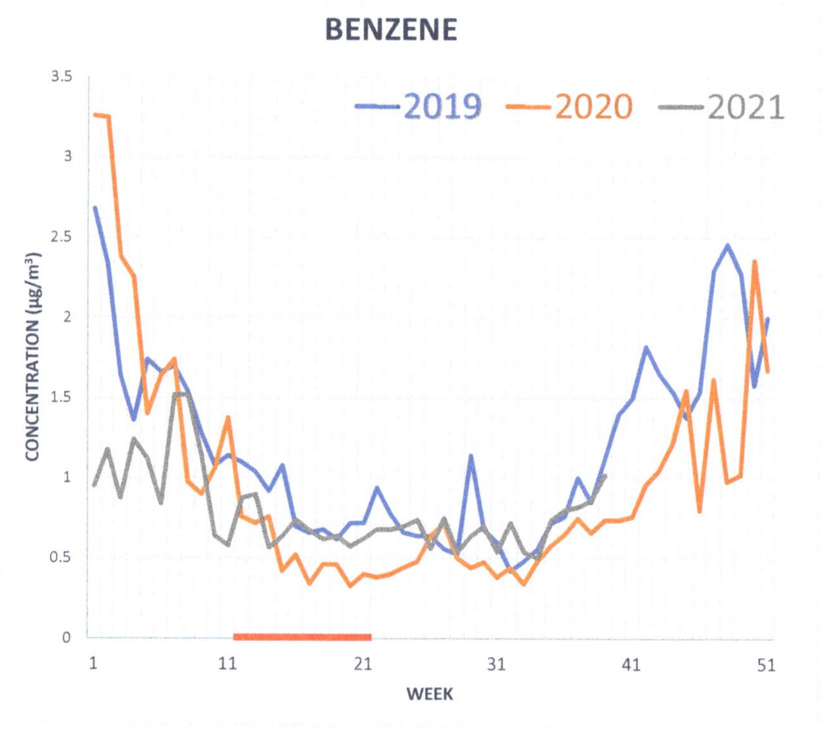

Figure 2. Weekly trend of the concentration of airborne benzene in the years 2019, 2020 and 2021. The weeks 11–21, marked in red, are the lockdown weeks in year 2020.

It can be observed that, while the reduction in benzene concentration is lower but consistent with the decrease in road traffic, for PM2.5 there was not a similar reduction, as other important emission sources have to also be considered, such as civil heating. No decrease in PM2.5 could also result from enhanced atmospheric oxidation during the COVID-19 lockdown [7,10].

The results of the targeted metabolomics, obtained using HPLC-MS/MS, are reported in Table 1 both for the years 2020 and 2021. The values are expressed as a ratio including the urinary creatinine concentrations of the same samples.

3.1. Targeted Metabolomics: Determination of Urinary Biomarkers

The urinary concentration values found for the volunteers for each metabolite in 2020 and 2021 were log-transformed and compared using a *t*-test. The statistically significant differences were indicated with an asterisk ($p < 0.01$).

The analysis of the concentrations of the metabolites SPMA and 6-OHNPy shows that their average value in 2020 is about 80% of that found in 2021, while the metabolite of pyrene, 1-OHPy, is about 70% in 2020 compared to 2021, all with a statistically significant difference.

The reduction in the exposure biomarkers in 2020 compared to 2021 reflects, albeit to a lesser extent, both that measured by ARPA Lazio for benzene (−40%) and by ANAS in relation to road traffic in Lazio (−36%). In this regard, it is important to remember that both benzene and PAHs, being produced by the combustion of organic products, can also be generated from indoor sources, such as the use of incense, cooking food and active and passive cigarette smoke, contributing to the production of the studied metabolites.

Examining the effect biomarkers, a highly significant difference was found for all the three measured biomarkers, with a reduction of about 40% for 8-oxoGuo and 8-oxodGuo in 2020 compared to 2021. As regards 8-oxoGua, this biomarker seems to be more sensitive to the exposure variations than the other two, with a reduction of around 60% in 2020 compared to 2021. However, all the biomarkers of oxidative stress were affected by the lockdown more strongly than the dose biomarkers of the considered pollutants.

3.2. Determination of Urinary Elements

Supplementary Materials Table S1 shows summary statistics for the creatinine-adjusted urinary metal concentrations at the two monitoring periods. Be, Bi and Mn levels were found to be lower than those of LOD in most or all cases (50% for Bi and 100% of Be and Mn). These elements were excluded from the statistical elaboration. The highest concentrations (>80 mg/g creatinine) were observed for essential elements, such as Ca, K, Mg and Na, while among the toxic or potentially toxic elements, the highest concentrations were obtained for As (mean = 122 or 54 µg/g creatinine in 2020 or 2021, respectively), Cu (mean = 30 or 40 µg/g creatinine in 2020 or 2021, respectively) and Li (mean = 27 or 21 µg/g creatinine in 2020 or 2021, respectively). Comparing the cohort at the two time measurements, the analysis showed a generally significant reduction during lockdown year for B, Co, Cu and Sb; Tl and Zn also decreased, but not significantly. We found that B was positively associated with Ni (rho = 0.713, $p < 0.01$) and Rb (rho = 0.706, $p < 0.01$). Cobalt does not show high correlations with any of the analyzed elements. Cu, Ni and Sb were highly correlated to each other (Spearman correlation ranging from 0.712 to 0.797, $p < 0.01$). Sb was also positively associated with Cr (rho = 0.723, $p < 0.01$) and Fe (rho = 0.704, $p < 0.01$), indicating a possible common source of exposure. Metals of concern for brake lining emissions include Cd, Cr, Cu, Ni, Pb, Sb and Zn [70]. Previous studies have shown significant contributions of road traffic to the Sb and Cu content of PM due to metal emissions from brake linings [34]. Brake lining wear has been considered to be responsible for 99% and 90% of airborne Sb and Cu, respectively [71]. Additionally, the link between physical activity and increased metal excretion was studied [72,73]. Although sweat is the most important way of the excretion of trace metals during physical exercise [74], it is possible to assume that more intense breathing during physical activity may lead to a greater inhalation of elements present in the air and, consequently, to an increase in their urinary levels. A sedentary lifestyle and reduced vehicular traffic may have caused lower urinary Cu and Sb levels during the 2020 lockdown. However, the difference in B and Co levels between the periods during and after the lockdown could be linked to different eating habits. Fruit-based beverages and products, tubers, legumes, and water could contribute a major portion of the dietary B vitamin [75]. Immune function and the metabolism of vitamin D, Ca, Cu, Mg and estrogen have been proposed as being related to the functioning of B vitamin in humans [75]. Cobalt is an essential trace element that is mainly known as a component of vitamin B12, which serves as a cofactor in the synthesis of methionine and the metabolism of folates and purines [75–77]. Good food sources of Co include fish, green leafy vegetables and cereals [78,79].

3.3. Untargeted Metabolomics and Multivariate Analysis

A representative urinary ^{1}H NMR spectrum is shown in Supplementary Materials Figures S1–S3. Thirty-five metabolites, including creatinine (used as a normalizing factor), were identified and quantified from the ^{1}H NMR urine spectra of a subgroup of 34 matched subjects in 2020 and 2021. The list of the identified metabolites with the relative chemical shifts of the resonances is reported in Supplementary Materials Table S2.

The data matrix included, as variables (in columns), 35 urinary NMR metabolites, 6 urinary effect biomarkers (8-oxoGua, 8-oxoGuo, 8-oxodGuo, 3-NO_2Tyr, 5-MeCyt and Cotinine), 6 urinary dose biomarkers (namely SPMA, 1-OHPy, 6-OHNPy, 3-OHBaP, 1-OHNAP and 2-OHNAP) and 28 urinary elements. All the concentrations were log-transformed before applying the multivariate and univariate analysis performed on 29 pairs of matched samples (mean age of 48.76 $\pm$ 16.16 years), among which 14 were females and 15 were males.

Firstly, we performed an explorative PCA for all the subjects (data not shown) to appreciate the possible spontaneous separation between urinary metabolic profiles in 2020 and 2021. Since no spontaneous separation was observed, we decided to analyze the data matrix by PLS-DA in order to identify the effects of the lockdown on the urinary metabolic profiles, as shown in Figure 3.

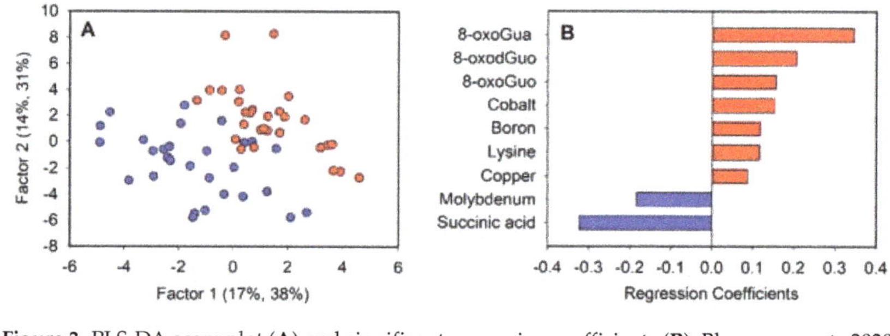

Figure 3. PLS-DA score plot (**A**) and significant regression coefficients (**B**). Blue represents 2020 and red 2021.

The PLS-DA (Figure 3A) showed a good discrimination model: the validation results indicate a $R^2 = 0.87$ and $Q^2 = 0.49$. Furthermore, eight metabolites were identified as significantly relevant for the discrimination based on the regression coefficient values (Figure 3B). In particular, the Lys levels of 8-oxoGuo, 8-oxoGua, 8-oxodGuo, B, Co and Cu were higher in 2021 (red color), while succinic acid and Mo were higher in 2020 (blue color).

We performed a paired Student's t-test on the metabolites that were significant for the PLS-DA model. We report only the metabolites that have statistically significant differences between 2020 and 2021, as indicated in the boxplots in Figures 4 and 5.

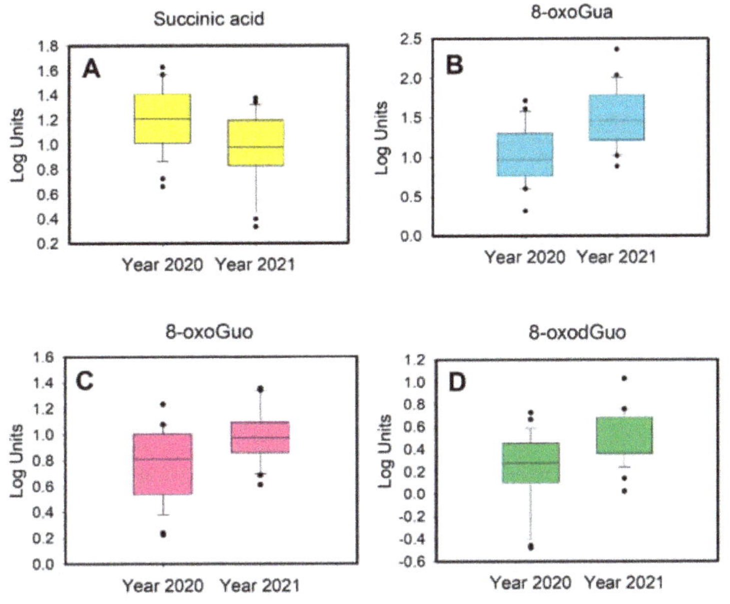

Figure 4. Boxplots of (**A**) succinic acid, (**B**) 8-oxoGua, (**C**) 8-oxoGuo and (**D**) 8-oxodGuo.

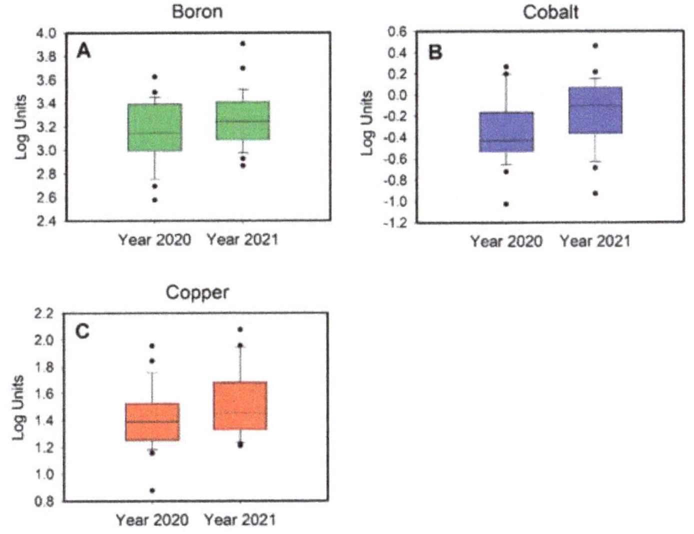

Figure 5. Boxplots of (**A**) boron, (**B**) cobalt and (**C**) copper.

The untargeted metabolomic analysis confirms the results of the targeted HPLC-MS tandem analysis regarding oxidative stress biomarkers, which were higher in 2021 than in 2020. In addition, the increase in succinic acid excretion in 2020 compared to 2021 is highlighted. Succinic acid is an intermediate product of the Krebs cycle. It has recently been observed that succinic acid is able to fuel inflammation and modify cell metabolism [80]. This acid is excreted by macrophages in the extracellular fluids of inflamed tissues; inflammation is favored by a diet rich in sugars, salt and foods containing succinic acid.

Furthermore, according to a further study, an increase in urinary levels of succinic acid is due not only to diet but also to a sedentary lifestyle [81].

The increase in the urinary concentration of succinic acid in 2020 is probably attributable to a different diet and a more sedentary lifestyle during the lockdown period.

This study has some limitations. Firstly, the low number of subjects based on the voluntary participation due to the lockdown situation. Secondly, the same subjects had to be contacted one year later for the second sample, causing some dropouts (n = 9). Another limitation is the fact that it was not possible to perform the indoor air monitoring in the houses of the volunteers, due to the reduced mobility of all the Rome residents. In addition, more metabolomic studies with larger sample sizes are needed to identify air pollution components linked to adverse health effects.

4. Conclusions

The lockdown scenario was a unique condition, which made possible to directly measure the effects of a reduced exposure to traffic pollution for a long period, on the same group of subjects.

The reduction in the dose exposure biomarkers of benzene and PAHs in 2020 compared to 2021 reflects the trend registered both by ARPA Lazio for airborne benzene concentrations and by ANAS in relation to road traffic in Lazio, Italy.

Lower levels of the urinary oxidative stress biomarkers 8-oxoGuo, 8-oxoGua and 8-oxodGuo were also observed in 2020 with respect to 2021. These three biomarkers were more strongly affected by the lockdown than the benzene and PAH biomarkers: in fact, these pollutants are also produced by the combustion of organic products and, therefore, can be generated from indoor sources that were not influenced by the lockdown, such as wood heating, use of incense, food cooking and active and passive cigarette smoke, contributing to their release in the atmosphere.

Elementary analysis showed a generally significant reduction in B, Co, Cu and Sb during the lockdown year. Indeed, the higher airborne levels of Sb and Cu during 2021 could be associated with a greater vehicular traffic, particularly brake lining wear. A sedentary lifestyle and less vehicular traffic may have also caused the lower urinary Cu and Sb levels during the 2020 lockdown.

The untargeted metabolomic analysis identified succinate, an intermediate of the Krebs cycle, as being lower in 2021, possibly due to a different diet and a more sedentary lifestyle during the 2020 lockdown period.

The results show a significant reduction in the traffic pollution biomarkers in Rome citizens during the lockdown period, and an even stronger reduction in oxidative stress, linked not only to the reduced traffic exposure, but also to a different lifestyle.

This study demonstrates that urban pollution due to traffic is only partly responsible for oxidative stress in the citizens, and that other factors also have a role. We think that it would be important to include the determination of effect biomarkers in the chemical exposure studies. In addition, our results enhance the importance of information and communication about a healthy lifestyle for the prevention of cancer diseases.

Supplementary Materials: The following supporting information can be downloaded at: https://www.mdpi.com/article/10.3390/toxics10050267/s1, Table S1: Descriptive statistic of 28 urinary elements (µg/g creatinine) on 29 pairs of matched samples; Figure S1: ^{1}H NMR spectrum of urine t; Figure S2: ^{1}H NMR spectrum of urine, spectral region of 1–4 ppm; Figure S3: ^{1}H NMR spectrum of urine, spectral region of 6.5–9.5 ppm; Table S2: ^{1}H NMR assignment of urinary metabolites.

Author Contributions: Conceptualization, G.T. and F.B. (Flavia Buonaurio); methodology, F.B. (Flavia Buonaurio), F.B. (Francesca Borra), D.P., A.M., M.S., S.C. and F.S.; investigation, F.B. (Flavia Buonaurio), F.B. (Francesca Borra), D.P., E.P., M.S., F.S., O.G., M.L.A. and C.A.; data curation, D.P., F.B. (Flavia Buonaurio), F.B. (Francesca Borra), M.S. and O.G.; writing—original draft preparation, G.T., F.B. (Flavia Buonaurio), F.B. (Francesca Borra), D.P., M.S., O.G. and M.L.A.; writing—review and editing, G.T., M.S., O.G., M.L.A., A.M. and S.C. All authors have read and agreed to the published version of the manuscript.

Funding: This research received no external funding.

Institutional Review Board Statement: Ethical review and approval were waived for this study as it was a non-interventional/observational study based on the definitions of the European Clinical Trials Directive 2001/20/EC, for which the approval of an ethics committee was not requested. It was conducted according to the Declaration of Helsinki and followed the International Code of Ethics for Occupational Health Professionals of the International Committee of Occupational Health.

Informed Consent Statement: Informed consent was obtained from all subjects involved in the study.

Data Availability Statement: The data presented in this study are available within this article. Further inquiries may be directed to the authors.

Acknowledgments: The authors would like to warmly thank all the volunteers who participated to this study.

Conflicts of Interest: The authors declare no conflict of interest.

References

1. Samanta, P.; Ghosh, A.R. Environmental perspectives of COVID-19 outbreaks: A review. *World J. Gastroenterol.* **2021**, *27*, 5822–5850. [CrossRef] [PubMed]
2. World Health Organization (WHO). WHO Coronavirus (COVID-19) Dashboard. Available online: https://covid19.who.int/ (accessed on 12 May 2022).
3. Giovanetti, M.; Benvenuto, D.; Angeletti, S.; Ciccozzi, M. The first two cases of 2019-nCoV in Italy: Where they come from? *J. Med. Virol.* **2020**, *92*, 518–521. [CrossRef]
4. *Decreto del Presidente del Consiglio dei Ministri 8 Mzo 2020*; Gazzetta Ufficiale: Rome, Italy, 2020.
5. *Decreto del Presidente del Consiglio dei Ministri 11 Giugno 2020*; Gazzetta Ufficiale: Rome, Italy, 2020.
6. Mishra, N.P.; Das, S.S.; Yadav, S.; Khan, W.; Afzal, M.; Alarifi, A.; Kenawy, E.R.; Ansari, M.T.; Hasnain, M.S.; Nayak, A.K. Global impacts of pre- and post-COVID-19 pandemic: Focus on socio-economic consequences. *Sens. Int.* **2020**, *1*, 100042. [CrossRef] [PubMed]
7. Wang, Y.; Wu, R.; Liu, L.; Yuan, Y.; Liu, C.G.; Hang Ho, S.S.; Ren, H.; Wang, Q.; Lv, Y.; Yan, M.; et al. Differential health and economic impacts from the COVID-19 lockdown between the developed and developing countries: Perspective on air pollution. *Environ. Pollut.* **2022**, *293*, 118544. [CrossRef] [PubMed]
8. Massimi, L.; Pietrodangelo, A.; Frezzini, M.A.; Ristorini, M.; De Francesco, N.; Sargolini, T.; Amoroso, A.; Di Giosa, A.; Canepari, S.; Perrino, C. Effects of COVID-19 lockdown on PM10 composition and sources in the Rome Area (Italy) by elements' chemical fractionation-based source apportionment. *Atmos. Res.* **2022**, *266*, 105970. [CrossRef]
9. Collivignarelli, M.C.; Abbà, A.; Bertanza, G.; Pedrazzani, R.; Ricciardi, P.; Carnevale Miino, M. Lockdown for CoViD-2019 in Milan: What are the effects on air quality? *Sci. Total Environ.* **2020**, *732*, 139280. [CrossRef]
10. Le, T.; Wang, Y.; Liu, L.; Yang, J.; Yung, Y.L.; Li, G.; Seinfeld, J.H. Unexpected air pollution with marked emission reductions during the COVID-19 outbreak in China. *Science* **2020**, *369*, 702–706. [CrossRef]
11. Winkler, A.; Amoroso, A.; Di Giosa, A.; Marchegiani, G. The effect of Covid-19 lockdown on airborne particulate matter in Rome, Italy: A magnetic point of view. *Environ. Pollut.* **2021**, *291*, 118191. [CrossRef]
12. Bai, X.; Chen, H.; Oliver, B.G. The health effects of traffic-related air pollution: A review focused the health effects of going green. *Chemosphere* **2022**, *289*, 133082. [CrossRef]
13. Shima, M. Health Effects of Air Pollution: A Historical Review and Present Status. *Nihon Eiseigaku Zasshi.* **2017**, *72*, 159–165. [CrossRef]
14. Wong, J.Y.Y.; Jones, R.R.; Breeze, C.; Blechter, B.; Rothman, N.; Hu, W.; Ji, B.; Bassig, B.A.; Silverman, D.T.; Lan, Q. Commute patterns, residential traffic-related air pollution, and lung cancer risk in the prospective UK Biobank cohort study. *Environ. Int.* **2021**, *155*, 106698. [CrossRef] [PubMed]
15. Astolfi, M.L.; Di Filippo, P.; Gentili, A.; Canepari, S. Semiautomatic sequential extraction of polycyclic aromatic hydrocarbons and elemental bio-accessible fraction by accelerated solvent extraction on a single particulate matter sample. *Talanta* **2017**, *174*, 838–844. [CrossRef] [PubMed]
16. Yuan, Y.; Wu, Y.; Ge, X.; Nie, D.; Wang, M.; Zhou, H.; Chen, M. In vitro toxicity evaluation of heavy metals in urban air particulate matter on human lung epithelial cells. *Sci. Total Environ.* **2019**, *678*, 301–308. [CrossRef] [PubMed]
17. Marconi, E.; Canepari, S.; Luisa Astolfi, M.; Perrino, C. Determination of Sb(III), Sb(V) and identification of Sb-containing nanoparticles in airborne particulate matter. *Procedia Environ. Sci.* **2011**, *4*, 209–217. [CrossRef]
18. Vogel, C.F.A.; Van Winkle, L.S.; Esser, C.; Haarmann-Stemmann, T. The aryl hydrocarbon receptor as a target of environmental stressors—Implications for pollution mediated stress and inflammatory responses. *Redox Biol.* **2020**, *34*, 101530. [CrossRef] [PubMed]
19. Guo, H.; Ahn, S.; Zhang, L. Benzene-associated immunosuppression and chronic inflammation in humans: A systematic review. *Occup. Environ. Med.* **2021**, *78*, 377–384. [CrossRef]

20. Tranfo, G.; Pigini, D.; Paci, E.; Bauleo, L.; Forastiere, F.; Ancona, C. Biomonitoring of urinary benzene metabolite SPMA in the general population in Central Italy. *Toxics* **2018**, *6*, 37. [CrossRef]
21. Tranfo, G.; Pigini, D.; Paci, E.; Marini, F.; Bonanni, R.C. Association of exposure to benzene and smoking with oxidative damage to nucleic acids by means of biological monitoring of general population volunteers. *Environ. Sci. Pollut. Res.* **2017**, *24*, 13885–13894. [CrossRef]
22. Carrieri, M.; Pigini, D.; Martinelli, A.; Paci, E.; Maratini, F.; Salamon, F.; Tranfo, G. Effect of benzene exposure on the urinary biomarkers of nucleic acid oxidation in two cohorts of gasoline pump attendants. *Int. J. Environ. Res. Public Health* **2019**, *16*, 129. [CrossRef]
23. Bolden, A.L.; Kwiatkowski, C.F.; Colborn, T. New Look at BTEX: Are Ambient Levels a Problem? *Environ. Sci. Technol.* **2015**, *49*, 5261–5276. [CrossRef]
24. Mchale, C.M.; Zhang, L.; Smith, M.T. Current understanding of the mechanism of benzene-induced leukemia in humans: Implications for risk assessment. *Carcinogenesis* **2012**, *33*, 240–252. [CrossRef] [PubMed]
25. European Parliament. *European Council Directive 2008/50/EC on Ambient Air Quality and Cleaner Air for Europe*; European Parliament: Strasbourg, France, 2008.
26. IARC Publications Website. Certain Polycyclic Aromatic Hydrocarbons and Heterocyclic Compounds. Available online: https://publications.iarc.fr/Book-And-Report-Series/Iarc-Monographs-On-The-Identification-Of-Carcinogenic-Hazards-To-Humans/Certain-Polycyclic-Aromatic-Hydrocarbons-And-Heterocyclic-Compounds-1973 (accessed on 13 May 2022).
27. Tombolini, F.; Pigini, D.; Tranfo, G.; Paci, E.; Carosi, I.; Marini, F.; Bauleo, L.; Ancona, C.; Forastiere, F. Levels of urinary metabolites of four PAHs and cotinine determined in 1016 volunteers living in Central Italy. *Environ. Sci. Pollut. Res.* **2018**, *25*, 28772–28779. [CrossRef]
28. Kim, K.H.; Jahan, S.A.; Kabir, E.; Brown, R.J.C. A review of airborne polycyclic aromatic hydrocarbons (PAHs) and their human health effects. *Environ. Int.* **2013**, *60*, 71–80. [CrossRef] [PubMed]
29. Adamiec, E.; Jarosz-Krzemińska, E.; Wieszała, R. Heavy metals from non-exhaust vehicle emissions in urban and motorway road dusts. *Environ. Monit. Assess.* **2016**, *188*, 1–11. [CrossRef] [PubMed]
30. Yuen, J.Q.; Olin, P.H.; Lim, H.S.; Benner, S.G.; Sutherland, R.A.; Ziegler, A.D. Accumulation of potentially toxic elements in road deposited sediments in residential and light industrial neighborhoods of Singapore. *J. Environ. Manag.* **2012**, *101*, 151–163. [CrossRef] [PubMed]
31. Astolfi, M.L.; Canepari, S.; Cardarelli, E.; Ghighi, S.; Marzo, M.L. Chemical Fractionation of Ellements in Airborne Particulate Matter: Primary Results on PM10 and PM2.5 Samples in the Lazio Region (Central Italy). *Ann. Chim.* **2006**, *96*, 183–194. [CrossRef]
32. Shi, G.; Chen, Z.; Bi, C.; Wang, L.; Teng, J.; Li, Y.; Xu, S. A comparative study of health risk of potentially toxic metals in urban and suburban road dust in the most populated city of China. *Atmos. Environ.* **2011**, *45*, 764–771. [CrossRef]
33. Zgłobicki, W.; Telecka, M.; Skupiński, S. Assessment of short-term changes in street dust pollution with heavy metals in Lublin (E Poland)—levels, sources and risks. *Environ. Sci. Pollut. Res.* **2019**, *26*, 35049–35060. [CrossRef]
34. Fort, M.; Grimalt, J.O.; Querol, X.; Casas, M.; Sunyer, J. Evaluation of atmospheric inputs as possible sources of antimony in pregnant women from urban areas. *Sci. Total Environ.* **2016**, *544*, 391–399. [CrossRef]
35. Ferreira-Baptista, L.; De Miguel, E. Geochemistry and risk assessment of street dust in Luanda, Angola: A tropical urban environment. *Atmos. Environ.* **2005**, *39*, 4501–4512. [CrossRef]
36. Kong, A.; Frigge, M.L.; Masson, G.; Besenbacher, S.; Sulem, P.; Magnusson, G.; Gudjonsson, S.A.; Sigurdsson, A.; Jonasdottir, A.; Jonasdottir, A.; et al. Rate of de novo mutations and the importance of father's age to disease risk. *Nature* **2012**, *488*, 471–475. [CrossRef] [PubMed]
37. Ali, M.U.; Liu, G.; Yousaf, B.; Abbas, Q.; Ullah, H.; Munir, M.A.M.; Fu, B. Pollution characteristics and human health risks of potentially (eco)toxic elements (PTEs) in road dust from metropolitan area of Hefei, China. *Chemosphere* **2017**, *181*, 111–121. [CrossRef] [PubMed]
38. Padoan, E.; Romè, C.; Ajmone-Marsan, F. Bioaccessibility and size distribution of metals in road dust and roadside soils along a peri-urban transect. *Sci. Total Environ.* **2017**, *601–602*, 89–98. [CrossRef]
39. Knasmüller, S.; Parzefall, W.; Sanyal, R.; Ecker, S.; Schwab, C.; Uhl, M.; Mersch-Sundermann, V.; Williamson, G.; Hietsch, G.; Langer, T.; et al. Use of metabolically competent human hepatoma cells for the detection of mutagens and antimutagens. *Mutat. Res. Mol. Mech. Mutagen.* **1998**, *402*, 185–202. [CrossRef]
40. Duzgoren-Aydin, N.S.; Wong, C.S.C.; Aydin, A.; Song, Z.; You, M.; Li, X.D. Heavy metal contamination and distribution in the urban environment of Guangzhou, SE China. *Environ. Geochem. Health* **2006**, *28*, 375–391. [CrossRef]
41. Protano, C.; Manigrasso, M.; Avino, P.; Vitali, M. Second-hand smoke generated by combustion and electronic smoking devices used in real scenarios: Ultrafine particle pollution and age-related dose assessment. *Environ. Int.* **2017**, *107*, 190–195. [CrossRef]
42. Kamunda, C.; Mathuthu, M.; Madhuku, M. Health Risk Assessment of Heavy Metals in Soils from Witwatersrand Gold Mining Basin, South Africa. *Int. J. Environ. Res. Public Health* **2016**, *13*, 663. [CrossRef]
43. Zhao, L.; Xu, Y.; Hou, H.; Shangguan, Y.; Li, F. Source identification and health risk assessment of metals in urban soils around the Tanggu chemical industrial district, Tianjin, China. *Sci. Total Environ.* **2014**, *468–469*, 654–662. [CrossRef]
44. Domingo-Relloso, A.; Grau-Perez, M.; Galan-Chilet, I.; Garrido-Martinez, M.J.; Tormos, C.; Navas-Acien, A.; Gomez-Ariza, J.L.; Monzo-Beltran, L.; Saez-Tormo, G.; Garcia-Barrera, T.; et al. Urinary metals and metal mixtures and oxidative stress biomarkers in an adult population from Spain: The Hortega Study. *Environ. Int.* **2019**, *123*, 171–180. [CrossRef]

45. Tranfo, G.; Paci, E.; Carrieri, M.; Marchetti, E.; Sisto, R.; Gherardi, M.; Costabile, F.; Bauleo, L.; Ancona, C.; Pigini, D. Levels of urinary biomarkers of oxidatively generated damage to DNA and RNA in different groups of workers compared to general population. *Int. J. Environ. Res. Public Health* **2019**, *16*, 2995. [CrossRef]
46. Hassan, A.M.; Alam, S.S.; Abdel-Aziem, S.H.; Ahmed, K.A. Benzo-a-pyrene induced genotoxicity and cytotoxicity in germ cells of mice: Intervention of radish and cress. *J. Genet. Eng. Biotechnol.* **2011**, *9*, 65–72. [CrossRef]
47. Riley, E.A.; Carpenter, E.E.; Ramsay, J.; Zamzow, E.; Pyke, C.; Paulsen, M.H.; Sheppard, L.; Spear, T.M.; Seixas, N.S.; Stephenson, D.J.; et al. Evaluation of 1-Nitropyrene as a Surrogate Measure for Diesel Exhaust. *Ann. Work Expo. Health* **2018**, *62*, 339–350. [CrossRef] [PubMed]
48. Toriba, A.; Kitaoka, H.; Dills, R.L.; Mizukami, S.; Tanabe, K.; Takeuchi, N.; Ueno, M.; Kameda, T.; Tang, N.; Hayakawa, K.; et al. Identification and Quantification of 1-Nitropyrene Metabolites in Human Urine as a Proposed Biomarker for Exposure to Diesel Exhaust. *Chem. Res. Toxicol.* **2007**, *20*, 999–1007. [CrossRef] [PubMed]
49. Paci, E.; Pigini, D.; Bauleo, L.; Ancona, C.; Forastiere, F.; Tranfo, G. Urinary Cotinine Concentration and Self-Reported Smoking Status in 1075 Subjects Living in Central Italy. *Int. J. Environ. Res. Public Health* **2018**, *15*, 804. [CrossRef]
50. Darrall, K.G.; Figgins, J.A.; Brown, R.D.; Phillips, G.F. Determination of benzene and associated volatile compounds in mainstream cigarette smoke. *Analyst* **1998**, *123*, 1095–1101. [CrossRef] [PubMed]
51. Hayes, J.D.; Dinkova-Kostova, A.T.; Tew, K.D. Oxidative Stress in Cancer. *Cancer Cell* **2020**, *38*, 167–197. [CrossRef]
52. Jay Forman, H.; Zhang, H. Targeting oxidative stress in disease: Promise and limitations of antioxidant therapy. *Nat. Rev. Drug Discov.* **2021**, *20*, 689–709. [CrossRef]
53. Chao, M.R.; Evans, M.D.; Hu, C.W.; Ji, Y.; Møller, P.; Rossner, P.; Cooke, M.S. Biomarkers of nucleic acid oxidation—A summary state-of-the-art. *Redox Biol.* **2021**, *42*, 101872. [CrossRef]
54. Tanaka, M.; Chock, P.B. Oxidative Modifications of RNA and Its Potential Roles in Biosystem. *Front. Mol. Biosci.* **2021**, *8*, 407. [CrossRef]
55. Jin, L.; Godri Pollitt, K.J.; Liew, Z.; Rosen Vollmar, A.K.; Vasiliou, V.; Johnson, C.H.; Zhang, Y. Use of Untargeted Metabolomics to Explore the Air Pollution-Related Disease Continuum. *Curr. Environ. Health Rep.* **2021**, *8*, 7–22. [CrossRef]
56. Ancona, C.; Bauleo, L.; Biscotti, G.; Bocca, B.; Caimi, S.; Cruciani, F.; Di Lorenzo, S.; Petrolati, M.; Pino, A.; Piras, G.; et al. A survey on lifestyle and level of biomarkers of environmental exposure in residents in Civitavecchia (Italy). *Ann. Ist. Super. Sanita* **2016**, *52*, 488–494. [CrossRef] [PubMed]
57. EU Directive. EU—Directive 2002/73/EC of the European Parliament and of the Council. *Off. J. Eur. Communities* **2002**, *269*, 15–20.
58. ICOH. *International Code of Ethics for Occupational Health Professionals*; International Commission on Occupational Health: Rome, Italy, 2014.
59. Kroll, M.H.; Chesler, R.; Hagengruber, C.; Blank, D.W.; Kestner, J.; Rawe, M. Automated determination of urinary creatinine without sample dilution: Theory and practice. *Clin. Chem.* **1986**, *32*, 446–452. [CrossRef] [PubMed]
60. Raponi, F.; Bauleo, L.; Ancona, C.; Forastiere, F.; Paci, E.; Pigini, D.; Tranfo, G. Quantification of 1-hydroxypyrene, 1- and 2-hydroxynaphthalene, 3-hydroxybenzo[a]pyrene and 6-hydroxynitropyrene by HPLC-MS/MS in human urine as exposure biomarkers for environmental and occupational surveys. *Biomarkers* **2017**, *22*, 575–583. [CrossRef] [PubMed]
61. Andreoli, R.; Manini, P.; De Palma, G.; Alinovi, R.; Goldoni, M.; Niessen, W.M.A.; Mutti, A. Quantitative determination of urinary 8-oxo-7,8-dihydro-2′- deoxyguanosine, 8-oxo-7,8-dihydroguanine, 8-oxo-7,8-dihydroguanosine, and their non-oxidized forms: Daily concentration profile in healthy volunteers. *Biomarkers* **2010**, *15*, 221–231. [CrossRef] [PubMed]
62. Protano, C.; Canepari, S.; Astolfi, M.L.; D'Onorio De Meo, S.; Vitali, M. Urinary reference ranges and exposure profile for lithium among an Italian paediatric population. *Sci. Total Environ.* **2018**, *619–620*, 58–64. [CrossRef] [PubMed]
63. Astolfi, M.L.; Vitali, M.; Marconi, E.; Martellucci, S.; Mattei, V.; Canepari, S.; Protano, C. Urinary Mercury Levels and Predictors of Exposure among a Group of Italian Children. *Int. J. Environ. Res. Public Health* **2020**, *17*, 9225. [CrossRef]
64. Astolfi, M.L.; Protano, C.; Marconi, E.; Massimi, L.; Brunori, M.; Piamonti, D.; Migliara, G.; Vitali, M.; Canepari, S. A new rapid treatment of human hair for elemental determination by inductively coupled mass spectrometry. *Anal. Methods* **2020**, *12*, 1906–1918. [CrossRef]
65. ANAS. *Bollettini Mensili dell'Osservatorio del Traffico*; ANAS: Rome, Italy; Available online: https://www.stradeanas.it/sites/default/files/SMI4-SintesiOsservatorioMarzo2020.pdf (accessed on 18 May 2022).
66. ANAS. *Bollettini Mensili dell'Osservatorio del Traffico/Archivio Osservatorio del Traffico*; ANAS: Rome, Italy; Available online: https://www.stradeanas.it/sites/default/files/SMI4-SintesiOsservatorioMarzo2021.pdf (accessed on 18 May 2022).
67. ARPA Lazio. *Bollettini Qualità dell'Aria*; ARPA Lazio: Rieti, Italy; Available online: http://www.arpalazio.net/main/aria/sci/basedati/bollettini/bs.php?year=2020 (accessed on 18 May 2022).
68. ARPA Lazio. *Bollettini Qualità dell'Aria*; ARPA Lazio: Rieti, Italy; Available online: http://www.arpalazio.net/main/aria/sci/basedati/bollettini/bs.php?year=2019 (accessed on 18 May 2022).
69. ARPA Lazio. *Bollettini Qualità dell'Aria*; ARPA Lazio: Rieti, Italy; Available online: http://www.arpalazio.net/main/aria/sci/basedati/bollettini/bs.php?year=2021 (accessed on 18 May 2022).
70. Hjortenkrans, D.; Bergbäck, B.O.; Bergbäck, B.; Aggerud, A.H. New Metal Emission Patterns in Road Traffic Environments. *Environ. Monit. Assess.* **2006**, *117*, 85–98. [CrossRef]
71. Thorpe, A.; Harrison, R.M. Sources and properties of non-exhaust particulate matter from road traffic: A review. *Sci. Total Environ.* **2008**, *400*, 270–282. [CrossRef]

72. Campbell, W.W.; Anderson, R.A. Effects of Aerobic Exercise and Training on the Trace Minerals Chromium, Zinc and Copper. *Sports Med.* **1987**, *4*, 9–18. [CrossRef]
73. Kovacs, L.; Zamboni, C.B.; Nunes, L.A.S.; Lourenço, T.F.; Macedo, D.V. Concentrations of ions and metals in blood of amateur and elite runners using NAA. *J. Radioanal. Nucl. Chem.* **2013**, *3*, 393–398. [CrossRef]
74. Genuis, S.J.; Birkholz, D.; Rodushkin, I.; Beesoon, S. Blood, urine, and sweat (BUS) study: Monitoring and elimination of bioaccumulated toxic elements. *Arch. Environ. Contam. Toxicol.* **2011**, *61*, 344–357. [CrossRef]
75. *Dietary Reference Intakes for Vitamin A, Vitamin K, Arsenic, Boron, Chromium, Copper, Iodine, Iron, Manganese, Molybdenum, Nickel, Silicon, Vanadium, and Zinc*; The National Academic Press: Washington, DC, USA, 2001. [CrossRef]
76. Barceloux, D.G.; Barceloux, D.D. Cobalt. *J. Toxicol. Clin. Toxicol.* **1999**, *37*, 201–216. [CrossRef] [PubMed]
77. Varela-Moreiras, G.; Murphy, M.M.; Scott, J.M. Cobalamin, folic acid, and homocysteine. *Nutr. Rev.* **2009**, *67*, S69–S72. [CrossRef] [PubMed]
78. *Toxicological Profile for Cobalt*; U.S. Department of Health and Human Services: Washington, DC, USA, 2004.
79. Expert Group on Vitamins and Minerals. *Safe Upper Levels for Vitamins and Minerals Expert Group on Vitamins and Minerals Contents*; Food Standards Agency: London, UK, 2003.
80. Mills, E.; O'Neill, L.A.J. Succinate: A metabolic signal in inflammation. *Trends Cell Biol.* **2014**, *24*, 313–320. [CrossRef] [PubMed]
81. Benetti, E.; Liberto, E.; Bressanello, D.; Bordano, V.; Rosa, A.C.; Miglio, G.; Haxhi, J.; Pugliese, G.; Balducci, S.; Cordero, C. Sedentariness and urinary metabolite profile in type 2 diabetic patients, a cross-sectional study. *Metabolites* **2020**, *10*, 205. [CrossRef]

toxics

Article

Risk Assessment of Dietary Exposure to Organophosphorus Flame Retardants in Children by Using HBM-Data

Veronika Plichta [1,*], Johann Steinwider [1], Nina Vogel [2], Till Weber [2], Marike Kolossa-Gehring [2]Lubica Palkovičová Murínová [3], Soňa Wimmerová [3], Janja Snoj Tratnik [4], Milena Horvat [4], Gudrun Koppen [5], Eva Govarts [5], Liese Gilles [5], Laura Rodriguez Martin [5], Greet Schoeters [5,6], Adrian Covaci [7], Clémence Fillol [8], Loïc Rambaud [8], Tina Kold Jensen [9] and Elke Rauscher-Gabernig [1]

1 Austrian Agency for Health and Food Safety(AGES), Division of Integrative Risk Assessment, Data & Statistics, Department of Risk Assessment, 1220 Vienna, Austria; johann.steinwider@ages.at (J.S.); elke.rauscher-gabernig@ages.at (E.R.-G.)
2 German Environment Agency (UBA), 06844 Dessau-Roßlau, Germany; nina.vogel@uba.de (N.V.); till.weber@uba.de (T.W.); marike.kolossa@uba.de (M.K.-G.)
3 Faculty of Public Health, Slovak Medical University, 833 03 Bratislava, Slovakia; lubica.murinova@szu.sk (L.P.M.); sona.wimmerova@szu.sk (S.W.)
4 Department of Environmental Sciences, Jožef Stefan Institute, 1000 Ljubljana, Slovenia; janja.tratnik@ijs.si (J.S.T.); milena.horvat@ijs.si (M.H.)
5 VITO Health, Flemish Institute for Technological Research (VITO), 2400 Mol, Belgium; gudrun.koppen@vito.be (G.K.); eva.govarts@vito.be (E.G.); liese.gilles@vito.be (L.G.); laura.rodriguezmartin@vito.be (L.R.M.); greet.schoeters@vito.be (G.S.)
6 Department of Biomedical Sciences, University of Antwerp, 2610 Wilrijk, Belgium
7 Toxicological Center, University of Antwerp, 2610 Wilrijk, Belgium; adrian.covaci@uantwerpen.be
8 Santé Publique France, French Public Health Agency (ANSP), 94415 Saint-Maurice, France; clemence.fillol@santepubliquefrance.fr (C.F.); loic.rambaud@santepubliquefrance.fr (L.R.)
9 Department of Environmental Medicine, Institute of Public Health, University of Southern Denmark, 5000 Odense, Denmark; tkjensen@health.sdu.dk
* Correspondence: veronika.plichta@ages.at

Citation: Plichta, V.; Steinwider, J.; Vogel, N.; Weber, T.; Kolossa-Gehring, M.; Murínová, L.P.; Wimmerová, S.; Tratnik, J.S.; Horvat, M.; Koppen, G.; et al. Risk Assessment of Dietary Exposure to Organophosphorus Flame Retardants in Children by Using HBM-Data. *Toxics* **2022**, *10*, 234. https://doi.org/10.3390/toxics10050234

Academic Editor: Sunmi Kim

Received: 31 March 2022
Accepted: 27 April 2022
Published: 3 May 2022

Publisher's Note: MDPI stays neutral with regard to jurisdictional claims in published maps and institutional affiliations.

Copyright: © 2022 by the authors. Licensee MDPI, Basel, Switzerland. This article is an open access article distributed under the terms and conditions of the Creative Commons Attribution (CC BY) license (https://creativecommons.org/licenses/by/4.0/).

Abstract: Due to their extensive usage, organophosphorus flame retardants (OPFRs) have been detected in humans and in the environment. Human are exposed to OPFRs via inhalation of indoor air, dust uptake or dietary uptake through contaminated food and drinking water. Only recently, few studies addressing dietary exposure to OPFRs were published. In this study, we used human biomonitoring (HBM) data of OPFRs to estimate how much the dietary intake may contribute to the total exposure. We estimated by reverse dosimetry, the daily intake of tris (2-chloroethyl) phosphate (TCEP), tris (1-chloro-2-propyl) phosphate (TCIPP), tris (1,3-dichloro-2-propyl) phosphate (TDCIPP) for children using HBM data from studies with sampling sites in Belgium, Denmark, France, Germany, Slovenia and Slovakia. For estimating the dietary exposure, a deterministic approach was chosen. The occurrence data of selected food categories were used from a published Belgium food basket study. Since the occurrence data were left-censored, the Lower bound (LB)—Upper bound (UB) approach was used. The estimated daily intake (EDI) calculated on the basis of urine metabolite concentrations ranged from 0.03 to 0.18 μg/kg bw/d for TDCIPP, from 0.05 to 0.17 μg/kg bw/d for TCIPP and from 0.02 to 0.2 μg/kg bw/d for TCEP. Based on national food consumption data and occurrence data, the estimated dietary intake for TDCIPP ranged from 0.005 to 0.09 μg/kg bw/d, for TCIPP ranged from 0.037 to 0.2 μg/kg bw/d and for TCEP ranged from 0.007 to 0.018 μg/kg bw/d (summarized for all countries). The estimated dietary intake of TDCIPP contributes 11–173% to the EDI, depending on country and LB-UB scenario. The estimated dietary uptake of TCIPP was in all calculations, except in Belgium and France, above 100%. In the case of TCEP, it is assumed that the dietary intake ranges from 6 to 57%. The EDI and the estimated dietary intake contribute less than 3% to the reference dose (RfD). Therefore, the estimated exposure to OPFRs indicates a minimal health risk based on the current knowledge of available exposure, kinetic and toxicity data. We were able to show that the dietary exposure can have an impact on the general exposure based on our underlying exposure scenarios.

Keywords: HBM4EU; organophosphorus flame retardants; dietary exposure; children

1. Introduction

Flame retardants are chemicals, which are added to materials to prevent and to delay the fire hazards. Due to their persistence and accumulation potency and because of toxicological concerns, brominated flame retardants e.g., polybrominated diphenyl ethers (PBDEs), such as penta-, octa-, and deca-BDE, and hexabromocyclododecanes (HBCDD) are regulated under the Stockholm Convention on Persistent Organic Pollutants. This has led to a surge in the usage of alternative flame retardants with similar technical characteristics, such as the organophosphorus flame retardants (OPFRs). OPFRs are not chemically bound to the product and can be released into the surroundings by leaching, evaporation, migration and abrasion from the polymer matrix during usage. Due to the extensive usage, OPFRs have been detected in humans and the environment. Humans are exposed to OPFRs via inhalative uptake especially in the indoor environment, and orally by uptake of dust or from the diet through contaminated food and drinking water [1–3]. In the human, OPFRs are rapidly absorbed and excreted via urine, the half-lives of OPFRs was assumed to be short (<1 day), but recent findings indicate a half-life of 3–53 days depending on the OPFRs [4]. Exposure to OPFRs may cause various adverse effects to the human body such as kidney toxicity, neurotoxicity, reproductive toxicity, carcinogenicity and endocrine disruption [5]. Within the European Union, some OPFRs are classified and labeled as Carc. 2 and, therefore, pose a health hazard [6]. In this study we focused on tris (2-chloroethyl) phosphate (TCEP), tris (1-chloro-2-propyl) phosphate (TCIPP) and tris (1,3-dichloro-2-propyl) phosphate (TDCIPP). TCIPP represents approximately 80% of the chlorinated OPFRs in Europe and is considered to be potentially carcinogenic. TCEP was replaced by TCIPP because of its lower toxicity. Studies indicate that TDCIPP might be more neurotoxic than TCEP and TCIPP [3]. Food can be contaminated with OPFRs by two routes. Firstly, crops and animal-derived and aquatic products can be contaminated with OPFRs via soil and water. Secondly, foodstuffs can be contaminated by OPFRs during the production process, packaging and storage, since OPFRs are also present in several materials used in food processing [6,7]. Human biomonitoring (HBM) studies, which address the exposure to OPFRs, had their focus mainly on the indoor environment as an exposure source, since OPFRs are detected in high amounts in indoor air and dust. Only recently, few studies focused on the dietary exposure to OPFRs [2,8–10] and concluded that the dietary exposure should be considered in future exposure assessments. Therefore, the aim of this risk assessment was to estimate the dietary exposure of TDCIPP, TCIPP, TCEP and their individual contribution to the general exposure. The latter was derived by using HBM data.

2. Materials and Methods

2.1. Human Biomonitoring Data

The used human biomonitoring data are part of the HBM4EU Aligned studies which have been described in detail by Gilles et al. [11]. Biobanked urine samples of children (6–12 years) from Belgium (3xG study), Denmark (OCC study), France (Esteban study), Germany (GerES V sub sample), Slovenia (SLO CRP study) and Slovakia (PCB cohort study) were collected and analysed for its bis (2-chloroethyl) phosphate (BCEP), bis (1-chloro-2-propyl) phosphate (BCIPP) and bis (1,3-dichloro-2-propyl) phosphate (BDCIPP) concentration to determine the exposure to TCEP, TCIPP and TDCIPP. A detailed description of the used HBM studies is given in the Supplementary Materials Table S1. In this risk assessment, when referring to a country, the HBM-study listed above is meant. However, the HBM- study does not always reflect the general exposure of the whole population of the country, since some of the studies were not population representative or only performed on a regional level.

Estimation of the Daily Intake (EDI)

The HBM data which were used for calculating the daily intake were not stratified and were volume based [μg/L] and non-creatinine adjusted. The estimated daily intake (*EDI*) from the urinary excretion of the OPFR was estimated by using the following equation:

$$EDI = \left(\frac{C_{Urine} * UV_{excr}}{Fue}\right) \times \left(\frac{MW_d}{MW_m}\right)$$

where C_{Urine} is the concentration of the metabolite in μg/L and UV_{excr} is the daily excreted urinary volume of 22.2 mL/kg bw/d for children [12]. F_{ue} is the molar fraction of the urine excreted metabolite with respect to its parent compound, and MW_d and MW_m are the molecular weights of the parent compound and its metabolite. The following molecular weights were used [g/mol]: TCEP (285.49), BCEP (222.99), TCIPP (327.55), BCIPP (251.04), TDCIPP (430.9) and BDCIPP (319.93). For estimating the daily intake the arithmetic mean and the 95th Percentile (P95) of the urinary metabolite concentration were considered. The data on kinetics of the metabolism of OPFRs are limited, therefore the F_{ue} of 0.63 for TDCIPP [13] was also used for TCIPP and TCEP. This equation is often used for substances with similar half–lives as e.g., for plasticizers [14].

2.2. Dietary Exposure

A deterministic approach was used to estimate the dietary intake of Belgian, Danish, French, German, Slovenian and Slovakian children. Since the toxicokinetic data are limited, it was assumed that 100% of the detected OPFRs concentration in the contaminated food is absorbed through the dietary intake. The provided occurrence data (see Table 1) originated from a Belgium food basket study published by Poma et al. [7], where samples were drawn from supermarkets, discount retailers and specialized meat and fish stores in two Belgian cities Brussels and Antwerp and in one village (Dessel) from spring 2015 to autumn 2016. In total, 165 samples belonging to 14 food categories were collected and analysed on its TDCIPP, TCIPP and TCEP concentration. The food category "animal and vegetable fats and oils and primary derivatives thereof" also included fish oil food supplement, which was the highest contaminated sample. This sample was excluded, since it is not regularly consumed by children and could distort the outcome. The edited occurrence data in regard to Lower Bound (LB), Medium Bound (MB) and Upper Bound (UB) were provided by Giulia Poma (2021; personal communication) and were not modified in this risk assessment. Since, the occurrence data were left-censored, the LB-, MB- and UB approach was used [15]. Occurrence data below the limit of quantification (LOQ) were set equal to zero (LB), to the half of the LOQ (MB) and equal to the LOQ (UB).

Table 1. Occurrence data of TDCIPP, TCIPP and TCEP concentration in the analysed food categories in μg/kg provided by Poma et al. (2018).

Food Category	TDCIPP				TCIPP				TCEP			
	df%	LB	MB	UB	df%	LB	MB	UB	df%	LB	MB	UB
Animal and vegetable fats and oils and primary derivatives thereof	0	5.7 *	16.1	34.7	0	11 *	31	66.8	11	1.55 *	2.5	3.6
Whole Eggs	0	0	0.05	0.1	25	0.1	0.15	0.19	25	0.02	0.03	0.04
Fish (meat)	39	0.4	0.5	0.6	59	0.71	0.76	0.8	54	0.12	0.13	0.14
Crustaceans	0	0	0.22	0.44	20	0.18	0.22	0.26	0	0	0.04	0.07
Grains and grain-based products	0	0	0.16	0.31	71	3.58	3.65	3.73	57	0.6	0.61	0.63
Meat and meat products	0	0	0.8	1.61	26	0.19	0.29	0.39	34	0.15	0.2	0.25
Milk	0	0	0.21	0.43	0	0	0.8	1.59	0	0	0.23	0.45
Cheese	44	2.52	3.09	3.65	50	1.42	1.53	1.63	50	0.66	0.71	0.77

df: detection frequency [%], LB: Lower bound, MB: Medium bound, UB: Upper bound, * the LB data were provided by G. Poma (personal communication).

The used food consumption data, which were surveyed by food recording are given in Table 2. In general, the mean occurrence data were multiplied with the available mean food consumption data for children provided by the EFSA Comprehensive European Food Consumption Database [16]. This was conducted for the overall population (=all subjects) as well as for those who actually consumed the relevant food (=consumers) as reported in the respective national food consumption survey. Subsequently, the food category with the highest exposure of consumer was determined. For this food category, the P95 of the consumers and for the other food categories the mean of the overall population was used in order to calculate the exposure of high consumers. No food consumption data for Slovenian and Slovakian children were available, therefore the Austrian data were used, because these were the closest geographically data available.

Table 2. Food consumption data presented by country and food groups [g/kg bw/d] (EFSA, 2021).

Food Category	Consumption Data Germany [1]		Consumption Data Austria [2,*]		Consumption Data Belgium [3]		Consumption Data Denmark [4]		Consumption Data France [5]	
	Total Mean Population	High Level Consumer Only	Total Mean Population	High Level Consumer Only	Total Mean Population	High Level Consumer Only	Total Mean Population	High Level Consumer Only	Total Mean Population	High Level Consumer Only
Animal and vegetable fats and oils and primary derivatives thereof	0.58	1.34	0.92	1.98	0.48	1.29	1.14	2.2	0.62	1.5
Whole Eggs	0.67	2.15	0.47	1.59	0.01	1.47	0.5	1.44	0.05	0.89
Fish (meat)	0.21	2.59	0.38	2.62	0.35	4.36	0.4	1.52	0.54	4.01
Crustaceans	0.0004	1	0.01	1.54	0.05	2.52	0.03	0.43	0.04	1.82
Grains and grain-based products	8.16	14.03	9.4	15.76	7.31	12.65	7.72	11.85	8.04	13.97
Meat and meat products	2.75	6.01	2.86	6.67	4.14	9.17	3.83	6.6	4.38	9.14
Milk	8.46	20.98	6.68	16.54	9.16	27.11	17.26	35.65	10	26.69
Cheese	0.74	2.28	0.89	2.25	1.09	4.37	0.73	1.88	2.05	6.07

[1] national food consumption survey from 2006 [17], [2] national food consumption survey from 2010, * used for Slovenia and Slovakia [18] [3] national food consumption survey from 2014 [19] [4] national food consumption survey from 2005 [20] and [5] national food consumption survey from 2014 [21].

2.3. Hazard and Risk Characterization

As no HBM4EU HBM-GV, neither a HBM-I nor a HBM-II value from the German Human biomonitoring Commission nor a biomonitoring equivalent (BE) were available, a health-based guidance value (HBGV) was derived on the basis of data for a toxicologically acceptable uptake. For each study population and exposure scenario (LB—UB), the estimated daily intake (EDI) and dietary intake were set in relation to the available HBGV which was derived from external uptake data. In the European Union Risk Assessment Report of TCIPP, a lowest observed adverse effect level (LOAEL) of 52 mg/kg bw/d was derived from a 13-week rat study. The LOAEL was based on the increased liver weight observed in male rats. A minimal margin of safety (MOS) for repeated dose toxicity of 50 was considered sufficient [22]. Furthermore, a reference dose (RfD) of 80 µg/kg bw/d was derived by Ali et al. [23], where the sensitive endpoint is unknown.

The Agency for Toxic Substances and Disease Registry (ATSDR) derived a minimal risk level (MRL) for TCEP of 0.2 mg/kg bw/d for chronic exposure (>365 days) based on renal tubule lesions in female rats [24]. The US-EPA derived a chronic provisional reference dose (p-RfD) of 0.007 mg/kg bw/d, based on an increased relative kidney weight ($BMDL_{1SD}$ of 6.9 mg/kg bw/d) [25]. Another RfD of 22 µg/kg bw/d was derived by Ali et al. [23], where the sensitive endpoint is increased relative liver and kidney weights in female rats, as was reported by WHO [26]. For TDCIPP, an MRL of 0.2 mg/kg bw/d was derived for chronic exposure (>365 days) based on renal tubule lesions in male rats [24]. An RfD of 15 µg/kg bw/d was derived by Ali et al. [23], where the sensitive endpoint is increased liver weight in female mice, as was reported by WHO [26]. Additionally, a cumulative risk assessment by the use of the hazard index (HI) was performed for the urinary EDI and the estimated dietary exposure. The HI is calculated by summing up the hazard quotient (HQ) of the individual substances, whereby the HQ is the ratio of the HBGV and the estimated exposure of the respective substances. If the calculation yields a HI < 1, the combined risk is considered acceptable [27].

3. Results

3.1. Estimated Daily Intake (EDI)

The estimated daily intake (EDI) summarized for all countries (mean and high intake) ranged from 0.03 to 0.18 µg/kg bw/d for TDCIPP, from 0.05 to 0.17 µg/kg bw/d for TCIPP and from 0.02 to 0.2 µg/kg bw/d for TCEP. In Table 3, the urinary metabolites and the EDIs of their parent compounds are given in detail. BDCIPP had the highest detection frequency (df) across the studies (65–97.7%) except for Slovakia with 37% df. The df of BCIPP was below 55% across all studies, therefore only the P95 could be used. For BCEP the df ranged between 19 and 63.3%, so only for the German study was an arithmetic mean available and for Slovenia and Slovakia only the P95.

Table 3. Urinary BDCIPP, BCIPP and BCEP concentration levels and the estimated daily intakes (EDI) of their parent compounds TDCIPP, TCIPP and TCEP.

Country	OPFR-Metabolite	n	Df [%]	Urinary Concentration [µg/L]		EDI of Parent Compound [µg/kg bw/d]	
				Mean	High	Mean	High
Belgium (3xG study)	BDCIPP	133	98	1.03	3.08	0.05	0.14
	BCIPP	133	13.5	-	1.4	-	0.063
Denmark (OCC study)	BDCIPP	291	97	0.72	2.8	0.03	0.13
	BCIPP	291	6.5	-	0.48		0.022
France (Esteban study)	BDCIPP	299	65	1.34	3.82	0.06	0.18
	BCIPP	299	31	-	3.71		0.17
Germany (GerES V study)	BDCIPP	300	80	1.03	3.09	0.05	0.15
	BCIPP	300	53	0.11 [1]	0.74	0.05	0.034
	BCEP	300	63	0.48	0.96	0.022	0.043
Slovenia (SLO CRP study)	BDCIPP	147	84	0.86	2.32	0.04	0.11
	BCIPP	147	18	-	0.6	-	0.027
	BCEP	147	20	-	4.79	-	0.214
Slovakia (PCB cohort study)	BDCIPP	300	17	-	1.46	-	0.07
	BCIPP	300	29	-	1.71	-	0.078
	BCEP	300	20	-	3.81	-	0.17

df: detection frequency in percent, arithmetic mean and high (P95), [1] the median has been used, since no arithmetic mean was available.

3.2. Dietary Exposure

TDCIPP was only detected in fish and cheese, whereas TCIPP and TCEP were present in almost all food groups [7].

The estimated dietary intake of TDCIPP ranged from 0.006 to 0.06 µg/kg bw/d (Belgium), 0.008 to 0.095 µg/kg bw/d (Denmark), 0.009 to 0.074 µg/kg bw/d (France), 0.005 to 0.06 µg/kg bw/d (Germany) and 0.008 to 0.08 µg/kg bw/d (Slovenia/Slovakia). For TCIPP the estimated dietary intake was 0.03 to 0.13 µg/kg bw/d for Belgium, 0.04 to 0.2 µg/kg bw/d for Denmark, 0.04 to 0.15 µg/kg bw/d for France, 0.04 to 0.14 µg/kg bw/d for Germany and 0.05 and 0.18 µg/kg bw/d for Slovenia/Slovakia. In regards to TCEP the dietary intake ranged from 0.008 and 0.018 µg/kg bw/d and 0.007 to 0.016 µg/kg bw/d for Slovenia/Slovakia and Germany respectively. As shown in Figure 1, food categories "animal and vegetable fats and oils and primary derivatives thereof", "grains and grain-based products" "cheese" and "milk" were the main contributors to the dietary exposure. A detailed overview of the dietary intake is given in Supplementary Materials Tables S2–S13.

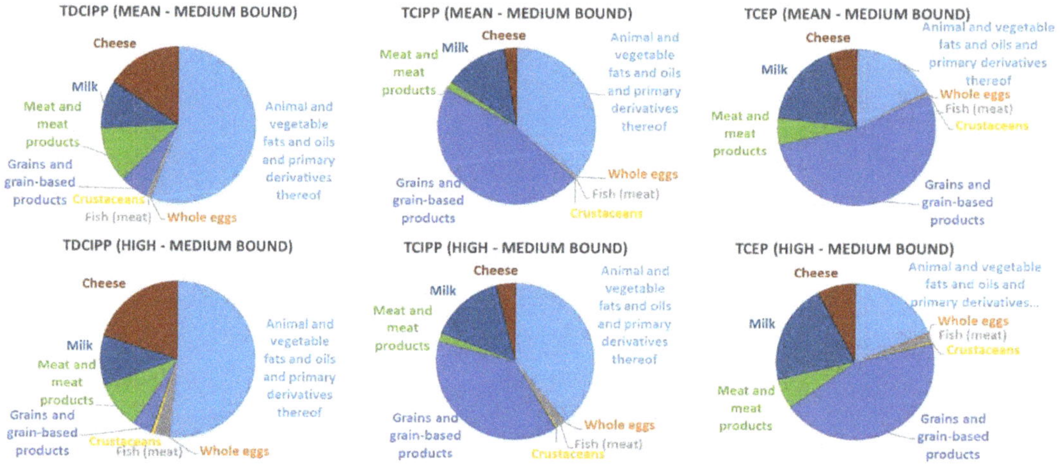

Figure 1. The contribution of the analysed food categories to the mean and high dietary exposure—an exemplary presentation of the medium bound.

3.3. *Risk Characterization*

Figure 2 shows the dietary intake and its contribution to the EDI. In the calculated worst-case scenario (upper bound, high exposure), the estimated dietary intake of TDCIPP contributes 43% (Belgium), 72% (Denmark), 41% (France), 41% (Germany), 76% (Slovenia) and 120% (Slovakia). In the best-case scenario (lower bound, mean/high exposure) the estimated dietary intake contributes, 12% (Belgium), 25% (Denmark), 14% (France), 11% (Germany), 19% (Slovenia) and 20% (Slovakia, high exposure) to the EDI. The estimated dietary intake of TCIPP was in all scenarios above 100%, except for the Belgian (83%, scenario lower bound, high exposure) and the French (90%, scenario upper bound, high exposure) study populations. In the case of TCEP, it was estimated that the dietary intake ranges from 6 to 57% for the German, Slovenian and Slovakian study populations respectively.

Neither the EDI of TDCIPP and TCEP nor the estimated dietary exposure exceeded the RfD of 15 µg/kg bw/d or the p-RfD of 0.007 mg/kg bw/d. Its maximum contribution to the RfD is below 1.2% (TDCIPP) and 3% (TCEP). The EDI and dietary intake for TCIPP contributes only 0.01–0.21% to the RfD of 80 µg/kg bw/d. In the European Union Risk Assessment, a MOS approach for TCIPP was used, which we also considered. The estimated intake and the calculated dietary exposure were above the MOS of 50 (range 307,619–10,375,137; data not shown). The estimated HI is below 0.1. The EDI and dietary exposure of TDCIPP, TCIPP and TCEP are not likely to cause adverse health effects based on current knowledge from available exposure, kinetic and toxicity data. A detailed overview is given in Supplementary Materials Table S14.

In Table 4, the Risk Characterisation Ratio (RCR) was estimated with the used HBM data (P95) and the different published HBGVs. Although, the RCR is well below 1 and indicates no adverse health effects, the values differ depending on which HBGV was chosen. The available HBGVs ranged from 15–200 µg/kg bw/d for TDCIPP, from 80 to 52,000 µg/kg bw/d (LOAEL used for MOS) for TCIPP and from 7–200 µg/kg bw/d for TCEP. Anyways, it should be kept in mind, that in this risk assessment different HBGVs for one OPFR also might have different sensitive endpoints as it is the case for TDCIPP e.g., renal tubule lesions in male rats vs. increased liver weight in female mice. However, if more than one HBGV value is available, the lowest HBGV is normally chosen to conduct an evidence-based single substance risk assessment.

Figure 2. The contribution of the dietary exposure to the estimated daily intake in µg/kg bw/d. marked as * means the dietary intake exceeded the estimated daily intake, therefore, only the dietary intake is shown.

Table 4. Risk Characterisation Ratio (RCR) of the used HBM data (P95) with the different HBGVs and the estimation of the Margin of Safety.

	TDCIPP		TCIPP		TCEP		
	MRL	RfD	RfD	MOS *	MRL	p-RfD	RfD
	(200 µg/kg bw/d)	(15 µg/kg bw/d)	(80 µg/kg bw/d)	(52 mg/kg bw/d)	(200 µg/kg bw/d)	(7 µg/kg bw/d)	(22 µg/kg bw/d)
Belgium	0.001	0.01	0.001	823,423	-	-	-
Denmark	0.001	0.01	0.0003	2,360,912	-	-	-
France	0.0009	0.01	0.002	307,619	-	-	-
Germany	0.0007	0.01	0.0004	1,538,093	0.0002	0.006	0.002
Slovenia	0.00054	0.0073	0.0003	1,908,470	0.001	0.03	0.01
Slovakia	0.0003	0.004	0.001	667,016	0.001	0.024	0.008

* Margin of Safety (MOS) is the ratio of NOAEL (or LOAEL) obtained from animal toxicological studies to the predicted or estimated human exposure. In this case a MOS over 50 is considered to be protective.

4. Discussion

4.1. Estimated Daily Intake

Only in the German study, BDCIPP, BCIPP and BCEP were detected frequently to estimate an average and high daily intake. The detected levels are in line with a previously published HBM study from Germany [8]. For the other HBM studies, no representative literature data were available to compare the detected levels in children within an acceptable sampling period. Anyways, it should also be considered, that only in the German, Slovenian, Belgian and French study population was the first morning urine sampled, this urine is normally more concentrated than spot urine. In a study conducted by Bastiaensen et al. [28], it was shown that the detection rate in a 24-h pooled urine sample is higher compared to spot urine samples. Furthermore, it was also shown that for certain OPFR metabolites such as BDCIPP, BCIPHIPP and DPHP, the first morning urine samples have the same representativeness as 24-h pooled samples. The authors concluded that the first morning urine samples might be a sensible alternative for biomonitoring studies to reduce exposure misclassification if the collection of several spot urine samples is not an option. The used metabolite might also have an impact on the EDI. The provided metabolite BDCIPP is known as an appropriate biomarker for TDCIPP since no other OPFR is metabolized to BDCIPP [29]. In the case of TCIPP, the provided metabolite BCIPP may not be the most representative biomarker. Several studies have reported that bis (1-chloro-2-propyl) 1-hydroxyl-2-propyl phosphate (BCIPHIPP) has a higher detection frequency than BCIPP and might be a more suitable biomarker to reflect the TCIPP body burden. Besides the biomarker BCEP, TCEP should also be considered as a biomarker, because of its resistance to be metabolized [30].

4.2. Dietary Intake

The estimated dietary intake of the sum of OPFRs ranged from 0.04 and 0.3 µg/kg bw/d (or 40–300 ng/kg bw/d), which is slightly lower compared to previous published data from Xu et al. [10]. They estimated a median dietary intake of 87 ng/kg bw/d and a high intake of 340 ng/kg bw/d of ∑OPFR (EHDPHP, TCEP, TPHP and TCPP) in a duplicate diet study. For the adult population, Poma et al. [2,7] estimated a mean dietary intake of 31 ng/kg bw/d (medium bound, Belgium food basket study) and mean dietary intake of 26.4 ng/kg bw/d (medium bound, Swedish food basket study). In all the above mentioned studies, EHDPHP was identified as the main contributor to the OPFR exposure. In this exposure assessment, the mean dietary intake (medium bound) was between 70 and 100 ng/kg bw/d for children.

The dietary exposure of TDCIPP contributes depending on the LB-MB-UB scenario between 7–173% to the EDI. In the case of TCEP, it is assumed that the dietary intake contributes 6–37% to the EDI. Except for France and the Belgian best case scenario (LB and high exposure), the estimated dietary exposure of TCIPP was in all scenarios above 100% to the EDI. In our study, the estimated total TCIPP intake in the German study was around 5 ng/kg bw/d. It was assumed that the total concentration in food was around 17.2 to 75.4 ng/g wet weight (ww; LB to UB), which might be an overestimation because it is unlikely that children in the used HBM studies consumed all the presented food categories in the time when the urine samples were collected. In a study performed by Xu et al. [10], they estimated that the total TCIPP exposure (inhalation, dust ingestion, dermal absorption and dietary intake) is around 8.6 ng/kg bw/d, which is comparable to our estimation. In their duplicate diet study, only 0.39 ng/g ww TCIPP was measured in the consumed food. In that study, air, dust and hand wipes were also analysed and based on the low concentration in food, they concluded the main exposure source was inhalation of personal ambient air [10,31]. Moreover, the provided biomarker might not reflect a realistic TCIPP exposure. This can also explain the higher estimated dietary exposure compared to the EDI.

The occurrence data were left-censored and for e.g., TDCIPP the detection frequency was quite low, it was only frequently detected in fish (df 39%) and cheese (df 44%). TCIPP and TCEP were more frequently detected in the range of 25–71% and 11–57% across the analysed food categories. In the LB-MB-UB approach, the food category "animal fats and vegetable fats and oils and primary derivatives of" contributes most to the total OPFRs exposure, which might not reflect a realistic exposure scenario because it was not detected at all. So it can either be not present in the samples or it was not detectable because of their high limit of quantification (LOQ) of 32.17 ng/g ww (TDCIPP) and of 61.97 ng/g ww (TCIPP) [7]. Based on the detection frequency of OPFRs in certain food categories, we conclude that a realistic exposure scenario occurs rather in the LB to MB range than in the UB. Despite the uncertainties by using aggregated occurrence data from a different country our estimations for the dietary intake in children are slightly higher than the data for adults in the literature. This is in line with the general observations that children have a higher uptake per kg body weight than adults due to their lower body weight.

Most of the published OPFRs human biomonitoring studies focused on the indoor exposure. The concentration in house dust is approximately three magnitudes higher than its occurrence in food. In contrast to this, a child ingests approximately 1.2 kg of food per day compared to 30 mg of dust [32]. Gbadamosi et al. [33] reviewed that for adults, food ingestion contributes 75%, 41% and 36% to the total TCEP, TDCIPP and TCIPP exposure. It can be assumed that the dietary intake might be equal or even a greater contributor to the total human exposure to OPFRs [6,7]. Similar to the few previous studies, we were able to show that the dietary exposure of OPFR can have an impact on the general exposure.

4.3. Uncertainties and Limitations

The following sources of uncertainties have been considered and are summarised in Table 5. Since we used the urinary molar excretion fraction (F_{ue}) of TDCIPP also for TCIPP and TCEP, we are not able to determine if the EDIs are over- or underestimated. Further, we used non-creatinine adjusted urinary HMB-data, since we had no detailed information on the anthropometric parameters. Some studies indicated that creatinine adjustment is not the appropriate approach to address the urine dilution for some OPFR metabolites, since they can be conjugated in the liver as glucuronides or sulfates and be actively excreted by the renal tubes [28]. It shows that adjusting it to the specific gravity, the ratio between the density of the urine and pure water, is more accurate to use. Nevertheless, in this risk assessment we did not use adjusted urinary values, and do not believe that it has a high impact on the outcome but it is still an uncertainty that can cause an over- or underestimation of the EDI. The provided biomarker BCIPP might not reflect the TCIPP exposure appropriately because of its low detection frequency. Since we have no information on the urinary BCIPHIPP concentration, an underestimation of the TCIPP exposure is likely.

Table 5. Qualitative evaluation of influence of uncertainties on the dietary exposure estimate and the estimated daily intake.

Sources of Uncertainty	Direction
Dietary exposure estimates	
The use of aggregated occurrence and food consumption data	+
Occurrence data	
The used occurrence data were from a single study and from only one country	+/−
Concentration data are considered applicable for all items within the entire food category	+
Food consumption data	
The used food consumption data were at a low hierarchy level	+
The used food consumption data were collected between 2006–2014 and might be outdated	+/−
For Slovenia and Slovakia no food consumption data were available, data from the geographically closest region was used	+/−
Use of data from food consumption surveys of a few days to estimate long-term (chronic) exposure for high percentiles (95th percentiles)	+
Hazard data	
Ali et al., 2012 did not provide information on the toxicological study and endpoints from which the reference dose for TCIPP was derived	+/−
Health-Based Guidance Values are only based on limited toxicological data	+/−
Uncertainty factors ranged from 100 to 1000	+/−
Estimated daily intake by using HBM data	
Extrapolation from single non-creatinine adjusted urine sample (spot or morning urine sample) to a 24 h urine sample	+/−
Slovakia and Denmark provided spot urine samples (less concentrated as first morning urine)	-
The estimated urinary molar excretion fraction (F_{ue}) from TDCIPP was used for TCIPP and TCEP	+/−
Absence of individual urine excretion volume data	+/−
BCIPP was the only provided biomarker for TCIPP	-

+: uncertainty with potential to cause overestimation of exposure; -: uncertainty with potential to cause underestimation of exposure; +/−: uncertainty can cause either an over- or underestimation of exposure.

Based on the uncertainty of aggregated occurrence data and the low hierarchy in the food consumption data, the used approach is rather conservative and an overestimation of the dietary exposure is likely. Further the occurrence data originated from Belgium, which might not represent the contamination on a European level.

The toxicological data are limited so we are not able to estimate if the used HBGVs are leading to an over- or underestimation of the risk characterisation.

The limitations in this risk assessment and by using HBM data were the lack of information on toxicokinetics as urinary molar excretion fraction or the oral bioavailability, which is still assumed to be 100% and can result in an overestimation of the exposure or a health risk. All HBGV were published between 2009 and 2012, there are no new data available, which is also a limiting factor in the hazard assessment. Additionally, for TCIPP, Ali et al. [23] did not provide any further information from which toxicological endpoints the RfD was derived, it is only known that chronic NOAELs and an uncertainty factor of 1000 were used. Therefore, it is not possible to evaluate this RfD on its plausibility and values like this one should be used carefully.

A refinement on analytical methods to achieve more sensitive LOQ and LOD is necessary to estimate the contamination, the LOQ in the food category "fats and oils" was rather high. Therefore, we actually do not know if an OPFR contamination in this food category is likely or not. The same applies to the OPFRs metabolites, a more sensitive LOQ and LOD would lead to a more accurate detection frequency of metabolites and reflects a more realistic human body burden.

5. Conclusions

We were able to show that the dietary exposure can have an impact on the general exposure. Under consideration of the current knowledge the estimated daily intake and dietary exposure indicate a minimal health risk. Moreover, we showed how HBM data can be integrated into a dietary risk assessment without performing a duplicate diet study. This kind of approach is expensive and time intense and are mainly conducted on a regional level. Anyways, further research is needed to overcome the data limitations e.g., occurrence data to provide a more comprehensive picture of the OPFRs exposure in children. Uncertainties especially information on toxicokinetics as urinary molar excretion fraction or the oral bioavailability should be addressed and reduced. Further, it is recommended to refine the analytical methods to achieve more sensitive LOQ and LOD to estimate the contamination in food and body burden accurately.

Supplementary Materials: The following supporting information can be downloaded at: https://www.mdpi.com/article/10.3390/toxics10050234/s1, Table S1. Detailed description of the used HBM studies, Table S2. Estimated dietary TDCIPP ex-posure for children in Slovenia and Slovakia (μg/kg bw/day), Table S3. Estimated dietary TCIPP exposure for children in Slovenia and Slovakia (μg/kg bw/day), Table S4. Estimated dietary TCEP exposure for children in Slovenia and Slovakia (μg/kg bw/day), Table S5. Estimated die-tary TDCIPP exposure for children in Germany (μg/kg bw/day), Table S6. Estimated dietary TCIPP exposure for children in Germany (μg/kg bw/day), Table S7. Estimated dietary TCEP ex-posure for children in Germany (μg/kg bw/day), Table S8. Estimated dietary TDCIPP exposure for children in Belgium (μg/kg bw/day), Table S9. Estimated dietary TCIPP exposure for children in Belgium (μg/kg bw/day), Table S10. Estimated dietary TDCIPP exposure for children in Den-mark (μg/kg bw/day), Table S11. Estimated dietary TCIPP exposure for children in Denmark (μg/kg bw/day), Table S12. Estimated dietary TDCIPP exposure for children in France (μg/kg bw/day), Table S13. Estimated dietary TCIPP exposure for children in France (μg/kg bw/day), Table S14. Dietary exposure estimation from each cohort in μg/kg bw/d and its contribution to the estimated intake and its relation to the RfD of 15 μg/kg bw/d (for TDCIPP) or 80 μg/kg bw/d (for TCIPP) and a p-RfD of 7 μg/kg bw/d (for TCEP) in percent.

Author Contributions: Conceptualization, V.P.; Data curation, E.G., L.G. and L.R.M.; Formal analysis, L.G. and L.R.M.; Investigation, V.P., N.V., T.W., M.K.-G., L.P.M., S.W., J.S.T., M.H., G.K., A.C., C.F., L.R. and T.K.J.; Project administration, M.K.-G. and G.S.; Supervision, J.S., E.G., G.S. and E.R.-G.; Writing—original draft, V.P.; Writing—review & editing, J.S., N.V., T.W., M.K.-G., L.P.M., E.G., G.S., A.C. and E.R.-G. All authors have read and agreed to the published version of the manuscript.

Funding: This project received funding from the European Union's Horizon 2020 research and innovation program under grant agreement No 733032 HBM4EU (www.HBM4EU.eu, accessed on 31 March 2022).

Institutional Review Board Statement: Not applicable.

Informed Consent Statement: Informed consent was obtained from all subjects involved in the study.

Data Availability Statement: Not applicable.

Acknowledgments: We would like to thank Giulia Poma for her cooperation regarding the occurrence data in food and Kirsten Baken, Lisa Melymuk, Selma Mahiout and Tiina Santonen for providing comments on this risk assessment. We would also like to thank Margaux Riou for her work on the ESTEBAN Study and in general we thank everyone who was involved in conducting the HBM studies.

Conflicts of Interest: The authors declare that they have no known competing financial interest or personal relationships that could have appeared to influence the work reported in this paper.

References

1. Hou, R.; Xu, Y.; Wang, Z. Review of OPFRs in animals and humans: Absorption, bioaccumulation, metabolism, and internal exposure research. *Chemosphere* **2016**, *153*, 78–90. [CrossRef] [PubMed]
2. Poma, G.; Glynn, A.; Malarvannan, G.; Covaci, A.; Darnerud, P.O. Dietary intake of phosphorus flame retardants (PFRs) using Swedish food market basket estimations. *Food Chem. Toxicol.* **2017**, *100*, 1–7. [CrossRef] [PubMed]
3. van der Veen, I.; de Boer, J. Phosphorus flame retardants: Properties, production, environmental occurrence, toxicity and analysis. *Chemosphere* **2012**, *88*, 1119–1153. [CrossRef] [PubMed]
4. Wang, X.; Liu, Q.; Zhong, W.; Yang, L.; Yang, J.; Covaci, A.; Zhu, L. Estimating renal and hepatic clearance rates of organophosphate esters in humans: Impacts of intrinsic metabolism and binding affinity with plasma proteins. *Environ. Int.* **2020**, *134*, 105321. [CrossRef] [PubMed]
5. Blum, A.; Behl, M.; Birnbaum, L.S.; Diamond, M.L.; Phillips, A.; Singla, V.; Sipes, N.S.; Stapleton, H.M.; Venier, M. Organophosphate Ester Flame Retardants: Are They a Regrettable Substitution for Polybrominated Diphenyl Ethers? *Environ. Sci. Technol. Lett.* **2019**, *6*, 638–649. [CrossRef]
6. Li, J.; Zhao, L.; Letcher, R.J.; Zhang, Y.; Jian, K.; Zhang, J.; Su, G. A review on organophosphate Ester (OPE) flame retardants and plasticizers in foodstuffs: Levels, distribution, human dietary exposure, and future directions. *Environ. Int.* **2019**, *127*, 35–51. [CrossRef]
7. Poma, G.; Sales, C.; Bruyland, B.; Christia, C.; Goscinny, S.; Van Loco, J.; Covaci, A. Occurrence of Organophosphorus Flame Retardants and Plasticizers (PFRs) in Belgian Foodstuffs and Estimation of the Dietary Exposure of the Adult Population. *Environ. Sci. Technol.* **2018**, *52*, 2331–2338. [CrossRef]
8. Fromme, H.; Lahrz, T.; Kraft, M.; Fembacher, L.; Mach, C.; Dietrich, S.; Burkardt, R.; Völkel, W.; Göen, T. Organophosphate flame retardants and plasticizers in the air and dust in German daycare centers and human biomonitoring in visiting children (LUPE 3). *Environ. Int.* **2014**, *71*, 158–163. [CrossRef]
9. Wang, Y.; Sun, H.; Zhu, H.; Yao, Y.; Chen, H.; Ren, C.; Wu, F.; Kannan, K. Occurrence and distribution of organophosphate flame retardants (OPFRs) in soil and outdoor settled dust from a multi-waste recycling area in China. *Sci. Total Environ.* **2018**, *625*, 1056–1064. [CrossRef]
10. Xu, F.; Tay, J.H.; Covaci, A.; Padilla-Sánchez, J.A.; Papadopoulou, E.; Haug, L.S.; Neels, H.; Sellström, U.; de Wit, C.A. Assessment of dietary exposure to organohalogen contaminants, legacy and emerging flame retardants in a Norwegian cohort. *Environ. Int.* **2017**, *102*, 236–243. [CrossRef]
11. Gilles, L.; Govarts, E.; Rambaud, L.; Vogel, N.; Castaño, A.; Esteban López, M.; Rodriguez Martin, L.; Koppen, G.; Remy, S.; Vrijheid, M.; et al. HBM4EU combines and harmonises human biomonitoring data across the EU, building on existing capacity-The HBM4EU survey. *Int. J. Hyg. Environ. Health* **2021**, *237*, 113809. [CrossRef]
12. Miller, L.A.; Stapleton, F.B. Urinary volume in children with urolithiasis. *J. Urol.* **1989**, *141*, 918–920. [CrossRef]
13. Lynn, R.K.; Wong, K.; Gravie-Gould, C.; Kennish, J.M. Disposition of the flame retardant, tris (1,3-dichloro-2-propyl) phosphate, in the rat. *Drug Metab. Dispos.* **1981**, *9*, 434–441.
14. Wittassek, M.; Koch, H.M.; Angerer, J.; Brüning, T. Assessing exposure to phthalates-the human biomonitoring approach. *Mol. Nutr. Food Res.* **2011**, *55*, 7–31. [CrossRef]

15. World Health Organization. Principles and Methods for the Risk Assessment of Chemicals in Food, International Programme on Chemical Safety, Environmental Health Criteria 240. Chapter 6: Dietary Exposure Assessment of Chemicals in Food. 2009. Available online: http://www.who.int/ipcs/food/principles/en/index1.html (accessed on 20 March 2022).
16. European Food Safety Authority (EFSA). Comprehensive European Food Consumption Database. 2021. Available online: https://www.efsa.europa.eu/de/food-consumption/comprehensive-database (accessed on 25 April 2022).
17. Stahl, A.; Vohmann, C.; Richter, A.; Heseker, H.; Mensink, G.B.M. Changes in food and nutrient intake of 6-to 17-year-old Germans between the 1980s and 2006. *Public Health Nutr.* **2009**, *12*, 1912–1923. [CrossRef]
18. Elmadfa, I. Österreichischer Ernährungsbericht 2012. 2012. Available online: http://www.bmg.gv.at/cms/home/attachments/4/5/3/CH1048/CMS1348749794860/oeb12.pdf (accessed on 25 April 2022).
19. De Ridder, K.; Bel, S.; Brocatus, L.; Cuypers, K.; Lebacq, T.; Moyersoen, I.; Ost, C.; Teppers, E. De consumptie van voedingsmiddelen en de inname van voedingsstoffen. In *Voedselconsumptiepeiling 2014–2015. Rapport 4*; Bel, S., Tafforeau, J., Eds.; WIV-ISP: Brussels, Belgium, 2016.
20. Pedersen, A.; Fagt, S.; Groth, M.; Christensen, T.; Biltoft-Jensen, A.; Matthiessen, J.; Andersen, N.L.; Korup, K.; Hartkopp, H.; Ygil, K.; et al. *Danskernes Kostvaner 2003–2008*; DTU Fødevareinstituttet: Søborg, Denmark, 2009.
21. ANSES. Third Study on the Food Consumption and Eating Habits of the French Population. 2017. Available online: https://www.anses.fr/en/content/inca-3-changes-consumption-habits-and-patterns-new-issues-areas-food-safety-and-nutrition (accessed on 20 March 2022).
22. European Commission. *European Union Risk Assessment Report Tris (2-chloro-1-methlyethyl) phosphate (TCPP)*; Chemicals Policy and Services Health and Safety Authority: Dublin, Ireland, 2008; Volume 57, pp. 98–419.
23. Ali, N.; Dirtu, A.C.; Van den Eede, N.; Goosey, E.; Harrad, S.; Neels, H.; Coakley, J.; Douwes, J.; Covaci, A. Occurrence of alternative flame retardants in indoor dust from New Zealand: Indoor sources and human exposure assessment. *Chemosphere* **2012**, *88*, 1276–1282. [CrossRef]
24. Agency for Toxic Substances and Disease Registry. Toxicological Profile for Phosphate Ester Flame Retardants. Volume 153, pp. 451–459. 2012. Available online: https://www.atsdr.cdc.gov/toxprofiles/tp202.pdf (accessed on 20 March 2022).
25. U.S Environmental Protection Agency's (US-EPA). *Provisional Peer-Reviewed Toxicity Values for Tris(2-chloroethyl)phosphate (TCEP) EPA/690/ R-09/069F*; U.S Environmental Protection Agency's (US-EPA): Cinicinnazi, OH, USA, 2009.
26. WHO Environmental Health Criteria 209 Flame Retardants: Tris (chloropropyl) Phosphate and Tris (2-chloroetyhl) Phosphate. 2004. Available online: https://www.who.int/ipcs/publications/ehc/who_ehc_209.pdf (accessed on 20 March 2022).
27. Vejdovszky, K.; Mihats, D.; Griesbacher, A.; Wolf, J.; Steinwider, J.; Lueckl, J.; Jank, B.; Kopacka, I.; Rauscher-Gabernig, E. Modified Reference Point Index (mRPI) and a decision tree for deriving uncertainty factors: A practical approach to cumulative risk assessment of food contaminant mixtures. *Food Chem. Toxicol.* **2019**, *134*, 110812. [CrossRef]
28. Bastiaensen, M.; Gys, C.; Malarvannan, G.; Fotache, M.; Bombeke, J.; Ait Bamai, Y.; Araki, A.; Covaci, A. Short-term temporal variability of urinary biomarkers of organophosphate flame retardants and plasticizers. *Environ. Int.* **2021**, *146*, 106147. [CrossRef]
29. Krystek, P.; Beeltje, H.; Noteboom, M.; van den Hoeven, E.M.; Houtzager, M.M.G. Analytical human biomonitoring method for the identification and quantification of the metabolite BDCPP originated from the organophosphate flame retardant TDCPP in urine. *J. Pharm. Biomed. Anal.* **2019**, *170*, 169–175. [CrossRef]
30. Wang, X.; Zhu, Q.; Liao, C.; Jiang, G. Human internal exposure to organophosphate esters: A short review of urinary monitoring on the basis of biological metabolism research. *J. Hazard. Mater.* **2021**, *418*, 126279. [CrossRef]
31. Xu, F.; Giovanoulis, G.; Van Waes, S.; Padilla-Sanchez, J.A.; Papadopoulou, E.; Magnér, J.; Haug, L.S.; Neels, H.; Covaci, A. Comprehensive study of human external exposure to organophosphate flame retardants via air, dust, and hand wipes: The importance of sampling and assessment strategy. *Environ. Sci. Technol.* **2016**, *50*, 7752–7760. [CrossRef] [PubMed]
32. U.S. Environmental Protection Agencys (US-EPA). *U.S EPA Update for of the Exposure Factors Handbook Chapter 5 (Update)-Soil and Dust Ingestion*; EPA/600/R-17/384F; U.S EPA Office of Research and Development: Washington, DC, USA, 2017.
33. Gbadamosi, M.R.; Abdallah, M.A.E.; Harrad, S. A critical review of human exposure to organophosphate esters with a focus on dietary intake. *Sci. Total Environ.* **2021**, *771*, 144752. [CrossRef] [PubMed]

toxics

MDPI

Article

Human Biomonitoring Guidance Values (HBM-GVs) for Bisphenol S and Assessment of the Risk Due to the Exposure to Bisphenols A and S, in Europe

Matthieu Meslin [1], Claire Beausoleil [1], Florence Anna Zeman [2], Jean-Philippe Antignac [3], Marike Kolossa-Gehring [4], Christophe Rousselle [1,*] and Petra Apel [4]

1 French Agency for Food, Environmental and Occupational Health & Safety, Anses, 14 Rue Pierre et Marie Curie, 94701 Maisons-Alfort, France; matthieu.meslin@anses.fr (M.M.); claire.beausoleil@anses.fr (C.B.)
2 French National Institute for Industrial Environment and Risks (INERIS), Parc ALATA BP2, 60550 Verneuil en Halatte, France; florence.zeman@ineris.fr
3 Oniris, National Research Institute for Agriculture, Food and the Environment (INRAE), LABERCA, 44300 Nantes, France; jean-philippe.antignac@oniris-nantes.fr
4 German Environment Agency (UBA), Corrensplatz 1, 14195 Berlin, Germany; marike.kolossa@uba.de (M.K.-G.); petra.apel@uba.de (P.A.)
* Correspondence: christophe.rousselle@anses.fr

Citation: Meslin, M.; Beausoleil, C.; Zeman, F.A.; Antignac, J.-P.; Kolossa-Gehring, M.; Rousselle, C.; Apel, P. Human Biomonitoring Guidance Values (HBM-GVs) for Bisphenol S and Assessment of the Risk Due to the Exposure to Bisphenols A and S, in Europe. *Toxics* **2022**, *10*, 228. https://doi.org/10.3390/toxics10050228

Academic Editor: Sunmi Kim

Received: 25 March 2022
Accepted: 26 April 2022
Published: 29 April 2022

Publisher's Note: MDPI stays neutral with regard to jurisdictional claims in published maps and institutional affiliations.

Copyright: © 2022 by the authors. Licensee MDPI, Basel, Switzerland. This article is an open access article distributed under the terms and conditions of the Creative Commons Attribution (CC BY) license (https://creativecommons.org/licenses/by/4.0/).

Abstract: Within the European Joint Programme HBM4EU, Human Biomonitoring Guidance Values (HBM-GVs) were derived for several prioritised substances. In this paper, the derivation of HBM-GVs for the general population ($HBM\text{-}GV_{GenPop}$) and workers ($HBM\text{-}GV_{worker}$) referring to bisphenol S (BPS) is presented. For the general population, this resulted in an estimation of the total urinary concentration of BPS of 1.0 µg/L assuming a 24 h continuous exposure to BPS. For workers, the modelling was refined in order to reflect continuous exposure during the working day, leading to a total urinary concentration of BPS of 3.0 µg/L. The usefulness for risk assessment of the HBM-GVs derived for BPS and bisphenol A (BPA) is illustrated. Risk Characterisation Ratios (RCRs) were calculated leading to a clear difference between risk assessments performed for both bisphenols, with a very low RCR regarding exposure to BPA, contrary to that obtained for BPS. This may be due to the endocrine mediated endpoints selected to derive the HBM-GVs for BPS, whereas the values calculated for BPA are based on the temporary Tolerable Daily Intake (t-TDI) from EFSA set in 2015. A comparison with the revised TDI recently opened for comments by EFSA is also discussed. Regarding the occupational field, results indicate that the risk from occupational exposure to both bisphenols cannot be disregarded.

Keywords: human biomonitoring (HBM); HBM4EU; internal exposure; biomarkers; endocrine disruptors; bisphenol A (BPA); bisphenol S (BPS); bisphenols; human biomonitoring guidance value (HBM-GV); physiologically based pharmacokinetic modelling (PBPK); risk assessment

1. Introduction

Bisphenols are a group of chemical compounds with two hydroxyphenyl functionalities. BPA, the most commonly used compound of the bisphenol family, is a monomer mainly used in the production of polycarbonate plastics and epoxy resins [1]. Polycarbonates can be used in the manufacture of many consumer items, such as food containers, baby bottles and childcare articles, water pipes, etc. Epoxy resins are used to make protective coatings for cans and beverages, and to make coatings used on the metal lids of jars [2]. Any residual BPA present in the final material or article made from polycarbonate plastics or epoxy resins has the potential to migrate into the food or water with which it comes into contact [3]. BPA is also found in air and dust particles and is used in some paper products (thermal paper, e.g., sales receipts) as well as in many other products, such as medical devices, surface coatings, printing inks, dental sealants, toys, cosmetics and flame retardants [2,4–7]. Human exposure to BPA is thus widespread, and ingestion of contaminated food is a major contributor to overall internal exposure to BPA for all age groups [8].

The toxicity of BPA has been extensively characterised in several risk assessments carried out by different bodies such as EFSA (the European Food Safety Authority), ANSES (the French Agency for Food, Environmental and Occupational Health & Safety, Maisons-Alfort, France) and even the Food and Agriculture Organization and the US Food and Drug Administration [2,9–11]. The German HBM Commission also reviewed BPA in 2012 and established HBM-I values for children and adults to assess HBM results [12,13]. These values were updated in 2015 on the basis of the new EFSA temporary Tolerable Daily Intake (t-TDI) available at that time. The latest comprehensive reassessment of BPA exposure and toxicity (by EFSA in January 2015 leading to its t-TDI of 4 μg/kg bw/d [2]) is based on the increase in relative kidney weight in the F0 generation of a two-generation mouse study by Tyl et al. (2008) seen as a critical effect [14]. According to EFSA, the risk to consumers was controlled as the highest estimates for dietary exposure and for exposure from several sources, both oral and dermal, are 3–5 times lower than the new TDI [2]. In 2015, the Risk Assessment Committee (RAC) of the European Chemicals Agency (ECHA) also issued an opinion following a request for the restriction of BPA in thermal paper. The restriction dossier was submitted by the French authorities in May 2014, following the identification of health risks for unborn children linked to the exposure (occupational or not) of their mothers to BPA contained in these thermal papers [15]. The RAC followed the same approach as that used by EFSA to calculate an oral Derived No Effect Level (DNEL) for the general population. Since the proposed restriction targeted dermal exposure when handling thermal paper, DNELs for dermal exposure (assuming 50% bioavailability) were also calculated for workers and the general population [15]. Based on these dermal DNELs and reasonable "worst case" scenario modelling, the RAC concluded that the risk of exposure to BPA from handling thermal paper was controlled for consumers but insufficiently controlled for workers. On 12 December 2016, the European Commission amended Annex XVII of REACH to include a restriction which set a threshold limit of 0.02% (by weight) for BPA in thermal paper. This restriction came into force on 2 January 2020 [16].

Nevertheless, endocrine disrupting properties of BPA at low doses are strongly suspected, and concern especially exists, for particularly sensitive population groups such as pregnant women and young children [17]. In January 2017, BPA was included on the candidate list of substances of very high concern (SVHC) due to its reproductive toxicity. In June of the same year, the ECHA Member State Committee supported the French proposal to include its endocrine disrupting properties, which cause probable serious effects on human health and give rise to a level of concern equivalent to carcinogenic, mutagenic, toxic to reproduction (CMRs) category 1A or 1B. In 2018, an update added an additional reason for inclusion in the candidate list of SVHC due to its endocrine disrupting properties leading to adverse environmental effects, as proposed by Germany [18].

The EFSA Panel on Food Contact Materials, Enzymes and Processing Aids (CEP Panel) released for a draft opinion for public consultation in December 2021 on the re-evaluation of risks to public health due to the presence of BPA in foodstuffs [19]. In this draft opinion, the immune system has now been identified as the most sensitive health endpoint to BPA exposure. Specifically, an increase of Th17 cells, key players in cellular immune mechanisms and involved in the development of allergic lung inflammation, was identified as the critical effect. This work led the EFSA CEP Panel to reduce the previous t-TDI by an order of magnitude of 10^5, which leads to a value of 0.04 ng/kg bw/day. However, as this opinion remains at this stage a draft, the risk assessment performed in this paper will still consider the t-TDI set in 2015.

Bisphenol S (BPS) is used in a variety of industrial applications as a substitute for BPA, for example, as a wash fixative in cleaning products, or as a starting monomer for the synthesis of polyether sulfone specifically used in the manufacture of baby bottles and children's tableware. Available data on the presence of BPS in the environment are more limited than for BPA. In Europe, BPS was detected in the wastewater of industrial facilities [20] and canned food [21]. BPS is also used as a developer in thermal paper, including in products marketed as "BPA-free paper" [22]. A survey of manufacturers selling thermal paper in the EU conducted by ECHA indicates that the use of BPS as an alternative to BPA in the manufacture of thermal paper had not increased markedly for the period 2014–2016 [23]. However, it is important to continue to monitor which alternatives thermal paper manufacturers will prefer, following the restriction on the use of BPA in these papers. Indeed, a study in China examined the presence of thirteen bisphenol-related compounds in paper products (120 thermal papers and 81 non-thermal papers) collected in Beijing [24]. The results indicated that the replacement of BPA, by alternatives such as BPS, has progressed significantly in several types of thermal paper such as stickers, train tickets, aeroplane boarding passes, lottery tickets, etc. Due to the probably globalised production of these products, such an increase in Europe cannot be excluded.

According to the Classification, Labelling and Packaging (CLP) regulation, BPS is considered a presumed reproductive toxicant for fertility and development (Reproductive Toxicant 1B). No risk assessment has yet been published for BPS, but several initiatives are underway. The toxicological profile of BPS has however been studied by Beausoleil et al. (2022) [25]. There is a fairly large body of data on the toxicity of BPS, including repeated dose toxicity, reproductive toxicity and also specific effects on the endocrine system.

Human biomonitoring (HBM) is a tool of growing importance for estimating internal aggregated exposure to chemicals and can therefore be used to improve human health risk assessment. The European Joint Programme on Human Biomonitoring (HBM4EU) is a joint effort of 30 countries and the European Environment Agency (co-funded by the European Commission within the framework of Horizon 2020) with the aim of advancing and harmonising human biomonitoring in Europe [26]. One of the limitations for the use of HBM data in risk assessments (RA) is the lack of HBM guidance values (HBM-GV) [27] which represent the concentration of a biomarker of exposure in a human biological matrix at and below which no adverse effect for human health is expected according to current toxicological knowledge. Those values are useful to risk assessors and risk managers for comparing to the HBM data measured in biomonitoring studies. This paper aims to characterise the risks associated with BPS and BPA according to available HBM data in Europe. For this purpose, the derivation of HBM-GVs for exposure to BPS for both the general population ($HBM\text{-}GV_{GenPop}$) and for workers ($HBM\text{-}GV_{Worker}$) has also been performed, as has been done previously for BPA [28].

2. Materials and Methods

2.1. Characterisation of Risk

To estimate the risk due to BPA or BPS exposure, Risk Characterisation Ratios (RCRs) are calculated by comparing the 95th percentile (P95) of the levels of biomarker measured to the *HBM-GV* derived, as described in Equation (1).

$$RCR = \frac{\text{P95 (biomarker)}}{HBM - GV} \quad (1)$$

If the RCRs are lower than 1, the risk due to BPA or BPS exposures can be ruled out for the sampled population according to the *HBM-GV* derived. Conversely, if RCRs are higher than 1, the risk cannot be ruled out for the health of the sampled population and investigations and/or management measures are required.

2.2. Derivation of HBM-GVs: General Methodology

The HBM-GVs are derived according to a systematic methodology agreed within the HBM4EU project and published in 2020 by Apel et al. [29].

In brief, this methodology relies on a tiered approach depending on the knowledge and data available (Figure 1). Option 1 is followed when relevant human data proving a relationship between internal concentrations of an adequate biomarker and the occurrence of adverse health effects are available. If those data are insufficient or unavailable, the second option can be considered by translating an existing external toxicity reference value (TRV), such as a TDI, or an Occupational Exposure Limit set by EU or relevant non-EU bodies, into an internal value. As a final option, HBM-GVs can be derived on the basis of critical effects observed in animal toxicological studies. This methodology has been applied for deriving HBM-GVs for BPA and BPS and the best approach was selected.

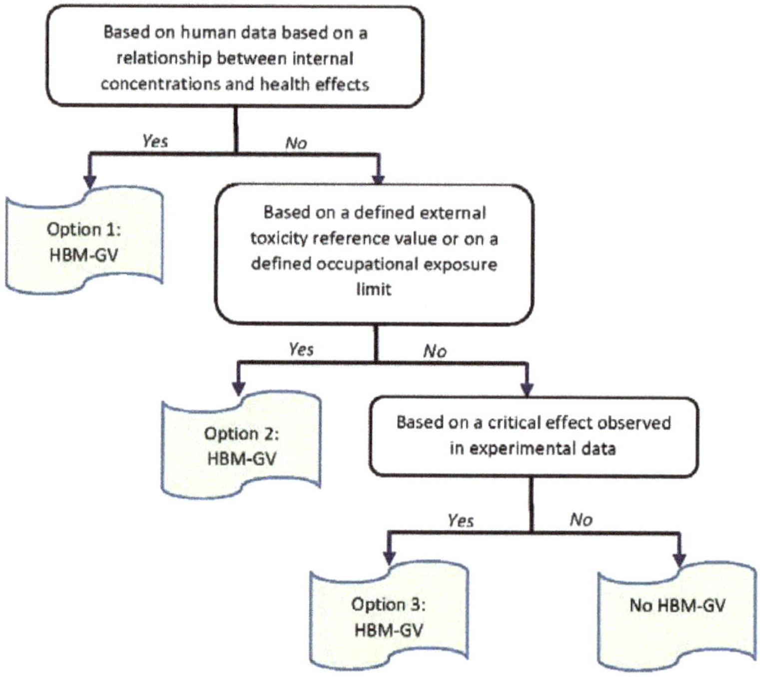

Figure 1. General methodology used to derive HBM-GVs as presented in Apel et al., 2020 [29].

2.3. *Characterisation of Internal Exposure to BPA and BPS*

The Information Platform for Chemical Monitoring (IPCheM) was consulted to obtain HBM data from European cohorts. IPCheM is a platform developed under the initiative of the European Commission for accessing chemical data collected and managed in Europe [30]. P95 for the selected biomarkers (total BPA and total BPS) in urine from aggregated HBM data included in the IPCheM platform was considered for the risk assessment of BPA and BPS. To respect data ownership, only results published in the literature were used in this paper. To document BPA exposure for the occupational population, data from occupational studies included in the review from Bousoumah et al. (2021) [31] were also used.

3. Results

3.1. *Assessment of Risk Due to BPA Exposure*

3.1.1. Derivation of HBM-GVs for BPA

Derivation of HBM-GV_{GenPop}

The HBM-GV_{GenPop} for adults and children was developed for total urinary BPA and published by Ougier et al. (2021) [28]. Human data proving a relationship between internal concentrations of a specific and sensitive biomarker of BPA and the occurrence of relevant adverse health effects were not sufficient to derive a HBM-GV. Thus, the second option for HBM-GV derivation, relying on the conversion of an external guidance value into a corresponding internal level, was adopted. For this purpose, total BPA concentrations in urine were estimated considering a steady-state exposure at the EFSA t-TDI of 4 μg/kg bw/d, using the PBPK model developed by Karrer et al. (2018) [32], assuming a 24 h averaged BPA exposure that is constant and occurs 100% via the oral route. The resulting values derived by Ougier et al. (2021) are 230 μg/L for adults and 135 μg/L for children [28] (Table 1).

Table 1. HBM-GVs calculated for the general population for total BPA in urine.

Key Study	Critical Effect	External Toxicity Reference Value	$HBM\text{-}GV_{GenPop}$	Key Study
Tyl et al., 2008 (Two-generation toxicity study in mice)	Increase in relative mean kidney weight in adult males F0	t-TDI (EFSA, 2015) 4 μg/kg bw/day	230 μg/L	135 μg/L

Derivation of $HBM\text{-}GV_{worker}$

For the occupationally exposed population, the absorption of BPA results mainly from dermal exposure, in particular for cashiers.

Therefore, the total BPA level in urine was also estimated by Ougier et al. (2021) for dermal absorption that would generate the same level of free BPA plasma concentration as an average 24 h oral absorption at the ECHA DNEL of 8 μg/kg bw/d for workers. A concentration of approximately 12 μg/L of total BPA in urine was calculated by Ougier et al. (2021) as a biological threshold value not to be exceeded in workers, assuming 100% dermal exposure in the workplace [28]. However, since this value for workers is much lower than the recommended $HBM\text{-}GV_{GenPop}$ (230 μg/L) and lower than the total BPA concentrations from some of the environmental (non-occupational) background exposures identified, no $HBM\text{-}GV_{worker}$ could be recommended. Thus, the theoretical threshold value of 12 μg/L corresponding to the worst-case scenario resulting from 100% dermal exposure was used only as an indication for comparison with available biomonitoring data.

3.1.2. Assessment of Exposure to BPA

For the assessment of risks due to BPA exposure for the general population, data of 16 HBM studies from 16 different European countries, representative of different parts of Europe, were compiled (Table 2). Data from the HELIX cohorts were creatinine adjusted. To be able to compare the creatinine adjusted levels to the reported HBM-GVs, data were converted in unadjusted levels (μg/L) using the mean value for creatinine obtained from studies by MacPherson et al. (2018) and Lau et al. (2018) for adults and children, respectively [33,34]. Moreover, reported levels of total BPA in urine corrected by urine specific gravity could not be converted, and were used as they were for the calculation of the RCR.

Regarding occupational exposure to BPA, HBM data are more limited and those identified in a review published by Bousoumah et al. (2021) from five European occupational studies are discussed for risk assessment purposes [31]. A Finnish study by Heinälä et al. published in 2017 and involving 49 workers was conducted in five different companies using BPA in the production of paints, composites, tractors and thermal papers [35]. Simonelli et al. (2017) collected urine samples from workers suffering endometriosis and a healthy control group, as well as information about their lifestyle, environment and work activities [36]. Lyapina et al. (2016) also conducted a study in Bulgaria in 55 students of dental medicine exposed to dental composite resins containing BPA as impurities due to their production process, and in 29 patients treated with those composite resins but not exposed occupationally as controls [37]. A Spanish study by Gonzales et al. (2019) has determined the levels of eight Bisphenols including BPA, BPS and BPF in biological samples from a controlled cohort of 29 workers in a hazardous waste incinerator located in Constantí in the

Catalonia region [38]. Lastly, a French study performed by Ndaw et al. (2016) evaluated the exposure to BPA in cashiers and in non-occupationally exposed workers from several workplaces [39].

3.1.3. Risk Characterisation Due to BPA Exposure

For the General Population

RCRs were calculated by comparing total BPA levels in urine from the HBM studies to the corresponding HBM-GVs derived for BPA (Table 2). The t-TDI based HBM-GV_{GenPop} for total BPA in urine for adults was used for calculating RCRs for teenagers. For studies reporting results for children and teenagers in combination, the HBM-GV for children was used with a conservative approach. In all cases the calculated RCRs are significantly lower than 1, both for adults and for children, and for each set of data available, meaning that exposure levels reported are far lower than the HBM-GV_{GenPop}. These results indicate that exposure to BPA for the populations sampled, according to current knowledge in a single substance–risk assessment approach, does not constitute a risk when taking into account the HBM-GV_{GenPop} derived from the t-TDI established by EFSA in 2015.

Table 2. Reported concentrations of total urinary BPA in the general population (P95) and RCR calculated using the HBM-GV_{GenPop} recommended by HBM4EU.

Cohorts	References	Country	Populations	P95 (µg/L)	HBM-GV_{GenPop} for Total Urinary BPA (µg/L)	RCR
IBS	Berman et al., 2014 [40]	Israel	Adults (40–59 years old)	20.48	230	0.09
GerES IV *	Becker et al., 2009 [41]	Germany	Children (3–14 years old)	14	135	0.10
GerES V	Tschersich et al., 2021 [42]	Germany	Children (3–5 years old)	7.79	135	0.06
			Children (6–10 years old)	5.13	135	0.04
			Children (11–13 years old)	9.95	135	0.07
			Adolescents (14–17 years old)	7.65	230	0.03
3xG	Study website [43]	Belgium	Adults (20–39 years old)	4.61	230	0.02
FLEHS II	Geens et al., 2014 [44]	Belgium	Adolescents (14–15 years old)	9.6	230	0.04
DEMOCOPHES	Covaci et al., 2015 [45]	Belgium, Denmark, Luxembourg, Slovenia, Spain, Sweden	Mothers (<45 years old)	11.1	230	0.05
			Children (5–12 years old)	13.1	135	0.10
		Belgium	Mothers (<45 years old)	11.6	230	0.05
			Children (5–12 years old)	13.4	135	0.10
		Denmark	Mothers (<45 years old)	11.5	230	0.05
			Children (5–12 years old)	7.9	135	0.06
		Luxembourg	Mothers (<45 years old)	7.4	230	0.03
			Children (5–12 years old)	8.3	135	0.06
		Slovenia	Mothers (<45 years old)	13.4	230	0.06
			Children (5–12 years old)	18.9	135	0.14
		Spain	Mothers (<45 years old)	12.2	230	0.05
			Children (5–12 years old)	9.8	135	0.07
		Sweden	Mothers (<45 years old)	5	230	0.02
			Children (5–12 years old)	6.2	135	0.05

Table 2. *Cont.*

Cohorts	References	Country	Populations	P95 (µg/L)	HBM-GV$_{GenPop}$ for Total Urinary BPA (µg/L)	RCR
PBAT	Hartmann et al., 2016 [46]	Austria	Children M (6–10 years old)	7.3	135	0.05
			Children F (6–10 years old)	5.8	135	0.04
			Teenagers M (11–15 years old)	2.7	230	0.01
			Teenagers F (11–15 years old)	5.2	230	0.02
			Adults M (18–64 years old)	3.6	230	0.02
			Adults F (18–64 years old)	1.3	230	0.01
			Senior M (65–79 years old)	1.3	230	0.01
			Senior M (65–79 years old)	4.6	230	0.02
HELIX **	Haug et al., 2018 [47]	France, Greece, Lithuania, Norway, Spain, United Kingdom.	Pregnant women (>18 years old)	22	230	0.10
			Children (6–12 years old)	15.7	135	0.12
Elfe	Dereumeaux et al., 2016 [48]	France	Pregnant women (>18 years old)	5.3	230	0.02
Esteban	Balicco et al., 2019 [49]	France	Children (6–10 years old)	7.3	135	0.05
			Adolescents (11–14 years old)	13.7	230	0.06
			Adolescents (15–17 years old)	6	230	0.03
			Adults M (18–74 years old)	10	230	0.04
			Adults F (18–74 years old)	6.9	230	0.03
Danish-HBM	Frederiksen et al., 2014 [50]	Denmark	Pregnant women (>18 years old)	7.52	230	0.03
RefLim 2011	Porras et al., 2014 [51]	Finland	Adults (22–67 years old)	7.9	230	0.03
RHEA	Myridakis et al., 2015 [52]	Greece	Pregnant women (>16 years old)	4.7	230	0.02
			Children (4 years old)	16.6	135	0.12
INMA	Casas et al., 2013 [53]	Spain	Pregnant women (≥16 years) first trimester	11.9	230	0.05
			Children (4 years old)	12.3	135	0.09
TH Pregnant women	Machtinger et al., 2018 [54]	Israel	Pregnant women	14.2	230	0.06
SLO-CRP	Tkalec et al., 2021 [55]	Slovenia	Children (6–9 years old)	9.5	135	0.07
			Adolescents (11–15 years old)	7.3	230	0.03

* The HBM-GV value for children was used to calculate the RCR using a protective approach. ** Conversion carried out with concentrations of 0.75 g creatinine/L for pregnant women (MacPherson et al., 2018) and 0.68 g creatinine/L for children (Lau et al., 2018).

For Workers

As explained above, no HBM-GV could be derived for workers exposed to BPA. However, the concentration of 12 µg/L of total BPA in urine calculated by Ougier et al. (2021) was used to compare the values measured in occupational studies [28]. The study by Heinälä et al. (2017) suggests an occupational exposure of concern, with median urinary concentrations (P50) of total BPA at the end of the shift much higher (130–150 µg/L) in an environment where the air levels of BPA are low [35]. Elevated levels of total urinary BPA were still observed on Monday morning after the weekend break. Taking into account this delay in the excretion of BPA, these results suggest significant skin exposure, for which absorption is generally slower in comparison with other routes of exposure. In addition, the highest reported concentrations of total BPA were in the order of 1000–1500 µg/L, which is well above the value of 12 µg/L, but also above the recommended HBM-GV for the general population (230 µg/L).

Simonelli et al. (2017) reported urinary concentrations of total urinary BPA in workers lower than in the study of Heinälä et al. (2017), with concentrations ranging from 1.17 μg/L to 12.68 μg/L, and an average value of 5.31 μg/L [36]. In the control group, the concentrations measured are between 1.28 and 2.35 μg/L, with an average value of 1.64 μg/L. These levels are 10 to 100 times lower than those reported by Heinälä et al. (2017) and are also below, or at the limit of, the value of 12 μg/L while also well below the HBM-GV_{GenPop} value (230 μg/L). In addition, the results from Simonelli et al. (2017) show that certain professional activities generally considered to be exposed to BPA, such as those carried out by housekeepers, are in reality slightly exposed or not exposed to BPA. These observations could be due to the wearing of personal protective equipment. On the contrary, other categories generally considered to have a rather small exposure to BPA through their professional activities, such as students or salespeople, would in reality be significantly exposed.

Lyapina et al. (2016) and Gonzales et al. (2019) also reported results below the theoretical limit value of 12 μg/L [37,38]. The urinary concentration of total BPA was on average 6.16 μg/L, with a standard deviation of 14.05 μg/L, in student dentists, and only 0.86 μg/L for employees working in a hazardous waste incinerator with a maximum value of 2.82 μg/L.

The study by Ndaw et al. published in 2016 reports median urinary concentrations of total BPA of 3.54 μg/L in controls and 8.82 μg/L in cashiers [39]. These levels are also below both the HBM-GV_{GenPop} and the value of 12 μg/L. However, the P95 values exceed this limit value for both exposed cashiers and for controls with 44 μg/L and 14.2 μg/L, respectively. However, the authors highlighted that no significant increase in the urinary concentrations of free BPA was observed.

3.2. Assessment of Risk Due to BPS Exposure

3.2.1. Derivation of HBM-GVs for BPS

Selection of the Methodological Approach for the Derivation of HBM-GVs for BPS

The human data analysed by Beausoleil et al. (2022) were considered insufficient for establishing a relationship between internal concentrations and relevant health effects [25]. In addition, neither toxicity reference values nor occupational exposure limits have been proposed by the EU or by any relevant non-EU organisations so far. Regarding workers, the scarcity of data on occupational exposure does not allow for assessment of the relationship between atmospheric concentrations of BPS and total urinary concentrations of BPS. Thus, the derivation method for HBM-GVs for BPS for both the general population and the workers was based on a point of departure (POD) identified from animal experimental studies, as tentatively proposed by Beausoleil et al., 2022 [25]. The HBM-GVs were then derived by translating the POD into the corresponding urinary levels in humans for the selected biomarker of exposure by using a PBPK model. Once the human equivalent dose was derived, different assessment factors were applied to estimate the HBM-GVs.

Karrer et al. (2018) extended to BPS the PBPK model developed by Yang et al. in 2015, used previously by EFSA in 2015 for calculating the Human Equivalent Dose Factor (HEDF) (ratio of AUC Animal/AUC Human) and deriving the t-TDI for BPA [2,32,56]. To validate the model, a comparison of predicted concentrations with measured concentrations of BPS and total BPS (BPS-G + free BPS) in two volunteer studies [57–59] has been performed. The blood flow was replaced by a plasmatic flow (BPS being mostly distributed in the plasma) and a value for the haematocrit of 0.45 was used. Outputs of excreted urinary BPS quantities were also replaced with urinary BPS concentrations by adding an estimated value for the daily 24 h urinary excretion. The same daily urinary excretion rate of 0.05 L/h previously used for BPA was considered. After oral administration, an absorption fraction of 100% was considered.

BPS is well absorbed by the oral route. However, dermal absorption of BPS is very limited. Based on an in vitro study by Champmartin et al. (2020) [60], it has been assumed that the dermal route will contribute very little to the overall exposure to BPS. Lastly, the relative contribution of the inhalation route may also be expected to be low, based on BPS physical-chemical properties (low vapor pressure) and by analogy with previous BPA risk assessments. It is thus anticipated that the 100% oral exposure scenario is appropriate for the general population, as well as for workers. Thus, the derivation method for HBM-GVs for BPS for both the general population and workers is based on the same constant and continuous exposure scenario to the LOAEL chosen as a POD via ingestion (100% oral exposure scenario).

Selection of the Biomarker of Exposure for BPS

Bisphenols are largely detoxified by phase II conjugating enzymes, including UDP-glucuronyl transferases (UGT) and sulfotransferases (SULT) (Figure 2) [61,62]. Similarly to what has been observed for BPA, BPS clearance is mainly driven by glucuronidation of BPS into BPS-G, principally excreted in urine [28]. Although the unconjugated form is known to be the active one able to induce an effect, very few data are available regarding the excretion levels of non-conjugated BPS in urine; however, these levels confirmed a very low proportion of free urinary BPS among all the excreted forms [49]. Conversely, the measurement of total BPS in urine allows for better compatibility with current methodological detection limits and is a protective approach for exposure assessment considering possible endogenous deconjugation. Therefore, in a large majority of available biomonitoring studies, the urinary exposure level is estimated after enzymatic hydrolysis of the conjugates so that the total of free and conjugated forms is finally determined. Total BPS in urine was then selected as the appropriate biomarker of exposure to BPS to be considered for the derivation of HBM-GV. It should be noticed that most of the existing studies have determined exposure levels in spot urine samples.

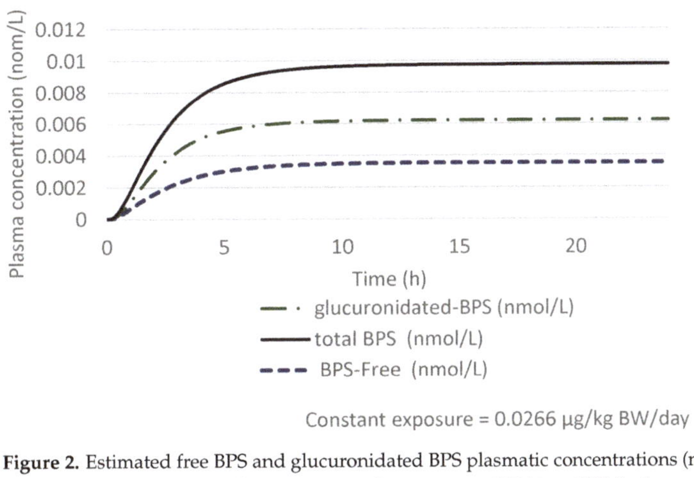

Figure 2. Estimated free BPS and glucuronidated BPS plasmatic concentrations (nmol/L) for a 70 kg adult after 24 h constant and continuous oral exposure to 0.0266 µg BPS/kg bw averaged over 24 h in the general population.

Selection of Key Studies and Choice of POD

The first two options that, according to methodology developed within the HBM4EU programme and published by Apel et al. (2020), should be used to derive the HBM-GV when the data available allows it, could not be followed for BPS. Therefore, the third option is to derive an HBM-GV based on critical effects observed in animal studies. The choice of POD is based on the recommendation formulated by Beausoleil et al. (2022) who reviewed regulatory toxicology studies conducted in accordance with OECD Test Guidelines, as well as numerous academic studies investigating more specific parameters dedicated to endocrine disrupting effects [25]. Adverse effects due to BPS exposure were reported in academic studies assessing male and female reproduction, mammary glands, neurobehavior and metabolism/obesity. LOAELs regarding developmental exposure to BPS were far lower in academic studies than in regulatory ones and were expressed, respectively, in μg/kg bw/day vs. mg/kg bw/day. Beausoleil et al. (2022) proposed to derive a HBM-GV from values determined in academic peer-reviewed experimental studies [25].

Converging studies by Kolla et al. published in 2018 and 2019, as well as by Catanese and Vandenberg published in 2017, were conducted by exposure via the oral route with at least two dose levels, and identified a common LOAEL of 2 μg/kg bw/day for mammary gland and neurobehavioral toxicity, respectively (Table 3) [63–65]. This LOAEL has been selected as a POD for the derivation of the HBM-GVs.

Table 3. Key studies and selection of the POD for the derivation of the HBM-GVs.

Species, Exposure Duration	Critical Endpoint	POD	References
CD-1 Mice, oral exposure, GD 9 to PND 20	Increased number of terminal end buds in the mammary gland	LOAEL = 2 μg/kg bw/day	Kolla et al., 2018 [63]
CD-1 Mice, oral exposure, GD 9 to GD 16 or lactation day 20	Dose dependent increase in ductal area in the mammary gland.		Kolla et al., 2019 [64]
Rat, oral exposure, GD8/9 to PND20/21	Neurobehavioral toxicity		Catanese and Vandenberg, 2017 [65]

Prediction of the Total Urinary BPS Concentration with a 100% Oral Exposure Scenario

The derivation method for HBM-GVs for BPS for both the general population and workers is based on the same LOAEL identified from animal experimental studies. In addition, dermal absorption of BPS is very limited, whereas BPS is well absorbed by the oral route. It is therefore anticipated that the dermal route will contribute very little to the overall exposure to BPS. The relative contribution of the inhalation route may similarly be expected to be low, based on its physical-chemical properties (low vapor pressure) and by analogy with previous BPA risk assessments. Therefore, a 100% oral exposure scenario was considered for the general population, as well as for workers.

The HBM-GVs were derived by translating the LOAEL of 2 μg/kg bw/day, chosen as a POD for the corresponding urinary levels in humans of total BPS, by using the modified PBPK model published by Karrer et al. in 2018 [32]. Moreover, an assessment factor of 3 is applied for extrapolating a LOAEL to a NOAEL. The interspecies differences in toxicokinetics between animal species and humans are determined by PBPK modelling, and an assessment factor of 2.5 is applied for remaining interspecies differences (mostly for toxicodynamic differences). A factor of 10, accounting for intra-species differences, is further applied to the human adjusted NOAEL. Generally, this factor is set to 5 when workers are the targeted population for which the HBM-GV is derived, in line with ECHA's R8 guidance for deriving DNELs for workers [66]. However, as the most sensitive endpoint(s) to be protected from are the effects on the unborn child, no differences can be assumed between the foetuses of the general population and those of workers.

Therefore, for an adult of 70 kg, applying assessment factors of 75 ($3 \times 2.5 \times 10$) to the POD and assuming a constant and continuous oral exposure throughout the day of BPS leads to a constant exposure to 0.0266 μg/kg bw/day. The estimated maximal concentration of plasmatic free and glucuronidated BPS is 9.7×10^{-3} nmol/L (2.42 ng/L) after 24 h constant oral exposure to 0.0266 μg/kg bw/day for an adult of 70 kg (Figure 2).

The estimated concentration of total BPS (sum of estimated urinary glucuronidated and free BPS) in urine is 4.13 nmol/L (1030 ng/L) after 24 h constant oral exposure to 0.0266 μg/kg bw/day for an adult of 70 kg (Figure 3). The HBM-GV_{GenPop} is then rounded to 1.0 μg/L.

Although a weight of 70 kg may not be representative of women, especially non-pregnant women, this weight gives more conservative values for lighter women and seems appropriate for calculating HBM-GVs for a population including men and women. Additionally, as the most sensitive endpoint(s) to be protected from are the effects on the unborn child, the calculated HBM-GVs for men are probably highly conservative.

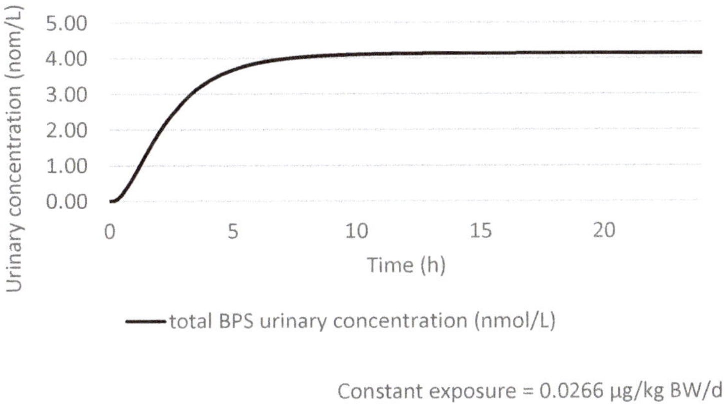

Figure 3. Estimated total urinary BPS concentration (nmol/L) for a 70 kg adult after 24 h constant and continuous oral exposure to 0.0266 µg of BPS/kg bw averaged over 24 h in the general population.

For workers, the same approach of using the POD and modified PBPK model by Karrer et al. (2018) was used to derive the HBM-GV$_{worker}$. Applying an assessment factor of 75 (3 × 2.5 × 10) and assuming a constant and continuous oral exposure of BPS throughout the day would lead to a constant exposure of 0.0266 µg/kg bw/day. However, as occupational exposure is not continuous throughout the day or week, the modelling was refined in order to reflect a continuous exposure of 8 h per day followed by a non-exposure period of 16 h. This 8 h exposure and 16 h non-exposure period was then repeated four times in order to mimic a working week. Thus, this 5 day occupational scenario includes an 8 h exposure period with 0.0798 µg/kg bw and a non-exposure period of 16 h which then corresponds to an exposure for the entire day (0–24 h) of 0.0266 µg/kg bw/day (Figures 4 and 5). The estimated concentration of total BPS (sum of estimated urinary BPS-G and free BPS) in urine is 12.1 nmol/L (3028 ng/L). The HBM-GV$_{worker}$ is then rounded to 3.0 µg/L.

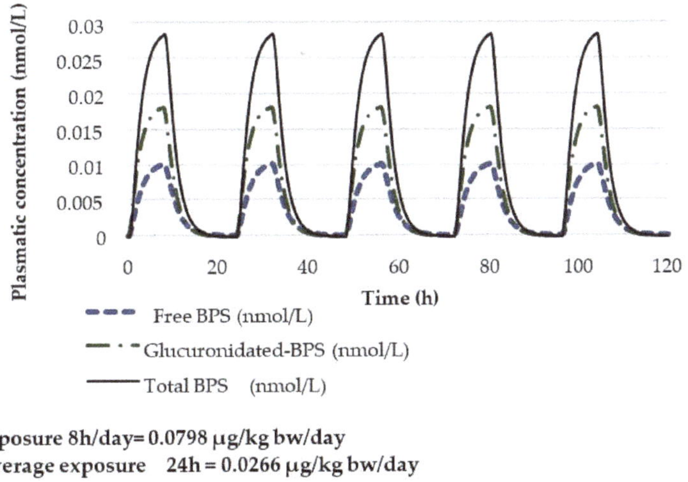

Figure 4. Estimated free BPS and glucuronidated BPS plasmatic concentrations (nmol/L) for an adult of 70 kg considering an occupational scenario lasting 5 days with daily exposure from 0 h to 8 h to 0.0798 µg/kg bw and from 8 h to 24 h to 0 µg/kg bw corresponding to an exposure for the entire day (0–24 h) of 0.0266 µg/kg bw/day.

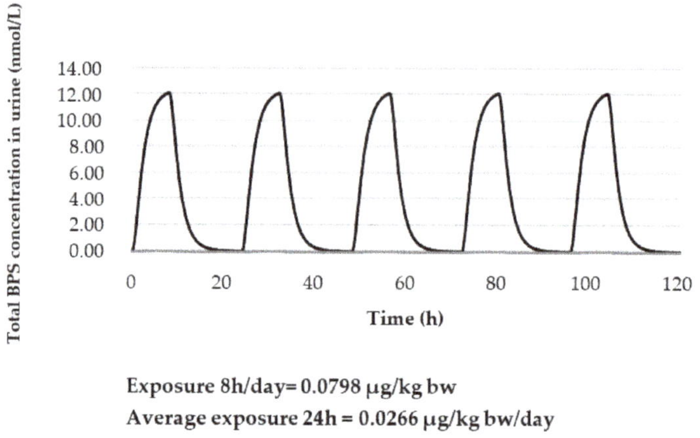

Figure 5. Estimated total urinary BPS concentrations (nmol/L) for an adult of 70 kg considering an occupational scenario lasting 5 days with daily exposure from 0 h to 8 h to 0.0798 μg/kg bw and from 8 h to 24 h to 0 μg/kg bw corresponding to an exposure for the entire day (0–24 h) of 0.0266 μg/kg bw/day.

3.2.2. Assessment of Exposure to BPS

The data from studies reporting total BPS levels in urine were used to calculate the RCRs with the corresponding HBM-GVs detailed above. For the risk assessment for the general population exposed to BPS, data from seven studies from six different European countries were compiled. RCRs were calculated for each available dataset, with the exception of studies for which P95 values were not reported. Some studies also reported results corrected by urine specific gravity. As with BPA, the results of those studies were used as is and even though they are close to the HBM-GVs this contributes to the uncertainty of the meaning of the corresponding RCRs calculated. In addition, unlike BPA, a specific HBM-GV_{GenPop} value for children was not developed for BPS and a general value calculated based on adult weight was used. Calculating such a value based on a lower weight that is more representative of a child population would have led to more conservative RCRs and the characterisation of a potentially increased risk for these populations.

Data on occupational exposure to BPS in Europe are more limited. In fact, given our literature search, only one of the retrieved studies describes the exposure assessment of European (French) workers to BPS and is related to thermal receipts.

3.2.3. Risk Characterisation Due to Exposure to BPS

For the General Population

Taking into account the HBM-GV_{GenPop} developed, and the internal exposure measured in the Esteban study, the RCRs obtained with the P95 are greater than 1 for all age groups (Table 4) [42]. Therefore, the health risk associated with exposure to BPS cannot be excluded. Looking at other percentiles than P95, we find that this risk exists for at least 10% of the sampled population. Regarding the Israeli study by Machtinger et al. (2018) [54], the RCR is less than 1, which means that the health risk due to BPS exposure of the studied population can be ruled out with respect to the HBM-GV_{GenPop}. According to the result of the Slovenian study by Tkalec et al. (2021) [55], BPS exposure is not a risk for children in contrast to the adolescents sampled in this study, for whom 10% of the samples equal or exceed the HBM-GV_{GenPop}. The RCRs calculated with the P95 results of Norwegian studies reported by Husøy et al. (2019) and Sakhi et al. (2018) [67,68] for adults are also lower than 1. However, the characterisation of risk is more uncertain because the results are reported as corrected by specific gravity. This makes it difficult to draw conclusions regarding risk because total BPS levels are very close to the HBM-GV. Similarly, for levels measured in children in the Sahki et al. (2018) study leading to a RCR that exceeds 1, the risk remains difficult to characterise, as these levels are adjusted for specific gravity and are close to the $\text{HBM-GV}_{\text{GenPop}}$.

Table 4. Summary of HBM data and calculated RCRs for total urinary BPS for the general population.

Cohort	Reference	Country	Population	$\text{HBM-GV}_{\text{GenPop}}$ (μg/L)	P25 (μg/L)	P50 (μg/L)	P75 (μg/L)	P90 (μg/L)	P95 (μg/L)	RCR
Esteban	Balicco et al., 2019 [49]	France	Adults (18–74 years old)	1.0	0.14	0.31	0.80	2.24	6.33	6.3
			Adult M (18–74 years old)	1.0	0.15	0.38	0.88	2.44	9.71	9.7
			Adult F (18–74 years old)	1.0	0.13	0.27	0.72	2.15	5.39	5.4
			Adults (18–29 years old)	1.0	0.16	0.44	0.99	3.38	7.13	7.1
			Adults (30–44 years old)	1.0	0.17	0.36	0.91	3.77	28.93	28.9
			Adults (45–59 years old)	1.0	0.14	0.29	0.7	1.97	3.33	3.3
			Adults (60–74 years old)	1.0	0.10	0.22	0.66	1.59	4.11	4.1
			Adults F (18–49 years old)	1.0	0.16	0.30	0.69	2.51	6.08	6.1
			Adults F (50 years and older)	1.0	0.11	0.23	0.78	1.98	3.49	3.5
Euromix	Husøy et al., 2019 * [67]	Norway	Adults (24–72 years old)	1.0	0.11	0.16	0.29	0.56	0.90	0.9
	Sakhi et al., 2018 * [68]	Norway	Mothers	1.0	<LOQ	<LOQ	0.26	0.44	0.59	0.6
			Children	1.0	<LOD	0.13	0.39	0.95	1.68	1.7
TH Pregnant women	Machtinger et al., 2018 [54]	Israel	Pregnant women	1.0		<LOD			0.40	0.4
SLO-CRP	Tkalec et al., 2021 [55]	Slovenia	Children (6–9 years old)	1.0		0.3		0.59	0.70	0.7
			Teenagers (11–15 years old)	1.0		<LOQ		1	1.80	1.8

* P95 reported as corrected by urine specific gravity.

For Workers

Based on the French study by Ndaw et al. (2018) on occupational biomonitoring for cashiers [69], RCRs were calculated considering the $\text{HBM-GV}_{\text{worker}}$ value of 3.0 μg/L (Table 5). We can see that the RCRs are greater than 1 for cashiers and control workers not exposed to BPS in their occupational activities. The two median values (P50) for cashiers and controls do not exceed the $\text{HBM-GV}_{\text{worker}}$ (3.0 μg/L). Therefore, according to the available results, although urinary total BPS levels are significantly higher in cashiers than in control workers, it appears that risk due to BPS exposure can be ruled out for at least 50% of the population in both groups, while for at least 5% of the same populations the risk exists. This also confirms that BPS exposure is ubiquitous in the general population and in workers.

Table 5. Summary of HBM data and calculated RCRs for total BPS for the occupational population.

Reference	Country	Population	$\text{HBM-GV}_{\text{GenPop}}$ (μg/L)	P50 (μg/L)	P95 (μg/L)	RCR
Ndaw et al., 2018 [69]	France	Professional (cashiers)	3.0	2.53	19.9	6.6
		Professional (controls)	3.0	0.67	12.6	4.2

4. Discussion

The expanding database on BPS provides convincing evidence of adverse effects on neurodevelopment or mammary glands following exposure during early life. The sensitivity to BPS is clearly dependent on the timing of exposure with specific periods of development being critical. Disruption of estrogenic signalling is likely central in the mediation of these effects although other modes of action may be involved [70–72]. The derived HBM-GV_{GenPop} is applicable to the whole general population and thereby is protective for all. Nevertheless, we are also aware of the current assessment of BPS at the EU level, which may be finalised during the coming months. Depending on the conclusion of this assessment, the derivation of the HBM-GVs for BPS could be updated accordingly.

Consideration of the type and time of sampling is crucial, because the proposed HBM-GVs should allow for direct comparison with urinary data on bisphenols collected from the general population or occupational biomonitoring studies. It has to be noted that exposure is likely to be overestimated if spontaneous urine is used because spot urine sampling includes more sources of variability (e.g., within-person variability, within-day variability) than 24 h urine sampling, and that generic values in that case are generally used for the urinary output rate and bodyweight.

Additionally, 24 h urine collections would be preferable as both the urinary concentration of total BPS or total BPA, and the urinary output rate (mL/day) are measured. The geometric mean (GM) (or median) can be directly used as an estimate for the average exposure of the population. The P95 might tend to overestimate high exposure as it includes not only the between-person variability, but also within-person and between-day variability [2]. However, collecting 24 h urine voids in large biomonitoring studies is not established and too challenging, mainly for reasons of logistics and cost. The collection of first morning urine is a non-random, single sampling that is not representative of daily variability, may introduce a bias and may result in an over- or underestimation of average exposure; however, the sparse evidence available from the literature does indicate comparability of the central tendency between first morning voids, spot urine samples and 24 h urine samples [2,73,74]. The total urinary concentration of BPS (or total BPA) in an individual spot urine sample cannot be used to arrive at a realistic estimate of daily exposure to bisphenols because of their non-persistent nature and short elimination half-life [2]. However, a set of spot urine samples can be used to obtain a reliable estimate of the average BPS exposure of a sufficiently large investigated population, provided that the sampling is at random in relation to meal ingestion and bladder-emptying times [2]. Vernet et al., who in 2018 characterised the within-day, between-day and between-week variability of phenol urinary biomarker concentrations (e.g., BPA) during pregnancy, came to the same conclusion that for biomonitoring purposes (and not aetiological studies) collecting spot samples was a good option if the population was large enough [74].

Regarding the assessment of risk due to exposure to BPA and BPS, the heterogeneity in terms of concentration units (i.e., µg/L, µg/g creatinine or µg/L corrected for urine specific gravity), statistical descriptors to document the distribution of exposure, as well as methodological sensitivity threshold limits of detection/limits of quantification and "quality assurance/quality control" provisions implemented to ensure consolidated consistency of the results generated, do not allow for direct and rapid comparability between different biomonitoring studies identified in the literature. The diversity of the subpopulations considered, the geographical sampling areas and the sampling years also contribute to the difficulty of comparing the reported data.

We find a clear difference between risk assessments for BPA and BPS exposure according to our current HBM-GVs. Indeed, according to current knowledge, while risk could be ruled out for BPA with very low RCRs for the general population, this is not the case for BPS where RCRs are high for a large portion of the sampled population, and it appears that protective measures need to be taken regarding BPS exposure in the general and occupational population. This is because the HBM-GVs for BPS were established on the basis of endocrine disruption effects in animals at very low doses, in contrast to the

HBM-GV$_{GenPop}$ for BPA. As mentioned in the introduction, the EFSA CEP Panel recently released a draft opinion for consultation regarding the re-evaluation of the t-TDI for BPA set in 2015 in light of the latest available data [19]. The new TDI would be established at 0.04 ng/kg bw/d of total BPA, which is 10^5 times lower than the current t-TDI. As the HBM-GV is calculated under a steady-state, the new HBM-GV$_{GenPop}$ for BPA based on the full new TDI using PBPK modelling would be 2.3 ng total BPA/L urine for adults and 1.4 ng total BPA/L urine for children assuming 100% oral intake in each case. Therefore, if this new TDI is confirmed after the consultation period, comparing the P95 for total BPA measured in the aligned studies to a HBM-GV derived from the new TDI proposed would result in RCRs far exceeding 1. Thus, the exposure to BPA would be of concern regarding this new value proposed, and protective measures would need to be taken. The possible concerns of citizens and stakeholders regarding their exposure over time to BPA doses would need to be considered. Such low levels of HBM-GVs may also be below the detection limits of most existing analytical methods, posing the challenge of further improving the sensitivity of detection.

The HBM-GVs developed in the HBM4EU joint programme are associated with confidence levels based on expert judgments regarding the reliability of the data and the calculation method used to derive the HBM-GVs, also taking into account: the nature and quality of the epidemiological and/or toxicological data, uncertainties regarding the modes of action, the choice of starting point, as well as confidence in the PBPK modelling and the exposure scenario assumptions. The confidence levels for the HBM-GVs developed for measuring BPS for the general population and for workers were considered to be medium to low. These HBM-GVs thus need further scientific evaluation using a weight-of-evidence approach to be validated or to improve their confidence level. Several initiatives are underway and, depending on their findings, the risk assessment proposed here may change.

With respect to occupational populations, more studies need to be conducted to assess the risk of exposure to bisphenols for the European population for various occupational activities. This is true for BPA for which data are still limited. This is even more important for BPS and other BPA analogues in a context of BPA replacement in some products following the restriction of BPA in thermal papers, but also the evolution of societal concerns. Furthermore, there is a lack of knowledge at the international level regarding the actual exposure to BPA according to occupational activities, which triggers the need to perform biomonitoring studies for each activity. Calculating the difference in total BPA in urine from pre- and post-work samples may help to assess exposure resulting from workplace BPA ingestion at the individual level. This recommendation is also valid for BPS.

Overall, the available biomonitoring data indicate that the risk from occupational exposure should not be overlooked, although in the absence of a larger data set and a recommendable HBM-GV$_{worker}$ for BPA it is difficult to draw a conclusion regarding risk to workers at this time.

5. Conclusions

In conclusion, this exercise exemplifies how biomonitoring data can be used in risk assessment associated to exposure to bisphenols, and also highlights a number of challenges and uncertainties related to this particular case study. The results point out the need to harmonise the process for collecting and expressing aggregated data, and thus to establish recommendations for conducting these biomonitoring studies, as well as the need to develop a network of laboratories with harmonised analytical capabilities to generate comparable exposure data. All these needs have been addressed in HBM4EU and should be further developed and improved in the future. There is also a need to bridge gaps in terms of knowledge of the toxicity of bisphenols A and S, particularly at low doses. Exposure assessment at the European level needs to be harmonised to better assess the risks. The EFSA Panel has recently reviewed new evidence of BPA toxicity, and if the effects of BPA at very low doses as suggested is confirmed in the near future, then most of the European population would be at risk. It is therefore necessary to continue to produce harmonised and accessible biomonitoring data both for the general population and according to different occupational activities. This is particularly relevant as regulatory action has been taken to reduce BPA, but an increase has been observed in the use and applications of BPS and other analogues.

All of this highlights the importance of developing and harmonising the use of biomonitoring to inform public policy, as is done in European research projects such as the HBM4EU. The new European Partnership for the assessment of risks from chemicals (PARC) is also moving in this direction and is even more broadly involved in the development of all aspects of chemical risk assessment.

Author Contributions: Conceptualisation, M.M. and C.R.; methodology, P.A., M.K, C.R. and M.M.; validation, C.R., P.A. and M.K.-G.; formal analysis, J.-P.A., F.A.Z. and M.M.; investigation, J.-P.A., F.A.Z. and M.M.; data curation, J.-P.A.; writing—original draft preparation, M.M., C.B., J.-P.A. and F.A.Z.; writing—review and editing, C.B., CR., P.A. and M.K.-G.; software, F.A.Z.; visualisation, F.A.Z. and M.M.; supervision, C.R. and P.A.; project administration, M.K.-G. All authors have read and agreed to the published version of the manuscript.

Funding: The HBM4EU project has received funding from the European Union's Horizon 2020 research and innovation programme under grant agreement No 733032 and received co-funding from the author's organisations.

Institutional Review Board Statement: Not applicable.

Informed Consent Statement: Not applicable.

Data Availability Statement: Not applicable.

Acknowledgments: The authors are grateful to Greet Schoeters and Eva Ougier for their support and discussions. The authors would like also to thank Céline Brochot from INERIS for having provided the PBPK model code developed by C. Karrer.

Conflicts of Interest: The authors declare no conflict of interest. The funders had no role in the design of the study; in the collection, analyses, or interpretation of data; in the writing of the manuscript; or in the decision to publish the results.

References

1. ANSES. *Analysis of the Most Appropriate Risk Management Option (RMOA) for Bisphenol A*; ECHA: Helsinki, Finland, 2017.
2. EFSA. Panel on Food Contact Materisals, Enzymes, Flavourings and Processing Aids (CEF) Scientific Opinion on the Risks to Public Health Related to the Presence of Bisphenol A (BPA) in Foodstuffs. *EFSA J.* **2015**, *13*, 3978. [CrossRef]
3. Geens, T.; Goeyens, L.; Covaci, A. Are Potential Sources for Human Exposure to Bisphenol-A Overlooked? *Int. J. Hydrogen Environ. Health* **2011**, *214*, 339–347. [CrossRef] [PubMed]
4. Vandenberg, L.N.; Hauser, R.; Marcus, M.; Olea, N.; Welshons, W.V. Human Exposure to Bisphenol A (BPA). *Reprod. Toxicol.* **2007**, *24*, 139–177. [CrossRef] [PubMed]
5. vom Saal, F.S.; Welshons, W.V. Evidence That Bisphenol A (BPA) Can Be Accurately Measured without Contamination in Human Serum and Urine, and That BPA Causes Numerous Hazards from Multiple Routes of Exposure. *Mol. Cell. Endocrinol.* **2014**, *398*, 101–113. [CrossRef]
6. Konieczna, A.; Rutkowska, A.; Rachoń, D. Health Risk of Exposure to Bisphenol A (BPA). *Rocz. Panstw. Zakl. Hig.* **2015**, *66*, 5–11.
7. Bernier, M.R.; Vandenberg, L.N. Handling of Thermal Paper: Implications for Dermal Exposure to Bisphenol A and Its Alternatives. *PLoS ONE* **2017**, *12*, e0178449. [CrossRef]
8. Rudel, R.A.; Gray, J.M.; Engel, C.L.; Rawsthorne, T.W.; Dodson, R.E.; Ackerman, J.M.; Rizzo, J.; Nudelman, J.L.; Brody, J.G. Food Packaging and Bisphenol A and Bis(2-Ethyhexyl) Phthalate Exposure: Findings from a Dietary Intervention. *Environ. Health Perspect.* **2011**, *119*, 914–920. [CrossRef]
9. ANSES. Évaluation Des Risques Du Bisphénol A (BPA) Pour La Santé Humaine. Available online: https://www.anses.fr/fr/system/files/CHIM2009sa0331Ra-0.pdf?msclkid=3c3a0147c61311eca4e97b466caa524b (accessed on 18 March 2022).
10. FAO; WHO. *Toxicological and Health Aspects of Bisphenol A*; WHO: Ottawa, ON, Canada, 2011.
11. US Food and Drug Administration Bisphenol A (BPA): Use in Food Contact Application. Available online: https://www.fda.gov/food/food-additives-petitions/bisphenol-bpa-use-food-contact-application (accessed on 18 March 2022).
12. Kommission Human-Biomonitoring des Umweltbundesamtes Stoffmonographie Bisphenol A (BPA)—Referenz- und Human-Biomonitoring-(HBM)-Werte für BPA im Urin. *Bundesgesundheitsblatt—Gesundh.—Gesundh.* **2012**, *55*, 1215–1231. [CrossRef]

13. Apel, P.; Angerer, J.; Wilhelm, M.; Kolossa-Gehring, M. New HBM Values for Emerging Substances, Inventory of Reference and HBM Values in Force, and Working Principles of the German Human Biomonitoring Commission. *Int. J. Hydrogen Environ. Health* **2017**, *220*, 152–166. [CrossRef]
14. Tyl, R.W.; Myers, C.B.; Marr, M.C.; Sloan, C.S.; Castillo, N.P.; Veselica, M.M.; Seely, J.C.; Dimond, S.S.; Van Miller, J.P.; Shiotsuka, R.N.; et al. Two-Generation Reproductive Toxicity Study of Dietary Bisphenol A in CD-1 (Swiss) Mice. *Toxicol. Sci.* **2008**, *104*, 362–384. [CrossRef]
15. ECHA; Committee for Risk Assessment (RAC). Opinion on an Annex XV Dossier Proposing Restrictions on Bisphenol A. Available online: https://echa.europa.eu/documents/10162/209030fc-ca4b-4745-97b6-98bfc4d6bdd3 (accessed on 18 March 2022).
16. *Commission Regulation (EU) 2016/2235 of 12 December 2016 Amending Annex XVII to Regulation (EC) No 1907/2006 of the European Parliament and of the Council Concerning the Registration, Evaluation, Authorisation and Restriction of Chemicals (REACH) as Regards Bisphenol A (Text with EEA Relevance)*; Official Journal of the European Union: 2016; Volume 337.
17. Danish, E.P.A. *Alternative Technologies and Substances to Bisphenol A (BPA) in Thermal Paper Receipts*; Danish Environmental Protection Agencey: Copenhagen, Denmark, 2014.
18. ECHA. Bisphenol A. Available online: https://echa.europa.eu/hot-topics/bisphenol-a (accessed on 18 March 2022).
19. EFSA. *Panel on Food Contact Materials, Enzymes and Processing Aids Re-Evaluation of the Risks to Public Health Related to the Presence of Bisphenol A (BPA) in Foodstuffs*; EFSA: Parma, Italy, 2021.
20. Česen, M.; Lenarčič, K.; Mislej, V.; Levstek, M.; Kovačič, A.; Cimrmančič, B.; Uranjek, N.; Kosjek, T.; Heath, D.; Dolenc, M.S.; et al. The Occurrence and Source Identification of Bisphenol Compounds in Wastewaters. *Sci. Total Environ.* **2018**, *616–617*, 744–752. [CrossRef] [PubMed]
21. Viñas, P.; Campillo, N.; Martínez-Castillo, N.; Hernández-Córdoba, M. Comparison of Two Derivatization-Based Methods for Solid-Phase Microextraction–Gas Chromatography–Mass Spectrometric Determination of Bisphenol A, Bisphenol S and Biphenol Migrated from Food Cans. *Anal. Bioanal. Chem.* **2010**, *397*, 115–125. [CrossRef] [PubMed]
22. Liao, C.; Liu, F.; Kannan, K. Bisphenol S, a New Bisphenol Analogue, in Paper Products and Currency Bills and Its Association with Bisphenol A Residues. *Environ. Sci. Technol.* **2012**, *46*, 6515–6522. [CrossRef]
23. ECHA. *Use of Bisphenol A and Its Alternatives in Thermal Paper in the EU 2014-16*; ECHA: Helsinki, Finland, 2017.
24. Yang, Y.; Yang, Y.; Zhang, J.; Shao, B.; Yin, J. Assessment of Bisphenol A Alternatives in Paper Products from the Chinese Market and Their Dermal Exposure in the General Population. *Environ. Pollut.* **2019**, *244*, 238–246. [CrossRef] [PubMed]
25. Beausoleil, C.; Le Magueresse-Battistoni, B.; Viguié, C.; Babajko, S.; Canivenc-Lavier, M.-C.; Chevalier, N.; Emond, C.; Habert, R.; Picard-Hagen, N.; Mhaouty-Kodja, S. Regulatory and Academic Studies to Derive Reference Values for Human Health: The Case of Bisphenol S. *Environ. Res.* **2022**, *204*, 112233. [CrossRef]
26. Ganzleben, C.; Antignac, J.-P.; Barouki, R.; Castaño, A.; Fiddicke, U.; Klánová, J.; Lebret, E.; Olea, N.; Sarigiannis, D.; Schoeters, G.R.; et al. Human Biomonitoring as a Tool to Support Chemicals Regulation in the European Union. *Int. J. Hydrogen Environ. Health* **2017**, *220*, 94–97. [CrossRef]
27. Louro, H.; Heinälä, M.; Bessems, J.; Buekers, J.; Vermeire, T.; Woutersen, M.; van Engelen, J.; Borges, T.; Rousselle, C.; Ougier, E.; et al. Human Biomonitoring in Health Risk Assessment in Europe: Current Practices and Recommendations for the Future. *Int. J. Hydrogen Environ. Health* **2019**, *222*, 727–737. [CrossRef]
28. Ougier, E.; Zeman, F.; Antignac, J.-P.; Rousselle, C.; Lange, R.; Kolossa-Gehring, M.; Apel, P. Human Biomonitoring Initiative (HBM4EU): Human Biomonitoring Guidance Values (HBM-GVs) Derived for Bisphenol A. *Environ. Int.* **2021**, *154*, 106563. [CrossRef]
29. Apel, P.; Rousselle, C.; Lange, R.; Sissoko, F.; Kolossa-Gehring, M.; Ougier, E. Human Biomonitoring Initiative (HBM4EU)—Strategy to Derive Human Biomonitoring Guidance Values (HBM-GVs) for Health Risk Assessment. *Int. J. Hydrogen Environ. Health* **2020**, *230*, 113622. [CrossRef]
30. IPCheM Portal. Available online: https://ipchem.jrc.ec.europa.eu/ (accessed on 18 March 2022).
31. Bousoumah, R.; Leso, V.; Iavicoli, I.; Huuskonen, P.; Viegas, S.; Porras, S.P.; Santonen, T.; Frery, N.; Robert, A.; Ndaw, S. Biomonitoring of Occupational Exposure to Bisphenol A, Bisphenol S and Bisphenol F: A Systematic Review. *Sci. Total Environ.* **2021**, *783*, 146905. [CrossRef]
32. Karrer, C.; Roiss, T.; von Goetz, N.; Gramec Skledar, D.; Peterlin Mašič, L.; Hungerbühler, K. Physiologically Based Pharmacokinetic (PBPK) Modeling of the Bisphenols BPA, BPS, BPF, and BPAF with New Experimental Metabolic Parameters: Comparing the Pharmacokinetic Behavior of BPA with Its Substitutes. *Environ. Health Perspect.* **2018**, *126*, 077002. [CrossRef] [PubMed]
33. MacPherson, S.; Arbuckle, T.E.; Fisher, M. Adjusting Urinary Chemical Biomarkers for Hydration Status during Pregnancy. *J. Expo. Sci. Environ. Epidemiol.* **2018**, *28*, 481–493. [CrossRef] [PubMed]
34. Lau, C.-H.E.; Siskos, A.P.; Maitre, L.; Robinson, O.; Athersuch, T.J.; Want, E.J.; Urquiza, J.; Casas, M.; Vafeiadi, M.; Roumeliotaki, T.; et al. Determinants of the Urinary and Serum Metabolome in Children from Six European Populations. *BMC Med.* **2018**, *16*, 202. [CrossRef] [PubMed]
35. Heinälä, M.; Ylinen, K.; Tuomi, T.; Santonen, T.; Porras, S.P. Assessment of Occupational Exposure to Bisphenol A in Five Different Production Companies in Finland. *Ann. Work Expo. Health* **2017**, *61*, 44–55. [CrossRef]
36. Simonelli, A.; Guadagni, R.; De Franciscis, P.; Colacurci, N.; Pieri, M.; Basilicata, P.; Pedata, P.; Lamberti, M.; Sannolo, N.; Miraglia, N. Environmental and Occupational Exposure to Bisphenol A and Endometriosis: Urinary and Peritoneal Fluid Concentration Levels. *Int. Arch. Occup. Environ. Health* **2017**, *90*, 49–61. [CrossRef]

37. Lyapina, M.; Dencheva, M.S.; Krasteva, A.Z.; Nikolov, G.; Cekova, M.P.; Deliverska, M.; Kisselova-Yaneva, A. Biomonitoring of Urinary Levels of Bisphenol A. *C. R. De L'académie Bulg. Des Sci.* **2016**, *69*, 807–813.
38. González, N.; Cunha, S.C.; Monteiro, C.; Fernandes, J.O.; Marquès, M.; Domingo, J.L.; Nadal, M. Quantification of Eight Bisphenol Analogues in Blood and Urine Samples of Workers in a Hazardous Waste Incinerator. *Environ. Res.* **2019**, *176*, 108576. [CrossRef]
39. Ndaw, S.; Remy, A.; Jargot, D.; Robert, A. Occupational Exposure of Cashiers to Bisphenol A via Thermal Paper: Urinary Biomonitoring Study. *Int. Arch. Occup. Environ. Health* **2016**, *89*, 935–946. [CrossRef]
40. Berman, T.; Goldsmith, R.; Göen, T.; Spungen, J.; Novack, L.; Levine, H.; Amitai, Y.; Shohat, T.; Grotto, I. Demographic and Dietary Predictors of Urinary Bisphenol A Concentrations in Adults in Israel. *Int. J. Hydrogen Environ. Health* **2014**, *217*, 638–644. [CrossRef]
41. Becker, K.; Güen, T.; Seiwert, M.; Conrad, A.; Pick-Fuß, H.; Müller, J.; Wittassek, M.; Schulz, C.; Kolossa-Gehring, M. GerES IV: Phthalate Metabolites and Bisphenol A in Urine of German Children. *Int. J. Hydrogen Environ. Health* **2009**, *212*, 685–692. [CrossRef]
42. Tschersich, C.; Murawski, A.; Schwedler, G.; Rucic, E.; Moos, R.K.; Kasper-Sonnenberg, M.; Koch, H.M.; Brüning, T.; Kolossa-Gehring, M. Bisphenol A and Six Other Environmental Phenols in Urine of Children and Adolescents in Germany—Human Biomonitoring Results of the German Environmental Survey 2014–2017 (GerES V). *Sci. Total Environ.* **2021**, *763*, 144615. [CrossRef] [PubMed]
43. 3xg Studie | 3xgstudie. Available online: https://studie3xg.be/nl (accessed on 18 March 2022).
44. Geens, T.; Bruckers, L.; Covaci, A.; Schoeters, G.; Fierens, T.; Sioen, I.; Vanermen, G.; Baeyens, W.; Morrens, B.; Loots, I.; et al. Determinants of Bisphenol A and Phthalate Metabolites in Urine of Flemish Adolescents. *Environ. Res.* **2014**, *134*, 110–117. [CrossRef] [PubMed]
45. Covaci, A.; Hond, E.D.; Geens, T.; Govarts, E.; Koppen, G.; Frederiksen, H.; Knudsen, L.E.; Mørck, T.A.; Gutleb, A.C.; Guignard, C.; et al. Urinary BPA Measurements in Children and Mothers from Six European Member States: Overall Results and Determinants of Exposure. *Environ. Res.* **2015**, *141*, 77–85. [CrossRef] [PubMed]
46. Hartmann, C.; Uhl, M.; Weiss, S.; Scharf, S.; König, J. Human Biomonitoring of Bisphenol A Exposure in an Austrian Population. *Biomonitoring* **2016**, *3*, 5–14. [CrossRef]
47. Haug, L.S.; Sakhi, A.K.; Cequier, E.; Casas, M.; Maitre, L.; Basagana, X.; Andrusaityte, S.; Chalkiadaki, G.; Chatzi, L.; Coen, M.; et al. In-Utero and Childhood Chemical Exposome in Six European Mother-Child Cohorts. *Environ. Int.* **2018**, *121*, 751–763. [CrossRef] [PubMed]
48. Dereumeaux, C.; Saoudi, A.; Pecheux, M.; Berat, B.; de Crouy-Chanel, P.; Zaros, C.; Brunel, S.; Delamaire, C.; le Tertre, A.; Lefranc, A.; et al. Biomarkers of Exposure to Environmental Contaminants in French Pregnant Women from the Elfe Cohort in 2011. *Environ. Int.* **2016**, *97*, 56–67. [CrossRef]
49. Balicco, A.; Bidondo, M.; Fillol, C.; Gane, J.; Oleko, A.; Saoudi, A.; Zeghnoun, A. *Imprégnation de la population française par les bisphénols A, S et F: Programme national de biosurveillance, Esteban 2014–2016*; Santé Publique France: Saint-Maurice, France, 2019.
50. Frederiksen, H.; Jensen, T.K.; Jørgensen, N.; Kyhl, H.B.; Husby, S.; Skakkebæk, N.E.; Main, K.M.; Juul, A.; Andersson, A.-M. Human Urinary Excretion of Non-Persistent Environmental Chemicals: An Overview of Danish Data Collected between 2006 and 2012. *Reproduction* **2014**, *147*, 555–565. [CrossRef]
51. Porras, S.P.; Heinälä, M.; Santonen, T. Bisphenol A Exposure via Thermal Paper Receipts. *Toxicol. Lett.* **2014**, *230*, 413–420. [CrossRef]
52. Myridakis, A.; Fthenou, E.; Balaska, E.; Vakinti, M.; Kogevinas, M.; Stephanou, E.G. Phthalate Esters, Parabens and Bisphenol-A Exposure among Mothers and Their Children in Greece (Rhea Cohort). *Environ. Int.* **2015**, *83*, 1–10. [CrossRef]
53. Casas, M.; Valvi, D.; Luque, N.; Ballesteros-Gomez, A.; Carsin, A.-E.; Fernandez, M.F.; Koch, H.M.; Mendez, M.A.; Sunyer, J.; Rubio, S.; et al. Dietary and Sociodemographic Determinants of Bisphenol A Urine Concentrations in Pregnant Women and Children. *Environ. Int.* **2013**, *56*, 10–18. [CrossRef]
54. Machtinger, R.; Berman, T.; Adir, M.; Mansur, A.; Baccarelli, A.A.; Racowsky, C.; Calafat, A.M.; Hauser, R.; Nahum, R. Urinary Concentrations of Phthalate Metabolites, Bisphenols and Personal Care Product Chemical Biomarkers in Pregnant Women in Israel. *Environ. Int.* **2018**, *116*, 319–325. [CrossRef] [PubMed]
55. Tkalec, Ž.; Kosjek, T.; Snoj Tratnik, J.; Stajnko, A.; Runkel, A.A.; Sykiotou, M.; Mazej, D.; Horvat, M. Exposure of Slovenian Children and Adolescents to Bisphenols, Parabens and Triclosan: Urinary Levels, Exposure Patterns, Determinants of Exposure and Susceptibility. *Environ. Int.* **2021**, *146*, 106172. [CrossRef] [PubMed]
56. Yang, X.; Doerge, D.R.; Teeguarden, J.G.; Fisher, J.W. Development of a Physiologically Based Pharmacokinetic Model for Assessment of Human Exposure to Bisphenol A. *Toxicol. Appl. Pharmacol.* **2015**, *289*, 442–456. [CrossRef]
57. Oh, J.; Choi, J.W.; Ahn, Y.-A.; Kim, S. Pharmacokinetics of Bisphenol S in Humans after Single Oral Administration. *Environ. Int.* **2018**, *112*, 127–133. [CrossRef] [PubMed]
58. Oh, J.; Choi, J.W.; Ahn, Y.-A.; Kim, S. Response to the Letter to the Editor. *Environ. Int.* **2018**, *115*, 395–396. [CrossRef] [PubMed]
59. Khmiri, I.; Côté, J.; Mantha, M.; Khemiri, R.; Lacroix, M.; Gely, C.; Toutain, P.-L.; Picard-Hagen, N.; Gayrard, V.; Bouchard, M. Toxicokinetics of Bisphenol-S and Its Glucuronide in Plasma and Urine Following Oral and Dermal Exposure in Volunteers for the Interpretation of Biomonitoring Data. *Environ. Int.* **2020**, *138*, 105644. [CrossRef]
60. Champmartin, C.; Marquet, F.; Chedik, L.; Décret, M.-J.; Aubertin, M.; Ferrari, E.; Grandclaude, M.-C.; Cosnier, F. Human in Vitro Percutaneous Absorption of Bisphenol S and Bisphenol A: A Comparative Study. *Chemosphere* **2020**, *252*, 126525. [CrossRef]

61. Zalko, D.; Soto, A.M.; Dolo, L.; Dorio, C.; Rathahao, E.; Debrauwer, L.; Faure, R.; Cravedi, J.-P. Biotransformations of Bisphenol A in a Mammalian Model: Answers and New Questions Raised by Low-Dose Metabolic Fate Studies in Pregnant CD1 Mice. *Environ. Health Perspect.* **2003**, *111*, 309–319. [CrossRef]
62. Gramec Skledar, D.; Peterlin Mašič, L. Bisphenol A and Its Analogs: Do Their Metabolites Have Endocrine Activity? *Environ. Toxicol. Pharmacol.* **2016**, *47*, 182–199. [CrossRef]
63. Kolla, S.; Morcos, M.; Martin, B.; Vandenberg, L.N. Low Dose Bisphenol S or Ethinyl Estradiol Exposures during the Perinatal Period Alter Female Mouse Mammary Gland Development. *Reprod. Toxicol.* **2018**, *78*, 50–59. [CrossRef]
64. Kolla, S.; McSweeney, D.B.; Pokharel, A.; Vandenberg, L.N. Bisphenol S Alters Development of the Male Mouse Mammary Gland and Sensitizes It to a Peripubertal Estrogen Challenge. *Toxicology* **2019**, *424*, 152234. [CrossRef] [PubMed]
65. Catanese, M.C.; Vandenberg, L.N. Bisphenol S (BPS) Alters Maternal Behavior and Brain in Mice Exposed during Pregnancy/Lactation and Their Daughters. *Endocrinology* **2017**, *158*, 516–530. [CrossRef] [PubMed]
66. ECHA. Guidance on Information Requirements and Chemical Safety Assessment Chapter R.8: Characterisation of Dose [Concentration]-Response for Human Health. Available online: https://echa.europa.eu/documents/10162/17224/information_requirements_r8_en.pdf/e153243a-03f0-44c5-8808-88af66223258?t=1353935239897 (accessed on 18 March 2022).
67. Husøy, T.; Andreassen, M.; Hjertholm, H.; Carlsen, M.H.; Norberg, N.; Sprong, C.; Papadopoulou, E.; Sakhi, A.K.; Sabaredzovic, A.; Dirven, H.A.A.M. The Norwegian Biomonitoring Study from the EU Project EuroMix: Levels of Phenols and Phthalates in 24-Hour Urine Samples and Exposure Sources from Food and Personal Care Products. *Environ. Int.* **2019**, *132*, 105103. [CrossRef] [PubMed]
68. Sakhi, A.K.; Sabaredzovic, A.; Papadopoulou, E.; Cequier, E.; Thomsen, C. Levels, Variability and Determinants of Environmental Phenols in Pairs of Norwegian Mothers and Children. *Environ. Int.* **2018**, *114*, 242–251. [CrossRef]
69. Ndaw, S.; Remy, A.; Denis, F.; Marsan, P.; Jargot, D.; Robert, A. Occupational Exposure of Cashiers to Bisphenol S via Thermal Paper. *Toxicol. Lett.* **2018**, *298*, 106–111. [CrossRef]
70. Almeida, D.L.; Pavanello, A.; Saavedra, L.P.; Pereira, T.S.; de Castro-Prado, M.A.A.; de Mathias, P.C.F. Environmental Monitoring and the Developmental Origins of Health and Disease. *J. Dev. Orig. Health Dis.* **2019**, *10*, 608–615. [CrossRef]
71. Le Magueresse-Battistoni, B.; Multigner, L.; Beausoleil, C.; Rousselle, C. Effects of Bisphenol A on Metabolism and Evidences of a Mode of Action Mediated through Endocrine Disruption. *Mol. Cell. Endocrinol.* **2018**, *475*, 74–91. [CrossRef]
72. Mhaouty-Kodja, S.; Belzunces, L.P.; Canivenc, M.-C.; Schroeder, H.; Chevrier, C.; Pasquier, E. Impairment of Learning and Memory Performances Induced by BPA: Evidences from the Literature of a MoA Mediated through an ED. *Mol. Cell. Endocrinol.* **2018**, *475*, 54–73. [CrossRef]
73. Christensen, K.L.Y.; Lorber, M.; Koch, H.M.; Kolossa-Gehring, M.; Morgan, M.K. Population Variability of Phthalate Metabolites and Bisphenol A Concentrations in Spot Urine Samples versus 24- or 48-h Collections. *J. Expo. Sci. Environ. Epidemiol.* **2012**, *22*, 632–640. [CrossRef]
74. Vernet, C.; Philippat, C.; Calafat, A.M.; Ye, X.; Lyon-Caen, S.; Siroux, V.; Schisterman, E.F.; Slama, R. Within-Day, Between-Day, and Between-Week Variability of Urinary Concentrations of Phenol Biomarkers in Pregnant Women. *Environ. Health Perspect.* **2018**, *126*, 037005. [CrossRef]

toxics

MDPI

Article

A Biomonitoring Pilot Study in Workers from a Paints Production Plant Exposed to Pigment-Grade Titanium Dioxide (TiO_2)

Enrico Bergamaschi [1], Valeria Bellisario [1], Manuela Macrì [1], Martina Buglisi [1], Giacomo Garzaro [1], Giulia Squillacioti [1], Federica Ghelli [1], Roberto Bono [1,*], Ivana Fenoglio [2], Francesco Barbero [2], Chiara Riganti [3], Antonella Marrocco [4], Sara Bonetta [1] and Elisabetta Carraro [1]

1 Department of Public Health and Pediatrics, University of Torino, Via Pietro Giuria, 10126 Torino, Italy; enrico.bergamaschi@unito.it (E.B.); valeria.bellisario@unito.it (V.B.); manuela.macri@unito.it (M.M.); martina.buglisi@unito.it (M.B.); giacomo.garzaro@unito.it (G.G.); giulia.squillacioti@unito.it (G.S.); federica.ghelli@unito.it (F.G.); sara.bonetta@unito.it (S.B.); elisabetta.carraro@unito.it (E.C.)

2 Department of Chemistry, University of Torino, Via Pietro Giuria, 10124 Torino, Italy; ivana.fenoglio@unito.it (I.F.); francesco.barbero@unito.it (F.B.)

3 Department of Oncology, University of Torino, 10124 Torino, Italy; chiara.riganti@unito.it

4 Department of Environmental Health, Harvard T.H. Chan School of Public Health, Boston, MA 02115, USA; amarrocco@hsph.harvard.edu

* Correspondence: roberto.bono@unito.it

Abstract: Among particulate matter composing paints, titanium dioxide (TiO_2) forms about 20% of the final suspension. Although TiO_2 is broadly used in many applications, TiO_2 powders represent an established respiratory hazard for workers with long-term exposure. In 35 workers of a paints production plant (15 exposed and 20 not exposed), we assessed pro-inflammatory cytokines (IL-1β, TNF-α, IL-10, IL-17), surfactant protein D (SP-D) and Krebs von den Lungen-6 glycoprotein (KL-6) in exhaled breath condensate (EBC). In urine samples, we measured 8-isoprostane (Isop) and Malondialdehyde (MDA) as biomarkers of oxidative stress, and Titanium (Ti-U) as a biomarker of exposure. Health status, habits and occupational history were recorded. Airborne respirable dusts and Ti were quantified. Particle number concentration and average diameter (nm) were detected by a NanoTracer™ monitoring device. Ti was measurable in filters collected at the respiratory breathing zone (0.11–0.44 $\mu g/m^3$ 8-h TWA). IL-1β and IL-10 values were significantly higher in exposed workers, whereas SP-D was significantly lower ($p < 0.001$). KL-6 was significantly higher in workers than in controls ($p < 0.01$). MDA levels were significantly increased in exposed workers and were positively correlated with Ti-U. Exposure to TiO_2 in paint production is associated with the subtle alterations of lung pathobiology. These findings suggest the need for an integrated approach relying on both personal exposure and biomarker assessment to improve the hazard characterisation in occupational settings.

Keywords: paints production; TiO_2 powders; biological monitoring; exhaled breath condensate; oxidative stress; occupational health

Citation: Bergamaschi, E.; Bellisario, V.; Macrì, M.; Buglisi, M.; Garzaro, G.; Squillacioti, G.; Ghelli, F.; Bono, R.; Fenoglio, I.; Barbero, F.; et al. A Biomonitoring Pilot Study in Workers from a Paints Production Plant Exposed to Pigment-Grade Titanium Dioxide (TiO_2). *Toxics* **2022**, *10*, 171. https://doi.org/10.3390/toxics10040171

Academic Editor: Christophe Rousselle

Received: 20 February 2022
Accepted: 29 March 2022
Published: 31 March 2022

Publisher's Note: MDPI stays neutral with regard to jurisdictional claims in published maps and institutional affiliations.

Copyright: © 2022 by the authors. Licensee MDPI, Basel, Switzerland. This article is an open access article distributed under the terms and conditions of the Creative Commons Attribution (CC BY) license (https://creativecommons.org/licenses/by/4.0/).

1. Introduction

Paint manufacturing involves different operations, such as powder handling, pouring, mixing, dispersing, bin filling and cleaning [1]. Huge volumes of the materials added in conventional paints are handled as dry powders, and some studies have revealed that large amounts of ultrafine particles (UFPs) can be generated during the handling of conventional materials [2,3], which may fall under the EU definition of nanomaterials or contain a fraction of nanoparticles. As a result, during material processing, workers are exposed to a heterogeneous mixture of different particles, which can make the quantitative exposure characterisation and risk assessment very complex. Recent studies have shown that the release of UFPs originating from the handling of conventional micron-sized materials may be substantial [4].

Many of the pigments and fillers used in paints can cause diverse negative health effects after lung exposure, e.g., [5–7], even though both the technological development and the compliance of companies to safety requirements have contributed to dramatically lowering exposure levels.

Among particulate matter composing paints, titanium dioxide (TiO_2) may represent more than 20% of the final suspension by weight. TiO_2 has been used for decades in a wide range of applications, such as photocatalysis, cosmetics and pigments for paints, mainly due to its ability to confer whiteness and opacity to these products. Its high technological attractiveness originates from its light-scattering properties and very high refractive index, which means that relatively low levels of the pigment are required to achieve a white, opaque coating, with the range of light scattered depending on particle size [8].

Although for many years TiO_2 has been considered as an inert and safe material, belonging to low-solubility and low-toxicity particles (LSLTP), with the increasing production of ultrafine TiO_2 powders, concern about the possible consequences of workplace exposure has emerged [9].

The European Chemical Agency (ECHA) has recently classified some TiO_2 powders, and powder mixtures containing TiO_2, as carcinogenic by inhalation (Category 2, H351) when supplied on its own or in mixtures, where the substance or mixture contains 1% or more TiO_2 particles with an aerodynamic diameter <10 μm. Liquid mixtures containing TiO_2 are not classified as carcinogenic, but if they contain particles with an aerodynamic diameter of <10 μm, they need to be labelled as "hazardous respirable droplets may be formed when sprayed" (EUH211) [10].

The hazard of nanoscale or ultrafine TiO_2 in occupational settings has been investigated in some studies concerning early health effects in workers manufacturing TiO_2 [11–14]. Pelclova et al. [13,15] performed a study of 36 workers exposed to NP-TiO_2 pigment and 45 controls. The median total masses of TiO_2 concentrations measured in manufacturing laboratories in 2012 and 2013 were 0.65 and 0.40 mg/m^3, respectively. The median of the concentrations measured by the scanning mobility particle sizer (SMPS) and aerodynamic particle sizer (APS) were 1.98×10^4 and 2.32×10^4 particles/cm^3, respectively, with about 80% of particles smaller than 100 nm in diameter. Most of the oxidative stress biomarkers, assessed in the exhaled breath condensate (EBC) samples, taken before the start and at the end of the shift, were higher in production workers than in the control groups. A multiple linear regressions analysis confirmed the association between the production of TiO_2 and the levels of the biomarkers. The Ti concentration was determined in EBC as a direct biomarker of the exposure of workers involved in the production of the TiO_2 pigment, while the biomarkers of oxidative stress reflected the biological effects of NP-TiO_2 on the respiratory tract. In a further study, Pelclova et al. [13] focused on lipid peroxidation biomarkers both in EBC and urine to identify the most suitable oxidative stress markers for the non-invasive routine biological monitoring of workers occupationally exposed to NP-TiO_2. Malondialdehyde (MDA), 4-hydroxy-trans-hexenal (HHE), 4-hydroxynonenal (HNE), 8-isoprostaglandin F2α (8-ISO) and linear aldehydes (C6-C12) were determined in the EBC and urine of 34 workers and 45 unexposed controls. All lipid peroxidation biomarkers were significantly higher in workers than in controls. Moreover, a dose-dependent association was found between TiO_2 and lipid peroxidation biomarkers in EBC, but not in urine samples. These results confirm the oxidative stress hypothesis as the main mechanism of action of TiO_2 and give a consistent explanation of the lung damage occurring after long-term exposure.

Tumor necrosis factor alpha (TNF-α) and interleukins 1-beta (IL-1 β) are cytokines activated in the presence of oxidative stress, as they are directly up-regulated by the redox-sensitive transcription factor nuclear factor-κB (NFκB) and/or increased upon the activation of inflammosome [16]. An increase in the level of TNF-α and IL-1 β is indicative of the recruitment and activation of pro-inflammatory Th1 lymphocytes [17]. In addition, IL-17 is produced upon pro-inflammatory stimuli to indicate the recruitment of macrophages and Th2 lymphocytes to promote the maturation of B lymphocytes in plasma cells [18]. By contrast, IL-10 is an immuno-suppressive cytokine, which reduces the recruitment of effector T cells and counteracts the effects of TNF-α and IL-1β. It is common that the exposure to pro-oxidant agents simultaneously increases biomarkers of oxidative stress such as MDA, HHE, HNE and 8-ISO, and inflammatory cytokines [16]. These mechanisms may trigger a vicious circle that amplifies the initial damage on the lung tissue, leading to necro-apoptosis or, in the case of prolonged oxidative stress and chronic inflammation, fibrosis [16].

Zhao et al. [14] carried out a cross-sectional study in a nano-TiO_2 manufacturing plant in eastern China. Besides TiO_2 exposure, the authors assessed some biomarkers of cardiopulmonary effects in exposed workers and in control subjects. In the packaging workshop, the total mass concentration of particles (measured by a micro-orifice uniform deposit impactor in a range 10–18,000 nm) was 3.17 mg/m^3, with 39% of nanoparticles accounting for the whole mass. Lung damage markers (namely, surfactant protein D—SP-D); cardiovascular disease markers (VCAM-1, ICAM-1, LDL, and TC); oxidative stress markers Superoxide Dismutase (SOD) and MDA; and inflammation markers (IL-8, IL-6, IL-1β, TNF-α, and IL-10) showed association with TiO_2 nanoparticles' concentration. Similarly, SP-D showed a (dose)-response relationship within the exposed group.

The above studies were carried out in production facilities, where high levels of exposure are expected, whereas no studies are available among TiO_2 end-users, which usually experience lower exposure levels. Moreover, the subtle changes observed in previous studies need to be further confirmed at lower exposure concentrations.

One existing gap is the lack of information regarding the physico-chemical and morphological characterisations of the TiO_2 used in the various production processes. In fact, the most recent epidemiological studies suggest the need to adequately characterise TiO_2 powders to understand which of them can cause harmful effects [19].

With the aim to fill in these gaps, we carried out a pilot study aimed at characterising the exposure to TiO_2 in an occupational setting and testing a battery of biomarkers of local pulmonary or systemic effects in a small group of occupationally exposed workers during the production of paints.

2. Materials and Methods

2.1. Subjects and Study Design

The study was carried out in a company producing paints in the province of Turin, north western Italy. A workplace site visit was carried out to assess the exposure to powders, including TiO_2, and to identify the critical production phases. Therefore, 35 workers were recruited: 15 exposed workers assigned to the production departments and 20 not-exposed workers (control group: administration, design, and marketing). All the workers were informed about the study through an information brochure on the aims, objectives and the various phases of the project aims. Each subject was identified and anonymised with a unique numeric code, and all data, in accordance with current EU legislation, have been processed in an aggregate and anonymous manner. The study was conducted according to the guidelines of the Declaration of Helsinki and approved by the Bioethics Committee of the University of Turin (protocol number 256219/2019). A format to gather the informed consent to participate in the study was signed by all the participants upon the approval by the Bioethical Committee. A questionnaire was administered to all workers involved to collect information about working exposure/habits and on some possible confounding factors (e.g., age, personal and work history, drug use, smoking and alcohol habits, etc.).

2.2. Characterisation of Materials Containing TiO_2

The two TiO_2 powders that were mainly used as pigment in the factory are herein indicated as (i) T-PS: a pure, untreated TiO_2; (ii) T-PR: a TiO_2 with traces of Al and Si and an organic treatment. The commercial names are not reported so as to not identify the supplier.

All samples were characterised for elemental composition, size distribution, morphology, crystal structure, surface charge and surface reactivity.

Size distribution in the micrometric range. An analysis was performed by using a Sysmex FPIA3000 analyser. Before the measurements, the samples were suspended in water (0.5 mg/mL) and sonicated for 2 min with a probe sonicator (100 W, 60 kHz, Sonoplus, Bandelin, Berlin, Germany). A high-power field (2× secondary lens) was applied allowing us to measure particles from 1 to 40 μm.

Size distribution in the nanometric range. An analysis was performed by using a ZetaView® PMX-120 (Particle Metrix GmbH, Inning am Ammersee, Germany) nanoparticle tracking analyser (NTA), equipped with a light source wavelength of 488 nm. Before the measurements, the samples were suspended in double filtered milli-Q water (5 mg/mL) and well vortexed, then stock dispersions were further diluted in double filtered milli-Q water (final concentration 5×10^{-5} mg/mL); the concentration was found suitable for the NTA analysis. After the optimisation of the instrumental parameters, the sensitivity and the shutter were set at 65 and 100, respectively; 33 videos of 1 s for each sample were recorded analysing ~50 particles/video.

Crystalline phase. X-ray diffraction (XRD) spectra measurements were performed by means of a diffractometer (PW1830, Philips, Amsterdam, The Netherlands) using CoKa radiation, in the (20–90) 2q range, with step width 2q ¼ 0.05. Diffraction peaks have been indexed according to the ICDD database (International Centre for Diffraction Data). The spectra have been elaborated (X'pert Highscore 1.0c, PANalytical B.V.) in order to assess the crystalline phase of the different specimens.

Specific surface area (BET). The surface area of the particles was measured by means of the Brunauer, Emmett, and Teller (BET) method based on N_2 adsorption at 77 K (Micrometrics ASAP 2020).

Surface chemistry. The ζ-potential was evaluated by means of electrophoretic light scattering (ELS) (Zetasizer Nano-ZS, Malvern Instruments, Worcestershire, UK). TiO_2 particles were suspended in ultrapure water and then sonicated for 2 min with a probe sonicator (100 W, 60 kHz, Sonoplus, Bandelin, Berlin, Germany). The ζ-potential was measured at different pH (2–9) by adding 0.1 M HCl or NaOH to the suspension.

Surface reactivity. The NMs surface reactivity was monitored by EPR spectroscopy (Miniscope 100 EPR spectrometer, Magnettech, Berlin, Germany) using TEMPONE-H (1-hydroxy-2,2,6,6-tetramethyl-4-oxo-piperidine, Enzo Life Sciences, Inc., New York, NY, USA) as the spin probe. An amount of powder, corresponding to an exposed surface area of 1.4 m^2, was suspended in 2 mL of water or phosphate buffer containing TEMPONE-H 50 mM (1-hydroxy-2,2,6,6-tetramethyl-4-oxo-piperidine, Enzo Life Sciences, Inc., New York, NY, USA). The suspension was constantly stirred under illumination in a quartz vial equipped with a 500 W mercury/xenon lamp (Oriel Instruments, Stratford, CT, USA) and an IR water filter, to avoid the overheating of the suspensions and a 400 nm cut-off filter. The EPR spectra were recorded on 50 μL of the suspension.

EDX analysis. A qualitative elemental analysis was performed by scanning electron microscopy (SEM—COXEM EM30AX Plus EDAX, Daejeon, Korea).

2.3. Environmental and Personal Monitoring

Airborne dust concentrations were measured at specific workstations identified during the site visit. In particular, departments corresponding to specific operations have been identified: (a) water-based paints and storage; (b) enamels and solvents; (c) administrative offices; (d) quality control laboratory; (e) packaging; (f) warehouse. Among these areas, those characterised by greater production activity were chosen for environmental and personal sampling.

The exposure of workers was assessed by environmental and personal sampling, determining the mass, number of fine/ultrafine particles and physicochemical characteristics (shape, size, crystalline form and reactivity) of the powders containing TiO_2. Air Check Touch fixed cyclone head samplers (SKC, PA) equipped with polycarbonate filters (0.2 μm cut-off) were used to sample the air directly breathed by workers (personal sampling) as well as near the main processing areas (environmental sampling).

The measurements were integrated with air sampling in the major production areas of the company, where TiO_2 is usually used, using a Nanotracer™ PNT1000 particle counter (Oxility, Bilthoven, The Netherlands). The instrument, based on the electric charge of the airborne particles, allows the collection of information concerning these particles in the 10–300 nm dimensional range (nanoparticles and ultrafine particles). Data stored were analysed by the NanoReporter™ software. On each of the two days of the study, we measured the particle number concentration (#particles/cm^3), the mean particle diameter (nm) and the surface area of lung deposition, i.e., an estimate of the surface area of the particles deposited in the various compartments of the respiratory tract per unit of inhaled air volume ($\mu m^2/cm^3$).

2.4. Biological Sampling

Biological sampling involved the collection of the exhaled breath condensate (EBC) for the measurement of both oxidative stress and inflammation biomarkers, following the recommendations of the American Thoracic Society and the European Respiratory Society Task Force [19,20]. The EBC was collected for each subject by means of a portable condenser (TURBO-DECCS™—Medivac, Parma, Italy). The TURBO is a refrigeration device and incorporates an aluminium Peltier cooling unit, which houses the EBC collection tube. DECCS refers to the "disposable medical plastic respiratory system"—medical polyethylene—consisting of a mouthpiece connected to a one-way suction valve with a saliva trap and an EBC collection tube inserted into the cooling unit set at −10 °C. Subjects included in the study were asked to breathe at tidal volume into a tube through a mouthpiece with a two-way breathing valve to separate inspiratory and expiratory air and saliva, while wearing a nose clip. On average, 1.7 to 2.3 mL of condensate was collected per participant. Internal standardisation was carried out by collecting a predetermined volume of air (90 L) within about 15 min of breathing at rest by VOLMET 20, suitable for measuring the total volume of air exhaled during an EBC collection session, placed on the DECCS disposable collection circuits. Immediately after the sampling, the EBC samples were divided into 200 µL aliquots in sterile Eppendorf™ tubes and stored at −20 °C for short-term storage (<30 days) and at −80 °C for long-term storage (>30 days).

In parallel, several compounds were analysed as biomarkers of early and subclinical effects in EBC. Tumour necrosis factor alpha (TNF-α), interleukins 1-beta (IL-1β), 17 (IL-17) and 10 (IL-10) were determined as biomarkers of inflammation and immunosuppression. Surfactant protein D (SP-D) and Krebs von den Lungen 6 (KL-6) glycoprotein were determined as potential biomarkers of interstitial lung disease. All analyses were performed with specific immune-enzymatic methods (ELISA).

Finally, before starting the morning work shift, the volunteers provided a spot urine sample for the determination of urinary Titanium (Ti-U), as a biomarker of exposure, and of oxidative stress biomarkers, namely 8-Isoprostane (IsoP) and Malondialdehyde (MDA). Ti-U was determined by inductively coupled plasma mass spectrometry (ICP-MS). Urinary 15-F2t-IsoP concentrations were measured by a competitive enzyme-linked immunoassay (ELISA, Oxford, MI, USA), while MDA concentrations were measured by a colorimetric assay (Oxford, MI, USA), according to the manufacturer's instructions and as already described in our previous work [21]. The 15-F2t-Isop concentrations were normalized by creatinine. Urinary creatinine was determined by the kinetic Jaffé method.

2.5. Statistical Analysis

A statistical analysis was conducted using SPSS for Windows®, Version 25.0 (Statistical Package for Social Sciences, Chicago, IL, USA). The normality of the distribution was assessed by the Kolmogorov–Smirnov test. The differences between the central tendency values of the two subgroups examined were evaluated by the Mann–Whitney test and Student's *t* test. Values are here reported as mean ± standard deviation (SD) and as median and interquartile range (25–75th percentile) for the parametric and non-parametric variables, respectively. The correlations between the variables were evaluated using the Pearson correlation coefficient (r) and Spearman's rho for the normally distributed and non-normally distributed variables, respectively. Values of $p < 0.05$ were considered statistically significant. A multiple linear regression analysis model, adjusted by smoke and working age, was applied to assess the association between the biomarkers and workplace exposure.

3. Results

3.1. Characterisation of the Exposure

Environmental sampling: the time course of particle number concentrations in the 10–300 nm dimensional range was measured by a NanotracerTM PNT1000 particle counter to estimate the emission profile over two standard working days (Figure 1).

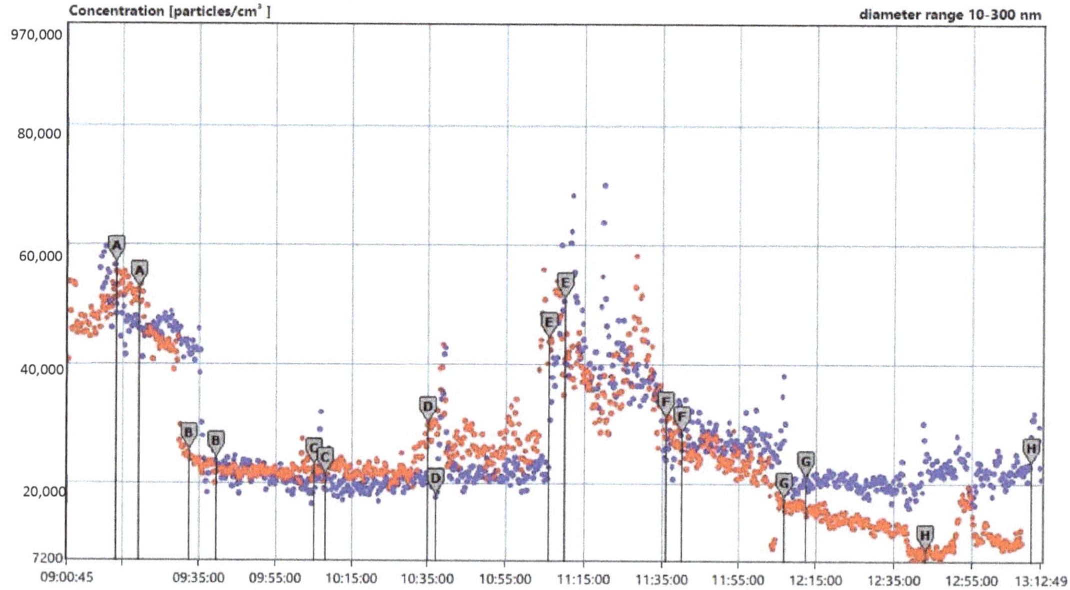

Figure 1. Time course (*X*-axis) of particle number concentrations (*Y*-axis) in the 10–300 nm dimensional range: first day (blue bullet) and second day (red bullet) of the sampling. A–H markers refer to the different working areas assessed in two subsequent days. A = TiO_2 storage area (confined area used as positive control); B = water-based paint system; C = water-based paint system during automatic bin filling; D = water-based paint system during manual bin filling; E = manual handling of powders, mixing and dispersion; F = manual handling, mixing and dispersion; G = office building; H = outdoor environment (NB: the first day there was one truck to supply the silos with TiO_2).

The median value of the respirable fraction of dusts was 0.064 mg/m^3 in the water-based paint production area, while the highest values were found in the enamel production area (0.112–0.137 mg/m^3). The values of the respirable fraction of airborne TiO_2 were lower, with peaks of 0.114 µg/m^3 as the weighted average concentration over an 8 h working time (8 h-TWA) in stationary samples. On the contrary, the TiO_2 concentrations (8 h-TWA) measured in the workers' breathing zone showed slightly higher values, with a minimum of 0.011 µg/m^3 up to values of 0.462 µg/m^3 detected during the production of water-based paints.

The number of particles in the nanometer range was on average five times greater than the environmental background measured outside the company, during the operations of dumping bags into containers and mixing wet powders. The time course of particle number concentration was quite similar in the two days, owing to quite standardised activities and working tasks.

Table 1 summarises the concentration values in mass units and numbers measured in the various areas during the production process.

Table 1. Results of the sampling of airborne dust in the different production environments. PBZ: personal breathing one.

Company Area/ Type of Samplig	Water-Based Paint System	Automatic Bin Filling	Mixing and Dispersion	Administrative Office	Outdoor (Day 1)
Area monitoring respirable dusts (mg/m^3 8 h-TWA)	0.064	0.013	0.112; 0.137	0.033	
Area monitoring respirable Ti ($\mu g/m^3$ 8 h-TWA)	0.018	0.018; 0.114	0.012; 0.024	0.013	
PBZ-Ti ($\mu g/m^3$ 8 h-TWA)	0.104; 0.462	0.011; 0.012	0.07; 0.014	0.012	
Particle number concentrations $\times 10^3$ (average aerodynamic diameter, nm)	24.98; 54.68 (64–73)	20.8; 27.68 (78–93)	40.72; 46.40 (67–95)	16.97 (71)	8.16 (72)

3.2. *Physico-Chemical Characterisation of the TiO_2 Powders Handled by the Workers*

Physico-chemical properties are known to modulate the hazard of particulate materials. Crystallinity [22], surface reactivity [23] and particles' size and shape [24] have been reported to modulate the toxicity of TiO_2 powders. To predict the possible properties of the airborne TiO_2, two TiO_2 powders used in the factory were analysed. Samples were characterised for composition (EDX), size and dimensional distribution in both nano- and micron-range (NTA, FPIA), morphology (FPIA), crystal structure (X-ray diffractometry), specific surface area (BET), and surface charge (ELS) and surface reactivity (EPR). Table 2 summarises the results.

Table 2. Elemental composition, crystallinity, average size (hydrodynamic diameter), dimensional distribution and polydispersion index (PdI) of the various samples of TiO_2.

TiO_2 Sample	Elemental Composition (% *w/w*)	Crystal Phase	Particles Size (FPIA) (µm)	Size (Hydrodynamic Diameter, NTA) (nm)	Surface Area BET m^2/g	ζ Potential (Water) (mV)
T-PS (untreated TiO_2)	Ti 58.7, O 40.2, Al 1.1	**100% rutile**	1–4	187.1 ± 118.6	56.2 ± 0.27	-22.7 ± 0.642
T-PR (Al, Si, organic treatment)	Ti 55.0, O 40.9, Al 2.3, Si 1.8	**100% rutile**	1–10	169.9 ± 96.6	14.66 ± 0.11	11.9 ± 1.13

Both powders were very electrostatic and able to disperse extremely quickly in the air at minimal contact. This characteristic undoubtedly increases the possibility that they will be absorbed through the respiratory tract by the workers who handle them. The size distribution of the powders was analysed in the micrometric and nanometric range by integrating FPIA (Figure 2) and NTA (Figure S1) techniques.

As expected, both powders contained particles in the micrometric range (Figure 2). However, Ti-PR appeared to have particles in the respirable range (1–4 µm). In total, 50% of the particles had a diameter lower than 2 µm. Ti-PS exhibited particles with a diameter in the whole detectable range (1–10 µm), and, similarly to Ti-PR, 50% of the particles had a diameter lower than 2 µm. Particles appeared to be mainly isometric (50% of the particles had a circularity >0.9), but with an irregular shape, as expected by the powder obtained by grinding. Nanometric particles were detected for both samples (Figure S1) in similar abundance. The nanoparticles covered a wide range of size distribution, and had a mean hydrodynamic diameter of 187.1 and 169.9 nm for Ti-PS and Ti-PR, respectively. The measurement of the specific surface area (SSA) gave complementary information on the particles' size for non-porous materials. Ti-PS exhibited an SSA of 56.2 + 0.27 m^2/g typical of a nanometric powder, suggesting that micrometric particles were formed by aggregates of smaller particles. Conversely, Ti-PR had an SSA equal to 14.66 + 0.11 m^2/g, typical of a micrometric powder.

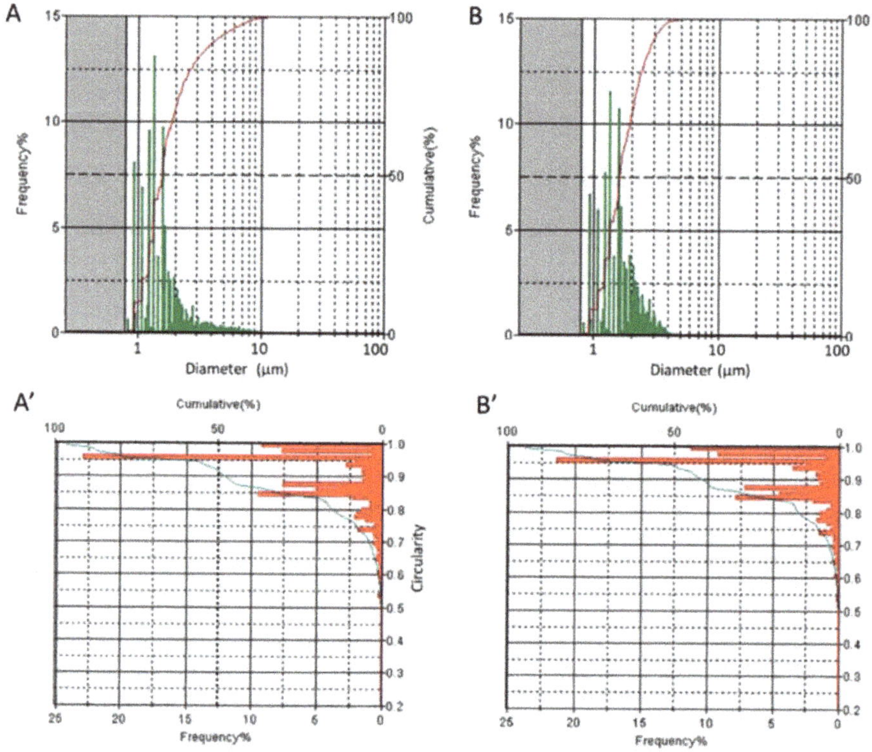

Figure 2. Particle size and shape distribution evaluated by flow particles images analyser (FPIA). Particle size distribution (frequency and cumulative) of (**A**) Ti-PS and (**B**) Ti-PR; circularity (frequency and cumulative) of (**A′**) Ti-PS and (**B′**) Ti-PR.

An XRD analysis revealed rutile as the single crystalline phase for both samples (Figure S2). The properties of the surface of the particles were investigated by electrophoretic light scattering (ELS) measuring the ζ-potential values at different pH (Figure S3A). The ζ-potential curves had the typical features of TiO_2 powder, with positive ζ-potential values at acidic pH and negative pH at basic pH. However, the two samples had a different point of zero charge (PZC), suggesting the presence at the surface of coating/contaminants. These data suggest that TiO_2 was exposed at the surface in both samples. This has been confirmed by EPR spectroscopy (Figure S3B). The powders have been put in contact with a probe [25] and activated with a UV lamp. In this condition, the probe reacts with the surface of TiO_2 generating a radical species that is detected by the EPR spectrometer. Both materials exhibited an intense EPR signal, due to the activation of the TiO_2 phase in both cases exposed to the surface. The signals were lower than those observed for an anatase/rutile photo-reactive material (Aeroxide P25). This was expected because of the crystalline phase (rutile), and less reactive with respect to anatase. T-PR appeared more reactive than T-PS, despite the lower specific surface area and the presence of an organic treatment. This might be due to the presence of surface contaminants possibly modifying the TiO_2 reactivity.

3.3. *General Descriptive of the Population under Study*

Table 3 summarises the main general characteristics of the enrolled groups. The two groups were homogeneous for age and for some characteristics (e.g., BMI) (Levene's test of homogeneity of variance = N.S.). Smokers reported 2–10 cigarettes/day.

Table 3. General descriptive of the population under study. No differences between groups were detectable (Levene's test of homogeneity of variance). Values are mean ± SD.

Main General Characteristics of the Population Investigated							
	General Sample		Exposed (No. 15)		Controls (No. 20)		Levene's Test
Height (cm) Mean ± SD	174 ± 8.1		175 ± 8.9		173 ± 8		0.4
Weight (Kg) Mean ± SD	80.7 ± 14.5		85.7 ± 14.5		82.3 ± 13.6		0.8
BMI Mean ± SD	26.6 ± 4.1		27.9 ± 3.5		26 ± 4.4		0.5
Age (years) Mean ± SD	47.3 ± 11.5		48.7 ± 10.05		45.9 ± 12.18		0.2
Working exposure (years) Mean ± SD	13.8 ± 10.9		14.3 ± 10.8		13.5 ± 12		0.4
Smoke habits (N.)	YES	NO	YES	NO	YES	NO	0.1
	9	26	6	9	3	17	

Table 4 reports the main general values of oxidative and inflammatory biomarkers, also distinguished by TiO_2 exposure. The table also reports the results of statistics (Mann–Whitney U test).

Table 4. General descriptive biomarkers measured in EBC and urine sample with Mann–Whitney U test results.

EBC				
	Total	Exposed	Not Exposed	Mann-Whitney U Test
TNF-α [pg/mL] Mean ± SD, median, range, IQ	1.6 ± 1.1 0.7; [0.5–1.7]; 1.2	0.6 ± 0.2 0.6; [0.5–0.7]; 0.3	1.6 ± 1.3 1.1; [0.6–2.6]; 2	0.03
IL-1β [pg/mL] Mean ± SD, median, range, IQ	0.4 ± 0.1 0.45; [0.3–0.5]; 0.1	0.5 ± 0.05 0.48; [0.4–0.5]; 0.07	0.3 ± 0.1 0.35; [0.25–0.45]; 0.2	<0.001
IL-10 [pg/mL] Mean ± SD, median, range, IQ	1.6 ± 1.1 0.7; [0.5–1.7]; 1.2	0.8 ± 0.07 0.8; [0.7–0.8]; 0.09	0.6 ± 0.1 0.65; [0.5–0.7]; 0.2	<0.001
IL 17 [pg/mL] Mean ± SD, median, range, IQ	1.1 ± 1.1 1.12; [1–1.2]; 1.12	1.1 ± 0.07 1.1; [1–1.1]; 0.09	1.1 ± 1.4 1.15; [1.1–1.2];0.12	0.11
KL-6 [U/L] Mean ± SD, median, range, IQ	990 ± 670 820; [620–1210]; 590	1370 ± 780 1190; [780–1470]; 690	690 ± 390 750; [300–1050]; 750	0.003
SPD [pg/mL] Mean ± SD, median, range, IQ	23.7 ± 2.7 22.8; [21.9–24.4]; 2.4	22.2 ± 0.8 22; [21.7–22.9]; 1.2	24.9 ± 3.2 23.4; [22.2–28.6]; 6.4	0.011
URINE				
MDA [μM] Mean ± SD, median, range, IQ	6.6 ± 3.8 5.7; [3.1–9.3]; 6.2	8.8 ± 3.9 9; [6–10.6]; 4.6	4.8 ± 2.8 4.4; [2.8–5.5]; 2.7	0.002
ISOP [ng/mg crea] Mean ± SD, median, range, IQ	3.7 ± 1.2 3.8; [2.7–4.4]; 1.8	3.9 ± 1.5 3.9; [2.6–4.6]; 2	3.6 ± 1 3.5; [2.7–4.2]; 1.6	0.5
Ti-U Mean ± SD, median, range, IQ	20.4 ± 16.2 17.9; [1.72–90.4]; 17.2	25.9 ± 19.3 23.3; [10.6–90.4]; 15.7	14.5 ± 10.5 11.9; [1.72–44.4]; 17.9	0.02

Among the urinary biomarkers of exposure and oxidative stress, Ti and MDA concentrations were significantly higher in exposed workers than in control subjects (Mann–

Whitney: $p = 0.02$ and $p = 0.002$, respectively) (Figure 3), while IsoP was not significantly different between the two groups.

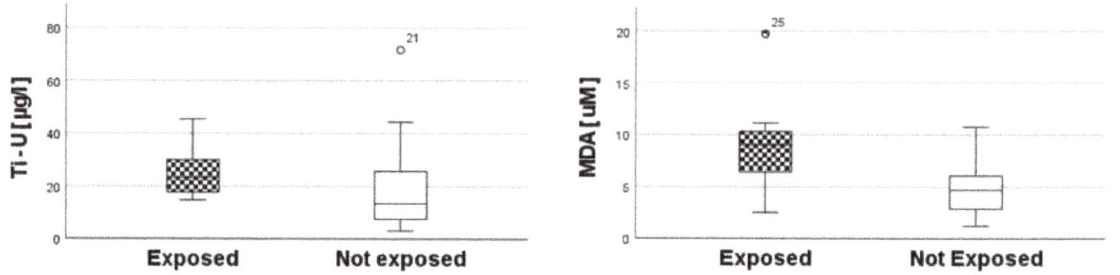

Figure 3. Differences in exposed vs. not exposed of Ti and MDA, as biomarkers of exposure and oxidative stress, respectively.

Figure 4 shows the results of EBC biomarkers in exposed and control workers. Pro-inflammatory cytokines TNF-α, IL-1 β and IL-10 were significantly different between groups (Mann–Whitney: $p = 0.003$, $p < 0.001$, $p < 0.001$, respectively). Conversely, IL-17 showed no significant difference between groups. SP-D values were significantly lower in exposed workers than in controls (Mann–Whitney test: $p = 0.011$). Conversely, the KL-6 glycoprotein values were significantly higher in exposed workers than in controls (Mann–Whitney test: $p = 0.003$).

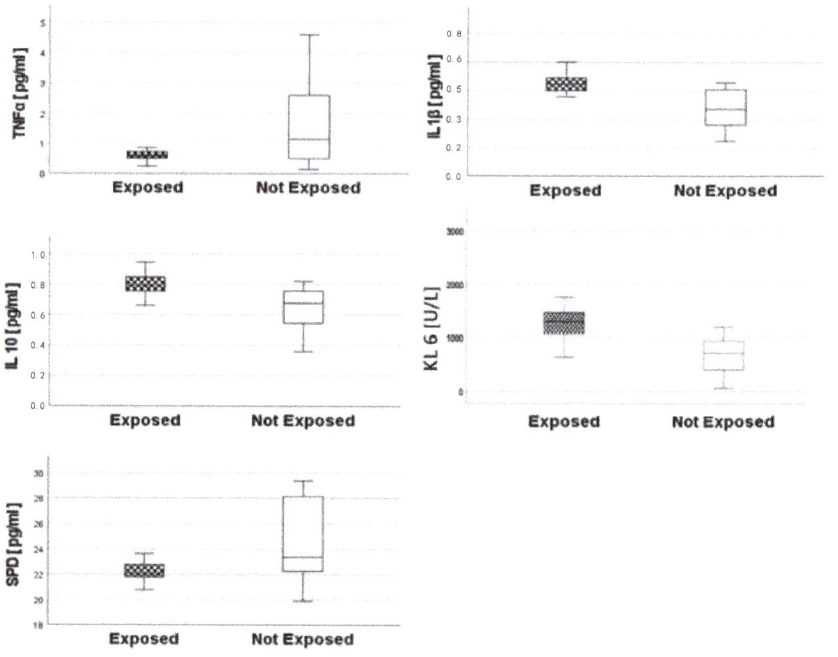

Figure 4. Descriptive statistics of biomarkers determined in EBC (Mann–Whitney test). Pro-inflammatory cytokines TNF-α, IL-1 β and of IL-10 were significantly different between groups ($p = 0.003$, $p < 0.001$, $p < 0.001$, respectively). KL-6 glycoprotein values were significantly higher in exposed workers than in not exposed ($p = 0.003$), whereas SP-D values were significantly lower in exposed workers than in not exposed (p = 0.011).

Positive correlations were found between the urinary biomarkers of effects and oxidative stress. In particular, Ti-U concentrations were positively correlated to MDA levels (Spearman's rho = 0.841, $p < 0.001$). EBC biomarkers also were inter-correlated. In particular, TNF-α was negatively correlated with IL-1β (Spearman's rho = −0.428, $p < 0.001$), with IL-10 (Spearman's rho = −0.454, $p = 0.006$) and with KL-6 (Spearman's rho = −0.391, $p = 0.02$). IL-1β and IL-10 showed a positive and statistically high correlation (Spearman's rho = 0.705, $p < 0.001$), whereas IL-10 was correlated with KL-6 (Spearman's rho = −0.477, $p = 0.004$). In addition, IL-1β and KLF6 showed a positive correlation (Spearman's rho = 0.433, $p = 0.009$). Finally, significant correlations were found also between MDA and TNF-α (Spearman's rho = −0.402, $p = 0.02$), MDA and IL-1β (Spearman's rho = 0.449, $p = 0.007$) and MDA and IL-10 (Spearman's rho = 0.434, $p = 0.009$). Despite the exclusion of different not-significant biomarkers, MDA and IL-1β were chosen as the gold standard of the molecular mechanisms in urine or in the EBC matrix, respectively.

Regarding the urine matrix, the linear regression model, adjusted by smoke and working age, proved an inverse relationship between MDA levels and the workplace exposure (B = −2.5, $p = 0.04$, CI: −0.065/−0.088) (Figure 5A). In EBC, IL-1β showed a similar significant relationship (B = −0.46, $p = 0.04$, CI: −0.007–0.005) (Figure 5B).

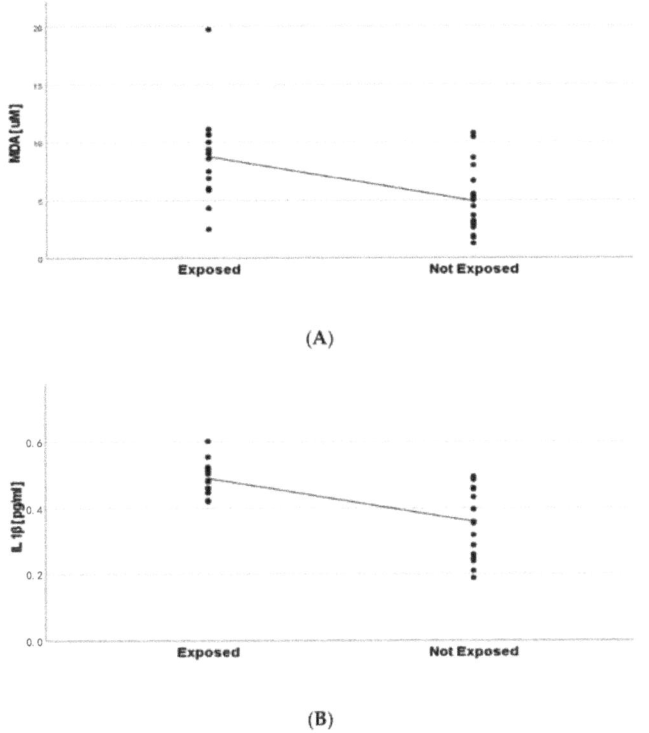

Figure 5. Regression model between working exposure and MDA (urine gold standard) (**A**) and IL1β (EBC gold standard) (**B**).

4. Discussion

Despite the large quantities of material (tons of pigment) handled daily for the production of paints, the present study revealed considerable compliance with safety procedures in the company under investigation. Indeed, the concentration of TiO_2 in the production areas was therefore very low, due to the presence of highly efficient capture systems located near the sources of potential emission, and the natural ventilation assured by the huge doors of the factory that were maintained open. As a result, the airborne concentrations of respirable dusts were always much lower than the current occupational exposure limit values proposed by various international agencies. Nevertheless, the concentration of TiO_2 in the workers' breathing zone was quantifiable to a modest extent and showed values higher than the airborne concentrations, probably because of the close proximity of workers to the emission sources. The dimensional characterisation of the TiO_2 powders handled by the workers confirmed the presence of particles in the micrometric range, but also the presence of an ultrafine fraction of particles, as inferred by NTA and, for Ti-PS, BET analyses.

Currently, in Europe, there are no harmonised limits regarding exposure to TiO_2 dusts neither for workers nor for the general population; however, the growing industrial application of NP-TiO_2 has led several countries to recommend threshold limit values valid for exposures of 8 h a day, for the entire duration of working life. The National Institute for Occupational Safety and Health (NIOSH) in 2011 recommended different occupational exposure limits (REL = recommended exposure level) for fine TiO_2 and for ultrafine TiO_2. The recommended airborne exposure limit (REL) is set at 2.4 mg/m^3 for fine TiO_2 and 0.3 mg/m^3 for ultrafine (including engineered nanoscale) TiO_2, as the time-weighted average (TWA) [9].

Some research programs have suggested occupational exposure limits, such as ENRHES (Engineered Nanoparticles: Review of Health and Environmental Safety), which defined a DNEL (derived no effect level), i.e., an exposure level above which humans should not be exposed equal to 0.017 mg/m^3 (17μg/m^3) for 8 working hours. This value was extrapolated from an NOAEC (no observed adverse effect concentration) corresponding to 5 mg/m^3 for repeated toxicity in a 13-week inhalation study (6 h a day, 5 days a week, exposed to particles from 21 nm) [26].

The French Agency for Food and the Environmental and Occupational Health and Safety (ANSES) has recommended an 8 h occupational exposure limit of 0.80 μg/m^3 [27]. Interestingly, the same agency has proposed a toxicological reference value of 0.12 μg/m^3 for nanosized TiO_2 used when conducting health risk assessments as part of the management of industrial facilities and sites in France [28]. The above value would represent a benchmark value for the prevention of fibro-proliferative alterations and the progressive alteration of the alveolar epithelium towards bronchiolysis [29,30].

However, this threshold can be applicable only to TiO_2-P25 powders (consisting of 80% anatase and 20% rutile with a diameter = 21 nm), which is the only TiO_2 in powder form tested in the study used to establish this value [28]. Considering the huge variety of TiO_2 powders on the European market (more than 350 different types) and the absence of exhaustive toxicological data for each of these, this value may not be applicable for the other forms of NP-TiO_2 (due to the different diameter, crystalline structure, surface coating, etc.) [19]. In the present study, the powders were in the rutile form, which is considered less reactive and hazardous compared to anatase [25]. Though one of the samples was declared by the provider to be treated with an organic coating, in both cases, the ELS and EPR measurements revealed TiO_2 exposed at the surface of the particles.

Over the two days of measurements, some airborne mass concentrations gave values very close to the limit value recommended by the ANSES (0.12 μg/m^3). Thus, in spite of the very low amount measured, we cannot rule out the possible effect occurring in the long time period. This is the main reason why we have assessed both exposure and biomarkers of effect, measured in accessible biological matrices.

Biological monitoring, a widely used practice in occupational and environmental toxicology, combines the use of exposure indicators that provide information on the level of exposure of the subjects, and biological indicators of the effect that aim to highlight potential early and reversible biological alterations associated with/caused by a given exposure.

Regarding biomarkers of exposure, we were able to quantify the urinary concentration of Titanium (Ti, element, not oxide) as a potential biomarker of exposure and internal dose. Interestingly, we found a substantial stability of the urinary Ti levels measured in spot urine samples of the subjects occupationally exposed at the beginning and at the end of the week (data not reported); the difference between the values at the end and the beginning of the week were negligible.

Toxicokinetic studies of ultrafine/nanoparticles suggest preferential absorption via the respiratory and oral routes with possible systemic translocation that can lead to accumulation in peripheral tissues or excretion via faeces and/or urine [31]. Our findings seem to be in agreement with information about the toxicokinetics of Ti evaluated in experimental animals following single exposure [32,33], suggesting that the kinetics of Ti (especially at low doses) are characterised by a persistence beyond 48 h in the lungs, with a gradual reduction of the pulmonary dose in about two weeks, with a slow urinary excretion. These findings suggest that even lower exposure levels (e.g., by occupation) can account for the difference we found between the exposed and the controls, thus mirroring a sort of "steady state" following the accumulation of Ti, which can also be taken up with the diet. However, a clear-cut meaning in toxicological terms to the amount of Ti measured is not predictable.

In the EBC samples collected from the subjects participating in the study, some biomarkers of effect reflecting morphological or functional alteration in the lung parenchyma have been measured. EBC consists mainly of water vapour condensed from the respiratory tract with small droplets of liquid lining the airways from the respiratory tract, including the bronchial and alveolar regions of the lungs, and it is supposed to mirror lung bio-pathology [20]. Within these droplets, there is an unknown fraction of volatile and non-volatile substances. Volatile substances occur in the form of water-soluble compounds in the condensed water of the EBC sample such as ammonia, hydrogen peroxide and ethanol, while non-volatile substances can include cytokines, lipids and salts, but also environmental contaminants, particles and viruses [20,34].

A common toxicity characteristic of the inhaled particles is their propensity to generate oxidative stress at the cellular level, in relation to their surface reactivity which inversely increases with particle size, and which is therefore exacerbated for nanoparticles. Among the many approaches to assess systemic oxidative stress, the determination of 8-isoprostane (IsoP), is of particular interest because it is a stable molecule, present in many tissues and biological fluids, reflecting lipid peroxidation [35–37]. Isoprostanes are prostaglandins generated by cyclooxygenase and by the peroxidation of arachidonic acid. They reliably reflect lipid peroxidation and are ubiquitous in the body, stable in biological fluids and not sensitive to dietary lipid intake. An increase in urinary Isop has been reported after exposure to fine and ultrafine wood smoke particles [37], and in a recent study that aimed to evaluate oxidative stress related to different degrees of urban pollution [17,38].

MDA is another biomarker of lipid peroxidation, which can be easily quantified after the controlled reaction with the reactive substance of thiobarbituric acid (TBARS). Modifications of TBARS concentrations are used to evaluate a wide range of diseases and MDA remains, to date, the most used assay to determine lipid peroxidation [39,40].

IsoP and MDA have been used as biomarkers reflecting systemic oxidative stress, i.e., the oxidative load supported by the organism, and have been found to increase in acute or chronic inflammation and indicate the recruitment and activation of neutrophils and macrophages [41]. However, in our study, urinary IsoP levels were not significantly different between the exposed workers and the control subjects. In spite of this result, we observed that urinary IsoP levels were generally higher if compared with the previous literature [42,43]. This is not surprising, as ELISA assays tend to overestimate IsoP concentrations, mainly due to interferences, and previous authors used liquid chromatography (LC) techniques. Our results are 10-fold higher than those reported by LC-quantified IsoP, falling into the expected range of ELISA overestimation in urine (0.4–61.9 folds) compared to LC [44]. We observed an increase in MDA among the exposed compared to controls, and a positive correlation between the exposure biomarker (Ti-U) and MDA in both the overall group and in the group of exposed workers. This confirms previous observations about an increase in urinary biomarkers of oxidative stress in workers occupationally exposed in the production of pigments [13,45]; although, we found generally lower levels of MDA compared to other studies [43].

In the EBC samples collected from the subjects participating in the study, we measured some biomarkers of effect to assess early functional alteration in the lung biopathology, such as TNF-α, IL-1 β, IL-17 and IL-10 as biomarkers of inflammation and immu-no-suppression. While the increase in IL-1β we detected can be interpreted in a pro-inflammatory sense [17], the significant decrease in TNF-α concentration in the exposed subjects is not easy to interpret. An imbalance between pro-inflammatory TNF-α and anti-inflammatory IL-10 in plasma has been considered a biomarker of suppression of Th1 cell-mediated immunity and exacerbation of Th2 humoral immune response, potentially leading to the development of allergic and autoimmune diseases [46]. We also know that TiO_2 can increase the cytokine secretion by macrophages in vitro [47]. Whether a similar pattern can occur in lung lining fluids, as reflected by EBC determination, should be further elucidated. Overall, an immunosuppressive effect seems to prevail, probably linked to a reduced macrophage activity in the airways of the exposed subjects.

Surfactant protein D (SP-D) values were significantly lower in exposed workers than in controls. Although no studies have considered the SP-D in EBC, it is reasonable that these pneumoproteins may be detectable and measurable. SP-A and SP-D belong to a subgroup of type C lectin (molecular weight: 26–36 and 43 kDa, respectively), and are involved in the regulation of pulmonary host defense and inflammation [48,49]. Our findings are consistent with those of Zhao et al. [14], who found a slight but significant decrease in SP-D levels in serum samples of workers occupationally exposed to nano-TiO_2. These changes were interpreted as the results of a cell injury and/or decrease in number of type II alveolar epithelial cells caused by nanomaterial-induced oxidative stress [14]. In general, it is known that the involvement of the distal airways caused by different types of ultrafine particulates and gases (such as ozone) can induce a reduction in the pneumocyte population, which can explain a reduction in lung and circulating levels of lung pneumoproteins [50].

In our study, conversely, EBC values of the Krebs von den Lungen-6 (KL-6) glycoprotein were significantly higher in exposed workers than in controls. Glycoprotein KL-6 is a high molecular weight mucin (1200 kDa) which has been classified as MUC1—human mucin [51]. Recent clinical studies have suggested that surfactant protein D (SP-D) and Krebs von den Lungen-6 (KL-6) glycoprotein are potential biomarkers of interstitial lung disease. For instance, in Japan they both are used in the clinical setting during the diagnostic tree for the identification of idiopathic pulmonary fibrosis and other types of interstitial lung diseases [52]. The negative correlation between KL-6 and SP-D found in particular in the exposed group seems to suggest a relationship between the depletion of the population of specialized epithelial cells and the activation of a pro-fibrotic cascade. Moreover, these findings are consistent with insights gathered from the mineralogical analyses of BAL samples of patients with lung fibrosis, suggesting a role of titanium nanoparticles in idiopathic pulmonary fibrosis [53].

The overall picture resulting from the combination of the different biomarkers reflecting subtle changes in lung physiology assessed in this study suggests a mild inflammatory status and the activation of a pro-fibrotic cascade. Most likely, these alterations are not due to recent exposure, but could represent a condition due to previous exposure, due to the cumulative effects of such low doses or to the higher exposure occurring in the past. The results also reveal an association between biomarkers of systemic oxidative stress and the exposure of concern. Although not specific, these biomarkers appear to be related to the specific type of exposure investigated.

This biomarker-based approach to assess early effects in workers exposed to nano-objects has been recommended as a component of a strategy to risk assessment and management in the workplace [54–56]. In order to enable feasible and acceptable routine screenings in populations at risk, biomarkers that could be analysed in biological matrices collected by non-invasive procedures are of particular relevance. From this perspective, exhaled breath condensates (EBC), exhaled air (EA) and urine are three preferred biological matrices for the non-invasive investigation of biomarkers in workers exposed to UFPs and nanomaterials, as they give insights on local (pulmonary) and systemic oxidative stress and inflammatory response [13,20,21,54–59]. Although we recognise the higher performance of the LC-MS/MS technique for the quantification of some biomarkers of oxidative stress as compared to ELISA assays [60,61], we acknowledge that ELISA are less time-consuming and more cost-effective techniques, which do not require particular laboratory equipment and highly-trained operators. This may be considered both a limitation and a strength of this study aimed at supporting an extensive quantification of oxidative stress and inflammatory biomarkers in non-invasively collected specimens to monitor workers' health in contexts other than research settings, e.g., periodical health surveillance.

5. Conclusions

Exposure to TiO_2 containing dusts well below the occupational exposure limits (OELs), but close to the threshold for preventing fibro-proliferative and progressive alteration of epithelium, can result in subtle lung changes, as reflected by changes in KL-6 and SP-D and an increase in biomarkers of oxidative stress. Interestingly, the lower threshold proposed by ANSES for TiO_2 represents the benchmark dose for preventing fibro-proliferative and progression alteration of epithelium and alveolar bronchiolisation.

With regard to the exposure parameter, the setting investigated is probably representative of a series of small companies that adopt similar production processes, suggesting a widespread micro-pollution mainly linked to operational but temporary tasks. At the concentrations found in such an environment, however, no short- or long-term effects are expected, owing to the compliance with the threshold limit values adopted or proposed for workplaces. We could argue that just quantifiable aerosols in the nanometer range could affect some health endpoints for which the current standards are not (yet) applicable.

This study suggests the need for a combined approach relying on both exposure assessment and biomarkers of effect to improve the risk assessment in occupational settings in which TiO_2 is handled, though under strict and effective control measures.

Thus, in spite of several drawbacks in implementing an integrated strategy [54], biomonitoring enables regular exposure and health assessment and can support effective risk management [54–56]. Owing to the small number of subjects evaluated and the intrinsic variability of biomarkers, the observed changes along with their health significance must be assessed in a long-term perspective.

Supplementary Materials: The following supporting information can be downloaded at: https://www.mdpi.com/article/10.3390/toxics10040171/s1. Figure S1: Particles size distribution evaluated by Nanoparticle Tracking Analysis (NTA). (A) Aqueous dispersion of T-PS. (B) Aqueous dispersion of T-PR. Figure S2. Crystalline phase. XRD patterns of (a) T-PS and (b) T-PR. Figure S3. ζ-potential and surface reactivity. (A) ζ-potential values measured at different pH of T-PS and T-PR; (B) Surface reactivity of T-PS and T-PR evaluated by EPR/spin trapping technique. Ctrl−. no powder; ctrl+ Aeroxide P25.

Author Contributions: Conceptualization, E.B., S.B., E.C., R.B.; methodology, E.B., V.B., G.S., I.F., F.B., C.R., A.M., S.B.; validation, M.M., G.G., F.G., F.B., C.R., A.M.; formal analysis, V.B., G.S., F.G., G.G., I.F., C.R., M.B.; data curation, V.B.; writing—original draft preparation, E.B., V.B., G.S.; writing—review and editing, R.B., I.F., C.R., S.B., E.C.; supervision, E.B., R.B., I.F., E.C. All authors have read and agreed to the published version of the manuscript.

Funding: This study has received funds from Fondazione Cassa di Risparmio di Torino (CRT) and EU LIFE17 NanoExplore (ENV/GR/000285).

Institutional Review Board Statement: The study was conducted according to the guidelines of the Declaration of Helsinki, and approved by the Bioethics Committee of the University of Turin (protocol number 256219/2019).

Informed Consent Statement: A format to gather the informed consent to participate in the study was signed by all the participants upon the approval of the format by the Bioethical Committee.

Data Availability Statement: Not applicable.

Acknowledgments: Many thanks are due to all workers for their willing cooperation, and to all the volunteers enrolled as controls.

Conflicts of Interest: The authors declare no conflict of interest.

References

1. Paint and Coating Manufacture. Available online: https://www.iloencyclopaedia.org/part-xii-57503/chemical-processing/item/380-paint-and-coating-manufacture (accessed on 23 March 2022).
2. Fonseca, A.S.; Viitanen, A.K.; Kanerva, T.; Säämänen, A.; Aguerre-Chariol, O.; Fable, S.; Dermigny, A.; Karoski, N.; Fraboulet, I.; Koponen, I.K.; et al. Occupational Exposure and Environmental Release: The Case Study of Pouring TiO_2 and Filler Materials for Paint Production. *Int. J. Environ. Res. Public Health* **2021**, *18*, 418. [CrossRef] [PubMed]
3. Van Broekhuizen, P. *Applicability of Provisional NRVs to PGNPs and FCNPs*; Bureau KLB: Den Haag, The Netherlands, 2017.
4. Viitanen, A.K.; Uuksulainen, S.; Koivisto, A.J.; Hämeri, K.; Kauppinen, T. Workplace measurements of ultrafine particles—A literature review. *Ann. Work Expo. Health* **2017**, *61*, 749–758. [CrossRef]
5. Saber, A.T.; Jensen, K.A.; Jacobsen, N.R.; Birkedal, R.; Mikkelsen, L.; Moller, P.; Loft, S.; Wallin, H.; Vogel, U. Inflammatory and genotoxic effects of nanoparticles designed for inclusion in paints and lacquers. *Nanotoxicology* **2012**, *6*, 453–471. [CrossRef] [PubMed]
6. Yanamala, N.; Farcas, M.T.; Hatfield, M.K.; Kisin, E.R.; Kagan, V.E.; Geraci, C.L.; Shvedova, A.A. In Vivo Evaluation of the Pulmonary Toxicity of Cellulose Nanocrystals: A Renewable and Sustainable Nanomaterial of the Future. *ACS Sustain. Chem. Eng.* **2014**, *2*, 1691–1698. [CrossRef] [PubMed]
7. Smulders, S.; Luyts, K.; Brabants, G.; Van Landuyt, K.; Kirschhock, C.; Smolders, E.; Golanski, L.; Vanoirbeek, J.; Hoet, P.H.M. Toxicity of nanoparticles embedded in paints compared with pristine nanoparticles in mice. *Toxicol. Sci.* **2014**, *141*, 132–140. [CrossRef] [PubMed]
8. Braun, J.H. Titanium Dioxide—A Review. *J. Coat. Technol.* **1997**, *69*, 59–72.
9. CDC-NIOSH. Occupational exposure to titanium dioxide: Current Intelligence Bulletin. *DHHS (NIOSH) Publ.* **2011**, *63*, 1–140. [CrossRef]
10. European Chemicals Agency. *ECHA Guide on the Classification and Labelling of Titanium Dioxide*; European Chemicals Agency: Helsinki, Finland, 2021. [CrossRef]
11. Pelclova, D.; Barosova, H.; Kukutschova, J.; Zdimal, V.; Navratil, T.; Fenclova, Z.; Vlckova, S.; Schwarz, J.; Zikova, N.; Kacer, P.; et al. Raman microspectroscopy of exhaled breath condensate and urine in workers exposed to fine and nano TiO_2 particles: A cross-sectional study. *J. Breath Res.* **2015**, *9*, 036008. [CrossRef]
12. Pelclova, D.; Zdimal, V.; Kacer, P.; Fenclova, Z.; Vlckova, S.; Syslova, K.; Navratil, T.; Schwarz, J.; Zikova, N.; Barosova, H.; et al. Oxidative stress markers are elevated in exhaled breath condensate of workers exposed to nanoparticles during iron oxide pigment production. *J. Breath Res.* **2016**, *10*, 016004. [CrossRef]
13. Pelclova, D.; Zdimal, V.; Kacer, P.; Zikova, N.; Komarc, M.; Fenclova, Z.; Vlckova, S.; Schwarz, J.; Makeš, O.; Syslova, K.; et al. Markers of lipid oxidative damage in the exhaled breath condensate of nano TiO_2 production workers. *Nanotoxicology* **2017**, *11*, 52–63. [CrossRef]
14. Zhao, L.; Zhu, Y.; Chen, Z.; Xu, H.; Zhou, J.; Tang, S.; Xu, Z.; Kong, F.; Li, X.; Zhang, Y.; et al. Cardiopulmonary effects induced by occupational exposure to titanium dioxide nanoparticles. *Nanotoxicology* **2018**, *12*, 169–184. [CrossRef]
15. Pelclova, D.; Zdimal, V.; Fenclova, Z.; Vlckova, S.; Turci, F.; Corazzari, I.; Kacer, P.; Schwarz, J.; Zikova, N.; Makes, O.; et al. Markers of oxidative damage of nucleic acids and proteins among workers exposed to TiO_2 (nano) particles. *Occup. Environ. Med.* **2016**, *73*, 110–118. [CrossRef]
16. Vietti, G.; Lison, D.; van den Brule, S. Mechanisms of lung fibrosis induced by carbon nanotubes: Towards an Adverse Outcome Pathway (AOP). *Part. Fibre Toxicol.* **2015**, *13*, 11. [CrossRef]
17. Yazdi, A.S.; Guarda, G.; Riteau, N.; Drexler, S.K.; Tardivel, A.; Couillin, I.; Tschopp, J. Nanoparticles activate the NLR pyrin domain containing 3 (Nlrp3) inflammasome and cause pulmonary inflammation through release of IL-1α and IL-1β. *Proc. Natl. Acad. Sci. USA* **2010**, *107*, 19449–19454. [CrossRef]
18. Kuwabara, T.; Ishikawa, F.; Kondo, M.; Kakiuchi, T. The Role of IL-17 and Related Cytokines in Inflammatory Autoimmune Diseases. *Mediat. Inflamm.* **2017**, *2017*, 3908061. [CrossRef]
19. Guseva Canu, I.; Fraize-Frontier, S.; Michel, C.; Charles, S. Weight of epidemiological evidence for titanium dioxide risk assessment: Current state and further needs. *J. Expo. Sci. Environ. Epidemiol.* **2020**, *30*, 430–435. [CrossRef]
20. Horváth, I.; Hunt, J.; Barnes, P.J.; Alving, K.; Antczak, A.; Baraldi, E.; Becher, G.; Van Beurden, W.J.C.; Corradi, M.; Dekhuijzen, R.; et al. Exhaled breath condensate: Methodological recommendations and unresolved questions. *Eur. Respir. J.* **2005**, *26*, 523–548. [CrossRef]
21. Bellisario, V.; Mengozzi, G.; Grignani, E.; Bugiani, M.; Sapino, A.; Bussolati, G.; Bono, R. Towards a formalin-free hospital. Levels of 15-F2t-isoprostane and malondialdehyde to monitor exposure to formaldehyde in nurses from operating theatres. *Toxicol. Res.* **2016**, *5*, 1122–1129. [CrossRef]
22. Gerloff, K.; Fenoglio, I.; Carella, E.; Kolling, J.; Albrecht, C.; Boots, A.W.; Förster, I.; Schins, R.P.F. Distinctive toxicity of TiO_2 rutile/anatase mixed phase nanoparticles on Caco-2 cells. *Chem. Res. Toxicol.* **2012**, *25*, 646–655. [CrossRef]

23. Fenoglio, I.; Ponti, J.; Alloa, E.; Ghiazza, M.; Corazzari, I.; Capomaccio, R.; Rembges, D.; Oliaro-Bosso, S.; Rossi, F. Singlet oxygen plays a key role in the toxicity and DNA damage caused by nanometric TiO_2 in human keratinocytes. *Nanoscale* **2013**, *5*, 6567–6576. [CrossRef]
24. Johnston, H.J.; Hutchison, G.R.; Christensen, F.M.; Peters, S.; Hankin, S.; Stone, V. Identification of the mechanisms that drive the toxicity of TiO_2 particulates: The contribution of physicochemical characteristics. *Part. Fibre Toxicol.* **2009**, *6*, 33. [CrossRef]
25. Marucco, A.; Carella, E.; Fenoglio, I. A comparative study on the efficacy of different probes to predict the photo-activity of nano-titanium dioxide toward biomolecules. *RSC Adv.* **2015**, *5*, 89559–89568. [CrossRef]
26. Bermudez, E.; Mangum, J.B.; Wong, B.A.; Asgharian, B.; Hext, P.M.; Warheit, D.B.; Everitt, J.I. Pulmonary responses of mice, rats, and hamsters to subchronic inhalation of ultrafine titanium dioxide particles. *Toxicol. Sci.* **2004**, *77*, 347–357. [CrossRef]
27. Dioxyde de Titane Sous Forme Nanoparticulaire: Recommandation de Valeurs Limites D'exposition Professionnelle | Anses—Agence Nationale de Sécurité Sanitaire de L'alimentation, de L'environnement et du Travail. Available online: https://www.anses.fr/fr/content/dioxyde-de-titane-sous-forme-nanoparticulaire-recommandation-de-valeurs-limites-d\T1\textquoterightexposition (accessed on 24 March 2022).
28. Anses—Agence Nationale de Sécurité Sanitaire de L'alimentation, de L'environnement et du Travail Valeurs Toxicologiques de Référence le Dioxyde de Titane Sous Forme Nanoparticulaire. Available online: https://www.anses.fr/fr/content/dioxyde-de-titane-sous-forme-nanoparticulaire-l%E2%80%99anses-d%C3%A9finit-une-valeur-toxicologique-de (accessed on 30 March 2022).
29. Driscoll, K.E.; Deyo, L.C.; Carter, J.M.; Howard, B.W.; Hassenbein, D.G.; Bertram, T.A. Effects of particle exposure and particle-elicited inflammatory cells on mutation in rat alveolar epithelial cells. *Carcinogenesis* **1997**, *18*, 423–430. [CrossRef] [PubMed]
30. Friemann, J.; Albrecht, C.; Breuer, P.; Grover, R.; Weishaupt, C. Time-course analysis of type II cell hyperplasia and alveolar bronchiolization in rats treated with different particulates. *Inhal. Toxicol.* **1999**, *11*, 837–854. [CrossRef] [PubMed]
31. Rinaldo, M.; Andujar, P.; Lacourt, A.; Martinon, L.; Raffin, M.C.; Dumortier, P.; Pairon, J.C.; Brochard, P. Perspectives in Biological Monitoring of Inhaled Nanosized Particles. *Ann. Occup. Hyg.* **2015**, *59*, 669–680. [CrossRef] [PubMed]
32. Tang, M.; Zhang, T.; Xue, Y.; Wang, S.; Huang, M.; Yang, Y.; Lu, M.; Lei, H.; Kong, L.; Yuepu, P. Dose dependent in vivo metabolic characteristics of titanium dioxide nanoparticles. *J. Nanosci. Nanotechnol.* **2010**, *10*, 8575–8583. [CrossRef]
33. Pujalté, I.; Dieme, D.; Haddad, S.; Serventi, A.M.; Bouchard, M. Toxicokinetics of titanium dioxide (TiO_2) nanoparticles after inhalation in rats. *Toxicol. Lett.* **2017**, *265*, 77–85. [CrossRef]
34. Mutlu, G.M.; Garey, K.W.; Robbins, R.A.; Danziger, L.H.; Rubinstein, I. Collection and analysis of exhaled breath condensate in humans. *Am. J. Respir. Crit. Care Med.* **2001**, *164*, 731–737. [CrossRef]
35. Cracowski, J.L. Isoprostanes: A putative key role in vascular diseases. *Rev. Med. Interne* **2004**, *25*, 459–463. [CrossRef]
36. Roberts, L.J.; Morrow, J.D. Measurement of F2-isoprostanes as an index of oxidative stress in vivo. *Free Radic. Biol. Med.* **2000**, *28*, 505–513. [CrossRef]
37. Bono, R.; Capacci, F.; Cellai, F.; Sgarrella, C.; Bellisario, V.; Trucco, G.; Tofani, L.; Peluso, A.; Poli, C.; Arena, L.; et al. Wood dust and urinary 15-F2t isoprostane in Italian industry workers. *Environ. Res.* **2019**, *173*, 300–305. [CrossRef]
38. Bono, R.; Tassinari, R.; Bellisario, V.; Gilli, G.; Pazzi, M.; Pirro, V.; Mengozzi, G.; Bugiani, M.; Piccioni, P. Urban air and tobacco smoke as conditions that increase the risk of oxidative stress and respiratory response in youth. *Environ. Res.* **2015**, *137*, 141–146. [CrossRef]
39. Del Rio, D.; Stewart, A.J.; Pellegrini, N. A review of recent studies on malondialdehyde as toxic molecule and biological marker of oxidative stress. *Nutr. Metab. Cardiovasc. Dis.* **2005**, *15*, 316–328. [CrossRef]
40. Tsikas, D. Assessment of lipid peroxidation by measuring malondialdehyde (MDA) and relatives in biological samples: Analytical and biological challenges. *Anal. Biochem.* **2017**, *524*, 13–30. [CrossRef]
41. Gong, J.; Zhu, T.; Kipen, H.; Wang, G.; Hu, M.; Ohman-Strickland, P.; Lu, S.E.; Zhang, L.; Wang, Y.; Zhu, P.; et al. Malondialdehyde in exhaled breath condensate and urine as a biomarker of air pollution induced oxidative stress. *J. Expo. Sci. Environ. Epidemiol.* **2013**, *23*, 322–327. [CrossRef]
42. Wu, W.-T.; Liao, H.-Y.; Chung, Y.-T.; Li, W.-F.; Tsou, T.-C.; Li, L.-A.; Lin, M.-H.; Ho, J.-J.; Wu, T.-N.; Liou, S.-H.; et al. Effect of Nanoparticles Exposure on Fractional Exhaled Nitric Oxide (FENO) in Workers Exposed to Nanomaterials. *Int. J. Mol. Sci.* **2014**, *15*, 878–894. [CrossRef]
43. Pelclova, D.; Zdimal, V.; Komarc, M.; Schwarz, J.; Ondracek, J.; Ondrackova, L.; Kostejn, M.; Vlckova, S.; Fenclova, Z.; Dvorackova, S.; et al. Three-Year Study of Markers of Oxidative Stress in Exhaled Breath Condensate in Workers Producing Nanocomposites, Extended by Plasma and Urine Analysis in Last Two Years. *Nanomaterials* **2020**, *10*, 2440. [CrossRef]
44. Klawitter, J.; Haschke, M.; Shokati, T.; Klawitter, J.; Christians, U. Quantification of 15-F2t-isoprostane in human plasma and urine: Results from enzyme-linked immunoassay and liquid chromatography/tandem mass spectrometry cannot be compared. *Rapid Commun. Mass Spectrom.* **2011**, *25*, 463–468. [CrossRef]
45. Buonaurio, F.; Astolfi, M.L.; Canepari, S.; Di Basilio, M.; Gibilras, R.; Mecchia, M.; Papacchini, M.; Paci, E.; Pigini, D.; Tranfo, G. Urinary Oxidative Stress Biomarkers in Workers of a Titanium Dioxide Based Pigment Production Plant. *Int. J. Environ. Res. Public Health* **2020**, *17*, 9085. [CrossRef]
46. Dobreva, Z.G.; Kostadinova, G.S.; Popov, B.N.; Petkov, G.S.; Stanilova, S.A. Proinflammatory and anti-inflammatory cytokines in adolescents from Southeast Bulgarian cities with different levels of air pollution. *Toxicol. Ind. Health* **2015**, *31*, 1210–1217. [CrossRef]

47. Bianchi, M.G.; Allegri, M.; Costa, A.L.; Blosi, M.; Gardini, D.; Del Pivo, C.; Prina-Mello, A.; Di Cristo, L.; Bussolati, O.; Bergamaschi, E. Titanium dioxide nanoparticles enhance macrophage activation by LPS through a TLR4-dependent intracellular pathway. *Toxicol. Res.* **2015**, *4*, 385–398. [CrossRef]
48. Kuroki, Y.; Takahashi, M.; Nishitani, C. Pulmonary collectins in innate immunity of the lung. *Cell. Microbiol.* **2007**, *9*, 1871–1879. [CrossRef]
49. Tomonaga, T.; Izumi, H.; Yoshiura, Y.; Nishida, C.; Yatera, K.; Morimoto, Y. Examination of Surfactant Protein D as a Biomarker for Evaluating Pulmonary Toxicity of Nanomaterials in Rat. *Int. J. Mol. Sci.* **2021**, *22*, 4635. [CrossRef]
50. Hermans, C.; Bernard, A. Lung epithelium-specific proteins: Characteristics and potential applications as markers. *Am. J. Respir. Crit. Care Med.* **1999**, *159*, 648–678. [CrossRef]
51. Ohyabu, N.; Hinou, H.; Matsushita, T.; Izumi, R.; Shimizu, H.; Kawamoto, K.; Numata, Y.; Togame, H.; Takemoto, H.; Kondo, H.; et al. An essential epitope of anti-MUC1 monoclonal antibody KL-6 revealed by focused glycopeptide library. *J. Am. Chem. Soc.* **2009**, *131*, 17102–17109. [CrossRef]
52. Hirasawa, Y.; Kohno, N.; Yokoyama, A.; Inoue, Y.; Abe, M.; Hiwada, K. KL-6, a human MUC1 mucin, is chemotactic for human fibroblasts. *Am. J. Respir. Cell Mol. Biol.* **1997**, *17*, 501–507. [CrossRef]
53. Forest, V.; Pourchez, J.; Pélissier, C.; Durand, S.A.; Vergnon, J.M.; Fontana, L. Relationship between Occupational Exposure to Airborne Nanoparticles, Nanoparticle Lung Burden and Lung Diseases. *Toxics* **2021**, *9*, 204. [CrossRef]
54. Bergamaschi, E.; Poland, C.; Guseva Canu, I.; Prina-Mello, A. The role of biological monitoring in nanosafety. *Nano Today* **2018**, *10*, 274–277. [CrossRef]
55. Bergamaschi, E.; Guseva Canu, I.; Prina-Mello, A.; Magrini, A. Biomonitoring. In *Adverse Effect of Engeneered Nanomaterials: Exposure, Toxicology and Impact on Human Health*, 2nd ed.; Elsevier: Amsterdam, The Netherlands, 2017; pp. 125–158. [CrossRef]
56. Schulte, P.; Leso, V.; Niang, M.; Iavicoli, I. Biological monitoring of workers exposed to engineered nanomaterials. *Toxicol. Lett.* **2018**, *298*, 112–124. [CrossRef] [PubMed]
57. Canu, I.G.; Schulte, P.A.; Riediker, M.; Fatkhutdinova, L.; Bergamaschi, E. Methodological, political and legal issues in the assessment of the effects of nanotechnology on human health. *J. Epidemiol. Community Health* **2018**, *72*, 148–153. [CrossRef]
58. Graczyk, H.; Lewinski, N.; Zhao, J.; Sauvain, J.J.; Suarez, G.; Wild, P.; Danuser, B.; Riediker, M. Increase in oxidative stress levels following welding fume inhalation: A controlled human exposure study. *Part. Fibre Toxicol.* **2016**, *13*, 31. [CrossRef]
59. Marie-Desvergne, C.; Dubosson, M.; Touri, L.; Zimmermann, E.; Gaude-Môme, M.; Leclerc, L.; Durand, C.; Klerlein, M.; Molinari, N.; Vachier, I.; et al. Assessment of nanoparticles and metal exposure of airport workers using exhaled breath condensate. *J. Breath Res.* **2016**, *10*, 036006. [CrossRef]
60. Shoman, Y.; Wild, P.; Hemmendinger, M.; Graille, M.; Sauvain, J.-J.; Hopf, N.B.; Canu, I.G. Reference Ranges of 8-Isoprostane Concentrations in Exhaled Breath Condensate (EBC): A Systematic Review and Meta-Analysis. *Int. J. Mol. Sci.* **2020**, *21*, 3822. [CrossRef]
61. Graille, M.; Wild, P.; Sauvain, J.J.; Hemmendinger, M.; Guseva Canu, I.; Hopf, N.B. Urinary 8-isoprostane as a biomarker for oxidative stress. A systematic review and meta-analysis. *Toxicol. Lett.* **2020**, *328*, 19–27. [CrossRef]

toxics

MDPI

Article

Health Risk Assessment of Ortho-Toluidine Utilising Human Biomonitoring Data of Workers and the General Population

Pasi Huuskonen [1,*], Spyros Karakitsios [2,3], Bernice Scholten [4], Joost Westerhout [4], Dimosthenis A. Sarigiannis [2,3] and Tiina Santonen [1]

1 Finnish Institute of Occupational Health, 00032 Työterveyslaitos, Finland; tiina.santonen@ttl.fi
2 Environmental Engineering Laboratory, Department of Chemical Engineering, Aristotle University of Thessaloniki, 54124 Thessaloniki, Greece; spyrosk@auth.gr (S.K.); sarigiannis@auth.gr (D.A.S.)
3 HERACLES Research Center on the Exposome and Health, Center for Interdisciplinary Research and Innovation, 57001 Thessaloniki, Greece
4 The Netherlands Organisation for Applied Scientific Research (TNO), 3508 TA Utrecht, The Netherlands; bernice.scholten@tno.nl (B.S.); joost.westerhout@tno.nl (J.W.)
* Correspondence: pasi.huuskonen@ttl.fi

Citation: Huuskonen, P.; Karakitsios, S.; Scholten, B.; Westerhout, J.; Sarigiannis, D.A.; Santonen, T. Health Risk Assessment of Ortho-Toluidine Utilising Human Biomonitoring Data of Workers and the General Population. *Toxics* **2022**, *10*, 217. https://doi.org/10.3390/toxics10050217

Academic Editors: Alison K. Bauer and Christophe Rousselle

Received: 31 March 2022
Accepted: 23 April 2022
Published: 25 April 2022

Publisher's Note: MDPI stays neutral with regard to jurisdictional claims in published maps and institutional affiliations.

Copyright: © 2022 by the authors. Licensee MDPI, Basel, Switzerland. This article is an open access article distributed under the terms and conditions of the Creative Commons Attribution (CC BY) license (https://creativecommons.org/licenses/by/4.0/).

Abstract: The aim of this work was to demonstrate how human biomonitoring (HBM) data can be used to assess cancer risks for workers and the general population. Ortho-toluidine, OT (CAS 95-53-4) is an aniline derivative which is an animal and human carcinogen and may cause methemoglobinemia. OT is used as a curing agent in epoxy resins and as intermediate in producing herbicides, dyes, and rubber chemicals. A risk assessment was performed for OT by using existing HBM studies. The urinary mass-balance methodology and generic exposure reconstruction PBPK modelling were both used for the estimation of the external intake levels corresponding to observed urinary levels. The external exposures were subsequently compared to cancer risk levels obtained from the evaluation by the Scientific Committee on Occupational Exposure Limits (SCOEL). It was estimated that workers exposed to OT have a cancer risk of 60 to 90:10^6 in the worst-case scenario (0.9 mg/L in urine). The exposure levels and cancer risk of OT in the general population were orders of magnitude lower when compared to workers. The difference between the output of urinary mass-balance method and the general PBPK model was approximately 30%. The external exposure levels calculated based on HBM data were below the binding occupational exposure level (0.5 mg/m^3) set under the EU Carcinogens and Mutagens Directive.

Keywords: ortho-toluidine; biomonitoring; urinary mass-balance; PBPK modelling; occupational exposure; general population

1. Introduction

Aniline and many of its derivatives are known or suspected human carcinogens. Ortho-toluidine, OT (CAS 95-53-4), also known as 2-aminotoluene, is one of these aniline derivatives which is manufactured and/or imported in the European Economic Area in quantities of 10,000–100,000 tonnes per annum [1]. OT is mainly used as a curing agent in epoxy resins, an intermediate in producing azo dyes and pigments, acid-fast dyestuffs, triarylmethane dyes, sulphur dyes, indigo compounds, photographic dyes, synthetic rubber, and rubber vulcanising chemicals, and an intermediate in the manufacture of herbicides [1,2].

OT is an animal and human carcinogen, classified as a category Carcinogenic 1B according to harmonised classification and labelling in European Union, EU (Classification, Labelling and Packaging (CLP) regulation 1272/2008). Other CLP classifications of OT are Acute Toxicity 3 for oral and inhalation toxicity, Eye Irritation 2 for irritative properties and Aquatic Acute 1 for environmental toxicity. In addition, OT may cause methemoglobinemia in humans [1,3]. Several aniline derivatives can be found on the candidate list of substances of very high concern (SVHCs) and the list of substances restricted under Registration, Evaluation, Authorisation and Restriction of Chemicals (REACH) EU regulation. OT has been added to the candidate list for eventual inclusion in Annex XIV to REACH [4]. Inclusion in annex XIV to REACH means that future use of OT requires authorisation in the EU. For authorisation purposes, users of OT need to demonstrate the adequate control of exposure by providing reliable and representative exposure information from their uses. Human biomonitoring (HBM) data could be a useful data source if measured biomarker levels can be related to cancer risk assessment.

OT is rapidly absorbed via the gastrointestinal tract, and is rapidly distributed, metabolised, and excreted (mainly via urine) in rats. The saturation of metabolic pathways is not apparent in the urinary excretion data. Absorption via skin as well as via respiratory tract has been demonstrated by acute toxicity studies and in occupationally exposed humans [1,2]. Metabolism studies have showed that N-acetylation and hydroxylation of the aromatic ring of OT as well as sulphate and glucuronide conjugation are the major metabolic pathways in rats. The human metabolic pathways are expected to be similar to those reported in experimental animals based on studies of other aromatic amines and knowledge of the key metabolising enzymes [5].

Occupational exposure to OT can be measured either by measuring air concentrations or by biomonitoring. Since dermal uptake may contribute significantly to systemic exposure to OT, biomonitoring can give a better picture on the total exposure than air monitoring. Both urinary OT and OT haemoglobin adduct analyses have been used to monitor occupational exposure to OT. However, the measurement of urinary excretion of total (free and conjugated) OT after hydrolysis is currently the most used method [1]. Since OT is a genotoxic carcinogen, there is no health-based biological limit value. In the EU, a binding occupational exposure limit value (BOELV) of 0.1 ppm (0.5 mg/m^3) has been set for OT [6]. This is not a purely health-based value but considers also socio-economic aspects.

Data on urinary OT levels of the general population have been used to set a biological guidance value (BGV) for urinary OT [1]. Based on the studies by Kütting et al. [7] and Weiss and Angerer [8], the 95th percentile of urinary OT (free and conjugated) among the non-smoking general population is approximately 0.2 µg/L which has been set as a biological guidance value (BGV) for workers in Germany for OT measured from post-shift urinary samples in the end of the working week [9]. Since smoking increases urinary OT levels, this value applies only for non-smokers or can be considered only after abstaining from smoking.

The aim of this work was to demonstrate how biomonitoring data can be used to assess cancer risks for workers and the general population. Using available information on the dose–response of OT carcinogenicity and published human biomonitoring (HBM) data, a risk assessment was conducted by applying a simple urinary mass-balance-based approach and a more refined exposure reconstruction algorithm coupled with a physiologically based pharmacokinetic (PBPK) model to convert urinary total OT levels as external daily intake.

2. Materials and Methods

The hazard and dose–response assessment of OT was based on the existing risk assessment from the Scientific Committee on Occupational Exposure Limits, SCOEL [1]. Since the SCOEL assessment covers only occupational exposure, cancer dose response for the general population was derived using a published benchmark dose lower limit (BMDL) level for OT as a starting point and using the general approach for cancer dose response setting described in REACH guidance R.8 [10]. A PubMed literature search was conducted for the past 20 years on OT in combination with occupation*, biomonitoring, and smoking, since OTs exist in cigarette smoke, to retrieve available occupational and general population biomonitoring data. Since the available HBM studies on the occupational exposure to OT were very limited in number, all relevant occupational HBM studies were included in the risk assessment. In the case of general population exposure assessment, only European HBM studies were included.

External exposure levels of OT were calculated from biomonitoring data of the general population and workers using a urinary mass-balance approach and by PBPK exposure reconstruction model to evaluate the cancer risk using the dose-response data published by SCOEL [1].

2.1. Urinary Mass-Balance Approach

The urinary mass-balance approach, as described by Angerer et al. [11], was used as a first-tier approach for the calculation of corresponding external intake levels from biomonitoring data. For this, the following formulas were used:

$$Css = (D \times BW \times FUE)/V24 \text{ or } D = (Css \times V24)/(FUE \times BW)$$

Here, D is the external dose as mg/kg bw, Css is the urinary level of the substance at steady state (mg/L), V24 is the estimated average 24 h urinary volume (L), and FUE is the mass of OT, including hydrolysed conjugate metabolites, excreted in urine during 24 h per mass of parent compound ingested (percentage).

In the calculations, OT urinary levels (Css) were assumed to represent a steady-state level, and 75% of the dose was assumed to be excreted in urine as measured parent compound, including hydrolysed conjugate metabolites (FUE), which was based on urine excretions observed in s.c. dosed rats after 24 h [2,5]. A value of 1.5 L/day was used as the 24 h urinary volume (V24) and 70 kg as the average body weight (BW).

As an example, the estimated urinary OT level corresponding to the BOELV 0.1 ppm (0.5 mg/m^3 = 0.07 mg/kg) assuming default 10 m^3 inhalation volume during working day, [10] is 2.5 mg/L (0.07 × 70 × 0.75/1.5). Conversely, the urinary OT BGV of 0.2 µg/L can be estimated to 5.7 ng/kg of OT by using parameters presented above (0.2 × 1.5/0.75 × 70).

2.2. PBPK Modelling

A generic PBPK model implemented in the INTEGRA platform for integrative exposure and risk modelling [12,13] was used to reconstruct external exposures of OT from HBM urinary data. Generic PBPK models are well-defined compartmental models capturing toxicokinetics of xenobiotics, accounting for real-life anatomy and physiology, independently of the compound considered. The applicability of the model to the different substances is ensured by parametrising the model for each compound. The generic human PBPK model developed on INTEGRA covers major ADME processes occurring in the human body at different stages of life. The model describes in detail the absorption, distribution, metabolism, and excretion (ADME) process of both OT and its metabolite OH-o-toluidine. This data was used as an input for the toxicokinetic model of INTEGRA, properly parameterised for the assessment of internal dose of OT. Non-compound specific parameters of the model have been presented in Sarigiannis et al. [13], while further parameterisation of the compound specific parameters illustrated in Table 1 was derived on quantitative structure–activity relationship (QSAR) modelling as described in the literature [14,15]. Validation of the estimates was done by comparing the predicted OT levels with the urinary levels obtained from occupational studies reporting urinary OT concentrations, and the results have been presented in the European Human Biomonitoring Initiative (HBM4EU) deliverable "AD12.3 Exposure model testing results" [16].

Regarding the exposure reconstruction, human biomonitoring data assimilation and their conversion into intake distributions is defined formally as a computational inversion problem. In such mathematical problems, the goal is to identify the specific input distributions that best explain the observed outputs while minimising the residual error. In our case, inputs comprise spatial and temporal information on micro-environmental media concentrations of xenobiotics and ancillary information on human activities, food intake patterns, or consumer product use that results in intakes; outputs are the observed levels of biomarkers in human biospecimens. In this study, we started from the urinary levels of OT and its hydrolysed conjugate metabolites; then, exposure levels were estimated for both occupational groups and the general population.

Table 1. Compound specific PBPK parameters for ortho-toluidine and the considered metabolite OH-o-toluidine.

	Ortho-Toluidine	**OH-o-Toluidine**
Tissue: Blood Partition Coefficients		
GI: Blood	3.8	0.7
Liver: Blood	1.9	0.65
Kidney: Blood	1.8	0.68
Fat: Blood	8.3	0.18
Bone: Blood	1.7	0.42
Brain: Blood	3.6	0.75
Gonads: Blood	0.79	0.83
Heart: Blood	1.6	0.57
Muscle: Blood	2.2	0.74
Skin: Blood	5.7	0.69
Lung: Blood	2.4	0.58
Fractions		
Fraction bound to plasma proteins	0.035	0.95
Fraction bound to red blood cells		0.005
Fraction of transformation from o-toluidine to OH-o-toluidine		1
Kinetic parameters		
Km (μmol/L)	27.2	
Vmax (μmol/h)	2835.3	
Clearances		
Kidney clearance rate (L/min)	0	0.17
Absorption GI tract		
Absorption fraction from GI tract	1	
Absorption rate in GI tract	1	

Variability and uncertainty of the overall model were addressed through the implementation of a Bayesian Markov Chain Monte Carlo [17,18] probabilistic framework. Markov chain Monte Carlo (MCMC) techniques are numerical approximation algorithms. They are used in Bayesian inference models to sample from probability distributions through the construction of Markov chains. In Bayesian inference processes, the target distribution of each Markov chain is a marginal posterior distribution. Each Markov chain begins with a seed value; the algorithm attempting to maximise the logarithm of the non-normalised joint posterior distribution eventually arrives at each target distribution by multiple iterations. Each iteration is considered a state. A Markov chain is a random process with a finite state-space. In Markov chains, the next state depends only on the current, not past, states. The method requires defining the prior distributions, the biomonitoring data upon which it will be applied, as well as a likelihood function, defining the likelihood of the data being correct given a set of forward model parameters. The MCMC approach takes into account an acceptance criterion that considers the likelihood of the data, given parameters. The MCMC process we have used samples using algorithms based on the Metropolis Hastings (M-H) (simulated annealing) or on differential evolution algorithms. Several studies have used MCMC techniques combined with PBPK models for inverse modelling [19–22]. Key factors that introduce variability are the anthropometric parameters (i.e., bodyweight) and the metabolic constants, for which a coefficient of uncertainty equal to 30% is considered as adequate [23]. In this study, 10,000 MCMC iterations were used in the model.

2.3. Risk Assessment

The risk assessment was based on the SCOEL [1] OT urinary bladder cancer risk estimation using the results from a two-year rat feeding study [24]. Since no adequate epidemiological data were available, SCOEL identified data on the formation of the urinary bladder transitional-cell carcinomas in female rats as being the most relevant for hazard characterisation, and derived a benchmark dose (BMD) causing 10% tumour incidence above background level (BMD_{10}) of 42.2 mg/kg bw per day. This BMD_{10} of 42.2 mg/kg bw per day was estimated to correspond to an inhaled dose of OT about 840 mg/m^3 as an 8 h time-weighted average (TWA) in occupational exposure [1]. This assumes a body weight of 70 kg, a default inhaled volume of 10 m^3 for an 8 h working day, and exposure of 48 weeks/year and 5 days/week. Absorption via inhalation and oral exposure was assumed the same. Allometric scaling of this dose level of 840 mg/m^3 from rat to human using a default value of 4 [10] provided a point of departure (POD) for 10% increase in tumour risk of 210 mg/m^3 (48 ppm). This resulted in linear OT occupational cancer risk estimations presented in Table 2 columns one and two. Using the same default numbers for a body weight and inhaled volume and mass-balance approach described in Section 2.1, we calculated the intake as mg/kg bw/d and steady-state urinary levels corresponding to the defined tumour risk levels (Table 2, columns 3 and 4).

Table 2. Ortho-toluidine linear cancer risk estimations, corresponding to occupational exposure to specific air levels, calculated daily occupational intake, and calculated urinary steady-state levels.

Tumour Risk	Ortho-Toluidine Concentration, mg/m^3 (ppm)	Occupational Intake (mg/kg bw/d) [1]	Steady-State Urinary Level (mg/L) [2]
1:10	210 (48)	30	1000
1:1000	2.1 (0.48)	0.3	10
1:10,000	0.21 (0.048)	0.03	1
1:100,000	0.021 (0.0048)	0.003	0.1
1:1,000,000	0.0021 (0.00048)	0.0003	0.01

[1] Converted to mg/bw/working day (as occupational exposure 8 h/day, 5 d/week, 70 kg worker, inhaling 10 m^3 of air during the working day). [2] Calculated urinary steady-state levels with mass-balance equation (75% of parent compound excreted to the urine including hydrolysed conjugate metabolites (FUE), 1.5 L/day of 24 h urinary volume (V24) and 70 kg body weight (BW).

The same BMD_{10} of 42.2 mg/kg bw per day was used as a starting point for the general population cancer risk assessment. After allometric scaling of this dose level from rat to human using a default value of 4, this provided a 10% increase in tumour risk of 10.6 mg/kg bw/d. Since the cancer data were based on the continuous oral feeding study in rats, no further dose adjustments were made. Linear OT lifetime cancer risks are presented in Table 3.

Table 3. Ortho-toluidine linear cancer risk estimations for general population, corresponding to specific lifetime cancer risk, and calculated urinary steady-state levels.

Tumour Risk	Ortho-Toluidine Intake (mg/kg bw/d)	Steady-State Urinary Level (mg/L) [1]
1:10	10.6	371
1:1000	0.106	3.71
1:10,000	0.0106	0.371
1:100,000	0.00106	0.0371
1:1,000,000	0.000106	0.00371

[1] Calculated urinary steady-state levels with mass-balance equation (75% of parent compound excreted to the urine including hydrolysed conjugate metabolites (FUE), 1.5 L/day of 24 h urinary volume (V24) and 70 kg body weight (BW).

3. Results

3.1. Summary of Exposure Biomonitoring Data

The literature search resulted in four relevant papers with occupational OT exposure biomonitoring data and three relevant papers for the general population. All studies measured OT after hydrolysis of the conjugated metabolites from urinary samples. The studies concerning occupational and the general population HBM data selected in the risk assessment are presented in Table 4.

Table 4. Human biomonitoring studies for ortho-toluidine in occupational settings and background concentrations of the general populations.

Study Origin	Urine Sample Type (*n*)	Urine Ortho-Toluidine (Range)	Reference
Workers, Liquid SO_2 plant, France	Post-shift (13)	Mean: 523 µg/L, P95: 962 µg/L (<LOD-984.1 µg/L)	[25]
Workers, Rubber industry, Germany	Post-shift (51)	Mean: 38.6 µg/L, P95: 292.4 µg/L (<LOD-292.4 µg/L)	[26]
Workers, Rubber industry, Sweden	Post-shift (157)	Median: 0.46 µg/L, (0.03–108 µg/L)	[27]
Workers, Pigment industry, Japan	Post-shift (36)	Mean: 55.5 µg/L, (<LOD-129.12 µg/L)	[28]
General population, Germany	24 h urine (81)	Median: 61.8 ng/24 h, Max: 401 ng/24 h	[29]
General population, Germany	24 h urine (20)	Mean (non-smokers): 167 ± 199.4 ng/24 h Mean (smokers): 204.2 ± 59.1 ng/24 h.	[30]
General population, Germany, Switzerland, United Kingdom	24 h urine (1631)	Mean (non-smokers): 64 ± 128 ng/24 h Mean (smokers): 179 ± 497 ng/24 h	[31]

LOD = limit of detection, *n* = number of study participants, P95 = 95th percentile.

Labat et al. [25] measured urine concentrations of OT among individuals (*n* = 13) who worked in a French liquid SO_2-plant polluted with OT. Pre-shift urine concentrations of OT had a range of 1.7 ± 1.5 µg/L, while the mean post-shift levels were 523 ± 321.6 µg/L, 95th percentile being 962 µg/L. Korinth et al. [26] assessed occupational exposure to aniline and OT among people (*n* = 51) working in the manufacturing of rubber products in Germany. The measured total urinary OT concentrations in post-shift samples were 38.6 µg/L (mean) and 292.4 µg/L (95th percentile). Li et al. [27] examined the relationship between DNA-damaging chemicals and average telomere length in 157 workers working in the Swedish rubber industry. OT levels were measured in urine samples collected during the last four hours of an eight-hour shift. Measured concentration range was 0.03–108 µg/L with a median of 0.46 µg/L. Eitaki et al. [28] studied OT-exposed workers (*n* = 36) in the Japanese pigment industry. Post-shift urinary OT measurements showed a mean concentration of 55.5 µg/L and a range of 16.5–129.12 µg/L.

Seidel [29] studied urinary excretion of aromatic amines including OT in general population in Germany. 24 h urine samples were collected from 81 non-smoking individuals aged 20–61 years. The median and maximum OT levels were found to be 61.8 and 401 ng/24 h, respectively. Another German study associated with exposure to carcinogenic aromatic amines was conducted by Riedel et al. [30]. Twenty-four urine samples were collected from 20 people (including 10 smokers). The non-smoking group had a mean OT concentration of 167 ± 199.4 ng/24 h while the concentration in the smoker group was 204.2 ± 59.1 ng/24 h. Lindner et al. [31] assessed the exposure of smokers and non-smokers to smoke constituents, including OT, by analysing 24 h urinary excretion of OT in 1631 adults (including 1223 smokers) from three countries (Germany, Switzerland, and United Kingdom). Exposure levels were lower in non-smokers than smokers, with mean OT urine concentrations of 64 ± 128 and 179 ± 497 ng/24 h respectively.

3.2. Reverse Calculation of External Exposure Based on Urinary Mass-Balance Approach and Generic PBPK Model

External exposures calculated for occupational and general population by using a urinary mass-balance method and a generic PBPK model for reverse dosimetry are presented in Tables 5 and 6. In addition, corresponding cancer risks estimated using linear extrapolation as described in the Methods are presented.

Table 5. Estimated occupational cancer risk values for o-toluidine using a urinary mass-balance method and a PBPK model in INTEGRA.

	Urinary Ortho-Toluidine (μg/L)	External Ortho-Toluidine Exposure (μg/kg bw/d)	Estimated Cancer Risk
Urinary mass-balance			
Labat et al. [25]	Mean: 523, P95: 962	Mean: 14.94, P95: 27.49	50–92:10^6
Korinth et al. [26]	Mean: 38.6, P95: 292.4	Mean: 1.10, P95: 8.35	4–28:10^6
Li et al. [27]	Median: 0.46, Max: 108	Median: 0.013, Max: 3.09	0.04–10:10^6
Eitaki et al. [28]	Mean: 55.5, Max: 129.1	Mean: 1.6, Max: 3.69	5–12:10^6
PBPK model			
Labat et al. [25]	Mean: 523, P95: 962	Mean: 9.96, P95: 18.74	33–62:10^6
Korinth et al. [26]	Mean: 38.6, P95: 292.4	Mean: 0.70, P95: 5.57	2–19:10^6
Li et al. [27]	Median: 0.46, Max: 108	Median: 0.012, Max: 2.06	0.04–7:10^6
Eitaki et al. [28]	Mean: 55.5, Max: 129.1	Mean: 1.1, Max: 2.5	4–8:10^6

P95 = 95th percentile.

Table 6. Estimated lifetime cancer risk values of o-toluidine for the general population using a urinary mass-balance method and a PBPK model in INTEGRA.

	Urinary Ortho-Toluidine (ng/24 h)	External Ortho-Toluidine Exposure (μg/kg bw/d)	Estimated Cancer Risk
Urinary mass-balance			
Seidel [29]	non-smokers: Mean: 61.8, Max: 401	non-smokers: Mean: 0.0012, Max: 0.008	0.0011–0.0072:10^6
Riedel et al. [30]	non-smokers: 167 ± 199.4	non-smokers: 0.0032 ± 0.004	0.001–0.01:10^6
	smokers: 204.2 ± 59.1	smokers: 0.0039 ± 0.001	0.004–0.007:10^6
Lindner et al. [31]	non-smokers: 64 ± 128	non-smokers: 0.0012 ± 0.0024	0.002–0.005:10^6
	smokers: 179 ± 497	smokers: 0.0034 ± 0.0095	0.009–0.018:10^6
PBPK model			
Seidel [29]	non-smokers: Mean: 61.8, Max: 401	non-smokers: Mean: 0.00078, range 0.00012–0.00514	0.0001–0.005:10^6
Riedel et al. [30]	non-smokers: 167 ± 199.4	non-smokers: Mean 0.00133, range 0.0001–0.0031	0.0001–0.003:10^6
Lindner et al. [31]	non-smokers: 64 ± 128	non-smokers: Mean 0.0008, range 0.00006–0.0019	0.0001–0.0018:10^6

3.3. Comparison of the Results

External dose estimations based on the urinary mass-balance method gave approximately 30% higher OT external exposure values both for workers and the general population when compared to corresponding values of the generic PBPK model. For example, in the study with the highest mean OT urinary concentration [25] the urinary mass-balance method gave a mean external exposure of 14.94 μg/kg whereas the PBPK reconstruction model gave a corresponding value of 9.96 μg/kg. The corresponding effect on OT cancer risk is in this example 50:10^6, compared to 33:10^6, respectively.

4. Discussion

A cancer risk assessment was performed for OT using existing HBM studies concerning workers and the general population. A urinary mass-balance method and a generic exposure reconstruction PBPK model were both used to estimate the external intake levels corresponding levels in urine.

In the case of OT, there are no measured correlation data on external exposure and urinary concentrations. Therefore, the urinary mass-balance method was selected, which is a rather rough method to correlate external exposure with the urinary levels. This can be further refined by using PBPK models, however, there is no specific model available for OT. A generic PBPK model was used to investigate the relation between external exposure and urinary values. The output of both models did not differ considerably; the urinary mass-balance method gave approximately 30% higher external exposure values. This difference results from the fact that the PBPK model considers exposure dynamics; hence, it can capture the intra-day variability of OT in urine and to attribute the higher concentration identified in the post-shift samples as the result of the 8 h shift. On the contrary, the biomonitored levels used for estimating intake with the urinary mass-balance method are assumed as steady-state values. If this assumption is applied on samples that have been taken from workers post-shift (representing a peak level rather than steady state level), this may result in an overestimation of external exposure. However, the 30% differences are minor when considering the uncertainties related to the cancer dose–response based on animal data.

4.1. Uncertainties in the Risk Assessment

The toxicokinetics of OT in humans were assumed to correspond to toxicokinetics observed in experimental animals. This is one source of uncertainty in this risk assessment. In the urinary mass-balance calculations, 75% of the OT dose (FUE) was assumed to be excreted in urine as parent compound or as hydrolysed conjugate metabolites. This was based on urinary excretion observed in s.c. dosed rats after 24 h [2]. The urinary mass-balance method used for reverse calculation assumes that the measured urinary levels represent steady-state levels. This must be considered especially in case of occupational exposure with rapidly eliminating chemicals because occupational exposure levels may vary during the working day. The half-life of OT has been reported in human plasma to be approximately four hours [32] and the available HBM data was based on post-shift urinary samples. OT is absorbed also through the human skin, which [26,33] can contribute significantly to the toxicokinetics of the compound. This should be considered when collecting/analysing HBM samples, since dermal absorption is slower when compared to inhalation exposure, resulting also in slower elimination. HBM data collected from 24 h urinary samples could better highlight the exposure. On the other hand, measurement of OT air concentrations does not consider the effect of personal protection of the worker and cannot inform us of the dermal exposure.

The carcinogenic properties of OT cause uncertainty in cancer risk assessments, since a threshold value for the cancer risk cannot be established. In this risk assessment, a linear extrapolation of cancer risk for humans was used from a BMD_{10} reference value derived for bladder tumours observed in rats. An allometric uncertainty scaling value of four was used for interspecies differences between rat and human. Linear extrapolation is considered as a conservative approach potentially resulting in the overestimation of risks at low levels. On the other hand, there is epidemiological evidence on the carcinogenicity of OT in occupationally exposed humans [3].

Individual variability in OT metabolism can have an influence on OT urinary levels. Genetic polymorphisms in human N-acetyltransferases can increase susceptibility to aromatic amine-induced cancer for slow-acetylators [34]. However, this can be considered to have a much smaller impact on the risk assessment than uncertainties in dose–response. Cigarette smoking can cause uncertainty to risk assessment since smoking elevates OT levels in the urine of smokers [31].

4.2. Recommendations for the Regulatory Risk Assessment

This assessment gives an example how HBM data can be used in occupational and general population risk assessments. The urinary mass-balance method may give a rough estimate of external exposure when using workers' post-shift urinary biomonitoring data. Since OT is a rapidly metabolised chemical, the best option would be to collect consecutive urinary samples covering a whole day, i.e., 24 h (i.e., pre-shift, during shift, post-shift, evening, and next morning). The data on consecutive periods of production of urine would be also very informative for parameterising a specific PBPK model in the future. Development of a PBPK model for the chemical group of arylamines would benefit future risk assessments, especially regarding exposure reconstruction modelling.

In the future, risk assessment would benefit from well-designed biomonitoring studies. In addition, proper toxicokinetic correlation studies for controlled OT air concentration and urine excretion in humans may benefit future modelling. Moreover, exposure assessment by biomonitoring OT haemoglobin adducts can highlight exposure for longer and cumulative time periods. Mean OT Hb-adduct levels have been 10 times higher in exposed versus unexposed workers and >100 times higher than the mean levels in unexposed populations previously studied [35]. Human metabolism and kinetics of OT should be studied further.

OT has been added to the candidate list of SVHC for eventual inclusion in Annex XIV to REACH [4]. This risk assessment can be beneficial for chemical authorities as well as manufacturers in the situation that the use of OT will require authorisation in EU.

5. Conclusions

In conclusion, by applying the urinary mass-balance methodology and the PBPK modelling based on four OT biomonitoring studies, we found that workers exposed to OT have a cancer risk of 60 to $90{:}10^6$ in the worst-case scenario (i.e., the biomonitoring study with the highest urinary levels). The exposure levels and cancer risk of OT in the general population were orders of magnitude lower when compared to workers. Although there is no generally agreed acceptable cancer risk levels, usually, risk levels in the order of $1{:}10^5$–10^6 for general population and around $1{:}10^5$ for workers are considered acceptable, but in case of workers higher levels have also been considered to be tolerable under certain circumstances [10]. The estimated external exposure levels for workers were also well below the intake levels, corresponding to the BOELV (0.5 mg/m^3) set under EU Carcinogens and Mutagens Directive. However, results should be considered carefully because of the limited number of HBM data. There is clearly a need for further biomonitoring data on OT exposure.

Author Contributions: Conceptualisation, P.H., T.S., S.K. and B.S.; methodology, P.H., T.S. and S.K.; software, S.K. and D.A.S.; investigation, P.H., B.S., S.K. and J.W.; writing—original draft preparation, P.H. and T.S.; writing—review and editing, B.S., S.K., D.A.S., J.W. and T.S.; supervision, D.A.S. and T.S. All authors have read and agreed to the published version of the manuscript.

Funding: This project has received funding from the European Union's Horizon 2020 research and innovation programme under grant agreement No. 733032 and received co-funding from the author's organizations and/or Ministries.

Data Availability Statement: The data presented in this study are available publicly.

Acknowledgments: The authors gratefully acknowledge the financial support of the European Union's Horizon 2020 research and innovation programme (Available online: www.HBM4EU.eu (accessed on 30 March 2022)).

Conflicts of Interest: The authors declare no conflict of interest.

References

1. Recommendation from the Scientific Committee on Occupational Exposure Limits. o-Toluidine, 2-methylaniline. SCOEL/REC/301. 2017. Available online: https://op.europa.eu/s/v5bh (accessed on 30 March 2022).
2. *SIDS Initial Assessment Report (SIAR) on O-Toluidine*; OECD: Paris, France; UN Environmental Programme Publications: Geneva, Switzerland, 2004.
3. A review of human carcinogens: Chemical agents and related occupations. *IARC Monogr. Eval. Carcinog. Risks Hum.* **2012**, *100*, 93–100.
4. ECHA—Candidate List of substances of Very High Concern for Authorisation. Available online: https://echa.europa.eu/candidate-list-table (accessed on 30 March 2022).
5. *Report on Carcinogens, Monograph on Ortho-Toluidine; National Toxicology Program*; U.S. Department of Health and Human Services: Washington, DC, USA, 2014. Available online: https://ntp.niehs.nih.gov/ntp/roc/thirteenth/monographs_final/otoluidine_508.pdf (accessed on 30 March 2022).
6. European Commission, Directive (EU) 2019/130 of the European Parliament and of the Council of 16 January 2019 Amending Directive 2004/37/EC on the Protection of Workers from the Risks Related to Exposure to Carcinogens or Mutagens at Work. Available online: https://eur-lex.europa.eu/legal-content/EN/TXT/?uri=uriserv:OJ.L_.2019.030.01.0112.01.ENG (accessed on 30 March 2022).
7. Kütting, B.; Göen, T.; Schwegler, U.; Fromme, H.; Uter, W.; Angerer, J.; Drexler, H. Monoarylamines in the general population—A cross-sectional population-based study including 1004 Bavarian subjects. *Int. J. Hyg. Environ. Health* **2009**, *212*, 298–309. [CrossRef] [PubMed]
8. Weiss, T.; Angerer, J. *Belastung der Bevölkerung der Bundesrepublik Deutschland durch nitroaromatischen Verbindungen—Der Einfluss von Ernährung und Bekleidung*; Institut für Arbeits-, Sozial- und Umweltmedizin der Friedrich-Alexander-Universität Erlangen-Nürnberg: Erlangen, Germany, 2005.
9. *MAK- und BAT-Werte-Liste. Grenzwerte in Biologischem Material zu o-Toluidin. Addendum zu o-Toluidin*; BAT Value Documentation in German language; Deutsche Forschungsgemeinschaft: Bonn, Germany, 2012.
10. ECHA—Guidance on Information Requirements and Chemical Safety Assessment Chapter R.8: Characterisation of Dose [Concentration]-Response for Human Health. ECHA-2010-G-19-EN. 2012. Available online: https://echa.europa.eu/documents/10162/13632/information_requirements_r8_en.pdf/e153243a-03f0-44c5-8808-88af66223258 (accessed on 30 March 2022).
11. Angerer, J.; Aylward, L.L.; Hays, S.M.; Heinzow, B.; Wilhelm, M. Human biomonitoring assessment values: Approaches and data requirements. *Int. J. Hyg. Environ. Health* **2011**, *214*, 348–360. [CrossRef] [PubMed]
12. Sarigiannis, D.; Karakitsios, S.; Gotti, A.; Loizou, G.; Cherrie, J.; Smolders, R.; De Brouwere, K.; Galea, K.; Jones, K.; Handakas, E.; et al. Integra: From global scale contamination to tissue dose. In Proceedings of the 7th International Congress on Environmental Modelling and Software: Bold Visions for Environmental Modeling, iEMSs, San Diego, CA, USA, 15–19 June 2014; pp. 1001–1008.
13. Sarigiannis, D.; Karakitsios, S.P.; Handakas, E.; Simou, K.; Solomou, E.; Gotti, A. INTEGRAted exposure and risk characterization of bisphenol-A in Europe. *Food Chem. Toxicol.* **2016**, *98*, 134–147. [CrossRef]
14. Papadaki, K.C.; Karakitsios, S.P.; Sarigiannis, D.A. Modeling of adipose/blood partition coefficient for environmental chemicals. *Food Chem. Toxicol.* **2017**, *110*, 274–285. [CrossRef]
15. Sarigiannis, D.A.; Papadaki, K.; Kontoroupis, P.; Karakitsios, S.P. Development of QSARs for parameterizing Physiology Based ToxicoKinetic models. *Food Chem. Toxicol.* **2017**, *106*, 114–124. [CrossRef]
16. Sarigiannis, D.A. AD12.3 Exposure Model Testing Results, HBM4EU Project. 2019. Available online: https://www.hbm4eu.eu/work-packages/additional-deliverable-12-3-exposure-model-testing-results/ (accessed on 30 March 2022).
17. Gelman, A.; Rubin, D.B. Markov chain Monte Carlo methods in biostatistics. *Stat. Methods Med. Res.* **1996**, *5*, 339–355. [CrossRef]
18. Gilks, W.; Spiegelhalter, D.; Richardson, S. *Markov Chain Monte Carlo in Practice*; Chapman and Hall/CRC Press: Boca Raton, FL, USA, 1996.
19. Chen, C.C.; Shih, M.C.; Wu, K.Y. Exposure estimation using repeated blood concentration measurements. *Stoch. Environ. Res. Risk Assess.* **2010**, *24*, 445–454. [CrossRef]
20. Georgopoulos, P.G.; Sasso, A.F.; Isukapalli, S.S.; Lioy, P.J.; Vallero, D.A.; Okino, M.; Reiter, L. Reconstructing population exposures to environmental chemicals from biomarkers: Challenges and opportunities. *J. Expo. Sci. Environ. Epidemiol.* **2009**, *19*, 149–171. [CrossRef]
21. Lyons, M.A.; Yang, R.S.; Mayeno, A.N.; Reisfeld, B. Computational toxicology of chloroform: Reverse dosimetry using Bayesian inference, Markov chain Monte Carlo simulation, and human biomonitoring data. *Environ. Health Perspect.* **2008**, *116*, 1040–1046. [CrossRef]
22. McNally, K.; Cotton, R.; Cocker, J.; Jones, K.; Bartels, M.; Rick, D.; Price, P.; Loizou, G. Reconstruction of exposure to m-xylene from human biomonitoring data using PBPK modelling, bayesian inference, and Markov chain Monte Carlo simulation. *J. Toxicol.* **2012**, *2012*, 760281. [CrossRef] [PubMed]
23. Clewell, R.; Clewell, H. Development and specification of physiologically based pharmacokinetic models for use in risk assessment. *Regul. Toxicol. Pharmacol.* **2008**, *50*, 129–143. [CrossRef] [PubMed]

24. *Bioassay of o-Toluidine Hydrochloride for Possible Carcinogenicity (CAS No. 636-21-5)*; Carcinogenesis Technical Report Series No. 153; NIH Publication No. 79-1709; US Department of Health, Education and Welfare, Public Health Service, National Institutes of Health, National Cancer Institute: Bethesda, MD, USA, 1979.
25. Labat, L.; Thomas, J.; Dehon, B.; Humbert, L.; Leleu, B.; Nisse, C.; Lhermitte, M. Evalution d'une exposition professionnelle à l'ortho-toluidine par chromatographie phase gazeuse couplée à la spectrométrie de masse. *Acta Clin. Belg.* **2006**, *61*, 61–65. [CrossRef]
26. Korinth, G.; Weiss, T.; Penkert, S.; Schaller, K.H.; Angerer, J.; Drexler, H. Percutaneous absorption of aromatic amines in rubber industry workers: Impact of impaired skin and skin barrier creams. *Occup. Environ. Med.* **2007**, *64*, 366–372. [CrossRef]
27. Li, H.; Jönsson, B.A.G.; Lindh, C.H.; Albin, M.; Broberg, K. N-nitrosamines are associated with shorter telomere length. *Scand. J. Work Environ. Health* **2011**, *37*, 316–324. [CrossRef]
28. Eitaki, Y.; Nakano, M.; Kawai, T.; Omae, K.; Takebayashi, T. Biological monitoring of o-toluidine in urine pretreated by an enzymatic deconjugation method. *J. Occup. Health* **2019**, *61*, 349–357. [CrossRef]
29. Seidel, A. *Ermittlung von Quellen für das Vorkommen von Nitro-/Aminoaromaten im Urin von Nichtrauchern*; Umweltbundesamt: Dessau, Germany, 2005.
30. Riedel, K.; Scherer, G.; Engl, J.; Hagedorn, H.W.; Tricker, A.R. Determination of three carcinogenic aromatic amines in urine of smokers and nonsmokers. *J. Anal. Toxicol.* **2006**, *30*, 187–195. [CrossRef]
31. Lindner, D.; Smith, S.; Leroy, C.M.; Tricker, A.R. Comparison of exposure to selected cigarette smoke constituents in adult smokers and nonsmokers in a European, multicenter, observational study. *Cancer Epidemiol. Biomark. Prev.* **2011**, *20*, 1524–1536. [CrossRef]
32. Richter, E. Biomonitoring of human exposure to arylamines. *Front. Biosci.* **2015**, *7*, 193–207. [CrossRef]
33. Lüersen, L.; Wellner, T.; Koch, H.M.; Angerer, J.; Drexler, H.; Korinth, G. Penetration of b-naphthylamine and o-toluidine through human skin in vitro. *Arch Toxicol.* **2006**, *80*, 644–646. [CrossRef]
34. Golka, K.; Prior, V.; Blaszkewicz, M.; Bolt, H.M. The enhanced bladder cancer susceptibility of NAT2 slow acetylators towards aromatic amines: A review considering ethnic differences. *Toxicol. Lett.* **2002**, *128*, 229–241. [CrossRef]
35. Hanley, K.W.; Viet, S.M.; Hein, M.J.; Carreón, T.; Ruder, A.M. Exposure to o-toluidine, aniline, and nitrobenzene in a rubber chemical manufacturing plant: A retrospective exposure assessment update. *J. Occup. Environ. Hyg.* **2012**, *9*, 478–490. [CrossRef] [PubMed]

toxics

MDPI

Article

Using Human Biomonitoring Data to Support Risk Assessment of Cosmetic Ingredients—A Case Study of Benzophenone-3

Christophe Rousselle [1], Matthieu Meslin [2,*], Tamar Berman [3], Marjolijn Woutersen [4], Wieneke Bil [4], Jenna Wildeman [4] and Qasim Chaudhry [5]

1 European and International Affairs Department, Anses, 94701 Maisons-Alfort, France; christophe.rousselle@anses.fr
2 Risk Assessment Department, Anses, 14 rue Pierre et Marie Curie, 94701 Maisons-Alfort, France
3 Ministry of Health, Jerusalem 9101002, Israel; tamar.berman@moh.gov.il
4 RIVM National Institute for Public Health and the Environment, 3721 MA Bilthoven, The Netherlands; marjolijn.woutersen@rivm.nl (M.W.); wieneke.bil@rivm.nl (W.B.); jennawildeman@outlook.com (J.W.)
5 Department of Clinical Sciences and Nutrition, University of Chester, Parkgate Road, Chester CH1 4BJ, UK; q.chaudhry@chester.ac.uk
* Correspondence: matthieu.meslin@anses.fr

Abstract: Safety assessment of UV filters for human health by the Scientific Committee on Consumer Safety (SCCS) is based on the estimation of internal dose following external (skin) application of cosmetic products, and comparison with a toxicological reference value after conversion to internal dose. Data from human biomonitoring (HBM) could be very useful in this regard, because it is based on the measurement of real-life internal exposure of the human population to a chemical. UV filters were included in the priority list of compounds to be addressed under the European Human Biomonitoring Initiative (HBM4EU), and risk assessment of benzophenone-3 (BP-3) was carried out based on HBM data. Using BP-3 as an example, this study investigated the benefits and limitations of the use of external versus internal exposure data to explore the usefulness of HBM to support the risk assessment of cosmetic ingredients. The results show that both approaches did indicate a risk to human health under certain levels of exposure. They also highlight the need for more robust exposure data on BP-3 and other cosmetic ingredients, and a standardized framework for incorporating HBM data in the risk assessment of cosmetic products.

Keywords: human biomonitoring; UV filters; benzophenone-3; HBM4EU; risk assessment; RCR; margin of safety

Citation: Rousselle, C.; Meslin, M.; Berman, T.; Woutersen, M.; Bil, W.; Wildeman, J.; Chaudhry, Q. Using Human Biomonitoring Data to Support Risk Assessment of Cosmetic Ingredients—A Case Study of Benzophenone-3. *Toxics* **2022**, *10*, 96. https://doi.org/10.3390/toxics10020096

Academic Editor: Giovanna Tranfo

Received: 10 January 2022
Accepted: 16 February 2022
Published: 19 February 2022

Publisher's Note: MDPI stays neutral with regard to jurisdictional claims in published maps and institutional affiliations.

Copyright: © 2022 by the authors. Licensee MDPI, Basel, Switzerland. This article is an open access article distributed under the terms and conditions of the Creative Commons Attribution (CC BY) license (https://creativecommons.org/licenses/by/4.0/).

1. Introduction

UV filters are used in sunscreen and other cosmetic products to protect the skin by absorbing or reflecting potentially harmful UVA and UVB rays.

UV filters in sunscreens are regulated in Europe in the Cosmetic Product Regulation (CPR), which provides a broad regulatory framework for restricting chemicals with potential risks to human health. The CPR does not include specific restrictions on endocrine disrupting (ED) chemicals, however, and as there was concern for these substances the European Commission started a review of cosmetic ingredients with ED properties in 2018 [1]. As a result of the review, the Commission established a "short" list of potential EDs in cosmetics, which included benzophenone-3 (BP-3). Subsequently, the Scientific Committee on Consumer Safety (SCCS), which provides opinions on health and safety risks (chemical, biological, mechanical and other physical risks) of non-food consumer products (cosmetics, personal-care products, toys, textiles, etc.), was mandated by the Commission to carry out a safety assessment of BP-3 in view of concerns related to potential ED properties [2].

The SCCS approach to assessing the safety of cosmetic ingredients to the consumers is based on first estimating the external dose that is in contact with the skin, followed by estimation of the internal dose that will be systemically available after crossing the skin. The internal exposure dose is then compared to a reference value that is either derived from human or experimental data. This toxicological reference value (TRV) is used by the SCCS to calculate a margin of safety (MoS) [3]. In this context, human biomonitoring (HBM) can provide a very useful tool because it provides real-life internal exposure of the human population to a given chemical in terms of measured concentrations in human samples, such as blood and urine. There has been an increasing use of HBM for exposure assessment in the European risk assessment schemes in recent years, although the level of use is still limited. Based on a survey of different risk assessment frameworks in the EU, less than 30% of respondents replied that HBM is regularly applied in the regulatory domain in their country [4]. The use of HBM in risk assessments has also been uneven across different frameworks; for example, compared to pesticides, biocides, and heavy metals, the use of HBM data has been scarce in cosmetics safety assessments.

The European Human Biomonitoring Initiative (HBM4EU), running from 2017 to 2022, has established a European-Union-wide human biomonitoring program to generate knowledge on human internal exposure to chemical pollutants and their potential health impacts in Europe, to support policy makers' efforts to regulate chemical safety and improve public health in Europe [5]. One of the goals of HBM4EU is to support policy making by providing evidence of the actual exposure of EU citizens to chemical substances and mixtures to inform risk assessment. The current study was aimed at exploring the potential usefulness of HBM data for supporting the risk assessment of cosmetic ingredients, using BP-3 as an example.

2. Methodological Approach

2.1. Risk Assessment Based on External Approach

The approach used by the SCCS to assess the safety of cosmetic ingredients, for which an opinion is required, is based on the calculation of a margin of safety. The method followed by the committee is described in detail in the Note of Guidance [3].

2.1.1. Margin of Safety Calculation

In the risk characterization of a cosmetic ingredient, the main focus is on systemic effects, with separate consideration of any local effects. The MoS is generally calculated based on oral toxicity studies unless robust dermal toxicity data are available. For this, a toxicological point of departure (PoD), derived from oral studies in test animals, has historically been used in the following Equation (1):

$$\text{MoS} = \text{PoDsys}/\text{SED} \quad (1)$$

where PoDsys (systemic PoD in mg/kg bw/day) is the potency descriptor, preferably a benchmark dose (BMD) or alternatively a NOAEL (no observed adverse effect level) or LOAEL (lowest observed adverse effect level) from a repeated-dose toxicity study. As these values are often based on oral studies, they are corrected for the oral bioavailability of the compounds to derive the PoDsys. It is considered that not more than 50% of an orally administered dose is systemically available. Thus, in the absence of data, 50% of the administered dose is used as the default oral absorption value for a cosmetic ingredient and the PoDsys is derived from the PoD by dividing by 2. SED (systemic exposure dose in mg/kg bw/day) is a dose descriptor for the systemic exposure to the substance calculated as the amount of the substance systemically absorbed from cosmetic products.

Taking into account the extrapolation uncertainties corresponding to differences in toxicokinetics and toxicodynamics between experimental animals and humans (a factor of 10), and toxicological sensitivity differences within the human population due to differences in biology, age, sex, health status, etc. (a factor of 10), an MoS equal to or higher than 100 (10×10) is considered an adequate indicator of safety of the substance to human health. Other default assessment factors (AFs) may also be applied to take into account for example lack of data, but this was not applicable in this case [6].

2.1.2. Hazard Assessment: Selection of an External Critical Dose

In this study, the selection of the critical effect, the key study, and the external critical dose of BP-3, was carried out following the methodology described in the SCCS Notes of Guidance [3]. Cosmetic safety assessments have historically used NOAEL as the toxicological PoD. The NOAEL is the highest dose or exposure level where no (adverse) treatment-related effects were observed. The external critical dose of the assessed ingredient (e.g., BP-3) is usually derived from a repeated-dose toxicity study that indicates an adverse effect, such as a chronic or a reproductive toxicity study [3].

The study selection process considered the assessment of all toxicological effect(s), their dose-response relationships, and possible threshold(s). Moreover, all periods of exposure including the most sensitive ones were considered, and in particular, exposure during pregnancy. The studies in which pregnant animals were exposed were therefore more thoroughly assessed.

2.1.3. Exposure Dose (SED) Calculation Based on Skin Application

The human exposure to BP-3 can arise from cosmetic products as well as from other sources. BP-3 is currently regulated in sunscreens as a UV-filter with a maximum allowed concentration of 6% and in other cosmetic products with a concentration limit of 0.5%. BP-3 is also used in other product groups, such as coating products, plasters, modeling clay and finger paints, machine washing liquids/detergents, automotive care products, paints and coating or adhesives, and fragrances and air fresheners [7]. Exposure to BP-3 used as a UV-filter in cosmetic products was calculated based on the exposure scenarios by multiplying the concentration of BP-3 at the maximum approved concentrations (6% in sunscreen products and 0.5% in other cosmetic products). The external dermal exposure (E_{dermal}) per day of BP-3 was then calculated taking into account the retention factor, concentration of BP-3 in the product, and amount of product applied daily, according to Equation (2):

$$E_{dermal\ BP\text{-}3} = C_{BP\text{-}3} \times q_x \times f_{ret} \tag{2}$$

where $E_{dermal\ BP\text{-}3}$ (mg/day): external exposure to BP-3 available for dermal uptake from a product category.

$C_{BP\text{-}3}$ (mg/g): concentration of BP-3 in the product category.

q_x (g/day): amount of product category that is applied per day.

f_{ret}: retention factor specific to the product category; for leave-on cosmetics, including UV-cream, a fraction of 1 (100%) is used, while for rinse-off cosmetics (e.g., shower gel, shampoo, etc.) a smaller fraction is used that depends on the type and use of the respective product.

Next, external dermal exposure was multiplied by skin penetration value to calculate the systemic exposure dose (SED). Skin penetration can be determined from in vivo or in vitro studies. The approach preferred by the SCCS is based on selecting in vitro studies using published criteria [8].

2.2. *Risk Assessment Based on Internal Approach*

In HBM the internal human exposure is determined by measuring a substance itself or its metabolites in biological samples, usually blood or urine, in a human study population. The risk is calculated by comparing this internal exposure to a human biomonitoring guidance value (HBM-GV). As the HBM-GV used for BP-3 has no legal status, and has not been derived as an official HBM4EU guidance value, it should be considered a provisional HBM-GV in the context of this study.

2.2.1. Risk Characterization Based on HBM Data

Following the internal approach, the risk characterization ratio (RCR) for BP-3 is calculated by comparing the typical case (TC) exposure estimate and the reasonable worst-case (RWC) exposure estimate in each study to the provisional HBM-GV, as described in Equations (3) and (4):

$$\text{RCR (TC)} = \frac{\text{Range of median values}}{\text{provisional HBM} - \text{GV}} \tag{3}$$

$$\text{RCR (RWC)} = \frac{\text{Range of P95 values}}{\text{provisional HBM} - \text{GV}} \tag{4}$$

RCRs greater than 1 imply that the population assessed is (partly) exposed above the derived guidance value, and is thus potentially at risk of adverse effects from the chemical (e.g., BP-3).

2.2.2. Provisional HBM-GV Derivation Using the Urinary Mass Balance Approach

A provisional HBM-GV is based on a PoD, reflecting the critical endpoint identified in hazard characterization and used as a reference point for the urinary concentration levels measured. For the derivation of the provisional HBM-GV, the urinary mass balance approach is applied [9]. This method can be applied to substances that are primarily eliminated through urinary excretion, and for which regular repeated exposure is likely. For such substances, the approach assumes a balance between the substance intake and the substance excretion, reaching a steady-state in the urine matrix. Under this assumption, the urinary excretion rate is a constant fraction of the intake rate [10].

Using this method, the critical dose from an experimental animal study (animal PoD) was transformed into a human biomarker concentration consistent with that dose (provisional HBM-GV) based on the urinary mass balance assumption. The animal PoD was first extrapolated to a human equivalent (external) TRV with application of assessment factors to account for extrapolation uncertainties (see Equation (5)). As for the external approach described above, the factors for allometric scaling and remaining toxicokinetic differences were applied to account for the differences in toxicokinetics and toxicodynamics between experimental animals and humans. An intraspecies factor was also applied [6].

$$\text{TRV} = \frac{\text{Animal PoD}}{\text{Overall AF}} \tag{5}$$

Next, a provisional HBM-GV was calculated from the TRV, using Equation (6), with parameters defined as follows:

- provisional HBM-GV: the human biomonitoring guidance value below which no adverse health effect should be expected, expressed as the substance concentration per gram creatinine (μg/g creatinine).
- TRV: toxicological reference value, the external human value corresponding to the animal PoD (mg/kg bw/day).
- Fue: substance-specific steady-state fraction of urinary excretion, the daily proportion of the intake dose excreted in urine.
- Creatinine excretion rate adjusted to BW: typical 24 h creatinine excreted, adjusted to default human bodyweight (g/kg bw/day), to compensate for differences in the volume of urine excreted.

$$\text{provisional HBM} - \text{GV} = \frac{\text{TRV} * \text{Fue (substance)}}{\text{Creatinine excretion rate adjusted to BW}} \tag{6}$$

2.2.3. Systemic Exposure Dose Calculation Based on an Internal Approach

A systematic review was performed to assess the levels of BP-3 exposure in the European general population using the search engine Embase.com (Amsterdam: Elsevier) to gather all available human biomonitoring studies. Articles were qualitatively considered for compiling the BP-UV filter exposure database with European exposure data. Selected articles were then quantitatively screened for meta-analysis of BP-3 exposure in the European general population. As older studies are considered less relevant for the current situation, only data after 2006 were included. The analysis included only European studies that applied enzymatic deconjugation of the samples and storage at <4 °C, as these are generally considered critical aspects of sample treatment to obtain reliable and comparable results [11].

After study selection based on the criteria, the overlapping cohorts amongst the multiple included articles were removed. In filtering out the overlapping cohorts, studies with larger sample sizes, more detailed data reporting, and analysis of multiple benzophenones, were prioritized. The risk of bias of the individual studies was evaluated and studies were excluded from the meta-analysis in case of a considerable risk of bias.

The summary statistics were calculated for a typical case exposure and a reasonable worst case exposure. The typical case exposure is described by the median and range (min-max) of median values from the included studies and the reasonable worst-case by the median and range (min-max) of P95 values. Descriptive statistics that were below LOD or LOQ were replaced with the corresponding LOD or LOQ value, conforming to the statistical analysis of the CONTAM panel [12]. All statistical analyses were performed in Excel.

3. Results

3.1. Risk Assessment Based on External Approach

3.1.1. Selection of External Critical Dose

Previous safety assessments of BP-3 by the European Scientific Committees [13–15] concluded that BP-3 can be safely used as a UV-filter up to 6% in cosmetic sunscreen products and up to 0.5% in all types of cosmetic products although it has the potential for contact allergy and photoallergy. The recent assessment by the SCCS [2] considered previous evaluations, as well as new information, with a particular focus on the potential ED properties of BP-3.

The information available under REACH has categorized BP-3 as non-sensitizing, non-irritant to eyes and skin, and suggests no acute adverse effects [7]. Both human epidemiological and animal experimental studies have associated BP-3 exposure to reproductive and developmental toxicity [2]. Within the CLP regulation (EC No. 1272/2008) (classification, labeling, and packaging) [16], BP-3 is not considered to meet the criteria for reproductive toxicity classification [7]. Systematic reviews have highlighted a varying relationship between BP-3 exposure and estrogenic, androgenic, and antiandrogenic activities frequently reported in vivo and in vitro [17,18]. However, the clinical relevance of the endocrine effects in these studies is not always clear, and uncertainty remains over when the observed endocrine effects can be interpreted as adverse. This led the SCCS to conclude that the currently available evidence on ED properties of BP-3—including in silico modeling and in vitro and in vivo studies—was inconclusive and at best equivocal when considered individually or taken together. Therefore, the SCCS did not derive a new PoD on ED properties to use in the risk assessment.

To calculate the MoS, the SCCS used the lowest NOAEL values (67.9 mg/kg bw/d) derived from a study by Nakamura et al. [19]. The Nakamura et al. (2015) study [19] determined the effects of maternal and lactational exposure to BP-3 (between 0–3448 mg/kg/day) on the development and reproductive organs of the offspring of time-mated female rats. There were no significant differences on the reproductive/ pregnancy parameters or sex ratio of the offspring between the control and BP-3-dosed groups. In the highest dose groups, there was a significant reduction in the normalized anogenital distance on post-natal day (PND) 23 in male pups, a decrease in serum ALT (albumin transferase) and a significant increase in cholesterol levels in both female and male offspring at PND 23. Body, ovarian and uterine weights were significantly reduced. Relative liver-to-body-weight ratios were also significantly increased and paired kidney weights were lower.

In addition, in the highest dose group, seminiferous tubules contained few or no spermatocytes compared to the control group. The SCCS considered 3000 ppm as the

LOAEL and 1000 ppm (67.9 mg/kg bw/day) as the NOAEL based on the decrease in spermatocytes. This value was used as a PoD for risk characterization.

3.1.2. Systemic Exposure Dose (SED) Calculation Based on an External Approach

- Skin penetration

Dermal absorption was determined using an in vitro study with 6% BP-3 that had been considered in the previous SCCP Opinion [15]. Due to some shortcomings in this study, the SCCS applied an additional 2 × SD (standard deviation) to the reported mean dermal absorption value. A dermal absorption of 9.9% was used for the calculation of SED. For other cosmetic products, a dermal absorption of 8% based on the results of the study performed with 2% BP-3 was used [2].

- Calculation of the SED

The calculation of the SED following application of a whole-body UV-cream containing up to 6% BP-3, the intended level of use requested by the cosmetic industry, was estimated at 1.78 mg/kg bw/day, as detailed in Table 1.

Table 1. Systemic exposure by dermal route from whole-body UV-cream containing up to 6% BP-3 (adapted from SCCS opinion [2]).

Description	Parameter	Value	Unit
Amount of sunscreen product used	A	18	g/day
Concentration of BP-3	C	6	%
Dermal absorption	DA_p	9.9	%
Human body weight	Bw	60	kg
Systemic exposure dose (SED)	$A \times 1000$ mg/kg $\times C/100 \times DA_p/100/bw$	1.78	mg/kg bw/day

Exposure via other types of cosmetic products (UV or non-UV) by dermal, inhalation or oral routes was also considered by SCCS (see Table 2 below).

Table 2. Calculation of margin of safety for different products containing BP-3 (adapted from SCCS opinion [2]).

Product Categories	Conc.	Surface	Total * Systemic Exposure Dose (SED) mg/kg bw/d	NOAEL **	MoS
UV cream	6%	whole body	1.78	67.9	38
UV aerosolized spray	6%	whole body	1.89	67.9	36
UV pump spray	6%	whole body	1.78	67.9	38
UV face cream	6%	face	0.15	67.9	447
UV hand cream	6%	hand	0.21	67.9	317
Non-UV ***	0.50%	whole body	0.12	67.9	585
Lipstick	6%	lips	0.05	67.9	1257

* Total = dermal + inhalation + oral, ** derived from prenatal and postnatal developmental study, oral, rat [19]. *** to protect cosmetic formulations.

3.1.3. MoS Calculation

Because of the evidence for rapid and almost complete absorption of BP-3 from the oral route, the SCCS did not apply any adjustment for bioavailability to this NOAEL value. Details of the calculation of systemic exposure dose (SED) are presented in the Tables in Annex 2 of the SCCS opinion. The calculation of MoS for different product types ranged between 1257 for lipstick and 36–38 for UV aerosolized spray and cream. MOS for hand cream and face cream were 317 and 447, respectively (see Table 2).

A margin of safety (MoS) of 38 was estimated for the use of BP-3 at a concentration of 6% as a UV-filter in sunscreens for the whole-body application (in the form of cream/lotion or pump spray), 36 for sunscreen propellant spray, and 447 for the face, 317 for the hand and 1257 for lipstick application. The MoS was estimated at 585 for the use of BP-3 at 0.5% to protect cosmetic formulations against sunlight. As mentioned before, a default value of 100 (10 × 10) accounting for inter- and intraspecies differences is generally acceptable, and as an MoS of at least 100 therefore indicates that a cosmetic ingredient is safe for use [3], the SCCS assessment of BP-3 concluded that:

- The use of BP-3 as a UV-filter up to a maximum concentration of 6% in sunscreen products is not safe for the consumer with the exception of face cream, hand cream, and lipsticks
- The use of BP-3 up to 0.5% in cosmetic products to protect the cosmetic formulation is safe for the consumer.

3.2. Risk Assessment Based on Internal Approach

3.2.1. Provisional HBM-GV Derivation Using the Urinary Mass Balance Approach

In this study, the same POD of 67.9 mg/kg bw/day as selected by the SCCS [2] was used as a starting point in the derivation of a provisional HBM-GV that was used to calculate the risk based on the biomonitoring data. In accordance with the ECHA guidance R.8, assessment factors (AFs) for allometric scaling, remaining interspecies differences, and intraspecies differences were applied [6]. The guidance additionally prescribes an AF for exposure duration extrapolation, where appropriate. This AF is generally applied to account for differences in exposure between the experimental toxicity study and real-life situation. However, as in the study from Nakamura et al. (2015) [19], animals were exposed during gestation, which corresponds to a sensitive window of exposure, and no other AF were considered necessary to extrapolate from sub-acute to chronic exposure duration. In view of this, an overall AF of 100 was constituted from the following default AFs:

- Allometric scaling (rat to human): 4
- Remaining interspecies differences: 2.5
- Intraspecies differences: 10
- Duration extrapolation: 1
- Overall AF: 4 * 2.5 * 10 * 1 = 100

The Fue was obtained from two human experimental exposure studies [20,21]. Hayden et al. (1997) reported BP-3 levels in urine to be 1–2% of the initial dose 10 h after dermal application to nine healthy adult volunteers [20]. Sarveiya et al. (2004) similarly reported up to 1% BP-3 of the initial dose dermally applied to three female volunteers and measured in urine after 48 h [21]. It was therefore considered appropriate to incorporate an Fue of 0.01 (1%) as a conservative value.

The HBM4EU default average daily creatinine excretion was incorporated in the derivation of the provisional HBM-GV. The HBM4EU default creatinine value is based on the calculation according to Aylward et al. (2009) [22], and amounts to an average total amount of creatinine excreted of 1.4 g/L per day for both sexes combined. Adjusted to the default body weight of 70 kg, as defined by ECHA R.8 [6] and followed by HBM4EU, this amounted to an average creatinine excretion rate of 0.02 g/kg bw/day. Table 3 presents an overview of the information and parameters incorporated in the provisional HBM-GV calculation.

Application of the overall AF to the animal PoD in Equation (7) provided the TRV:

$$\text{TRV} = \frac{67.9}{100} = 0.68\ \text{mg/kg bw/day} \tag{7}$$

Incorporating the TRV and remaining parameters (Table 3) in Equation (8) provided the provisional HBM-GV:

$$\text{provisional HBM} - \text{GV (GenPop)} = \frac{0.68 * 0.01}{0.02} = 0.34 \text{ mg/g creatine } = 340 \text{ µg/g creatinine} \quad (8)$$

Table 3. Overview of parameter values incorporated in the provisional HBM-GV calculation, with sources.

Parameter	Value	Source
Animal PoD	NOAEL = 67.9 mg/kg bw/day	SCCS opinion (2021) [2]
Overall AF	AF = 100	ECHA R.8 guidance, SCCS opinion [6]
Fue	1% or 0.01	Hayden et al. (1997); Sarveiya et al. (2004) [20,21]
24 h creatinine	0.02 g/kg bw/day	HBM4EU default from Aylward et al. (2009) [22]

3.2.2. Systemic Exposure Dose Calculation Based on an Internal Approach

The internal exposure was assessed with a meta-analysis of the biomonitoring data. Screening and selection resulted in 17 articles with biomonitoring data on BP-3 in the European cohorts. After application of the exclusion criteria and selection of the studies that used sufficient numbers of samples (>120), 8 studies were selected. Out of these, only 3 studies that had used creatinine correction were considered in the meta-analysis [23–25]. Table 4 presents the cohort characteristics and study-specific descriptive statistics of creatinine-corrected concentrations of BP-3 and BP-1. Frederiksen et al. (2013) [25] reported BP-3 measurements in first morning urine samples of mother–child pairs in Denmark. Dewalque et al. (2014) [24] reported BP-3 measurements in spot urine samples from the general population in Belgium. Adoamnei et al. (2018) [23] reported BP-3 and BP-1 measurements in first morning urine samples of male students in Spain. However, this study did not report maximum measured concentrations.

Table 4. Cohort characteristics and descriptive statistics reported by the final three studies included in the meta-analysis of exposure. All concentrations (P50, P75, P95, max) are creatinine-corrected and thus expressed as µg/g creatinine. Concentrations > provisional HBM-GV in bold.

Study	Sampling Period	N/Sex	Age Range	Sample	Chemical	LOD (%>LOD)	P50	P75	P95 (RCR)	Max
Frederiksen et al. (2013) Denmark	September 2011–December 2011	143 M/F	6–11	Morning	BP-3	0.07 (97.0%)	2.00	6.30	33.00 (0.1)	**408.00**
		154 F	31–52	Morning	BP-3	0.07 (98.0%)	4.40	15.00	**392.00 (1.15)**	**2139.00**
Dewalque et al. (2014) Belgium	January 2013–April 2013	123 M	2–75	Spot	BP-3	0.20 (82.1%)	0.60	2.00	28.80 (0.08)	**414.20**
		138 F	1–85	Spot	BP-3	0.20 (83.3%)	1.30	4.40	33.30 (0.1)	141.30
Adoamnei et al. (2018) Spain	October 2010–November 2011	215 M	18–23	Morning	BP-3	0.20 (65.6%)	0.96	4.60	16.30 (0.05)	NA
					BP-1	0.10 (97.2%)	1.60	3.10	9.90 (0.03)	NA

N: sample size, M: male; F: female; age range: minimum-maximum age in years; LOD: limit of detection; NA: not available.

The TC of BP-3 exposure (based on median values across the studies) ranged from 0.60 to 4.40 µg/g creatinine, with a median of 1.30 µg/g creatinine. The RWC BP-3 exposure (based on P95 values across the studies) varied between 16.30 and 392.00 µg/g creatinine, with a median of 33.00 µg/g creatinine.

3.2.3. Risk Characterization Ratio (RCR) Calculation

Comparing the TC exposure range (0.60 to 4.40 µg/g creatinine) and the RWC exposure range (16.30 to 392.00 µg/g creatinine) with the provisional HBM-GV (340 µg/g creatinine) in Equations (3) and (4) provided the following RCRs:

$$\text{RCR (TC)} = \frac{0.60 \text{ to } 4.40}{340} = 0.002 \text{ to } 0.01 = \text{RCR} < 1$$

$$\text{RCR (RWC)} = \frac{16.30 \text{ to } 392.00}{340} = 0.05 \text{ to } 1.15 = \text{RCR} > 1$$

For typical case exposure, the RCR was below 1, so median HBM results did not exceed the orientation value. However, for a reasonable worst-case exposure, the RCR slightly exceeded 1. Table 4 shows the HBM values exceeding the provisional HBM-GV in bold, as well as the RCR values at the RWC/P95 value. The P95 value reported by Frederiksen et al. (2013) [25] for Danish females exceeded the provisional HBM-GV. The maximum values for female participants and children in Frederiksen et al. (2013) [25] and for male participants in Dewalque et al. (2014) [24] exceeded the provisional HBM-GV.

In summary, HBM values did not exceed the provisional HBM-GV for a typical exposure scenario and were therefore within safe limits. They did, however, exceed the provisional HBM-GV slightly in the reasonable worst-case exposure scenario, indicating potential risk to the highly exposed part of the population.

4. Discussion

In most chemical risk assessment frameworks, the default approach is to assess external intake from different sources of exposure, as well as via different routes of exposure, which are often assessed separately. In the context of cosmetics risk assessment, dermal exposure is calculated using the estimated amount of a product applied, retention factor of the product, and skin penetration value of the chemical. This approach includes various uncertainties and often overestimates the real uptake because default, conservative estimates are used, e.g., for the amount of product used. At the same time, actual (real-life) exposure may be underestimated by not taking into account that exposure to a chemical substance may occur from different sources that may fall under different legislative frameworks [4]. The use of HBM studies in chemical risk assessment has not yet become common practice, and its value in risk assessment remains largely unexplored. Within the HBM4EU project, it is the aim to increase the quality of HBM data and provide policymakers with tools to use these data in risk assessment.

BP-3 is mainly used in cosmetics, and a recent safety evaluation has been performed by the SCCS, instigated by concerns on the ED properties of BP-3. As is usual practice, the exposure estimate in the opinion was based on the use patterns of sunscreen and other cosmetic products (SCCS/1625/20) and dermal absorption of BP-3. No risk assessment for exposure of the European population to BP-3 using HBM data has been performed to date. This study used BP-3 as an example to investigate the usefulness of HBM in risk assessment, and for informing decision-making throughout the risk assessment process—in particular to address the additional research questions.

This study has compared two parallel risk assessment approaches—one based on the external exposure (the SCCS approach), and the other on internal exposure data from three different countries in Europe (the HBM approach). The results showed a remarkably similar overall conclusion in that the highly exposed individuals would be at a (slightly) increased risk from the intended levels of BP-3 use in certain product categories.

It needs to be noted that the internal dose approach used in HBM includes all sources of exposure, including from cosmetics. Therefore, although the estimated RWC is only slightly above 1, it does not truly reflect the risk to consumers from cosmetics only. The RWC was based on the P95, which could reflect exposure to BP-3 via cosmetic products, whereas the RTC, which is based on the mean, could be more representative of the background exposure to BP-3 either via cosmetic products other than sun-care products or via other consumer products. BP-3 is indeed also used in consumer products to protect formulations from degradation by the sun. Unfortunately, in the biomonitoring studies that have measured BP-3 in Europe, there is no product use data to show where this exposure comes from. In the study from Zamoiski et al. (2016) [26], the authors observed based on urinary concentrations of BP-3 measured in NHANES 2003–2006 and 2009–2012 that BP-3 was positively associated with self-reported frequency of sunscreen use and that concentrations of BP-3 in the more frequent users were 116.8 μg/g creatinine, which is above the P95 reported in Belgium and Spain [23,24] but lower than the Danish values [25]. It was rather surprising to see the highest BP-3 exposure in a study with a sampling period only in the fall in a northern country [25]. A similarly unexpected outcome was seen in a recent study performed in Israel, where sunscreen products are widely used, that reported relatively low internal exposure [27]. Thus, more information on product use would be valuable to gain more insight on the products contributing to BP-3 exposure. Moreover, the study populations were relatively small, which adds uncertainty in the extrapolation to the entire European population. Vogel et al. (2019) [28], recommended samples of at least 72 to 120 individuals to be able to define reference values for the general population, which is in line with the 3 biomonitoring studies we have included in our assessment.

In contrast, the external approach, as currently used by the SCCS, is conservative as it considers that the concentration is at the maximum requested level for all the cosmetic products formulated with BP-3. In addition, for the aggregated exposure scenario, the SCCS does not take into account market penetration and assumes that all the products in the product category concerned contain BP-3. Neither of these assumptions are realistic and they make the SCCS approach conservative. Where data are available, the SCCS methodology incorporates a refinement in terms of the use of a probabilistic approach to exposure assessment to take into account the distribution of values for exposure parameters, such as concentration in the products, amounts of the products used, etc. However, this approach is subject to the availability of data on these distributions and therefore may not be practicable in all cases.

There are certain differences between the external and internal dose approaches that also warrant further attention.

Both external/internal exposure-based approaches have certain advantages and disadvantages that are illustrated in Table 5.

HBM data provide data on exposure from all routes and sources, which can be an advantage when there is a concern over aggregated exposure. In such situations, the availability of actual data is a distinct advantage because aggregate exposure measurements often suffer from high levels of uncertainty. However, to answer questions on specific product contributions, additional product-specific exposure estimations will remain a necessity. Based on the available information, it is difficult to have a clear understanding of the exposure profile to BP-3 and the contribution of different sources of exposure. BP-3 is used as a UV-filter not only in cosmetic sun-care products, but also in other consumer products such as washing and cleaning products, anti-freeze products, biocides (e.g., disinfectants, pest control products), and printing inks in food contact materials [7]. Intervention studies have shown that by avoiding the use of cosmetic products containing BP-3, internal exposure can be significantly reduced, which indicates that cosmetics are an important source of exposure to BP-3 [29,30].

Another consideration for biomonitoring studies is that they are by nature retrospective. They can only be used for retrospective risk management by showing the actual exposure under a certain regulatory regime, but not to predict exposure before placing a new product on the market, or authorizing a new ingredient. HBM data can, however, be used to assess the impact of new regulatory measures by measuring time trends. In fact, the HBM data used here were from before the previous decrease in maximum concentration of BP-3 in cosmetic products that went into effect in 2017. An update of this risk assessment will be performed within HBM4EU based on measurements from 2020, the outcome of which will give a better understanding of the current situation.

Table 5. Differences between current SCCS approach and HBM approach for risk assessment.

	Current SCCS Approach	HBM Approach
Dose estimation	Modeled/estimated	Measured, real-world conditions
Exposure pathways	Dermal exposure	Provides data on total exposure from all exposure pathways
Temporality	No time lag; can be used in a predictive approach	Time lag between exposure estimate and risk assessment; only retrospective; cannot be used in a prospective approach
Product specificity	Calculations per product type, combining several conservative parameters in a deterministic assessment may lead to overestimation	No product-specific data: aggregate exposure modeling needed to identify relative contribution of a product to the overall exposure
Consideration for toxicokinetic aspects	Generally uses in vitro studies for dermal absorption and historic animal studies for PoD and applies an AF to correct for animal–human differences	Considers biotransformation and elimination of the substance in humans, but requires appropriate timing of sampling
Conclusion of risk assessment for BP-3	Exposure at the intended use levels exceeds safe dose for whole-body cream and spray but not face or hand cream	Exposure exceeds safe dose in highly exposed individuals

Where there is a special interest in a specific, vulnerable groups within the population, such as children or pregnant woman, HBM data have the advantage in terms of providing an insight into the exposure of separate groups. This is equally possible through modeling of external exposure, but only if the required use data are available for these groups.

The SCCS guidance for the testing of cosmetic ingredients and their safety evaluation [3] noted that HBM data should be used to support risk assessment and risk management. For example, HBM can serve to evaluate whether the dose estimation model over- or under-estimates the real exposure. Especially in light of the prohibition of in vivo animal studies on cosmetic substances, HBM offers a means to inform risk assessment by providing real-life in vivo information from human studies, without the need for interspecies extrapolation or the limitation of a small number of subjects often used in human volunteer studies. Indeed, HBM data were also noted in the SCCS opinion on BP-3 (SCCS/1625/20), but not used as such in modeling the exposure, or supporting the safety assessment.

5. Conclusions

HBM has the potential to be a useful tool for regulatory risk assessment as it provides actual exposure levels, but it is currently rarely applied in regulatory practice. Within the HBM4EU project, it was proposed to collect and create successful examples on the use of HBM data in chemical risk assessment. The experience from these cases could be used as a basis for developing harmonized guidance on the use of biomonitoring data in risk assessment, which was indicated as critical to the use of HBM. The aim of this work was to compare an HBM case study on BP-3 with a 'traditional' risk assessment. The results have shown that both the current SCCS approach and the HBM approach have their own advantages for risk assessment of cosmetic products. In the case of BP-3, both approaches indicated similar levels of risk. The comparison also indicated that HBM data can be useful in supporting risk assessment by providing real-life data on exposure, and may also play an important role in post-approval assessment studies on exposure trends. However, before being adopted for use on a regular basis in regulatory risk assessments, more efforts are needed to better harmonize HBM surveys, and to obtain robust data that are representative of the exposure of the European population. This study has also highlighted the need for the development of a standardized framework for incorporation of HBM data in the current risk assessment of cosmetic products.

Author Contributions: Conceptualization, C.R., M.W. and M.M.; methodology, C.R., M.W., W.B., J.W. and T.B.; validation Q.C. and T.B.; investigation, C.R., M.W., W.B., J.W. and Q.C.; writing—original draft preparation, C.R., M.M. and M.W.; writing—review and editing, Q.C. and T.B.; Q.C. and C.R. are members of the SCCS. C.R., M.M., T.B., M.W., W.B. and J.W. are partners of the HBM4EU project. All authors have read and agreed to the published version of the manuscript.

Funding: Some of the authors (Christophe Rousselle, Matthieu Meslin, Tamar Berman, Marjolijn Woutersen, Wieneke Bil and Jenna Wildeman) received funding as part of the European Union's Horizon 2020 research and innovation program (grant agreement No 733032, HBM4EU), and also received co-funding from their respective organizations.

Institutional Review Board Statement: Not applicable.

Informed Consent Statement: Not applicable.

Data Availability Statement: Not applicable.

Acknowledgments: The authors are grateful to Members of the SCCS and Partners of HBM4EU project whose work contributed to this article.

Conflicts of Interest: The authors declare no conflict of interest. The funders had no role in the design of the study; in the collection, analyses, or interpretation of data; in the writing of the manuscript, or in the decision to publish the results.

Abbreviations

AF: assessment factor, ALT: albumin transferase, BMD: benchmark dose, BP-3: benzophenone-3, BW: body weight, C: concentration, CPR: Cosmetic Product Regulation, CLP: classification, labeling, and packaging, ED: endocrine disrupting, E_{dermal}: dermal exposure, EU: European Union, f_{ret}: retention factor, F_{ue}: steady-state fraction of urinary excretion, HBM: human biomonitoring, HBM4EU: European Human Biomonitoring Initiative, HBM-GV: human biomonitoring guidance value, LOAEL: lowest observed adverse effect level, LOD: limit of detection, LOQ: limit of quantification, MoS: margin of safety, NOAEL: no observed adverse effect level, PND: post-natal day, PoD: point of departure, PoDsys: systemic PoD, q_x: quantity applied per day, RCR: risk characterization ratio, RWC: reasonable worst-case, SCCS: Scientific Committee on Consumer Safety, SD: standard deviation, SED: systemic exposure dose, TC: typical case, TRV: toxicological reference value.

References

1. Report from the Commission to the European Parliament and the Council Review of Regulation (EC) No 1223/2009 of the European Parliament and of the Council on Cosmetic Products with Regard to Substances with Endocrine-Disrupting Properties. Available online: https://eur-lex.europa.eu/legal-content/EN/TXT/PDF/?uri=CELEX:52018DC0739 (accessed on 16 October 2021).
2. Scientific Committee on Consumer Safety (SCCS). Request for a Scientific Opinion on Benzophenone-3 (CAS No 131-57-7, EC No 205-031-5). Available online: https://ec.europa.eu/health/sites/default/files/scientific_committees/consumer_safety/docs/sccs2016_q_038.pdf (accessed on 16 October 2021).
3. Scientific Committee on Consumer Safety (SCCS). The SCCS Notes of Guidance for the Testing of Cosmetic Ingredients and Their Safety Evaluation 11th Revision, Adopted in March 2021. Available online: https://ec.europa.eu/health/sites/default/files/scientific_committees/consumer_safety/docs/sccs_o_250.pdf (accessed on 16 October 2021).
4. Louro, H.; Heinälä, M.; Bessems, J.; Buekers, J.; Vermeire, T.; Woutersen, M.; van Engelen, J.; Borges, T.; Rousselle, C.; Ougier, E.; et al. Human biomonitoring in health risk assessment in Europe: Current practices and recommendations for the future. *Int. J. Hyg. Environ. Health* **2019**, *222*, 727–737. [CrossRef] [PubMed]
5. HBM4EU–Science and Policy for a Healthy Future. Available online: https://www.hbm4eu.eu/ (accessed on 16 October 2021).

6. European Chemical Agency (ECHA). Guidance on Information Requirements and Chemical Safety Assessment. Chapter R.8: Characterisation of Dose [Concentration]-Response for Human Health. Available online: https://echa.europa.eu/documents/10162/13632/information_requirements_r8_en.pdf/e153243a-03f0-44c5-8808-88af66223258 (accessed on 16 October 2021).
7. European Chemical Agency (ECHA). Substance Information Benzophenone 3. Available online: https://echa.europa.eu/de/substance-information/-/substanceinfo/100.003.943 (accessed on 16 October 2021).
8. Scientific Committee on Consumer Safety (SCCS). Basic Criteria for the In Vitro Assessment of Dermal Absorption of Cosmetic Ingredients, Adopted June 2010. Available online: https://ec.europa.eu/health/scientific_committees/consumer_safety/docs/sccs_s_002.pdf (accessed on 5 January 2022).
9. Apel, P.; Rousselle, C.; Lange, R.; Sissoko, F.; Kolossa-Gehring, M.; Ougier, E. Human biomonitoring initiative (HBM4EU)-Strategy to derive human biomonitoring guidance values (HBM-GVs) for health risk assessment. *Int. J. Hyg. Environ. Health* **2020**, *230*, 113622. [CrossRef] [PubMed]
10. Angerer, J.; Aylward, L.L.; Hays, S.M.; Heinzow, B.; Wilhelm, M. Human biomonitoring assessment values: Approaches and data requirements. *Int. J. Hyg. Environ. Health* **2011**, *214*, 348–360. [CrossRef] [PubMed]
11. Frederiksen, H.; Nielsen, O.; Skakkebaek, N.E.; Juul, A.; Andersson, A.-M. UV Filters Analyzed by Isotope Diluted TurboFlow-LC–MS/MS in Urine from Danish Children and Adolescents. *Int. J. Hyg. Environ. Health* **2017**, *220*, 244–253. [CrossRef] [PubMed]
12. Knutsen, H.K.; Alexander, J. Risk to human health related to the presence of perfluorooctane sulfonic acid and perfluorooctanoic acid in food. *EFSA J.* **2018**, *16*, 5194.
13. Scientific Committee on Cosmetic Products and Non-Food Products Intended for Consumers (SCCNFP). Opinion on the Evaluation of Potentially Estrogenic Effects of UV-Filters Adopted by the SCCNFP during the 17th Plenary Meeting of 12 June 2001. Available online: https://ec.europa.eu/health/scientific_committees/consumer_safety/opinions/sccnfp_opinions_97_04/sccp_out145_en.htm (accessed on 5 January 2022).
14. Scientific Committee on Consumer Products (SCCP). Opinion on Benzophenone-3. (COLIPA N° S38), SCCS/1069/06, Adopted 19 December 2006. Available online: https://ec.europa.eu/health/ph_risk/committees/04_sccp/docs/sccp_o_078.pdf (accessed on 5 January 2022).
15. Scientific Committee on Consumer Products (SCCP). Opinion on Benzophenone-3. (COLIPA N° S38), SCCS/1201/08, Adopted 16 December 2008. Available online: https://ec.europa.eu/health/ph_risk/committees/04_sccp/docs/sccp_o_159.pdf (accessed on 5 January 2022).
16. Regulation (EC) No 1272/2008 of the European Parliament and of the Council of 16 December 2008 on Classification, Labelling and Packaging of Substances and Mixtures, Amending and Repealing Directives 67/548/EEC and 1999/45/EC, and Amending Regulation (EC) No 1907/2006. Available online: https://eur-lex.europa.eu/legal-content/en/TXT/?uri=CELEX%3A32008R1272 (accessed on 16 October 2021).
17. Ghazipura, M.; McGowan, R.; Arslan, A.; Hossain, T. Exposure to benzo-phenone-3 and reproductive toxicity: A systematic review of human and animal studies. *Reprod. Toxicol.* **2017**, *7*, 175–183. [CrossRef] [PubMed]
18. Kim, S.; Choi, K. Occurrences, toxicities, and ecological risks of benzophenone-3, a common component of organic sunscreen products: A mini-review. *Environ. Int.* **2014**, *70*, 143–157. [CrossRef] [PubMed]
19. Nakamura, N.; Inselman, A.; White, G.A.; Chang, C.-W.; Trbojevich, R.A.; Sephr, E.; Voris, K.L.; Patton, R.E.; Bryant, M.S.; Harrouk, W.; et al. Effects of maternal and lactational exposure to 2-hydroxy-4-methoxybenzone on development and reproductive organs in male and female rat offspring. *Birth Defects Res. Part B Dev. Reprod. Toxicol.* **2015**, *104*, 35–51. [CrossRef] [PubMed]
20. Hayden, C.; Roberts, M.; Benson, H.A. Systemic absorption of sunscreen after topical application. *Lancet* **1997**, *350*, 863–864. [CrossRef]
21. Sarveiya, V.; Risk, S.; Benson, H.A. Liquid chromatographic assay for common sunscreen agents: Application to in vivo assessment of skin penetration and systemic absorption in human volunteers. *J. Chromatogr. B* **2004**, *803*, 225–231. [CrossRef] [PubMed]
22. Aylward, L.L.; Hays, S.M.; Gagné, M.; Krishnan, K. Derivation of Biomonitoring Equivalents for di(2-ethylhexyl)phthalate (CAS No. 117-81-7). *Regul. Toxicol. Pharmacol.* **2009**, *55*, 249–258. [CrossRef] [PubMed]
23. Adoamnei, E.; Mendiola, J.; Moñino-García, M.; Vela-Soria, F.; Iribarne-Durán, L.M.; Fernández, M.F.; Olea, N.; Jørgensen, N.; Swan, S.; Torres-Cantero, A.M. Urinary concentrations of benzophenone-type ultra violet light filters and reproductive parameters in young men. *Int. J. Hyg. Environ. Health* **2018**, *221*, 531–540. [CrossRef] [PubMed]
24. Dewalque, L.; Pirard, C.; Charlier, C. Measurement of Urinary Biomarkers of Parabens, Benzophenone-3, and Phthalates in a Belgian Population. *BioMed Res. Int.* **2014**, *2014*, 649314. [CrossRef] [PubMed]
25. Frederiksen, H.; Nielsen, J.K.S.; Mørck, T.A.; Hansen, P.W.; Jensen, J.F.; Nielsen, O.; Andersson, A.-M.; Knudsen, L.E. Urinary excretion of phthalate metabolites, phenols and parabens in rural and urban Danish mother–child pairs. *Int. J. Hyg. Environ. Health* **2013**, *216*, 772–783. [CrossRef] [PubMed]
26. Zamoiski, R.D.; Cahoon, E.K.; Freedman, D.M.; Linet, M.S. Self-reported sunscreen use and urinary benzophenone-3 concentrations in the United States: NHANES 2003–2006 and 2009–2012. *Environ. Res.* **2015**, *142*, 563–567. [CrossRef] [PubMed]
27. Machtinger, R.; Berman, T.; Adir, M.; Mansur, A.; Baccarelli, A.A.; Racowsky, C.; Calafat, A.M.; Hauser, R.; Nahum, R. Urinary concentrations of phthalate metabolites, bisphenols and personal care product chemical biomarkers in pregnant women in Israel. *Environ. Int.* **2018**, *116*, 319–325. [CrossRef] [PubMed]

28. Vogel, N.; Conrad, A.; Apel, P.; Rucic, E.; Kolossa-Gehring, M. Human biomonitoring reference values: Differences and similarities between approaches for identifying unusually high exposure of pollutants in humans. *Int. J. Hyg. Environ. Health* **2018**, *222*, 30–33. [CrossRef] [PubMed]
29. Harley, K.G.; Kogut, K.; Madrigal, D.S.; Cardenas, M.; Vera, I.A.; Meza-Alfaro, G.; She, J.; Gavin, Q.; Zahedi, R.; Bradman, A.; et al. Reducing Phthalate, Paraben, and Phenol Exposure from Personal Care Products in Adolescent Girls: Findings from the HERMOSA Intervention Study. *Environ. Health Perspect.* **2016**, *124*, 1600–1607. [CrossRef] [PubMed]
30. Dodson, R.E.; Boronow, K.E.; Susmann, H.; Udesky, J.O.; Rodgers, K.M.; Weller, D.; Woudneh, M.; Brody, J.G.; Rudel, R.A. Consumer behavior and exposure to parabens, bisphenols, triclosan, dichlorophenols, and benzophenone-3: Results from a crowdsourced biomonitoring study. *Int. J. Hyg. Environ. Health* **2020**, *230*, 113624. [CrossRef] [PubMed]

toxics

Article

Relationship between Occupational Exposure to Airborne Nanoparticles, Nanoparticle Lung Burden and Lung Diseases

Valérie Forest [1,*], Jérémie Pourchez [1], Carole Pélissier [2,3], Sabyne Audignon Durand [4,5], Jean-Michel Vergnon [6,7] and Luc Fontana [2,3]

1 Centre CIS, Mines Saint-Etienne, Univ Lyon, Univ Jean Monnet, INSERM, U1059 Sainbiose, F-42023 Saint-Etienne, France; pourchez@emse.fr
2 Department of Occupational Medicine, University Hospital of Saint-Etienne, F-42055 Saint-Etienne, France; carole.pelissier@chu-st-etienne.fr (C.P.); luc.fontana@chu-st-etienne.fr (L.F.)
3 Univ Lyon, Univ Eiffel, Univ Lyon 1, Univ St Etienne, IFSTTAR, UMRESTTE, UMR_T9405, F-42005 Saint-Etienne, France
4 EPICENE Team, Inserm U1219, Bordeaux Population Health Research Center, University of Bordeaux, F-33076 Bordeaux, France; sabyne.audignon@u-bordeaux.fr
5 Department of Occupational and Environmental Medicine, Bordeaux Hospital, F-33400 Talence, France
6 Univ Lyon, Univ Jean Monnet, INSERM, U1059 Sainbiose, F-42023 Saint-Etienne, France; Jean.Michel.Vergnon@univ-st-etienne.fr
7 Department of Chest Diseases and Thoracic Oncology, University Hospital of Saint-Etienne, F-42055 Saint-Etienne, France
* Correspondence: vforest@emse.fr

Abstract: The biomonitoring of nanoparticles in patients' broncho-alveolar lavages (BAL) could allow getting insights into the role of inhaled biopersistent nanoparticles in the etiology/development of some respiratory diseases. Our objective was to investigate the relationship between the biomonitoring of nanoparticles in BAL, interstitial lung diseases and occupational exposure to these particles released unintentionally. We analyzed data from a cohort of 100 patients suffering from lung diseases (NanoPI clinical trial, ClinicalTrials.gov Identifier: NCT02549248) and observed that most of the patients showed a high probability of exposure to airborne unintentionally released nanoparticles (>50%), suggesting a potential role of inhaled nanoparticles in lung physiopathology. Depending on the respiratory disease, the amount of patients likely exposed to unintentionally released nanoparticles was variable (e.g., from 88% for idiopathic pulmonary fibrosis to 54% for sarcoidosis). These findings are consistent with the previously performed mineralogical analyses of BAL samples that suggested (i) a role of titanium nanoparticles in idiopathic pulmonary fibrosis and (ii) a contribution of silica submicron particles to sarcoidosis. Further investigations are necessary to draw firm conclusions but these first results strengthen the array of presumptions on the contribution of some inhaled particles (from nano to submicron size) to some idiopathic lung diseases.

Keywords: biomonitoring; nanoparticles; lung diseases; mineralogical analysis of broncho-alveolar lavages; occupational exposure

Citation: Forest, V.; Pourchez, J.; Pélissier, C.; Audignon Durand, S.; Vergnon, J.-M.; Fontana, L. Relationship between Occupational Exposure to Airborne Nanoparticles, Nanoparticle Lung Burden and Lung Diseases. *Toxics* **2021**, *9*, 204. https://doi.org/10.3390/toxics9090204

Academic Editors: Jeongim Park and Christophe Rousselle

Received: 13 July 2021
Accepted: 27 August 2021
Published: 30 August 2021

Publisher's Note: MDPI stays neutral with regard to jurisdictional claims in published maps and institutional affiliations.

Copyright: © 2021 by the authors. Licensee MDPI, Basel, Switzerland. This article is an open access article distributed under the terms and conditions of the Creative Commons Attribution (CC BY) license (https://creativecommons.org/licenses/by/4.0/).

1. Introduction

Nanoparticles are ubiquitous in nature, naturally occurring as by-products of wild fires, volcanic eruptions, and other natural processes, and are usually called ultra-fine particles (UFP). Nanoparticles can also be a result of human activities unintentionally produced and present in polluting emissions, such as welding fumes, cigarette smoke, aircraft waste gas, or diesel exhaust, also called UFP. In addition to these sources, a number of artificial nanoparticles, engineered nanomaterials (NM) which exhibit unique physical, chemical and/or biological characteristics associated with their nanostructure, have been developed and produced in a controlled, engineered manner to exploit their novel properties and functions. Due to the tremendous development of nanotechnologies during the last few decades and the subsequent potential exposure of humans to nanomaterials, nanotoxicology is a rapidly evolving research field. According to Stone et al. in a review [1], UFP and NM toxicology are not two distinct fields. Rather, they overlap extensively with the potential to extrapolate from one to the other in many respects. Furthermore, for these authors, ambient particulate matter research provided evidence of potential health impacts for UFPs, and NM toxicology has largely provided essential evidence of the mechanistic plausibility of these health effects. Finally, according to Stone et al., it seems safe to conclude that UFPs and NMs share the same general biological mechanisms of adverse effects.

In their review, Manno et al. defined biomonitoring as "the repeated, controlled measurement of chemical or biological markers in fluids, tissues, or other accessible samples from subjects exposed or exposed in the past or to be exposed to chemical, physical or biological risk factors in the workplace and/or the general environment" [2]. Consequently, in a context of health risk assessment, biomonitoring can be a particularly useful approach. Biomarkers used in human health studies typically fall within three categories: biomarkers of exposure, effect, and susceptibility [3]. They can bring critical information on the relationship between exposure to a harmful substance and biological/pathological effects.

Biomonitoring has been widely used in pulmonology, especially in the case of pneumoconiosis. One typical example is the assessment of asbestosis bodies in patient lung tissues or in broncho-alveolar lavage (BAL) fluids which has allowed defining values specific of diseases [4–6]. More recently, it has been suggested that the chemical composition of BAL from idiopathic pulmonary fibrosis patients had a specific profile that can be distinguished from that of patients with other interstitial lung diseases or healthy subjects [7]. The extension of this approach to the nanotoxicology field, although it has to face some technical challenges [8,9], could be very interesting especially to get new insights into the role of inhaled biopersistent nanoparticles in the etiology or development of some respiratory diseases. Indeed, although the impact of air pollution, including the contribution of nano-sized particles, on human health has been well documented [10], fewer data are available on the effects of nanoparticles, either engineered or unintentionally released, in the context of occupational exposure.

The biological monitoring of nanoparticles in human lung tissues or fluids could fill a gap and represents a promising way to investigate potential causal links between an exposure to inhaled nanoparticles and biological effects and even diseases [8,11–14] (Figure 1).

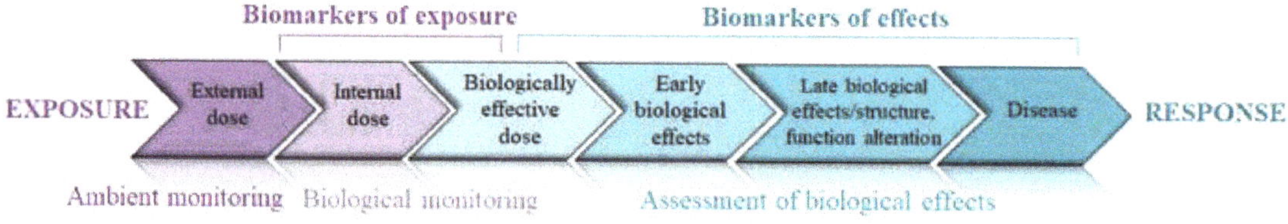

Figure 1. The mineralogical analysis of metal load extracted from pulmonary fluids could be used as an indicator of exposure to nanoparticles and could contribute to the assessment of potential causal links between the presence of inhaled biopersistent nanoparticles in the lungs and respiratory diseases.

Indeed, mineralogical analyses of BAL (i.e., biological monitoring of inhaled particles) allow quantifying the internal dose of inhaled biopersistent nanoparticles in a lung sample, which differs from the external dose that can be measured by ambient monitoring (i.e., atmospheric metrology). The assessment of the internal dose is a first step towards the characterization of persistent nanoparticles in tissues and the understanding of this potential source of adverse effects.

We adopted this approach to detect and quantify nanoparticles in various types of clinical samples. We developed optimized protocols for each kind of biological matrix to isolate the micro, sub-micro, and nano fractions of various types of inorganic particles and thus perform comprehensive mineralogical analyses. We were thus able to determine the nanoparticle load in patients' biological samples such as seminal and follicular fluids [15], colon [16], amniotic fluids [17], or BAL [18–20]. We especially focused our attention on these latter, as the biomonitoring of biopersitent nanoparticles in the lung could be particularly relevant in the case of respiratory diseases. Indeed, the respiratory tract represents the main route of entry for nanoparticles in the body and despite the lack of clear evidence it has been suggested that inhaled engineered nanoparticles accumulated in the lungs could be responsible, or at least could contribute, to idiopathic respiratory diseases.

We previously conducted a clinical trial on a cohort of 100 patients (NanoPI clinical trial, ClinicalTrials.gov Identifier: NCT02549248). We separated micron-sized particles (>1 µm) from submicron (100 nm–1 µm) and nano-sized particles (<100 nm) contained in BAL from patients who suffered from interstitial lung diseases (ILD). We then determined the metal load in each of these size-fractions. We evidenced a concentration of submicron silica particles higher in patients suffering from sarcoidosis than in patients suffering from other ILD, suggesting a potential role of these inhaled particles in the etiology and/or development of sarcoidosis [19]. Similarly, we observed a concentration of titanium nanoparticles higher in patients suffering from idiopathic fibrosis than in patients suffering from other ILD allowing to suspect a relationship between titanium nanoparticles and idiopathic pulmonary fibrosis even though in this case we had a too limited number of patients to reach a satisfactory statistical power to draw firm conclusions.

To complement mineralogical analyses of BAL and offer a comprehensive vision of the events from exposure to airborne nanoparticles to the biological response induced (Figure 1), we investigated associations between respiratory diseases and occupational exposures. To that purpose, we estimated the exposure to inhaled unintentionally released nanoparticles of the patients for each job held in their working life.

Thus, the objective of the present paper was to further investigate the relationship between the biological monitoring of nanoparticles in human BAL, interstitial lung diseases, and occupational exposure using a retrospective occupational exposure to unintentionally released nanoparticles assessment. Getting a complete picture from exposure to disease illustrates a comprehensive and useful approach in terms of human health risk assessment.

2. Materials and Methods

2.1. Patients

A prospective, monocentric and exploratory study called NanoPI (ClinicalTrials.gov Identifier: NCT02549248) was carried out during two years at the University Hospital of Saint-Etienne (Chest diseases and thoracic oncology Department). One-hundred patients exhibiting a clinical image of diffuse ILD and in need of a bronchoscopy associated with a BAL were included in this study after being fully informed and having given their written consent. Our protocol was in accordance with ethical principles defined by the World Medical Association declaration of Helsinki and subsequent amendments and was approved by an ethics committee (Comité de Protection des Personnes, Sud-Est I) as well as by the French agency regulating biomedical research (Agence Nationale de Sécurité du Médicament et des produits de santé, ANSM).

The following criteria of inclusion were applied: (i) patients with an ILD determined based on clinical signs and CT scan, requiring a flexible bronchoscopy associated with a BAL; (ii) patients older than 18; (iii) patients who had given their voluntary, informed and written consent; (iv) patients having a social insurance or beneficiary (mandatory for any French clinical study). Patients were excluded in the following cases: (i) patients who had not given their consent; (ii) when flexible bronchoscopy or BAL was not possible; (iii) patients under legal protection or pregnant women; (iv) patients with contagious disease (e.g., HIV infection, tuberculosis, viral hepatitis) for safety reasons.

2.2. Broncho-Alveolar Lavages

BAL were performed by injecting 50 mL of a warmed saline solution in the selected area of the patient lung. This solution was slowly aspirated, the collected sample constituting the bronchial wash (BW). Then, 2 to 4 additional 50 mL doses of warmed saline solution were injected and aspirated successively, the collected sample representing the BAL strictly speaking. After cytological analysis of the samples performed by the Histology-Cytology Department of the University Hospital of Saint-Etienne, the remaining samples (5 mL of BW and 20 mL of BAL) were added with an equivalent volume of sodium hypochlorite and stored at 4 °C until the mineralogical analyses were performed.

2.3. Sample Pre-Treatment and Analysis

We previously developed and validated a size fractionation protocol allowing to separate microparticles (>1 μm) from submicron particles (ranging from 100 nm to 1 μm) and nanoparticles and ions (particles < 100 nm), described extensively in our previous publications [18–20]. We applied this protocol to BAL and BW samples. Dynamic light scattering (DLS) method (Nanozetasizer®, Malvern Instrument, Orsay, France) allowed verifying the efficiency of the size fractionation. All fractions were also analyzed using inductively coupled plasma atomic emission spectroscopy (ICP-AES, Jobin-Yvon JY138 Ultrace) to assess for each metal the quantity of matter expressed in parts-per-billion (ppb), i.e., as ng/mL. The content of BAL and BW in aluminum (Al), beryllium (Be), cobalt (Co), chromium (Cr), copper (Cu), iron (Fe), nickel (Ni), silicon (Si), titanium (Ti), tungsten (W), zinc (Zn), and zirconium (Zr) was thus determined. "Blank" samples were also included to ensure samples were not contaminated with particles present in the environment, the materials, and solutions used. These control samples consisted of saline solution aspirated through the bronchoscope and that underwent exactly the same processes as clinical samples did.

2.4. Comparison to Clinical Data

More than 200 different conditions are grouped under the term interstitial lung disease, classified together because of similar clinical, radiographic, physiologic, or pathologic manifestations [21]. These diseases can be subdivided into those with a known origin (e.g., systemic disease, iatrogenic causes by drug, radiation, extrinsic allergic, pneumoconiosis, post-infectious) and those without, the latter usually called idiopathic interstitial pneumonias (mainly sarcoidosis, other granulomatous ILD and idiopathic pulmonary fibrosis). Patients from our cohort were thus first classified in two groups depending on the origin of the disease they suffer from, either with a known etiology or idiopathic. Details on the cohort are reported in Table 1. We then focused on sarcoidosis and idiopathic pulmonary fibrosis, two groups for which we suspected the contribution of Si submicron particles and Ti nanoparticles, respectively [19].

2.5. Retrospective Occupational Exposure to Nanoparticles Assessment

Patients were asked about all occupations held and industrial sectors, for at least six months since leaving school during their working life, up to the date of the diagnosis. For each occupation, the employer's sector was coded into the French classification of activities (NAF, Nomenclature d'Activités Françaises, 1999) of the National Institute for Statistics and Economics Studies (INSEE) [22], and the occupation was coded according to the International Standard Classification of Occupations (ISCO, 1968) of the International Labour Organisation, The International Labour Office [23].

Patients were also asked about working conditions such as exposure to chemicals, dusts, fumes, and the level of preventive measures used in occupational settings e.g., ventilation, and use of personal protective equipment. The assessment of occupational exposure to unintentionally released nanoparticles of each patient was independently and anonymously performed by an experienced industrial hygienist, follow-up on the review of occupational physicians, on the basis of these data. A probability of exposure to unintentionally produced nanoparticles from work-processes implemented in each occupation held by the patients was determined. Classes of probability were defined in a four scale: 0 when it was not found, 1 when it was possible (<10%), 2 when it was likely (10–50%), and 3 when it was very likely (>50%). Then they considered the final probability of exposure to unintentionally released nanoparticles as the highest probability of exposure observed in the career.

Table 1. Description of the cohort.

Disease	Number of Patients	Median Age (Min–Max)	Sex Ratio (M:F)	Smokers (Former Smokers)
With a known etiology	**58**	**70.5 (22–87)**	**43:15**	**10.3% (46.6%)**
Drug related ILD	15	71 (46–81)	14:1	6.7% (73.3%)
Infectious ILD	12	69.5 (42–85)	7:5	16.7% (33.3%)
Hypersensitivity pneumonitis	7	75 (34–78)	6:1	0% (14.3%)
Auto-immune pneumonitis	5	72 (22–87)	2:3	0% (40%)
Lymphangitis carcinomatosis/Neoplasia	4	75.5 (67–83)	2:2	0% (50%)
Desquamative interstitial pneumonia	3	71 (48–81)	2:1	66.7% (0%)
Pneumoconiosis	2	50 (41–59)	2:0	50% (50%)
Pulmonary veno-occlusive disease	2	73 (72–74)	2:0	0% (100%)
Antisynthetase syndrome	2	70.5 (70–71)	1:1	0% (50%)
Silicosis	1	55	1:0	0% (100%)
Microscopic polyangiitis	1	71	0:1	0% (0%)
Granulomatosis with polyangitis (Wegener's granulomatosis)	1	65	1:0	0% (0%)
Left heart failure	1	59	1:0	0% (100%)
Lipoid pneumonia	1	69	1:0	0% (100%)
Bronchiolitis obliterans	1	40	1:0	0% (0%)
Idiopathic	**34**	**67.5 (25–81)**	**22:12**	**14.7% (23.5%)**
Sarcoidosis	14	47 (25–80)	6:8	14.3% (7.1%)
Idiopathic nonspecific interstitial pneumonia	11	76 (46–81)	8:3	9.1% (27.3%)
Idiopathic pulmonary fibrosis	9	69 (61–81)	8:1	22.2% (36.4%)
Others	**8**	**61 (46–83)**	**3:5**	**0% (28.6%)**

3. Results

3.1. Relationship between Biomonitoring of Nanoparticles in Broncho-Alveolar Lavages and Lung Diseases

We previously separated micro from submicron and nanoparticles contained in BAL and BW samples from 100 patients suffering from ILD. We then assessed the metal load in each of these fractions. Results are fully detailed in our previous publications [18,19].

As shown by Figure 2A, we evidenced a concentration of submicron silica particles higher in patients suffering from sarcoidosis than in patients suffering from other ILD. Similarly, we observed a concentration of titanium nanoparticles higher in patients suffering from idiopathic fibrosis than in patients suffering from other ILD (Figure 2B).

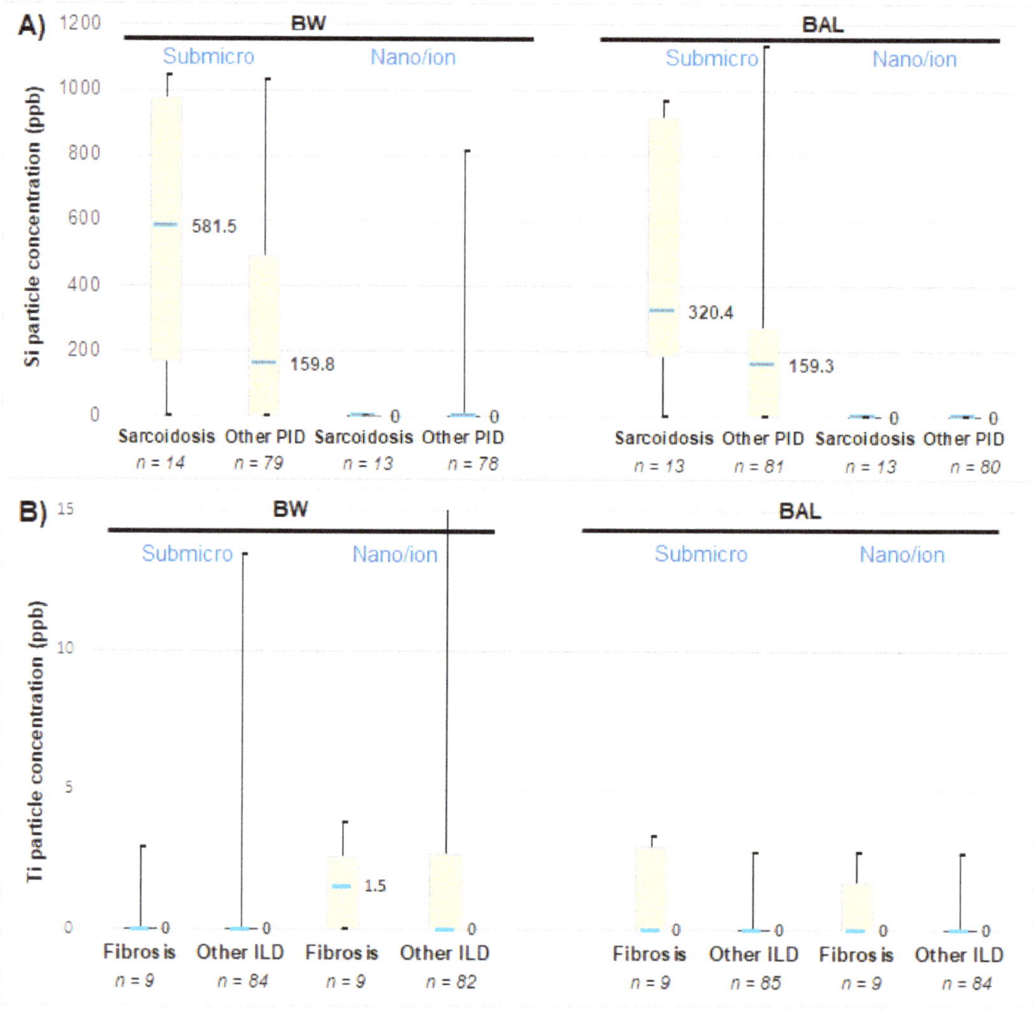

Figure 2. (**A**) Si particles concentration in bronchial wash (BW) and broncho-alveolar lavages (BAL) of patients suffering either from sarcoidosis or another type of ILD. (**B**) Ti particles concentration in BW and BAL of patients suffering either from idiopathic pulmonary fibrosis or another type of ILD. Comparison between the fractions containing either the submicron particles and the nanoparticles and ions. The median (in bold), minimal and maximal values, as well as the first and third quartiles are indicated. The number of patients (n) from each group is reported (please note that some data are missing because for technical reasons some BW and BAL samples could not be analyzed).

We further analyzed potential relationships between the particle load in BAL and BW and the patients' gender and their past or current smoking status by calculating the Pearson correlation coefficient. No correlation was observed (data not shown).

3.2. Relationship between Lung Diseases and Occupational Exposure

To explore possible associations between interstitial lung diseases and occupational exposure to airborne nanoparticles, we considered the highest probability of exposure to unintentionally released nanoparticles observed in the career. Results are reported Table 2.

Table 2. Description of the patients' cohort in terms of lung disease, occupations and probability of exposure to unintentionally released nanoparticles. The probability of exposure assessment resulted from the nanoparticle-release potential of the considered work process (determined from the literature) which was adjusted for each occupation according to the description of tasks and work-processes implemented as provided in the ISCO 1968 classification.

Patient Number	Lung Disease	Group (E: Disease of Known Etiology, I: Idiopathic Disease)	Occupations			Probability of Exposure to Nanoparticles: 0 Not Found, 1: Possible < 10%, 2: Likely 10–50%, 3: Very Likely > 50%			Final Exposure to Nanoparticles Probability: Highest Probability of Exposure to Nanoparticles in the Career
			1	2	3	Occupation 1	Occupation 2	Occupation 3	
1	Drug related ILD	E	Coachbuilder/painte	Welder	Printing machine operator	3	3	1	3
2	Idiopathic pulmonary fibrosis	I	Textile products machine operator			1			1
3	Drug related ILD	E	Farmer	Switching operator	Switching operator	3	1	1	3
4	Lymphangitis carcinomato-sis/Neoplasia	E	Farmer			2			2
5	Other		Mason	Refractory bricklayer	Ceramics operator	3	3	3	3
6	Drug related ILD	E	Market gardener	Farmer	Farm hands	0	2	1	3
7	Other		Seamstress	Cook	Childminder	1	3	1	3
8	Other		Floor sander	Truck driver	Pressman	3	3	1	3
9	Idiopathic pulmonary fibrosis	I	Miner	Miner	Train driver	3	3	3	3
10	Drug related ILD	E	Miner	Tile setter		3	3		3
11	Auto-immune pneumonitis	E	Domestic help	Domestic help		1	1		1
12	Drug related ILD	E	Farmer			3			3
13	Hypersensitivity pneumonitis	E	Bank employee			0			0
14	Infectious ILD	E	Accountant	Medical secretary		0	0		0
15	Other		/						
16	Infectious ILD	E	Printing machine operator	Printing machine operator		1	1		1
17	Lymphangitis carcinomato-sis/Neoplasia	E	Masseuse	Chocolate—products machine operator	Waitress/manageres	0	0	1	1
18	Infectious ILD	E	Teacher	Teacher		0	0		0
19	Idiopathic pulmonary fibrosis	I	Mason			3			3
20	Drug related ILD	E	Truck driver	Salesman	Company director	3	0	0	3
21	Drug related ILD	E	Pipe fitter/welder			3			3
22	Desquamative interstitial pneumonia	E	Coachbuilder/painter			3			3
23	Drug related ILD	E	Waitress	Waitress	Candle production machine operator	0	2	0	2
24	Other		Factory worker	Factory worker	Market gardener	0	3	0	3
25	Pulmonary veno-occlusive disease	E	Farmer			3			3
26	Idiopathic pulmonary fibrosis	I	Printing machine operator	Truck driver	Salesman	1	3	0	3
27	Idiopathic nonspecific interstitial pneumonia	I	Mason	Joiner		3	3		3
28	Idiopathic nonspecific interstitial pneumonia	I	Secretary	Secretary	Domestic help	0	0	1	1
29	Drug related ILD	E	Mason	Factory worker		3	1		3
30	Sarcoidosis	I	Seamstress	Childminder	Seamstress	1	0	0	1
31	Drug related ILD	E	Baker	Post officer	Foundry worker	3	0	3	3
32	Pneumoconiosis	E	Dental prosthetist			3			3
33	Auto-immune pneumonitis	E	Carer			0			0
34	Idiopathic nonspecific interstitial pneumonia	I	Baker			3			3
35	Bronchiolitis obliterans	E	Boilermaker/welder	Pipe fit-ter/boilermak		3	3		3
36	Other		Cleaner	Saleswoman	Waitress/manageress	0	0	1	1
37	Idiopathic nonspecific interstitial pneumonia	I	Domestic help	Food salesperson	Cashier	1	2	0	2
38	Sarcoidosis	I	Farmer			3			3
39	Antisynthetase syndrome	E							
40	Pulmonary veno-occlusive disease	E	Farmer			3			3
41	Idiopathic pulmonary fibrosis	I	Plant operator	Plant operator	Cleaner	1	3	1	3
42	Granulomatosis with polyangitis (Wegener's granulomatosis)	E	Joiner			3			3
43	Sarcoidosis	I	Seamstress	Cleaner	Childminder	1	1	0	1
44	Left heart failure	E	Boilermaker/welder	Carpenter/metal fitter	Carpenter/metal fitter	3	3	3	3
45	Hypersensitivity pneumonitis	E	Farmer			3			3

Table 2. *Cont.*

Patient Number	Lung Disease	Group (E: Disease of Known Etiology, I: Idiopathic Disease)	Occupations			Probability of Exposure to Nanoparticles: 0 Not Found, 1: Possible < 10%, 2: Likely 10–50%, 3: Very Likely > 50%			Final Exposure to Nanoparticles Probability: Highest Probability of Exposure to Nanoparticles in the Career
			1	2	3	Occupation 1	Occupation 2	Occupation 3	
46	Infectious ILD	E	Joiner			3			3
47	Idiopathic pulmonary fibrosis	I	Manufacturing labourer	Manufacturing labourer		3	3		3
48	Sarcoidosis	I	Salesman	Accountant		0	0		0
49	Idiopathic nonspecific interstitial pneumonia	I	Butcher			1			1
50	Idiopathic pulmonary fibrosis	I	Manufacturing labourer	Baker	Mason	3	3	3	3
51	Pneumoconiosis	E	Joiner			3			3
52	Microscopic polyangiitis	E	Saleswoman			0			0
53	Idiopathic nonspecific interstitial pneumonia	I	Machine-tool operator	Manufacturing labourer	Fiberglass plant operator	3	1	1	3
54	Desquamative interstitial pneumonia	E	Waitress	Fruit picker	Cleaner in a plastic products factory	2	0	2	2
55	Sarcoidosis	I	Animator in retirement home			0			0
56	Idiopathic nonspecific interstitial pneumonia	I	Foundry moulder			3			3
57	Idiopathic pulmonary fibrosis	I	Hospital caregiver			2			2
58	Sarcoidosis	I	Cleaner			2			2
59	Idiopathic pulmonary fibrosis	I							
60	Other		Speech therapist	Speech therapist	Speech therapist	0	0	0	0
61	Infectious ILD	E	Gym teacher			0			0
62	Auto-immune pneumonitis	E	Farmer	Textile products machine operator		3	1	0	3
63	Drug related ILD	E	Optical assembler	Butcher		0	0		0
64	Lipoid pneumonia	E	Steel materials handling	Machine-tool operator	Mason	3	3	3	3
65	Desquamative interstitial pneumonia	E	Metal carpenter			3			3
66	Sarcoidosis	I	Secretary	Policeman		0	0		0
67	Auto-immune pneumonitis	E	Office worker	Communicatic manager	Director	1	0	1	1
68	Infectious ILD	E	Plasterer/painter	Handler		3	2		3
69	Hypersensitivity pneumonitis	E	Cleaner	Quality manager	Windshield manufacturer	2	0	3	3
70	Auto-immune pneumonitis	E	Joiner			3			3
71	Other								
72	Infectious ILD	E	Carer			0			0
73	Hypersensitivity pneumonitis	E	Wood-products machine operator	Farmer		3	3		3
74	Drug related ILD	E	Metal polish			3			3
75	Drug related ILD	E	Baker	Salesman	Estate agent	3	1	1	3
76	Hypersensitivity pneumonitis	E	Mason	Train driver		3	3		3
77	Sarcoidosis	I	Manufacturing labourers	Construction sites truck driver	Manufacturing labourers	3	3	3	3
78	Infectious ILD	E	Mason	Rubber products (tyre) machine operator	Textile products machine operator	3	3	1	3
79	Drug related ILD	E	Metal industry operator	Machine-tool operator	Boilermaker	3	3	3	3
80	Idiopathic nonspecific interstitial pneumonia	I	Manufacturing labourers	Textile products machine operator		3	1		3
81	Sarcoidosis	I							
82	Lymphangitis carcinomatosis/Neoplasia	E	Meter reader	Storekeeper	Executive	0	0	0	0
83	Idiopathic nonspecific interstitial pneumonia	I	Textile products machine operator			1			1
84	Antisynthetase syndrome	E	Baker	Machine-tool operator	Security officer	3	3	0	3
85	Infectious ILD	E	Farmer			3			3
86	Silicosis	E	Miner	Mason		3	3		3
87	Sarcoidosis	I	Mason			3			3
88	Infectious ILD	E	Textile products machine operator			3			3
89	Infectious ILD	E	Chocolate—products machine operator			1			1
90	Hypersensitivity pneumonitis	E	Plumber			3			3

Table 2. *Cont.*

Patient Number	Lung Disease	Group (E: Disease of Known Etiology, I: Idiopathic Disease)	Occupations			Probability of Exposure to Nanoparticles: 0 Not Found, 1: Possible < 10%, 2: Likely 10–50%, 3: Very Likely > 50%			Final Exposure to Nanoparticles Probability: Highest Probability of Exposure to Nanoparticles in the Career
			1	2	3	Occupation 1	Occupation 2	Occupation 3	
91	Drug related ILD	E	Machine-tool operator	Machine-tool operator	Metal coating machine operator	3	3	3	3
92	Idiopathic nonspecific interstitial pneumonia	I	Boilermaker			3			3
93	Sarcoidosis	I	Electrician	Electrician in food industry	Electrician in mining plant	1	1	3	3
94	Idiopathic nonspecific interstitial pneumonia	I	Farmer	Machine finishing	Machine operator	3	3	2	3
95	Infectious ILD	E	Post officer			0			0
96	Lymphangitis carcinomato-sis/Neoplasia	E	Machine-tool operator	Policeman		3	0		3
97	Sarcoidosis	I	Train controller	Music teacher		0	0		0
98	Hypersensitivity pneumonitis	E	Welder/machine-tool operator			3			3
99	Sarcoidosis	I	Electrician			3			3
100	Sarcoidosis	I	Mason	Mover	Metal-heat-treating plant operator	3	1	3	3

Figure 3 reports the distribution of the patients, irrespective of the disease they suffer from, depending on their final probability of exposure to unintentionally released nanoparticles.

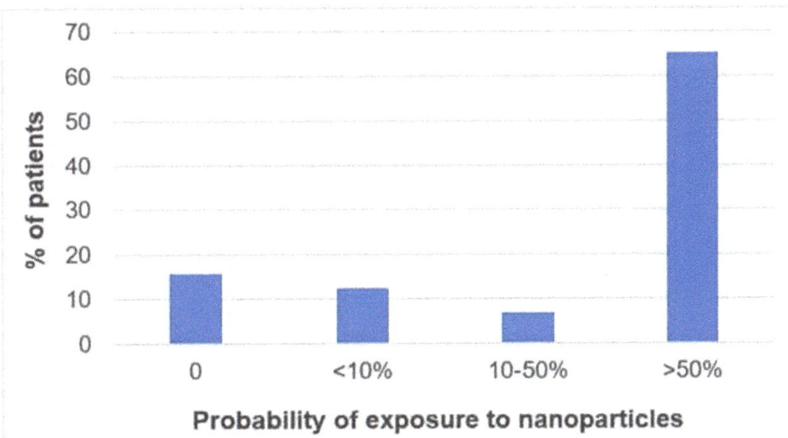

Figure 3. Distribution of the patients depending on their probability of exposure to unintentionally released nanoparticles.

We first observed that few patients (16%) had a null probability of exposure to unintentionally released nanoparticles during their occupational life. On the contrary, the vast majority of the patients (65%) exhibited a high probability of exposure to unintentionally released nanoparticles (>50%).

As shown by Figure 4A, we then reported the distribution of patients depending on their probability of exposure to unintentionally released nanoparticles (we grouped the 0–10% probability of exposure to unintentionally released nanoparticles on one hand and the 10–100% probability on the other hand) and the origin of their disease (either with a known etiology or idiopathic).

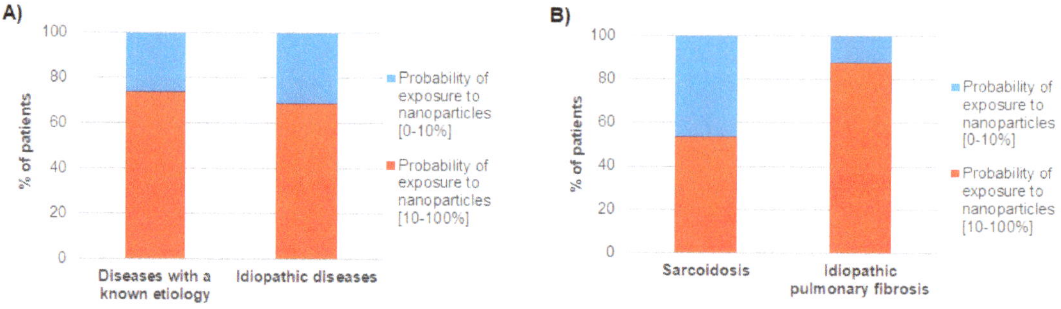

Figure 4. (**A**) Distribution of patients depending on the probability of exposure to unintentionally released nanoparticles and depending on the origin of their disease. (**B**) Distribution of patients depending on the probability of exposure to unintentionally released nanoparticles and depending on the nature of their disease.

The probability of exposure to unintentionally released nanoparticles was higher than 10% for a large majority of patients, either suffering from a disease with a known etiology or from an idiopathic disease (74% and 69% respectively).

We then focused our attention on sarcoidosis and idiopathic pulmonary fibrosis, two idiopathic diseases for which our mineralogical analyses had suggested correlations with the concentration of submicron silica particles and that of titanium nanoparticles, respectively (Figure 4B).

Interestingly, we observed different profiles between the two types of diseases. Regarding sarcoidosis, patients were almost equally distributed between the group of low probability of exposure to unintentionally released nanoparticles and that with a probability of exposure higher than 10%. On the contrary, for idiopathic pulmonary fibrosis, the probability of exposure to unintentionally released nanoparticles was higher than 10% for almost 88% of the patients.

4. Discussion

Besides the widely used in vivo and in vitro studies, mineralogical analyses of human biological samples can bring interesting and useful information, especially to investigate relationship between exposure to airborne nanoparticles and idiopathic lung diseases. For these reasons, here we go one step further and couple biomonitoring to exposure estimates based on expert judgments. Thus, we underwent a retrospective occupational exposure to unintentionally emitted nanoparticles assessment. This was mainly based on expert's judgment through job title, workplace conditions and their knowledge of occupations and similar documented situations, that may however lead to possible sources of errors in the estimates. We thus observed that most of the patients, whatever the type of disease they suffer from, showed a high probability of exposure to unintentionally released nanoparticles (Figure 3). This observation could suggest a potential contribution of inhaled nanoparticles to the development or exacerbation of lung diseases. This is particularly true for idiopathic pulmonary fibrosis where 88% of the patients exhibited a high probability of exposure to unintentionally released nanoparticles (Figure 4B). This finding is consistent with our mineralogical analyses that suggested a role of titanium nanoparticles in this disease. However, regarding sarcoidosis, only half (54%) of the patients were classified in the group of high probability of exposure to unintentionally released nanoparticles. Once again, this observation is in agreement with our mineralogical analyses that previously highlighted a potential contribution of silica submicron particles, i.e., particles bigger than nanoparticles, in this disease [19].

To the best of our knowledge, only two studies have established a clear relationship between exposure to unintentionally released nanoparticles and long-term negative effects in humans. Song et al. [24,25] found silica nanoparticles in clinical samples from seven patients suffering from lung injuries after an occupational exposure. However, these patients were also exposed to other toxic substances; consequently, no firm conclusion could be reached. Another study [26] had reported that pulmonary injuries were more severe in welders than in unexposed people suggesting that nanoparticles present in welding fumes could be responsible, at least in part, for the pulmonary inflammation. But this study was limited to the description of few clinical cases and did not have a significant statistical power. It was also restricted to a target population and conclusions can hardly be extrapolated.

In the literature, the investigation of inhaled nanoparticles' presence in patients' lungs is rare and when it exists it is limited to electron microscopy observations that do not allow a complete physicochemical characterization of the nanoparticles and is not suitable for large cohort analysis as it is a time-consuming and expensive technique. Laser-induced breakdown spectroscopy (LIBS) could appear as a promising alternative for the direct visualization of endogenous or exogenous elements within tissues but is still under development and not routinely used for biomedical applications [27]. Nevertheless, the biomonitoring of nanoparticles in biological samples appears as a promising approach to get new insights into the understanding of the genesis or evolution of lung diseases due to nanoparticle exposure.

In the present paper, we propose to couple biomonitoring to the assessment of unintentionally released nanoparticles exposure to get a larger picture on the relationship between exposure to airborne nanoparticles and interstitial lung diseases. Although this strategy has several advantages as previously discussed, it has also some limitations we have to take into account. First, results should be considered with caution as we have a small number of patients, especially in the idiopathic pulmonary disease group (nine patients). Moreover, no significant difference between the unintentionally released nanoparticle exposure assessment of different distinguished interstitial lung diseases groups was observed (results not shown). As we said above, further investigations are necessary to confirm the results observed. We may also remind the reader that we performed our analyses in a cohort of patients. It should be interesting to compare these data to those obtained with healthy control subjects. However, for ethical reasons, it is impossible to perform broncho-alveolar lavages in healthy persons due to the invasive nature of this exam. We should mention that we focused our analyses on occupational exposure to unintentionally released nanoparticles, to be complete, we should also consider other sources of exposure to nanoparticles, for instance environmental exposure (taking into account patients' living area, mode of transport, of heating, leisure or use of hygiene and cosmetics products. . .). However, these data are much more complex to collect with accuracy. Finally, it should be kept in mind that the presence of a given particle in a larger amount within biological samples is not sufficient to prove a causal link with a disease, and toxicity assessment is necessary to demonstrate a pathogenic effect as well as mechanistic studies to understand the underlying mechanisms. One perspective we propose to strengthen our array of presumptions is to couple the nanoparticle biomonitoring in lung clinical samples to the in vitro assessment of their toxicity [14]. For instance, by incubating cells with nanoparticles extracted from patients' BAL and then analyzing the cell response in terms of induction of cell death, pro-inflammatory response, oxidative stress, etc. The advantage of such strategy is that the in vitro assays are performed using nanoparticles which nature and dose are representative of real-life. From such approach benefits could be expected in the therapeutic and/or prevention fields. Indeed, the knowledge of the toxicity potential of nanoparticles can lead to preventive measures to limit the exposure to the harmful substances.

We determined a probability of exposure to unintentionally produced nanoparticles present in polluting emissions during each occupation held by the patients. We did not determine intensity and frequency of nanoparticle exposure. We also observed that the probability of exposure alone was not useful to discriminate some exposure differences between different distinguished interstitial lung diseases groups. Furthermore, our results did not determine the extent of occupational or daily life exposure and recent or lifetime exposure to unintentionally released nanoparticles. However, given the current state of knowledge on occupational exposure levels, which is patchy and based on heterogeneous methods, we have chosen to limit the assessment to probability [28]. The nanoparticle assessment used in this study was possible thanks to the knowledge acquired in the framework of the development of the job-exposure matrix MatPUF [29]. Such a job-exposure matrix might be useful to improve the quality and the accuracy of nanoparticle exposure assessment during a full occupational career, providing for each job chronological exposure data and main chemical families of released nanoparticles. MatPUF job-exposure matrix has already been used in epidemiological studies and has made it possible to highlight a relationship between occupational exposure to unintentionally released nanoparticles and small for gestational age and, cancers, such as lung cancer and brain nervous system tumors [30,31].

Job-exposure matrices have already been used to investigate associations between occupational exposure and idiopathic pulmonary fibrosis, however conflicting results were reported. Indeed, while Abramson et al. showed that occupational exposures to specific organic, mineral, or metal dusts were not associated with idiopathic pulmonary fibrosis [32], on the contrary Andersson et al. reported that occupational exposure to inorganic dusts, excluding silica and asbestos, was associated with increased risk of idiopathic pulmonary fibrosis [33]. Regarding sarcoidosis, both Graff et al. and Jonsson et al. concluded that occupational exposure to silica dust led to an increased risk of sarcoidosis [34,35]. These studies confirmed that lung diseases such as idiopathic pulmonary fibrosis or sarcoidosis could be caused by the inhalation of mineral particles [36–38]. Thus, such job-exposure matrix are interesting and useful tools for the assessment of occupational UFP exposure that can both contribute to the improvement of epidemiological knowledge of health risks and to the implementation of prevention in the workplace.

Nanoparticles possess nanostructure-dependent properties (e.g., chemical, physical, biological), which make them desirable for commercial or industrial applications. Workers are increasingly exposed to nanoparticles in occupational settings. However, these same properties may potentially lead to atypical toxicity and health risks are yet unknown. For these reasons, there is an unceasing scientific interest to the human toxicokinetics and toxicodynamics of nanoparticles and concerns about the potential risks of exposure to humans have been raised. Although toxic effects have not been really demonstrated in humans, there is accumulating evidence from experimental studies that exposure to some nanoparticles may be harmful [39]. However, it is mostly based on in vitro tests and animal experiments. Key questions, regarding the duration and level of exposure in humans, the toxic behavior of nanoparticles in humans, the physiological and chemical interaction with human body, the harmlessness of these interactions, and acute or chronic effects adverse effects need to be resolved [40]. As potential occupational exposure to unintentionally released nanoparticles becomes more prevalent, it is important that the principles of risk assessment and risk management be considered for workers. The risk assessment/risk management framework comprises three essential components: research, risk assessment, and risk management [40]. This exploratory study had the objective to characterize exposure to unintentionally released nanoparticles, using association of two approaches, biomonitoring and occupational exposure to unintentionally released nanoparticles assessment, in patients with ILD, to study possible links between with these health outcomes and exposure to unintentionally released nanoparticles and, thus, to contribute to the knowledge to risk factors of occupational and environmental lung ILD. Moreover, these approaches might be used to contribute to nanoparticles risk assessment. In fact, quantification of unintentionally released nanoparticles exposure remains the major challenge for prevention.

5. Conclusions

Large epidemiological studies are too personnel, require financial resources, and are time-consuming. By combining mineralogical analyses of human BAL samples and estimation of occupational nanoparticle exposure, the relationship between occupational exposure to airborne nanoparticles, nanoparticle lung burden, and implications on the etiology of lung diseases can be investigated. These complementary approaches appear as a promising strategy to get a comprehensive picture and could bring informative data for human health risk assessment and management.

Following this approach, it is evidenced that most of the patients from our cohort, whatever the type of disease they suffer from, showed a high probability of exposure to unintentionally released nanoparticles. This observation was consistent with the nanoparticle lung burden previously assessed, suggesting a potential role of inhaled nanoparticles to the development or exacerbation of lung diseases, although further experiments are necessary to draw firm conclusions.

Author Contributions: Conceptualization, L.F., J.-M.V., V.F. and J.P.; formal analysis, S.A.D. and C.P.; investigation, V.F., J.P. and S.A.D.; writing—original draft preparation, V.F.; writing—review and editing, S.A.D., L.F., J.-M.V., J.P. and C.P. All authors have read and agreed to the published version of the manuscript.

Funding: This research was funded by Fonds de Recherche en Santé Respiratoire de la Société de Pneumologie de Langue Française (NanoPI grant), the Canceropôle CLARA (Expo-Nano grant) and Santé-Environnement Rhône-Alpes (EnvitéRA).

Institutional Review Board Statement: The study was conducted according to the guidelines of the Declaration of Helsinki, and approved by an ethics committee (Comité de Protection des Personnes, Sud-Est I, identification code: 2012-A00798-35; date of approval: 27 August 2012) as well as by the French agency regulating biomedical research (Agence Nationale de Sécurité du Médicament et des produits de santé, ANSM, identification code: B121211-21; date of approval: 9 November 2012).

Informed Consent Statement: Informed consent was obtained from all subjects involved in the study. Written informed consent has been obtained from the patients.

Conflicts of Interest: The authors declare no conflict of interest. The funders had no role in the design of the study; in the collection, analyses, or interpretation of data; in the writing of the manuscript; or in the decision to publish the results.

References

1. Stone, V.; Miller, M.R.; Clift, M.J.D.; Elder, A.; Mills, N.L.; Møller, P.; Schins, R.P.F.; Vogel, U.; Kreyling, W.G.; Alstrup Jensen, K.; et al. Nanomaterials Versus Ambient Ultrafine Particles: An Opportunity to Exchange Toxicology Knowledge. *Environ. Health Perspect.* **2017**, *125*, 106002. [CrossRef]
2. Manno, M.; Viau, C.; Cocker, J.; Colosio, C.; Lowry, L.; Mutti, A.; Nordberg, M.; Wang, S. Biomonitoring for occupational health risk assessment (BOHRA). *Toxicol. Lett.* **2010**, *192*, 3–16. [CrossRef]
3. Silins, I.; Högberg, J. Combined Toxic Exposures and Human Health: Biomarkers of Exposure and Effect. *Int. J. Environ. Res. Public Health* **2011**, *8*, 629–647. [CrossRef]
4. De Vuyst, P.; Karjalainen, A.; Dumortier, P.; Pairon, J.-C.; Monsó, E.; Brochard, P.; Teschler, H.; Tossavainen, A.; Gibbs, A. Guidelines for mineral fibre analyses in biological samples: Report of the ERS Working Group. European Respiratory Society. *Eur. Respir. J.* **1998**, *11*, 1416–1426. [CrossRef] [PubMed]
5. Case, B.W.; Abraham, J.L.; Meeker, G.; Pooley, F.D.; Pinkerton, K.E. Applying Definitions of "Asbestos" to Environmental and "Low-Dose" Exposure Levels and Health Effects, Particularly Malignant Mesothelioma. *J. Toxicol. Environ. Health B Crit. Rev.* **2011**, *14*, 3–39. [CrossRef] [PubMed]
6. Mossman, B.T.; Lippmann, M.; Hesterberg, T.W.; Kelsey, K.T.; Barchowsky, A.; Bonner, J.C. Pulmonary Endpoints (Lung Carcinomas and Asbestosis) Following Inhalation Exposure to Asbestos. *J. Toxicol. Environ. Health B Crit. Rev.* **2011**, *14*, 76–121. [CrossRef] [PubMed]
7. Bargagli, E.; Lavorini, F.; Pistolesi, M.; Rosi, E.; Prasse, A.; Rota, E.; Voltolini, L. Trace metals in fluids lining the respiratory system of patients with idiopathic pulmonary fibrosis and diffuse lung diseases. *J. Trace Elem. Med. Biol.* **2017**, *42*, 39–44. [CrossRef]
8. Bergamaschi, E.; Poland, C.; Guseva Canu, I.; Prina-Mello, A. The role of biological monitoring in nano-safety. *Nano Today* **2015**, *10*, 274–277. [CrossRef]
9. Bitounis, D.; Pourchez, J.; Forest, V.; Boudard, D.; Cottier, M.; Klein, J.-P. Detection and analysis of nanoparticles in patients: A critical review of the status quo of clinical nanotoxicology. *Biomaterials* **2016**, *76*, 302–312. [CrossRef]
10. Rapport de l'Anses—Anses Particules de l'air Ambiant Extérieur—Impact Sur La Pollution Atmosphérique Des Technologies et de La Composition Du Parc de Véhicules Automobiles Circulant En France. 2019. Available online: https://www.anses.fr/fr/content/rapport-de-lanses-particules-de-l\T1\textquoterightair-ambiant-extérieur-impact-sur-la-pollution (accessed on 9 July 2021).

11. Groopman, J.D.; Kensler, T.W. The light at the end of the tunnel for chemical-specific biomarkers: Daylight or headlight? *Carcinogenesis* **1999**, *20*, 1–11. [CrossRef]
12. Angerer, J.; Ewers, U.; Wilhelm, M. Human biomonitoring: State of the art. *Int. J. Hyg. Environ. Health* **2007**, *210*, 201–228. [CrossRef] [PubMed]
13. Rinaldo, M.; Andujar, P.; Lacourt, A.; Martinon, L.; Canal Raffin, M.; Dumortier, P.; Pairon, J.-C.; Brochard, P. Perspectives in Biological Monitoring of Inhaled Nanosized Particles. *Ann. Occup. Hyg.* **2015**, *59*, 669–680. [CrossRef]
14. Forest, V.; Vergnon, J.-M.; Pourchez, J. Biological Monitoring of Inhaled Nanoparticles in Patients: An Appealing Approach To Study Causal Link between Human Respiratory Pathology and Exposure to Nanoparticles. *Chem. Res. Toxicol.* **2017**, *30*, 1655–1660. [CrossRef] [PubMed]
15. Bitounis, D.; Klein, J.-P.; Mery, L.; El-Merhie, A.; Forest, V.; Boudard, D.; Pourchez, J.; Cottier, M. Ex vivo detection and quantification of gold nanoparticles in human seminal and follicular fluids. *Analyst* **2018**, *143*, 475–486. [CrossRef]
16. Rinaldi, L.; Barabino, G.; Klein, J.-P.; Bitounis, D.; Pourchez, J.; Forest, V.; Boudard, D.; Leclerc, L.; Sarry, G.; Roblin, X.; et al. Metals distribution in colorectal biopsies: New insight on the elemental fingerprint of tumour tissue. *Dig. Liver Dis.* **2015**, *47*, 602–607. [CrossRef] [PubMed]
17. Raia-Barjat, T.; Prieux, C.; Leclerc, L.; Sarry, G.; Grimal, L.; Chauleur, C.; Pourchez, J.; Forest, V. Elemental fingerprint of human amniotic fluids and relationship with potential sources of maternal exposure. *J. Trace Elem. Med. Biol.* **2020**, *60*, 126477. [CrossRef]
18. Forest, V.; Vergnon, J.-M.; Guibert, C.; Bitounis, D.; Leclerc, L.; Sarry, G.; Pourchez, J. Metal load assessment in patient pulmonary lavages: Towards a comprehensive mineralogical analysis including the nano-sized fraction. *Nanotoxicology* **2017**, *11*, 1211–1224. [CrossRef] [PubMed]
19. Forest, V.; Pourchez, J.; Guibert, C.; Bitounis, D.; Leclerc, L.; Sarry, G.; Vergnon, J.-M. Nano to micron-sized particle detection in patients' lungs and its pathological significance. *Environ. Sci. Nano* **2019**, *6*, 1343–1350. [CrossRef]
20. Bitounis, D.; Barnier, V.; Guibert, C.; Pourchez, J.; Forest, V.; Boudard, D.; Hochepied, J.-F.; Chelle, P.; Vergnon, J.-M.; Cottier, M. A method for the quantitative extraction of gold nanoparticles from human bronchoalveolar lavage fluids through a glycerol gradient. *Nanoscale* **2018**, *10*, 2955–2969. [CrossRef] [PubMed]
21. Mikolasch, T.A.; Garthwaite, H.S.; Porter, J.C. Update in diagnosis and management of interstitial lung disease. *Clin. Med. (Lond.)* **2017**, *17*, 146–153. [CrossRef]
22. National Institute for Statistics and Economic Studies (INSEE). Nomenclature d'Activités Françaises. 2000. Available online: https://www.insee.fr/fr/information/2406147 (accessed on 7 July 2021).
23. International Labour Organization. *International Standard Classification of Occupations*; ILO: Geneva, Switzerland, 1968; Available online: https://ilo.primo.exlibrisgroup.com/discovery/fulldisplay/alma991141423402676/41ILO_INST:41ILO_V2 (accessed on 7 July 2021).
24. Song, Y.; Li, X.; Du, X. Exposure to nanoparticles is related to pleural effusion, pulmonary fibrosis and granuloma. *Eur. Respir. J.* **2009**, *34*, 559–567. [CrossRef]
25. Song, Y.; Li, X.; Wang, L.; Rojanasakul, Y.; Castranova, V.; Li, H.; Ma, J. Nanomaterials in Humans: Identification, Characteristics, and Potential Damage. *Toxicol. Pathol.* **2011**, *39*, 841–849. [CrossRef]
26. Andujar, P.; Simon-Deckers, A.; Galateau-Sallée, F.G.; Fayard, B.; Beaune, G.; Clin, B.; Billon-Galland, M.-A.; Durupthy, O.; Pairon, J.-C.; Doucet, J.; et al. Role of metal oxide nanoparticles in histopathological changes observed in the lung of welders. *Part. Fibre Toxicol.* **2014**, *11*, 23. [CrossRef] [PubMed]
27. Leprince, M.; Sancey, L.; Coll, J.-L.; Motto-Ros, V.; Busser, B. Elemental imaging using laser-induced breakdown spectroscopy: Latest medical applications. *Med. Sci. MS* **2019**, *35*, 682–688. [CrossRef]
28. Viitanen, A.-K.; Uuksulainen, S.; Koivisto, A.J.; Hämeri, K.; Kauppinen, T. Workplace Measurements of Ultrafine Particles-A Literature Review. *Ann. Work. Expo. Health* **2017**, *61*, 749–758. [CrossRef] [PubMed]
29. Audignon-Durand, S.; Gramond, C.; Ducamp, S.; Manangama, G.; Garrigou, A.; Delva, F.; Brochard, P.; Lacourt, A. Development of a Job-Exposure Matrix for Ultrafine Particle Exposure: The MatPUF JEM. *Ann. Work Expo. Health* **2021**, *65*, 516–527. [CrossRef] [PubMed]
30. Manangama, G.; Migault, L.; Audignon-Durand, S.; Gramond, C.; Zaros, C.; Bouvier, G.; Brochard, P.; Sentilhes, L.; Lacourt, A.; Delva, F. Maternal occupational exposures to nanoscale particles and small for gestational age outcome in the French Longitudinal Study of Children. *Environ. Int.* **2019**, *122*, 322–329. [CrossRef]
31. Manangama, G.; Gramond, C.; Audignon-Durand, S.; Baldi, I.; Fabro-Peray, P.; Gilg Soit Ilg, A.; Guénel, P.; Lebailly, P.; Luce, D.; Stücker, I.; et al. Occupational exposure to unintentionally emitted nanoscale particles and risk of cancer: From lung to central nervous system—Results from three French case-control studies. *Environ. Res.* **2020**, *191*, 110024. [CrossRef]
32. Abramson, M.J.; Murambadoro, T.; Alif, S.M.; Benke, G.P.; Dharmage, S.C.; Glaspole, I.; Hopkins, P.; Hoy, R.F.; Klebe, S.; Moodley, Y.; et al. Occupational and environmental risk factors for idiopathic pulmonary fibrosis in Australia: Case-control study. *Thorax* **2020**, *75*, 864–869. [CrossRef]
33. Andersson, M.; Blanc, P.D.; Torén, K.; Järvholm, B. Smoking, occupational exposures, and idiopathic pulmonary fibrosis among Swedish construction workers. *Am. J. Ind. Med.* **2021**, *64*, 251–257. [CrossRef]
34. Jonsson, E.; Järvholm, B.; Andersson, M. Silica dust and sarcoidosis in Swedish construction workers. *Occup. Med.* **2019**, *69*, 482–486. [CrossRef] [PubMed]

35. Graff, P.; Larsson, J.; Bryngelsson, I.-L.; Wiebert, P.; Vihlborg, P. Sarcoidosis and silica dust exposure among men in Sweden: A case–control study. *BMJ Open* **2020**, *10*, e038926. [CrossRef] [PubMed]
36. Rafnsson, V.; Ingimarsson, O.; Hjalmarsson, I.; Gunnarsdottir, H. Association between exposure to crystalline silica and risk of sarcoidosis. *Occup. Environ. Med.* **1998**, *55*, 657–660. [CrossRef]
37. Taskar, V.; Coultas, D. Exposures and Idiopathic Lung Disease. *Semin. Respir. Crit. Care Med.* **2008**, *29*, 670–679. [CrossRef]
38. Vincent, M.; Lievre, M. Sarcoidosis and pulmonary dust exposure, a plausible pathogenic link. *Rev. Mal. Respir.* **2002**, *19*, 103–104. [PubMed]
39. Nanoparticle Task Force ACOEM. Nanotechnology and Health. *J. Occup. Environ. Med.* **2011**, *53*, 687–689. [CrossRef]
40. Yokel, R.A.; MacPhail, R.C. Engineered nanomaterials: Exposures, hazards, and risk prevention. *J. Occup. Med. Toxicol.* **2011**, *6*, 7. [CrossRef]

toxics

Review

Partitioning of Persistent Organic Pollutants between Adipose Tissue and Serum in Human Studies

Meg-Anne Moriceau [1,†], German Cano-Sancho [1,†], MinJi Kim [2], Xavier Coumoul [3], Claude Emond [4], Juan-Pedro Arrebola [5,6,7], Jean-Philippe Antignac [1], Karine Audouze [3] and Christophe Rousselle [8,*]

1 Oniris, INRAE, LABERCA, 44300 Nantes, France
2 INSERM UMR-S 1124, Université Paris Sorbonne Nord, 93017 Bobigny, France
3 INSERM UMR-S 1124, Université Paris Cité, 45 rue des Saints-Pères, 75006 Paris, France
4 School of Public Health, Department of Environmental and Occupational Health, University of Montreal, Montreal, QC H3C 3J7, Canada
5 Department of Preventive Medicine and Public Health, Universidad de Granada, Campus de Cartuja s/n, 18071 Granada, Spain
6 Instituto de Investigación Biosanitaria (ibs.GRANADA), Avda. de Madrid, 15. Pabellón de Consultas Externas 2, 2a Planta, 18012 Granada, Spain
7 Consortium for Biomedical Research in Epidemiology and Public Health (CIBERESP), Instituto de Salud Carlos III, 28029 Madrid, Spain
8 ANSES, European and International Affairs Department, 14 rue Pierre et Marie Curie, 94701 Maisons-Alfort, France
* Correspondence: christophe.rousselle@anses.fr
† These authors contributed equally to this work.

Abstract: Blood is the most widely used matrix for biomonitoring of persistent organic pollutants (POPs). It is assumed that POPs are homogenously distributed within body lipids at steady state; however, the variability underlying the partitioning of POPs between fat compartments is poorly understood. Hence, the objective of this study was to review the state of the science about the relationships of POPs between adipose tissue and serum in humans. We conducted a narrative literature review of human observational studies reporting concentrations of POPs in paired samples of adipose tissue with other lipid-based compartments (e.g., serum lipids). The searches were conducted in SCOPUS and PUBMED. A meta-regression was performed to identify factors responsible for variability. All included studies reported high variability in the partition coefficients of POPs, mainly between adipose tissue and serum. The number of halogen atoms was the physicochemical variable most strongly and positively associated with the partition ratios, whereas body mass index was the main biological factor positively and significantly associated. To conclude, although this study provides a better understanding of partitioning of POPs to refine physiologically based pharmacokinetic and epidemiological models, further research is still needed to determine other key factors involved in the partitioning of POPs.

Keywords: adipose tissue; biomonitoring; meta-regression; partition coefficients; persistent organic pollutants

Citation: Moriceau, M.-A.; Cano-Sancho, G.; Kim, M.; Coumoul, X.; Emond, C.; Arrebola, J.-P.; Antignac, J.-P.; Audouze, K.; Rousselle, C. Partitioning of Persistent Organic Pollutants between Adipose Tissue and Serum in Human Studies. *Toxics* **2023**, *11*, 41. https://doi.org/10.3390/toxics11010041

Academic Editor: Jeongim Park

Received: 1 December 2022
Revised: 23 December 2022
Accepted: 27 December 2022
Published: 31 December 2022

Copyright: © 2022 by the authors. Licensee MDPI, Basel, Switzerland. This article is an open access article distributed under the terms and conditions of the Creative Commons Attribution (CC BY) license (https://creativecommons.org/licenses/by/4.0/).

1. Introduction

Persistent organic pollutants (POPs) are chemicals from a large class of substances heavily produced and released in the environment during the last century, peaking in the 1970s and switching to declining trends when regulations and bans began to be enforced [1]. These chemicals attracted international attention because of their high persistence in the environment and capacity to be transported over long distances and absorbed in the fat tissues of living organisms, prompting their magnification along trophic chains [2,3]. Additionally, many POPs may elicit adverse health effects on animals and humans, motivating global regulatory frameworks such as the Stockholm Convention, adopted in 2004, intended to reduce and monitor their worldwide production. Despite regulatory efforts, substantial concentrations of POPs can still be found in ecosystems, or in fatty tissues and specimens from animals and humans across the globe, decades after the banning period [1,4,5]. The vast family of POPs includes legacy chemicals intentionally produced for agricultural applications, like pesticides (e.g., organochlorinated pesticides), or for industrial uses, including polychlorinated biphenyls (PCBs), which can also be unintentionally released through thermal processes [6]. In addition, other classes, like polybrominated diphenyl ethers (PBDE) used as additive flame retardants, are among the substances of emerging concern [2]. Some POPs have shown the capacity for interacting with and disrupting a variety of molecular pathways, triggering pathogenic processes involved in human health, including impaired reproduction [7,8], carcinogenesis [9,10] or metabolic dysregulation [11,12], among others. Experimental and epidemiological studies on POPs have mainly relied on the effects of a few individual congeners, whereas the effect of low-dose mixtures of POPs resembling realistic background exposures remains scarcely explored [13]. Computational systems biology modeling also offers a new way of investigating potential toxicological effects of POPs [14].

Globally, POPs are characterized by their hydrophobic properties, dissolving in lipid-based foodstuffs like fatty fish or butter, favoring diet becoming a major entry pathway in humans [15]. The mechanism of absorption for many hydrophobic pollutants is the same as for dietary lipids, including the lymphatic pathway or the portal vein to the liver, and is determined by the octanol–water partition coefficients (log K_{ow}) of chemicals [16]. For instance, the absorption of the highly lipophilic pesticide hexachlorobenzene (HCB, log K_{ow} =5.7) mostly happens in the intestine; it is transported through the lymph to the tissues within chylomicrons, being mainly stored in adipose tissue [17]. POPs stored in adipocytes are mobilized with neutral lipids to systemic circulation upon energy demands and packed within lipoproteins with lipids and other hydrophobic chemicals. Thus, blood is often used as a convenient matrix for biomonitoring purposes, yet adipose tissue is considered the gold-standard matrix for long-term POP exposure assessment [18]. Elimination of POPs is conducted through bile acids into the intestine where they can ultimately be evacuated within feces, whereas they can be reabsorbed within the enterohepatic recycling process of conjugated bile acids [19]. The dynamics of lipids, and thus of POPs, may be impaired through metabolic disorders (e.g., obesity, diabetes), and POPs may disrupt lipid metabolism and energy balance [20]. In addition, the anthropometry of individuals and variations of fat pad volumes during the lifespan may modify the kinetics of POPs and the chemical exposure biomarkers in environmental health studies [21].

In the last decade, pharmacokinetic models, and especially physiologically based pharmacokinetic (PBPK) models, have been developed to refine risk assessment of chemicals [22]. PBPK models have also been used in epidemiological studies to refine the estimation of individual exposures at different critical life stages and/or in different target tissues [23]. They allow the simulation of toxicokinetic profiles for different sub-populations exposed to lipophilic pollutants from single-time-point measurements of POP concentrations in different matrices (e.g., serum, milk), potentially highlighting critical periods of vulnerability to the toxicity of POPs [24–26]. These models incorporate physiological parameters such as tissue-specific blood flows, metabolic processes, and tissue partitioning. An assumption usually made for those models is a homogeneous distribution of POPs within body lipids at steady state, thus a partitioning between adipose tissue and blood lipids close to the unit [27,28]. The same assumption has also been generalized to the partitioning of POPs between blood and breast milk if the concentrations are reported in lipid units, yet some authors have reported that this approach may be too simplistic [29,30]. Compound-specific physicochemical properties (e.g., log K_{ow}, molar volume) and individual physiological parameters (e.g., age, body mass index) may influence such assumptions in the case of breast milk [30]. Nonetheless, little attention has been paid to the partitioning of POPs between adipose tissue and blood lipids; hence, further characterization of POPs' dynamics through lipid compartments is therefore needed to refine the parametrization of PBPK models and right interpretation of blood biomarkers in epidemiological settings.

Recent publications have previously addressed the interest in adipose tissue for biomonitoring of environmental chemicals, with a special focus on women [31] and also in epidemiological studies [12], highlighting that adipocytes are also the target of POP. To the best of our knowledge, there are no published studies reporting biomonitoring data on POPs in matched lipid-based matrices (i.e., adipose tissues and blood). Another research gap concerns the biological and physicochemical factors influencing the variability in partitioning of POPs among lipid-based tissues reported in individual studies. Thus, the main objective of this study was to review the state of the science of interrelationships in POP concentrations in adipose tissue with other internal lipid-based compartments including blood lipids, excluding breast milk. Hence, we conducted a narrative literature review of primary articles reporting concentrations of POPs in paired samples of adipose tissue with blood (e.g., serum) or other measured internal compartments. For those articles with reported partition coefficients (e.g., adipose tissue: serum) in comparable metrics, a meta-regression was conducted to identify factors responsible for the variability, otherwise a qualitative synthesis is provided.

2. Methods

A literature search strategy was developed to identify recent human studies that measured POPs in serum and adipose tissue from the same individuals. Although the review is essentially narrative, we used the principles of systematic review methodology to ensure the transparency and reproducibility of the process and PRISMA guidelines for reporting [32]. Hence, the bibliographic search was carried out using harmonized standards [33] to develop the search protocol. In this regard, the search strategy was developed to reflect the main key elements of the PECO statement (Table 1), including search strings developed to cover the relevant articles in Scopus and PubMed. The search strings detailed in Table S1 were used in Scopus and PUBMED on 05/08/2022. A 'publication date' filter was applied from 2011 to July 2021 in order to focus on the most recent studies that use highly sensitive analytical methods. Manual searches were conducted to retrieve studies published before 2011, based on references cited in the selected studies and the expert knowledge of the panel.

Table 1. PECO statement.

PECO	Description
Population	Human populations (including men, women, pregnant women, foetuses exposed in utero, and children exposed).
Exposure	All type of persistent organic pollutants (e.g., PBDEs, PCBs, etc)
Comparator	Not applicable.
Outcome	Simultaneous determination in adipose tissues and serum (or different adipose tissue locations).

To be included, studies had to be conducted in human populations, measuring internal levels of any type of POP in at least one adipose tissue location and serum or another different adipose tissue location. Hence, articles were excluded when concentrations of POPs in human adipose tissue, either subcutaneous adipose tissue or visceral adipose tissue and serum or in different adipose tissue compartments, were not reported. The search was restricted to articles in English and primary studies (review and conference publications were excluded).

Synthesis and Meta-Analysis

Studies with comparative metrics (e.g., ratio of POP concentrations in adipose tissue: serum-determined lipid-based weight, p.e. ng/g lipid weight) were pooled in a meta-analysis. The effect sizes were the adipose tissue ratio means and the respective standard deviations. The means and standard deviations were estimated from the median and interquartile range using the Wan method [34] for two studies [35,36]. Meta-estimates were appraised using a random-effect meta-analysis approach implemented with the *metamean* function in the "meta" R package [37,38]. Higgins & Thompson's I^2 statistic was used to quantify the between-study heterogeneity. The influence of variables on meta-estimates was assessed by linear meta-regression using the restricted maximum likelihood method and implemented with the *metareg* function of the same package. In this regard, individual variables (e.g., age, body mass index), year of collection and physicochemical properties (summarized in Table 2) were assessed in single and multivariate models. A narrative synthesis displaying the study design and main outputs was conducted for the rest of studies excluded from the meta-analysis, i.e., those for which no ratio of POP concentrations in adipose tissue: serum was provided.

Table 2. Physicochemical properties of included chemicals retrieved from PUBCHEM database [39] and adapted from previous publications [40–42].

Name	CAS N°	Molecular Mass	Log K_{ow}	Half-Life (Year)	N° Halogen	Molar Volume
Dioxin-like Polychlorinated Biphenyls						
PCB77	32598-13-3	292	6.34	0.1	4	268.2
PCB81	70362-50-4	292	6.34	0.7	4	268.2
PCB126	57465-28-8	326.4	6.98	1.6	5	289.1
PCB169	32774-16-6	360.9	7.62	7.3	6	310
PCB105	32598-14-4	326.4	6.98	2.4	5	289.1
PCB114	74472-37-0	326.4	6.98	10	5	289.1
PCB118	31508-00-6	326.4	6.98	3.8	5	289.1
PCB123	65510-44-3	326.4	6.98	7.4	5	289.1
PCB156	38380-08-4	360.9	7.62	16	6	310
PCB157	69782-90-7	360.9	7.62	18	6	310
PCB167	52663-72-6	360.9	7.62	12	6	310
PCB189	39635-31-9	395.3	7.27	22	7	330.9
Non-dioxin-like Polychlorinated Biphenyls						
PCB28	7012-37-5	257.5	5.62	16	3	247.3
PCB52	35693-99-3	292	6.34	10	4	268.2
PCB101	37680-73-2	326.4	6.98	11	5	289.1
PCB138	35065-28-2	360.9	7.44	27	6	310
PCB153	35065-27-1	360.9	7.62	14.4	6	310
PCB180	35065-29-3	395.3	8.27	11.5	7	330.9
Organochlorinated pesticides						
HCB	118-74-1	284.8	5.73	6	6	221.4
p,p'-DDE	72-55-9	318	6.51	10	4	305.2
p,p'-DDT	50-29-3	354.5	6.91	7	5	333.5
Brominated flame retardants						
BDE28	41318-75-6	406.89	5.88	0.94	3	365.5
BDE47	5436-43-1	485.79	6.77	0.37	4	288.8
BDE99	60348-60-9	564.7	7.66	8.2	5	312.1
BDE100	189084-64-8	564.7	7.66	2	5	312.1
BDE153	68631-49-2	643.6	8.55	3.5	6	335.4
BDE154	207122-15-4	643.6	7.82	3.3	6	335.4
BDE183	207122-16-5	722.5	9.44	0.25	7	358.7
BDE209	1163-19-5	959.2	12.11	0.04	10	428.6
PBB153	59080-40-9	627.6	9.1	6.2	6	335.4

3. Results

3.1. Study Characteristics

Finally, 249 publications were initially gathered from PubMed and SCOPUS after duplicate removal, among which 212 did not match the selection criteria and thus were excluded based on title and abstract (Figure 1). The remaining 37 references were evaluated based on the full text, of which 18 studies were considered eligible articles. A final list of 19 relevant articles reporting concentrations of POPs in adipose tissue and serum was included, of which 6 provided comparable metrics to conduct a meta-analysis (Table 2).

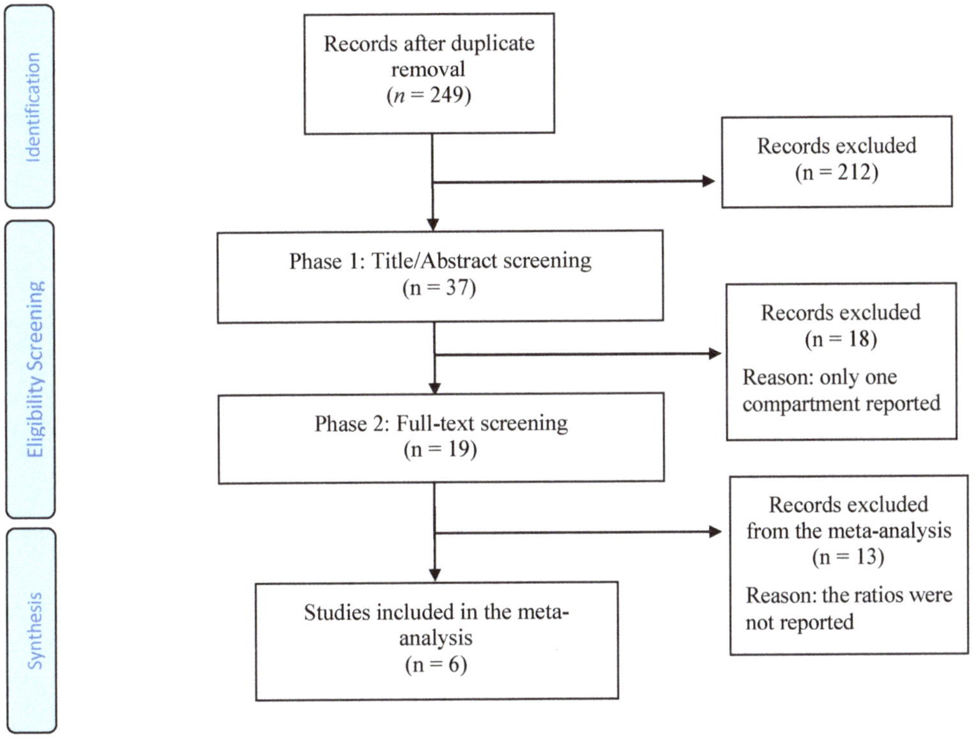

Figure 1. PRISMA flowchart of search strategy and study selection with exclusion criteria.

The main characteristics of 19 included studies are displayed in Table 3; they were published between 1999 [43] and 2021 [44]. Among these studies, 11 were performed in Europe and 8 outside Europe. It should be noticed that, for most of these studies, the sampling was conducted several years before publication, extending the exposure timeframe by several years. Different families of compounds belonging to POPs were measured depending on the study: the most frequently measured compounds were PCBs, either as individual congeners such as PCB 153, 138, or 180, or as a sum of PCBs, followed by 1,1,1-Trichloro-2,2-bis(4-chlorophenyl)ethane (*p,p′*-DDT), 1,1-dichloro-2,2-bis(p-chlorophenyl) ethylene (*p,p′*-DDE), HCB, beta hexachlorocyclohexane (β-HCH) and PBDE. Most studies represented a clinical population undergoing surgery exposed to background levels of POPs, but several involved occupational exposures [36] or highly contaminated areas [45]. The study participants provided adipose tissue biopsies for the surgical intervention during caesareans [36], breast cancer [10,43] or hepatocarcinoma interventions [45], for endometriosis [46,47], or for bariatric surgery [48,49]. Hence, the majority of the population was represented by females with a minor presence of males. Mean ages ranged between 29 and 59 years, and BMI was between 21 and 48 kg/m^2. Subcutaneous adipose tissue (SAT) was the most frequent location for biopsies, yet some other locations were also reported, including visceral adipose tissue (VAT), breast (BAT) and the gluteal region (GAT). A few studies measured concentrations of POPs in two different adipose tissue types/locations simultaneously [49–51]. Nine studies reported or provided the adipose tissue: serum ratios, of which one only reported ratios expressing serum concentrations in wet weight [47], three were either in lipid weight or wet weight, and six were in lipid weight.

Table 3. Summary of study characteristics.

Reference	Country	n	Condition	Gender	Age [a]	BMI [a]	Year Collection	Adipose Tissue	Serum		Ratio [c]	Meta [d]	Dioxin	PCB	OCP	BFR
					(Years)	(kg/m²)			Fasting	Units						
[52]	France	86	Caesarean	F	33 (20–46)	NA	2004–2006	SAT	Yes	LW						25
[35]	Spain	103	NA	F	54 (SD:12)	27	2012–2014	BAT	Yes	WW	X			3	2	
[35]	Spain	103	NA	F	54 (SD:12)	27	2012–2014	BAT	Yes	LW	X	X		3	2	
[51]	Bolivia	112	NA	M/F+	31 (18–70)	27	2010	VAT	NA	WW	X			3	3	
[51]	Bolivia	112	NA	M/F+	31 (18–70)	27	2010	VAT	NA	LW	X	X		3	3	
[53]	Spain	387	NA	M/F	52 [b] (37–63)	27	2003–2004	SAT	No	LW					1	
[54]	Italy	70	Caesarean	F	33 [b] (SD:4)	24	2006	SAT	NA	LW				30	3	
[55]	Sweden	20	Stillbirths	F	25–40	21–35	2015–2016	FAT	NA	WW	X			10	9	3
[50]	Qatar	34	IR	M/F	32 [b] (SD: 10)	45	NA	SAT, VAT	No blood							28
[48]	France	42	Obesity/H	M/F	44/40	48/22	2006–2008	SAT(VAT)	NA	W/LW			17	18		
[56]	Korea	50	T2D/N-T2D	M/F	66/62 (8/11)	23 (3)	NA	SAT,VAT	No blood					19	13	
[10]	France	48	BC	F	64	24	2015	BAT	No	LW	X	X		18		8
[36]	China	37–55	Caesarean	F	29	21	2011	SAT	Yes	LW	X	X		6		8
[36]	China	37–55	Caesarean	F	29	21	2011	SAT	Yes	WW	X			6		8
[43]	Mexico	198	BC/H	F	40 (19–78)	NA	1994–1996	BAT	No	LW				1		
[49]	Belgium	52	Obesity	M/F	40 (18–58)	42	2010–2012	SAT, VAT	No blood					28	2	
[57]	Belgium	101	Infertile	F	32 (24–42)		1996–1998	SAT		LW				7	7	
[58]	Portugal	189	Obesity	M/F+	43 (19–65)	45 (5)	2010–2011	SAT, VAT	No blood				1		11	
[46]	France	67	ENDO/H	M+/F	34 (48–80)	24	2013–2015	VAT, SAT	NA	LW	X	X		18		8
[47]	US	473	ENDO/H	F	33 (IQR:11)	26(IQR:10)	2007–2009	VAT	No	WW	X			13	5	5
[59]	India	29	BC	F	24–65	NA	1996–1997	BAT	Yes	WW	X			2	3	
[59]	India	29	BC	F	24–65	NA	1996–1997	GAT	Yes	WW	X			2	3	
[59]	India	29	BC	F	24–65	NA	1996–1997	BAT	Yes	LW	X			2	3	
[59]	India	29	BC	F	24–65	NA	1996–1997	GAT	Yes	LW	X			2	3	
[44]	Australia	32	NA	M/F	55 (22–89)	25	2016	SAT	Yes	LW	X			3		
[45]	Italy	59	HCC	M+/F	69 (48–80)	NA	NA	VAT	Yes	LW	X	X		7		

[a] Mean age or body mass index and range, standard deviation (SD) or interquartile range (IQR), [b] geometric mean or median, [c] identifier of studies that reported ratios, [d] identifier of studies included in the meta-regression. Abbreviations: BAT, breast adipose tissue; BC, breast cancer; BFR, brominated flame retardants; BMI, body mass index; FAT, fetal adipose tissue; GAT, gluteal adipose tissue; F, female; F+, overrepresentation of female; H, healthy subjects; HCC, hepatocellular carcinoma; IQR, interquartile range; IR, insulin resistance; LW, lipid weight; M, male; NA, not available; OCP, organochlorinated pesticides; PCBs, polychlorinated biphenyls; R, identifier for studies reporting calculated partition ratios; SAT, subcutaneous adipose tissue; SD, standard deviation; VAT, visceral adipose tissue; WW, wet weight.

3.2. Partition Coefficients of Persistent Organic Pollutants between Adipose Tissue and Serum

The studies tended to report the ratios considering both metrics on a lipid weight basis, but in some cases (Table 4) reported both lipid weight and wet weight biomarkers [35,36,51,59]. One study chose to present the values of ratios in wet weight [47]. Among PCBs, the congeners 138, 153 and 180 were the most reported ones in the literature because of their abundance (eight studies). In most cases, the congeners showed adipose tissue: serum ratios between 1.5 and 2; however, some exceptions were found. For instance, a few studies showed mean ratios of PCB 153 close to 5 [35,44], whereas others were close to 0.7–0.9 for the same congener [35,59]. In turn, the metabolite of the insecticide DDT, *p,p'*-DDE, was the organochlorinated pesticide (OCP) most widely reported in the six studies. As previously described for PCBs, the mean levels of ratios of OCP were mostly in the range 1.5–2, with several exceptions. The adipose tissue: serum ratios of brominated flame retardants (BFRs) were reported in three studies. Two French studies tended to show consistent values for most PBDEs [10,46], showing mean values in the range between two and three for most congeners with the exception of BDE209, which showed the highest values (4.6–8.0). Conversely, in one study conducted in an e-waste facility area with expectant women, the ratios tended to be below the unit for most congeners (0.4–1.1) [36]. In summary, for most studies, the reported adipose tissue: serum ratios displayed a similar pattern, with a 1.5-3-fold increased concentration of POPs in the adipose tissue, considering the measures based on lipid weight.

Table 4. Congener-specific summary of mean coefficient partitions of persistent organic pollutants between adipose tissue and serum, both normalized on the lipid weight basis.

Chemical	[52]	[51]	[35]	[10]	[43]	[36]	[46]	[44]	[59] BAT	[59] GAT	[45]
PCB101				1.8			1.5				
PCB105				2.1			2.1				
PCB114				2.1			2.3				
PCB118				1.8		1.4	1.8				1.2
PCB123				1.6			1.9				
PCB126				1.7			1.7				
PCB138		3.6	4.9	2.1		1.6	2.3	5.4			2.0
PCB153		0.9	4.7	2.1		1.7	2.3	5.4	0.7	0.7	2.0
PCB156				2.1			2.6				1.7
PCB157				1.6			1.8				
PCB167				1.5			1.7				
PCB169				1.8			2.0				
PCB169											2.2
PCB180		1.7	2.2	2.0		1.5	2.4	4.4	0.8	0.8	2
PCB189				2.6			2.6				
PCB194											2.1
PCB28				1.7		1.2	1.3				
PCB52				1.5			1.2				
PCB77				1.6			0.4				
PCB81				1.2			0.7				
PCB99						1.4					
p,p'-DDE		1.9	1.7		4.2			3.4	0.7	0.9	
p,p'-DDT		2							1.1	1.5	
HCB		1.9	2.2								
β-HCH									1.1	1.6	

Table 4. *Cont.*

Chemical	[52]	[51]	[35]	[10]	[43]	[36]	[46]	[44]	[59] BAT	[59] GAT	[45]
BDE100	0.91			2.3		0.7	3.2				
BDE153	2.12			2.4		1.0	3.0				
BDE154	1.24			1.3		0.9	2.3				
BDE183				3.0		1.1	3.2				
BDE209	0.16			4.6		0.4	7.9				
BDE28	1.00			2.7		0.6	2.5				
BDE47	0.97			2.5		0.5	3.7				
BDE99	0.36			3.0		0.5	4.1				
PBB153				2.5							

Abbreviations: BAT, brown adipose tissue; BDE, bromodiphenyl ether; β-HCH, β-hexachlorocyclohexane; GAT, gluteal adipose tissue; HCB, hexachlorobenzene; PCB, polychlorinated biphenyl; PBB153, 2,2′,4,4′,5,5′-hexabromobiphenyl; *p,p′*-DDT, 1,1,1-Trichloro-2,2-bis(4-chlorophenyl)ethane; *p,p′*-DDE, 1,1-dichloro-2,2-bis(p-chlorophenyl) ethylene.

3.3. Physicochemical Determinants of the Variability in Adipose Tissue: Serum Partition Coefficients

In order to gain insight into the physicochemical factors potentially affecting the variability in adipose tissue: serum ratios, we conducted a meta-regression based on available summary partitioning data reported in the selected studies. The meta-regression slopes, graphically depicted by bubble plots, are displayed in Figure 2A–D. The number of halogens was the variable most strongly associated with adipose tissue: serum ratios (β 0.25 95% CI (0.17–0.34), $p = 0.05$ (Figure 2B)). Log K_{ow}, molar weight (M_w) and molar volume showed null or mild associations with the ratios (Figure 2A,C,D).

3.4. Biological/Individual/Instrumental Determinants of the Variability in Adipose Tissue: Serum Partition Coefficients

Among individual variables, BMI was the most strongly associated with the adipose tissue: serum ratios for all POPs combined (Figure 3A), β 0.26 95% CI (0.17–0.34), $p < 0.001$, whereas age showed a more flattened slope than BMI, yet was statistically significant, β 0.02 95% CI (0.00–0.03), $p = 0.03$ (Figure 3B). Due to the close relationship between age and BMI, we also considered both variables in the same meta-regression model, suggesting a potential confounding effect of age, resulting in null associations for age with a minor impact on BMI coefficient. The year of sample collection (Figure 3C) also showed a positive association with the ratios β 0.26 95% CI (0.17–0.34), $p < 0.001$. For the congeners with at least three studies reporting the adipose tissue: serum ratios and BMI, we conducted a detailed meta-regression for individual congeners of PCBs and BDEs (Figure 4A–F). These stratified results supported the existence of congener-specific relationships between the ratios and BMI. For instance, BDE 209 was associated with BMI with the steepest slope, whereas PCB180 showed the most flattened one (Figure 4).

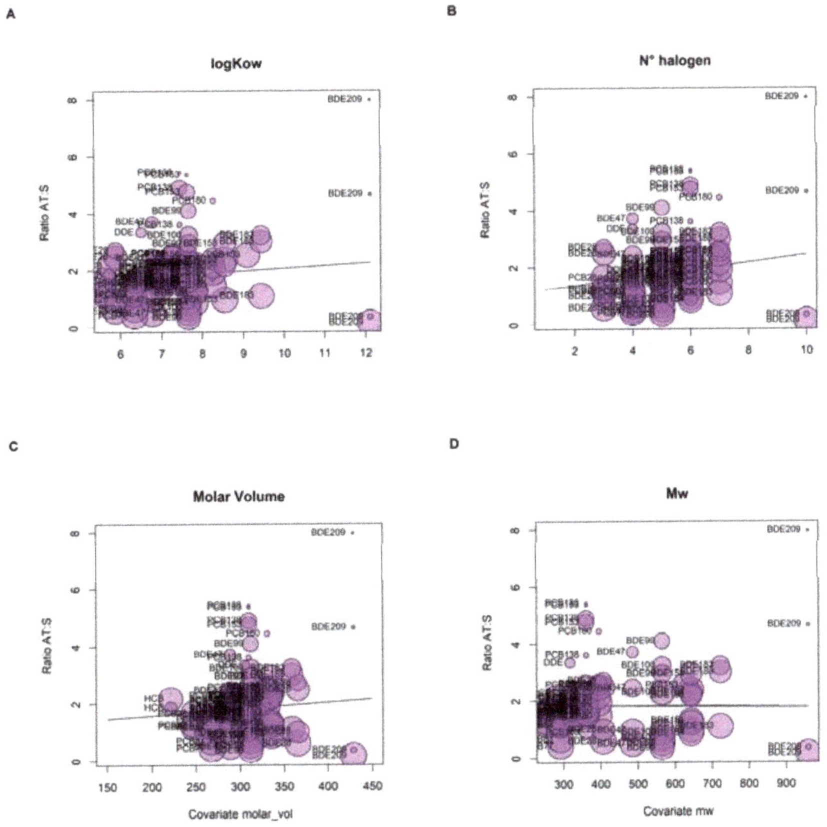

Figure 2. Bubble plots representing the associations between the mean partitioning ratios of persistent organic pollutants between adipose tissue and serum reported in the literature and physicochemical properties, including the partition octanol–water (Log K_{OW}, (**A**)), number of halogen atoms (**B**), molar volume (**C**) and molar weight (Mw, (**D**)).

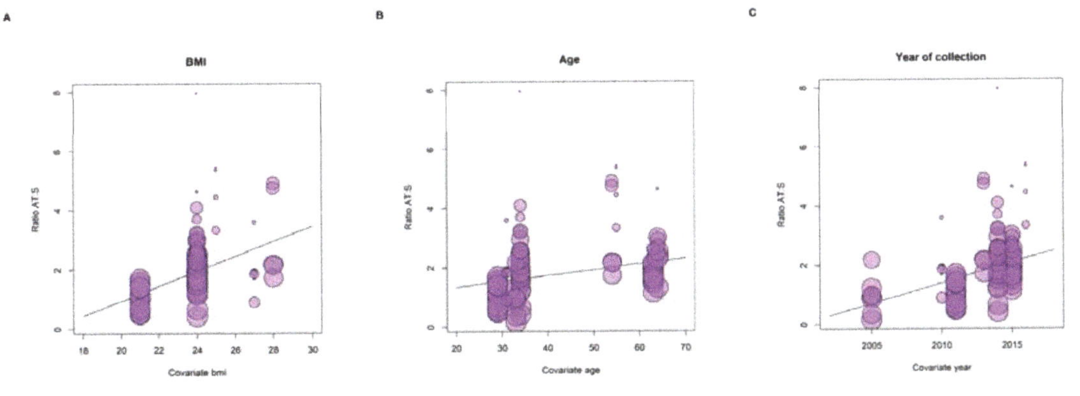

Figure 3. Bubble plots representing the associations between adipose tissue: serum ratios with body mass index (BMI, (**A**)), age (**B**), and year of collection (**C**) from the random effects meta-regression model. The study [52] was excluded from this analysis because the BMI was not reported.

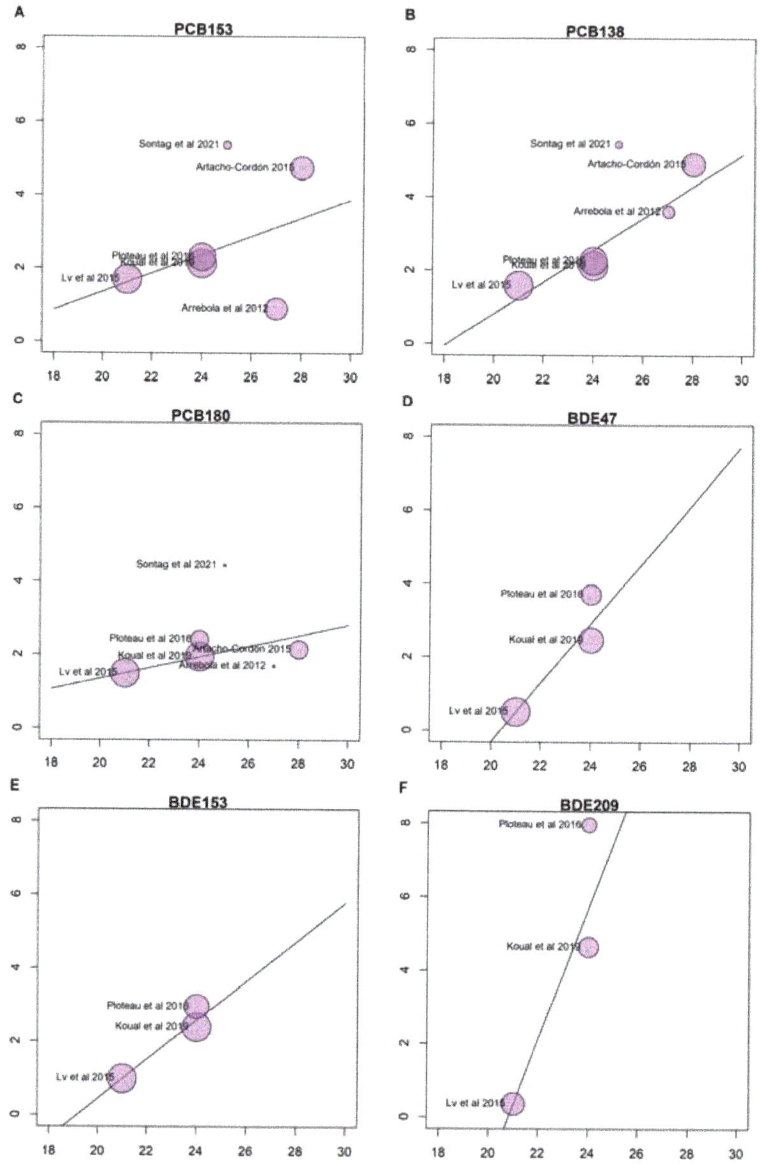

Figure 4. Bubble plots representing the associations between adipose tissue: serum ratios and body mass index (BMI) from the random effects meta-regression model for specific congeners including PCB 153 (**A**), PCB 138 (**B**), PCB 180 (**C**), BDE 47 (**D**), BDE 153 (**E**), and BDE 209 (**F**). The study [52] was excluded from this analysis because the BMI was not reported.

3.5. Variability in Levels of POPs between Adipose Tissue Types/Location

Seven studies analysed POPs in different adipose tissue locations (mainly visceral and subcutaneous) matched from the same individuals, including obese populations and women with endometriosis [46,48–50,56,58,59]. Globally, concentrations of organochlorinated and brominated chemicals analysed in paired VAT and SAT matrices showed high correlations, for instance with R Pearson correlation >0.9 ($p < 0.001$) for all POPs from

France [46] or R^2 0.7–0.6 ($p < 0.0001$) for PCBs from Belgium [49]. Congener-specific and strong sex-specific differences were noticed for the correlation slopes, showing lower correlations of POPs between VAT and SAT among females [49]. Noteworthy, two studies reported high correlations for individual POPs or ∑POPs between VAT and SAT locations but statistically different concentrations, with higher concentrations of individual POPs or ∑POPs in VAT than SAT [56,58]. Concentrations of organochlorinated pesticides (*p,p′*-DDE, *p,p′*-DDT, β-HCH) and PCBs (153 and 180) were also highly correlated between breast and gluteal fat pads (rho > 0.7, $p < 0.001$), showing comparable concentration levels [59]. The relationships between POPs in human foetal adipose tissue and maternal serum or placenta have been addressed by one study, showing in general high correlations for all PCBs and pesticides analysed (HCB, β-HCH, *p,p′*-DDE and transnonachlor) [55]. The authors reported sex-specific patterns of tissue accumulation, reporting higher tissue ratios among male foetuses than among females and lower among those with placental insufficiency.

4. Discussion

To the best of our knowledge, this is the first study attempting to summarize the available scientific evidence on partitioning of POPs between adipose tissue and serum from human studies. We have also aimed at characterizing and gaining insight into the determining factors of variability for partitions, as was already done for partitioning between serum and breast milk [30]. Data were retrieved mostly from studies conducted with non-occupationally exposed populations, reporting matched samples on lipid-based compartments, thus illustrating chronic low-dose exposure to POPs. Adipose tissue is acknowledged to be the gold-standard matrix for biomonitoring internal POP levels, yet blood (lipids) has been used as a surrogate for many years, assuming that blood POPs reflect their adipose tissue levels in equilibrated states [60]. Consequently, blood concentrations of POPs are extensively used for biomonitoring and epidemiological research. The widespread presence of low doses of POPs in internal biological matrices urges a comprehensive understanding of their kinetics for accurate use of exposure biomarkers in biomonitoring and epidemiological studies.

4.1. Overall Variability in Adipose Tissue: Serum Partition Coefficients among POPs

All individual studies included in this review reported high variability in the partition coefficients of POPs between adipose tissue and serum, at different levels of magnitude depending on the specific congeners. Whereas it has often been assumed that highly lipophilic chemicals would distribute homogeneously within lipid compartments of the human body, similarly to 2,3,7,8-tetrachlorodibenzo-p-dioxin [60], the studies included in this review illustrate that the mean ranges between 0.4 and 8 in a congener-specific base. Most ratios were above 1, suggesting a higher concentration of POPs in adipose tissue than in blood; nonetheless, in a few cases the ratios were inversed, reflecting the highest concentrations in serum. Among the different physicochemical properties investigated in the meta-regression, the number of halogen atoms was the most strongly and positively associated variable with the partition ratios, at the limit of statistical significance ($p = 0.05$), with the rest being statistically null (Log K_{ow}, Mw, and molar volume). The bubble plots reveal the strong dispersion of data for BDE209 and the underlying correlation of physicochemical variables (e.g., BDE209 showed the largest physicochemical properties: Log K_{ow} = 12.11, number of halogens = 10, molar volume = 428.6 m^3/mol). In addition, the bubble plots reflect that the linear meta-regression may be strongly influenced by two dominant studies reporting low ratios for BDE209, suggesting a non-monotonic relationship of the ratios with these physicochemical properties. These trends are consistent with the previous literature. For instance, positive associations between log K_{ow} and adipose tissue: serum ratios were reported for most POPs with the exception of PCB180 and BDE209, with higher log K_{ow} values (8.27 and 12.11, respectively) and a higher number of halogens [36]. Similar trends were reported for the blood/milk partitioning, which decreased among substances with larger log K_{ow} (>8), higher molecular weight (>400 for PCB/PCDD/Fs,

>700 for PBDEs), and more halogen compounds (>7) [30]. A bioaccumulation factor increase correlated with increased K_{ow} was reported for lipophilic compounds with $K_{ow} < 7$ in wild aquatic species [61]: the bioaccumulation factor would increase for $K_{ow} < 7$, then decline for $K_{ow} > 7$. A parabolic relationship was also observed between log biomagnification factor and log K_{ow}, with a peak for K_{ow} of approximately 8 [62]. Lv et al. (2015) also noted that compounds with similar log K_{ow} but from different families (PCB vs. PBDE) exhibited different partition ratios, lower for BDEs than PCBs [36]. It was assumed that deposition of PBDE in human adipose tissues was lower than deposition of PCB with similar K_{ow}; however, these trends were not consistent in other studies [10,46].

Globally, these data support that physicochemical parameters, such as number of halogens, may have an impact on POP partitioning between adipose tissue and serum, but likely in a non-monotonic manner and influenced by other parameters. For instance, the tridimensional structure and associated degree of rigidity of the different compounds could also have an impact on the diffusion-limited partition of POPs [30]

4.2. Congener-Specific Variability in Adipose Tissue: Serum Partition Coefficients between Studies

Table 4 summarizes the high variability in partition ratios determined for the same congener among different studies. For instance, PCB138's reported mean ratios ranged from 1.59 to 5.44, a three-fold difference for the same compound. Systematic differences were appreciated between studies, in some cases even switching the direction in the case of BDEs [36]. These differences may be influenced by specific study characteristics, such as timing of sampling, differences in analytical procedures introducing analytical uncertainties for POPs, or the different determinations of lipid content derived from different equations [15].

Among the individual biological factors analyzed, BMI was the main determinant, positively associated with the partition ratios (β 0.26 95% CI (0.17–0.34), $p < 0.001$), while age was associated to a weaker degree and non-associated when both BMI and age were considered in the multivariate meta-regression model. Globally, these findings suggest that the larger the fat mass, the higher the concentration of POPs in adipose tissue relative to blood lipids, independently of age. These findings would empirically support the hypothesis that populations with high adipose fractions may be sequestering more POPs in their adipocytes, preventing their release into circulation and thus protecting other organs [63]. Most studies that reported different types of adipose tissue deposits did not support differences between locations (e.g., VAT vs. SAT), as high or similar concentrations were reported. In any case, the number of studies was very limited to include the adipose tissue location as a co-variable in the meta-regression or to conduct a stratification analysis. One study showed different trends, with high correlations between ∑POP and AT, with the highest concentrations in VAT rather than SAT [56,58]. These differences may be due to the dissimilarities in lipid content [58], differences in blood flow, or in the density of blood vessels and/or differences in metabolism, with VAT being more metabolically active [64,65]. A PBPK model was proposed for γ-HBCD in mice [66], including an additional compartment corresponding to the deep fat fraction of the body. For POPs with log now >4, it was suggested that ratio of the neutral lipid-equivalent content (corresponding to neutral lipids and phospholipids) of adipose tissue and blood was the main determinant of POP partitioning between those tissues [67]: different types of lipids and compositions of non-lipid molecules might play a role in POP dynamics within AT, and interactions between POPs and lipid compounds at the molecular scale should be further investigated [46], especially for biomonitoring with blood. It was also suggested that extracellular and lymphatic circulation might contribute to POP storage [66]. Metabolic status and disease condition may also play a role in POP partitioning [15], but those parameters were not investigated in the meta-regression due to the limited number of studies. The level and the route of exposure could also have an impact on the metabolic disruption of adipocytes, and thus the kinetics of POPs. In this regard, the exposure scenario may also contribute to explain the variabilities of coefficients. Nonetheless, formal statistics were not applicable to compare background exposures with occupational or highly contaminated sites. For instance, one occupational e-waste study [36] showed serum levels of some BFRs 20 times higher than in populations exposed at background levels [10,46], which would explain ratios systematically lower, below the unit.

4.3. Limitations of the Review

First, the current review was essentially narrative, even if a robust methodology based on systematic review principles was applied. If our review is not exhaustive, we believe that most studies with AT and blood POP measurements published during the last decade have been included. Because of this criterion, only a limited number of studies were eligible in the meta-analysis. In addition, each individual study involved a modest sample size. In order to strengthen the statistical power, studies with larger sample size will be required. Furthermore, the selected studies explored different sets of congeners, limiting the ability to compare ratios among studies for the same compounds and to conclude about the influence of population biological characteristics. Actually, the studies provided few individual variables that could be integrated in the meta-regression, but other factors may likely impact on the kinetics of POPs and consequently on the ratios (e.g., parity, breastfeeding, physical activity). In addition, we pooled some summary metrics through approximations, because not all studies reported the means and standard deviations. Optimally, statistical analyses combining raw data from individual studies would improve the accuracy of estimates. Considering the invasive character of adipose tissue biopsies collected from the populations with cancer or obesity the data may not be representative of the general population. This review also stressed the need for further studies on dioxins scarcely reported in the literature. Per/polyfluorinated substances (PFAS) are a wide class of chemicals with some congeners listed as POPs in the Stockholm Convention commonly characterized in blood due to their amphiphilic properties; hence, they are not represented in the present review because they are more present in livers than in AT. Nonetheless, some studies have shown that even non-persistent chemicals like bisphenol A or parabens can be found in adipose tissue [68,69]. New approaches based on artificial intelligence, which used text mining to automatically screen large literature sources and databases, could also be considered in order to get as much information as possible. As an example, the AOP-helpFinder tool [70,71], which helps the development of adverse outcome pathways, could be adapted in future meta-analysis studies.

4.4. Impacts of Findings

We believe that the findings of this study are of high relevance due to the widespread use of exposure biomarkers of POPs for biomonitoring and epidemiological research. Biomonitoring of chemicals has been progressively integrated as a novel tool within the chemical risk assessment process, notably by using PBPK modelling and the development of biomonitoring equivalents [72–74]. PBPK models of POPs are often based on the assumption of a homogeneous distribution of POP in lipid compartments within the body (i.e., that adipose tissue: blood ratios are comparable). Thus, a better understanding of partitioning of POPs between fat compartments will help the refinement of those models and reduce uncertainty within chemical risk assessment and regulatory decision-making. Likewise, in epidemiological research, the extensive use of blood biomarkers of POPs has not been supported by a deep understanding of the complex relationship of POPs in blood and adipose tissue [15]. Nowadays, different statistical methods have been proposed to use the blood biomarkers, addressing the different causal structures depicted when the perturbation of lipid metabolism is in the causal pathway between the exposure and the health outcomes [75]. Although some methods may attenuate the statistical bias, none of the methods using a single-spot blood biomarker may fully recapitulate the complex dynamics of POPs. The problem becomes more complex from the perspective of POP mixtures, where individual congeners may impact the kinetics of the others. Hence, a better understanding of the dynamics of partitions between fat compartments is of key relevance for accurate use and interpretation of surrogates' measures in environmental health studies.

5. Conclusions

This review supports the fact that partitioning of POPs between fat compartments is highly variable. These findings also show that the partition ratios of POPs between fat compartments systematically depart from the unit, even at equilibrated states, as extensively assumed. BMI and number of halogens were the main determinants of the variability in the adipose tissue: serum ratio of POP among the selected studies. Finally, this study highlights that further research is still needed to better understand the behavior of POPs in the main matrices used for biomonitoring, especially for those scarcely studied congeners.

Supplementary Materials: The following supporting information can be downloaded at: https://www.mdpi.com/article/10.3390/toxics11010041/s1, Table S1. Search strings.

Author Contributions: Conceptualization, G.C.-S., M.K., X.C., J.-P.A. (Juan-Pedro Arrebola), C.E. and C.R.; methodology, G.C.-S., M.K., X.C., J.-P.A. (Jean-Philippe Antignac) and C.R.; software, G.C.-S.; validation, M.-A.M., G.C.-S. and C.R.; formal analysis, M.-A.M., G.C.-S. and C.R.; resources, K.A., C.R.; data curation, M.-A.M. and G.C.-S.; writing—original draft preparation, M.-A.M., G.C.-S. and C.R.; writing—review and editing, all authors.; visualization, G.C.-S.; supervision, C.R.; project administration, C.R., K.A.; funding acquisition, C.R., K.A. All authors have read and agreed to the published version of the manuscript.

Funding: The project has been funded by the French National Research Agency (ANR-18-CE34-0001-01, Creative Project). Juan Pedro Arrebola is under contract within the Ramón y Cajal Program (RYC-2016-20155, Ministerio de Economía, Industria y Competitividad, Spain).

Institutional Review Board Statement: Not applicable.

Informed Consent Statement: Not applicable.

Data Availability Statement: Not applicable.

Acknowledgments: The authors would like to acknowledge the CREATIvE project, as this project has received funding from the French National Research Agency (ANR-18-CE34-0001-01). This work was supported by the Université de Paris and INSERM.

Conflicts of Interest: The authors declare they have no actual or potential competing financial interests.

References

1. van den Berg, M.; Kypke, K.; Kotz, A.; Tritscher, A.; Lee, S.Y.; Magulova, K.; Fiedler, H.; Malisch, R. WHO/UNEP global surveys of PCDDs, PCDFs, PCBs and DDTs in human milk and benefit-risk evaluation of breastfeeding. *Arch. Toxicol.* **2017**, *91*, 83–96. [CrossRef] [PubMed]
2. Wu, Z.; He, C.; Han, W.; Song, J.; Li, H.; Zhang, Y.; Jing, X.; Wu, W. Exposure pathways, levels and toxicity of polybrominated diphenyl ethers in humans: A review. *Environ. Res.* **2020**, *187*, 109531. [CrossRef] [PubMed]
3. Long, M.; Wielsøe, M.; Bonefeld-Jørgensen, E.C. Time Trend of Persistent Organic Pollutants and Metals in Greenlandic Inuit during 1994–2015. *Int. J. Environ. Res. Public Health* **2021**, *18*, 2774. [CrossRef] [PubMed]
4. Pan, H.-Y.; Li, J.-F.; Li, X.-H.; Yang, Y.-L.; Qin, Z.-F.; Li, J.-B.; Li, Y.-Y. Transfer of dechlorane plus between human breast milk and adipose tissue and comparison with legacy lipophilic compounds. *Environ. Pollut.* **2020**, *265*, 115096. [CrossRef] [PubMed]
5. Nøst, T.H.; Berg, V.; Hanssen, L.; Rylander, C.; Gaudreau, E.; Dumas, P.; Breivik, K.; Sandanger, T.M. Time trends of persistent organic pollutants in 30 year olds sampled in 1986, 1994, 2001 and 2007 in Northern Norway: Measurements, mechanistic modeling and a comparison of study designs. *Environ. Res.* **2019**, *172*, 684–692. [CrossRef]
6. Quinete, N.; Schettgen, T.; Bertram, J.; Kraus, T. Occurrence and distribution of PCB metabolites in blood and their potential health effects in humans: A review. *Environ. Sci. Pollut. Res. Int.* **2014**, *21*, 11951–11972. [CrossRef]
7. Kahn, L.G.; Harley, K.G.; Siegel, E.L.; Zhu, Y.; Factor-Litvak, P.; Porucznik, C.A.; Klein-Fedyshin, M.; Hipwell, A.E. Persistent organic pollutants and couple fecundability: A systematic review. *Hum. Reprod. Update* **2021**, *27*, 339–366. [CrossRef]
8. Lefebvre, T.; Fréour, T.; Ploteau, S.; Le Bizec, B.; Antignac, J.P.; Cano-Sancho, G. Associations between human internal chemical exposure to Persistent Organic Pollutants (POPs) and In Vitro Fertilization (IVF) outcomes: Systematic review and evidence map of human epidemiological evidence. *Reprod. Toxicol.* **2021**, *105*, 184–197. [CrossRef]
9. Ennour-Idrissi, K.; Ayotte, P.; Diorio, C. Persistent Organic Pollutants and Breast Cancer: A Systematic Review and Critical Appraisal of the Literature. *Cancers* **2019**, *11*, 1063. [CrossRef]

10. Koual, M.; Cano-Sancho, G.; Bats, A.-S.; Tomkiewicz, C.; Kaddouch-Amar, Y.; Douay-Hauser, N.; Ngo, C.; Bonsang, H.; Deloménie, M.; Lecuru, F.; et al. Associations between persistent organic pollutants and risk of breast cancer metastasis. *Environ. Int.* **2019**, *132*, 105028. [CrossRef]
11. Mendes, V.; Ribeiro, C.; Delgado, I.; Peleteiro, B.; Aggerbeck, M.; Distel, E.; Annesi-Maesano, I.; Sarigiannis, D.; Ramos, E. The association between environmental exposures to chlordanes, adiposity and diabetes-related features: A systematic review and meta-analysis. *Sci. Rep.* **2021**, *11*, 14546. [CrossRef] [PubMed]
12. Mustieles, V.; Pérez-Carrascosa, F.M.; León, J.; Lange, T.; Bonde, J.-P.; Gómez-Peña, C.; Artacho-Cordón, F.; Barrios-Rodríguez, R.; Olmedo-Requena, R.; Expósito, J.; et al. Adipose Tissue Redox Microenvironment as a Potential Link between Persistent Organic Pollutants and the 16-Year Incidence of Non-hormone-Dependent Cancer. *Environ. Sci. Technol.* **2021**, *55*, 9926–9937. [CrossRef]
13. Lee, D.H.; Jacobs, D.R., Jr.; Park, H.Y.; Carpenter, D.O. A role of low dose chemical mixtures in adipose tissue in carcinogenesis. *Environ. Int.* **2017**, *108*, 170–175. [CrossRef] [PubMed]
14. Wu, Q.; Achebouche, R.; Audouze, K. Computational systems biology as an animal-free approach to characterize toxicological effects of persistent organic pollutants. *Altex* **2020**, *37*, 287–299. [CrossRef] [PubMed]
15. Cano-Sancho, G.; Labrune, L.; Ploteau, S.; Marchand, P.; Le Bizec, B.; Antignac, J.-P. The challenging use and interpretation of circulating biomarkers of exposure to persistent organic pollutants in environmental health: Comparison of lipid adjustment approaches in a case study related to endometriosis. *Chemosphere* **2018**, *200*, 388–396. [CrossRef]
16. Jandacek, R.J.; Rider, T.; Yang, Q.; Woollett, L.A.; Tso, P. Lymphatic and portal vein absorption of organochlorine compounds in rats. *Am. J. Physiol. Gastrointest. Liver Physiol.* **2009**, *296*, G226–G234. [CrossRef]
17. Jandacek, R.J.; Zheng, S.; Yang, Q.; Tso, P. Rapid clearance of hexachlorobenzene from chylomicrons. *Lipids* **2004**, *39*, 993–995. [CrossRef]
18. Vorkamp, K.; Castaño, A.; Antignac, J.P.; Boada, L.D.; Cequier, E.; Covaci, A.; Esteban López, M.; Haug, L.S.; Kasper-Sonnenberg, M.; Koch, H.M.; et al. Biomarkers, matrices and analytical methods targeting human exposure to chemicals selected for a European human biomonitoring initiative. *Environ. Int.* **2021**, *146*, 106082. [CrossRef]
19. Genuis, S.J.; Lane, K.; Birkholz, D. Human Elimination of Organochlorine Pesticides: Blood, Urine, and Sweat Study. *BioMed Res. Int.* **2016**, *2016*, 1624643. [CrossRef]
20. Lee, D.-H.; Porta, M.; Jacobs, D.R.; Vandenberg, L.N. Chlorinated persistent organic pollutants, obesity, and type 2 diabetes. *Endocr. Rev.* **2014**, *35*, 557–601. [CrossRef]
21. Li, L.; Wania, F. Mechanistic Pharmacokinetic Modeling of the Bioamplification of Persistent Lipophilic Organic Pollutants in Humans during Weight Loss. *Environ. Sci. Technol.* **2017**, *51*, 5563–5571. [CrossRef] [PubMed]
22. Desalegn, A.; Bopp, S.; Asturiol, D.; Lamon, L.; Worth, A.; Paini, A. Role of Physiologically Based Kinetic modelling in addressing environmental chemical mixtures—A review. *Comput. Toxicol.* **2019**, *10*, 158–168. [CrossRef] [PubMed]
23. Andersen, M.E.; Mallick, P.; Clewell, H.J., 3rd; Yoon, M.; Olsen, G.W.; Longnecker, M.P. Using quantitative modeling tools to assess pharmacokinetic bias in epidemiological studies showing associations between biomarkers and health outcomes at low exposures. *Environ. Res.* **2021**, *197*, 111183. [CrossRef]
24. Verner, M.A.; Ayotte, P.; Muckle, G.; Charbonneau, M.; Haddad, S. A physiologically based pharmacokinetic model for the assessment of infant exposure to persistent organic pollutants in epidemiologic studies. *Environ. Health Perspect.* **2009**, *117*, 481–487. [CrossRef] [PubMed]
25. Verner, M.A.; Hart, J.E.; Sagiv, S.K.; Bellinger, D.C.; Altshul, L.M.; Korrick, S.A. Measured Prenatal and Estimated Postnatal Levels of Polychlorinated Biphenyls (PCBs) and ADHD-Related Behaviors in 8-Year-Old Children. *Environ. Health Perspect.* **2015**, *123*, 888–894. [CrossRef] [PubMed]
26. Pruvost-Couvreur, M.; Béchaux, C.; Rivière, G.; Le Bizec, B. Impact of sociodemographic profile, generation and bioaccumulation on lifetime dietary and internal exposures to PCBs. *Sci. Total Environ.* **2021**, *800*, 149511. [CrossRef]
27. Emond, C.; Charbonneau, M.; Krishnan, K. Physiologically based modeling of the accumulation in plasma and tissue lipids of a mixture of PCB congeners in female Sprague-Dawley rats. *J. Toxicol. Environ. Health Part A* **2005**, *68*, 1393–1412. [CrossRef]
28. Emond, C.; Krishnan, K. A physiological pharmacokinetic model based on tissue lipid content for simulating inhalation pharmacokinetics of highly lipophilic volatile organic chemicals. *Toxicol. Mech. Methods* **2006**, *16*, 395–403. [CrossRef]
29. Needham, L.L.; Grandjean, P.; Heinzow, B.; Jørgensen, P.J.; Nielsen, F.; Patterson, D.G., Jr.; Sjödin, A.; Turner, W.E.; Weihe, P. Partition of environmental chemicals between maternal and fetal blood and tissues. *Environ. Sci. Technol.* **2011**, *45*, 1121–1126. [CrossRef]
30. Mannetje, A.; Coakley, J.; Mueller, J.F.; Harden, F.; Toms, L.M.; Douwes, J. Partitioning of persistent organic pollutants (POPs) between human serum and breast milk: A literature review. *Chemosphere* **2012**, *89*, 911–918. [CrossRef]
31. Sousa, S.; Maia, M.L.; Delerue-Matos, C.; Calhau, C.; Domingues, V.F. The role of adipose tissue analysis on Environmental Pollutants Biomonitoring in women: The European scenario. *Sci. Total Environ.* **2022**, *806*, 150922. [CrossRef] [PubMed]
32. Page, M.J.; McKenzie, J.E.; Bossuyt, P.M.; Boutron, I.; Hoffmann, T.C.; Mulrow, C.D.; Shamseer, L.; Tetzlaff, J.M.; Akl, E.A.; Brennan, S.E.; et al. The PRISMA 2020 statement: An updated guideline for reporting systematic reviews. *BMJ Clin. Res. Ed.* **2021**, *372*, n71. [CrossRef]
33. EFSA. Application of systematic review methodology to food and feed safety assessments to support decision making. *EFSA J.* **2010**, *8*, 1637. [CrossRef]

34. Wan, X.; Wang, W.; Liu, J.; Tong, T. Estimating the sample mean and standard deviation from the sample size, median, range and/or interquartile range. *BMC Med. Res. Methodol.* **2014**, *14*, 135. [CrossRef] [PubMed]
35. Artacho-Cordón, F.; Fernández-Rodríguez, M.; Garde, C.; Salamanca, E.; Iribarne-Durán, L.; Torné, P.; Expósito, J.; Papay-Ramírez, L.; Fernández, M.; Olea, N.; et al. Serum and adipose tissue as matrices for assessment of exposure to persistent organic pollutants in breast cancer patients. *Environ. Res.* **2015**, *142*, 633–643. [CrossRef]
36. Lv, Q.-X.; Wang, W.; Li, X.-H.; Yu, L.; Zhang, Y.; Tian, Y. Polychlorinated biphenyls and polybrominated biphenyl ethers in adipose tissue and matched serum from an E-waste recycling area (Wenling, China). *Environ. Pollut.* **2015**, *199*, 219–226. [CrossRef]
37. Luo, D.; Wan, X.; Liu, J.; Tong, T. Optimally estimating the sample mean from the sample size, median, mid-range, and/or mid-quartile range. *Stat. Methods Med. Res.* **2018**, *27*, 1785–1805. [CrossRef]
38. Balduzzi, S.; Rücker, G.; Schwarzer, G. How to perform a meta-analysis with R: A practical tutorial. *Evid.-Based Ment. Health* **2019**, *22*, 153–160. [CrossRef]
39. Kim, S.; Chen, J.; Cheng, T.; Gindulyte, A.; He, J.; He, S.; Li, Q.; Shoemaker, B.A.; Thiessen, P.A.; Yu, B.; et al. PubChem in 2021: New data content and improved web interfaces. *Nucleic Acids Res.* **2021**, *49*, D1388–D1395. [CrossRef]
40. Sjödin, A.; Mueller, J.F.; Jones, R.; Schütze, A.; Wong, L.-Y.; Caudill, S.P.; Harden, F.A.; Webster, T.F.; Toms, L.-M. Serum elimination half-lives adjusted for ongoing exposure of tri-to hexabrominated diphenyl ethers: Determined in persons moving from North America to Australia. *Chemosphere* **2020**, *248*, 125905. [CrossRef]
41. Bu, Q.; MacLeod, M.; Wong, F.; Toms, L.-M.L.; Mueller, J.F.; Yu, G. Historical intake and elimination of polychlorinated biphenyls and organochlorine pesticides by the Australian population reconstructed from biomonitoring data. *Environ. Int.* **2015**, *74*, 82–88. [CrossRef] [PubMed]
42. Milbrath, M.O.; Wenger, Y.; Chang, C.W.; Emond, C.; Garabrant, D.; Gillespie, B.W.; Jolliet, O. Apparent half-lives of dioxins, furans, and polychlorinated biphenyls as a function of age, body fat, smoking status, and breast-feeding. *Environ. Health Perspect.* **2009**, *117*, 417–425. [CrossRef] [PubMed]
43. López-Carrillo, L.; Torres-Sánchez, L.; López-Cervantes, M.; Blair, A.; Cebrián, M.E.; Uribe, M. The adipose tissue to serum dichlorodiphenyldichloroethane (DDE) ratio: Some methodological considerations. *Environ. Res.* **1999**, *81*, 142–145. [CrossRef] [PubMed]
44. Sontag, N.-J.; Banks, A.P.W.; Aylward, L.L.; O'Rourke, N.A.; Cavallucci, D.J.; Mueller, J.F.; Drage, D.S. Comparison of lipid-normalised concentrations of persistent organic pollutants (POPs) between serum and adipose tissue. *Int. J. Hyg. Environ. Health* **2021**, *236*, 113801. [CrossRef] [PubMed]
45. Zani, C.; Gelatti, U.; Donato, F.; Capelli, M.; Portolani, N.; Bergonzi, R.; Apostoli, P. Polychlorinated biphenyls in serum, liver and adipose tissue of subjects with hepatocellular carcinoma living in a highly polluted area. *Chemosphere* **2013**, *91*, 194–199. [CrossRef]
46. Ploteau, S.; Antignac, J.-P.; Volteau, C.; Marchand, P.; Vénisseau, A.; Vacher, V.; Le Bizec, B. Distribution of persistent organic pollutants in serum, omental, and parietal adipose tissue of French women with deep infiltrating endometriosis and circulating versus stored ratio as new marker of exposure. *Environ. Int.* **2016**, *97*, 125–136. [CrossRef] [PubMed]
47. Pollack, A.Z.; Krall, J.R.; Kannan, K.; Louis, G.M.B. Adipose to serum ratio and mixtures of persistent organic pollutants in relation to endometriosis: Findings from the ENDO Study. *Environ. Res.* **2021**, *195*, 110732. [CrossRef] [PubMed]
48. Kim, M.J.; Marchand, P.; Henegar, C.; Antignac, J.-P.; Alili, R.; Poitou, C.; Bouillot, J.-L.; Basdevant, A.; Le Bizec, B.; Barouki, R.; et al. Fate and complex pathogenic effects of dioxins and polychlorinated biphenyls in obese subjects before and after drastic weight loss. *Environ. Health Perspect.* **2011**, *119*, 377–383. [CrossRef]
49. Malarvannan, G.; Dirinck, E.; Dirtu, A.C.; Pereira-Fernandes, A.; Neels, H.; Jorens, P.G.; Van Gaal, L.; Blust, R.; Covaci, A. Distribution of persistent organic pollutants in two different fat compartments from obese individuals. *Environ. Int.* **2013**, *55*, 33–42. [CrossRef]
50. Helaleh, M.; Diboun, I.; Al-Tamimi, N.; Al-Sulaiti, H.; Al-Emadi, M.; Madani, A.; Mazloum, N.A.; Latiff, A.; Elrayess, M.A. Association of polybrominated diphenyl ethers in two fat compartments with increased risk of insulin resistance in obese individuals. *Chemosphere* **2018**, *209*, 268–276. [CrossRef]
51. Arrebola, J.; Cuellar, M.; Claure, E.; Quevedo, M.; Antelo, S.; Mutch, E.; Ramirez, E.; Fernandez, M.; Olea, N.; Mercado, L. Concentrations of organochlorine pesticides and polychlorinated biphenyls in human serum and adipose tissue from Bolivia. *Environ. Res.* **2012**, *112*, 40–47. [CrossRef] [PubMed]
52. Antignac, J.-P.; Cariou, R.; Zalko, D.; Berrebi, A.; Cravedi, J.-P.; Maume, D.; Marchand, P.; Monteau, F.; Riu, A.; Andre, F.; et al. Exposure assessment of French women and their newborn to brominated flame retardants: Determination of tri- to deca-polybromodiphenylethers (PBDE) in maternal adipose tissue, serum, breast milk and cord serum. *Environ. Pollut.* **2009**, *157*, 164–173. [CrossRef] [PubMed]
53. Arrebola, J.P.; Fernández, M.F.; Olea, N.; Ramos, R.; Martin-Olmedo, P. Human exposure to *p,p'*-dichlorodiphenyldichloroethylene (*p,p'*-DDE) in urban and semi-rural areas in southeast Spain: A gender perspective. *Sci. Total Environ.* **2013**, *458–460*, 209–216. [CrossRef]
54. Bergonzi, R.; Specchia, C.; Dinolfo, M.; Tomasi, C.; De Palma, G.; Frusca, T.; Apostoli, P. Distribution of persistent organochlorine pollutants in maternal and foetal tissues: Data from an Italian polluted urban area. *Chemosphere* **2009**, *76*, 747–754. [CrossRef] [PubMed]

55. Björvang, R.D.; Vinnars, M.-T.; Papadogiannakis, N.; Gidlöf, S.; Mamsen, L.S.; Mucs, D.; Kiviranta, H.; Rantakokko, P.; Ruokojärvi, P.; Lindh, C.H.; et al. Mixtures of persistent organic pollutants are found in vital organs of late gestation human fetuses. *Chemosphere* **2021**, *283*, 131125. [CrossRef] [PubMed]
56. Kim, K.-S.; Lee, Y.-M.; Kim, S.G.; Lee, I.-K.; Lee, H.-J.; Kim, J.-H.; Kim, J.; Moon, H.-B.; Jacobs, D.R.; Lee, D.-H. Associations of organochlorine pesticides and polychlorinated biphenyls in visceral vs. subcutaneous adipose tissue with type 2 diabetes and insulin resistance. *Chemosphere* **2014**, *94*, 151–157. [CrossRef]
57. Pauwels, A.; Schepens, P.J.; D'Hooghe, T.; Delbeke, L.; Dhont, M.; Brouwer, A.; Weyler, J. The risk of endometriosis and exposure to dioxins and polychlorinated biphenyls: A case-control study of infertile women. *Hum. Reprod.* **2001**, *16*, 2050–2055. [CrossRef]
58. Pestana, D.; Faria, G.; Sá, C.; Fernandes, V.C.; Teixeira, D.; Norberto, S.; Faria, A.; Meireles, M.; Marques, C.; Correia-Sá, L.; et al. Persistent organic pollutant levels in human visceral and subcutaneous adipose tissue in obese individuals–depot differences and dysmetabolism implications. *Environ. Res.* **2014**, *133*, 170–177. [CrossRef]
59. Rusiecki, J.A.; Matthews, A.; Sturgeon, S.; Sinha, R.; Pellizzari, E.; Zheng, T.; Baris, D. A correlation study of organochlorine levels in serum, breast adipose tissue, and gluteal adipose tissue among breast cancer cases in India. *Cancer Epidemiol. Biomark. Prev.* **2005**, *14*, 1113–1124. [CrossRef]
60. Patterson, D.G., Jr.; Needham, L.L.; Pirkle, J.L.; Roberts, D.W.; Bagby, J.; Garrett, W.A.; Andrews, J.S., Jr.; Falk, H.; Bernert, J.T.; Sampson, E.J.; et al. Correlation between serum and adipose tissue levels of 2,3,7,8-tetrachlorodibenzo-p-dioxin in 50 persons from Missouri. *Arch. Environ. Contam. Toxicol.* **1988**, *17*, 139–143. [CrossRef]
61. Wu, J.P.; Luo, X.J.; Zhang, Y.; Luo, Y.; Chen, S.J.; Mai, B.X.; Yang, Z.Y. Bioaccumulation of polybrominated diphenyl ethers (PBDEs) and polychlorinated biphenyls (PCBs) in wild aquatic species from an electronic waste (e-waste) recycling site in South China. *Environ. Int.* **2008**, *34*, 1109–1113. [CrossRef] [PubMed]
62. Wu, J.P.; Luo, X.J.; Zhang, Y.; Chen, S.J.; Mai, B.X.; Guan, Y.T.; Yang, Z.Y. Residues of polybrominated diphenyl ethers in frogs (Rana limnocharis) from a contaminated site, South China: Tissue distribution, biomagnification, and maternal transfer. *Environ. Sci. Technol.* **2009**, *43*, 5212–5217. [CrossRef] [PubMed]
63. La Merrill, M.; Emond, C.; Kim, M.J.; Antignac, J.-P.; Le Bizec, B.; Clément, K.; Birnbaum, L.S.; Barouki, R. Toxicological function of adipose tissue: Focus on persistent organic pollutants. *Environ. Health Perspect.* **2013**, *121*, 162–169. [CrossRef] [PubMed]
64. Freedland, E.S. Role of a critical visceral adipose tissue threshold (CVATT) in metabolic syndrome: Implications for controlling dietary carbohydrates: A review. *Nutr. Metab.* **2004**, *1*, 12. [CrossRef]
65. Preis, S.R.; Massaro, J.M.; Robins, S.J.; Hoffmann, U.; Vasan, R.S.; Irlbeck, T.; Meigs, J.B.; Sutherland, P.; D'Agostino, R.B., Sr.; O'Donnell, C.J.; et al. Abdominal subcutaneous and visceral adipose tissue and insulin resistance in the Framingham heart study. *Obesity* **2010**, *18*, 2191–2198. [CrossRef]
66. Emond, C.; DeVito, M.J.; Birnbaum, L.S. A PBPK model describing the pharmacokinetics of γ-HBCD exposure in mice. *Toxicol. Appl. Pharmacol.* **2021**, *428*, 115678. [CrossRef]
67. Haddad, S.; Poulin, P.; Krishnan, K. Relative lipid content as the sole mechanistic determinant of the adipose tissue:blood partition coefficients of highly lipophilic organic chemicals. *Chemosphere* **2000**, *40*, 839–843. [CrossRef]
68. Artacho-Cordón, F.; Arrebola, J.P.; Nielsen, O.; Hernández, P.; Skakkebaek, N.E.; Fernández, M.F.; Andersson, A.M.; Olea, N.; Frederiksen, H. Assumed non-persistent environmental chemicals in human adipose tissue; matrix stability and correlation with levels measured in urine and serum. *Environ. Res.* **2017**, *156*, 120–127. [CrossRef]
69. Venisse, N.; Cambien, G.; Robin, J.; Rouillon, S.; Nadeau, C.; Charles, T.; Rabouan, S.; Migeot, V.; Dupuis, A. Development and validation of an LC-MS/MS method for the simultaneous determination of bisphenol A and its chlorinated derivatives in adipose tissue. *Talanta* **2019**, *204*, 145–152. [CrossRef]
70. Jornod, F.; Jaylet, T.; Blaha, L.; Sarigiannis, D.; Tamisier, L.; Audouze, K. AOP-helpFinder webserver: A tool for comprehensive analysis of the literature to support adverse outcome pathways development. *Bioinformatics* **2021**, *38*, 1173–1175. [CrossRef]
71. Carvaillo, J.C.; Barouki, R.; Coumoul, X.; Audouze, K. Linking Bisphenol S to Adverse Outcome Pathways Using a Combined Text Mining and Systems Biology Approach. *Environ. Health Perspect.* **2019**, *127*, 47005. [CrossRef] [PubMed]
72. Nakayama, S.F.; St-Amand, A.; Pollock, T.; Apel, P.; Bamai, Y.A.; Barr, D.B.; Bessems, J.; Calafat, A.M.; Castaño, A.; Covaci, A.; et al. Interpreting biomonitoring data: Introducing the international human biomonitoring (i-HBM) working group's health-based guidance value (HB2GV) dashboard. *Int. J. Hyg. Environ. Health* **2022**, *247*, 114046. [CrossRef] [PubMed]
73. Hays, S.M.; Aylward, L.L. Interpreting human biomonitoring data in a public health risk context using Biomonitoring Equivalents. *Int. J. Hyg. Environ. Health* **2012**, *215*, 145–148. [CrossRef]
74. Audouze, K.; Sarigiannis, D.; Alonso-Magdalena, P.; Brochot, C.; Casas, M.; Vrijheid, M.; Babin, P.J.; Karakitsios, S.; Coumoul, X.; Barouki, R. Integrative Strategy of Testing Systems for Identification of Endocrine Disruptors Inducing Metabolic Disorders-An Introduction to the OBERON Project. *Int. J. Mol. Sci.* **2020**, *21*, 2988. [CrossRef] [PubMed]
75. O'Brien, K.M.; Upson, K.; Cook, N.R.; Weinberg, C.R. Environmental Chemicals in Urine and Blood: Improving Methods for Creatinine and Lipid Adjustment. *Environ. Health Perspect.* **2016**, *124*, 220–227. [CrossRef] [PubMed]

Disclaimer/Publisher's Note: The statements, opinions and data contained in all publications are solely those of the individual author(s) and contributor(s) and not of MDPI and/or the editor(s). MDPI and/or the editor(s) disclaim responsibility for any injury to people or property resulting from any ideas, methods, instructions or products referred to in the content.

toxics

Review

A Critical Scoping Review of Pesticide Exposure Biomonitoring Studies in Overhead Cultures

Christian Tobias Willenbockel [1], Julia Prinz [1], Stefan Dietrich [2], Philip Marx-Stoelting [1], Cornelia Weikert [2], Tewes Tralau [1] and Lars Niemann [1,*]

1 Department for Pesticide Safety, German Federal Institute for Risk Assessment, Max-Dohrn-Str. 8–10, 10589 Berlin, Germany; christian-tobias.willenbockel@bfr.bund.de (C.T.W.); julia.prinz@bfr.bund.de (J.P.); philip.marx-stoelting@bfr.bund.de (P.M.-S.); tewes.tralau@bfr.bund.de (T.T.)

2 Department for Food Safety, German Federal Institute for Risk Assessment, Max-Dohrn-Str. 8–10, 10589 Berlin, Germany; stefan.dietrich@bfr.bund.de (S.D.); cornelia.weikert@bfr.bund.de (C.W.)

* Correspondence: lars.niemann@bfr.bund.de

Abstract: The exposure of operators, workers, residents and bystanders to pesticides is of high potential concern. Yet, reports on pesticide residues in the environment and near treated fields often spark debates if such findings might indicate a health risk. Although the underlying models are considered conservative, there are only limited field data on systemic exposure available. As a first step to improve the situation, we conducted a scoping review of state-of-the-art pesticide exposure biomonitoring studies in operators, workers, residents or bystanders. In contrast to existing reviews, we focused on target cultures of potential high pesticide exposure such as tree-grown produce, vine or hops. The search was conducted in Web of Science, Scopus and PubMed. Out of 17 eligible articles, a total of 11 studies met our search criteria, and 6 of them quantified the systemic exposure of humans. The analysis revealed that exposure was mainly driven by application of pesticides and reentry work, resulting in a higher exposure of operators and workers than of residents and bystanders. In nearly all cases, the systemic exposure was below the relevant toxicological reference values. The studies were subsequently analyzed to identify key criteria for a reliable design of a biomonitoring study on pesticide exposure.

Keywords: human biomonitoring; pesticides; exposure; operators; workers; residents; bystanders; tree-grown produce; fruits; vine; systemic exposure

Citation: Willenbockel, C.T.; Prinz, J.; Dietrich, S.; Marx-Stoelting, P.; Weikert, C.; Tralau, T.; Niemann, L. A Critical Scoping Review of Pesticide Exposure Biomonitoring Studies in Overhead Cultures. *Toxics* **2022**, *10*, 170. https://doi.org/10.3390/toxics10040170

Academic Editor: Christophe Rousselle

Received: 1 March 2022
Accepted: 24 March 2022
Published: 31 March 2022

Publisher's Note: MDPI stays neutral with regard to jurisdictional claims in published maps and institutional affiliations.

Copyright: © 2022 by the authors. Licensee MDPI, Basel, Switzerland. This article is an open access article distributed under the terms and conditions of the Creative Commons Attribution (CC BY) license (https://creativecommons.org/licenses/by/4.0/).

1. Introduction

In recent decades, there has been growing public concern about the effects of pesticide use on health and the environment. Consequently, there is an increasing number of epidemiological and toxicological studies addressing potential adverse health effects of pesticides in humans under field conditions in which the reliable measurement of exposure is crucial. Unfortunately, this latter point is often not sufficiently fulfilled, particularly in epidemiological studies. This represents a major weakness and is a frequent cause for scientific dispute and criticism [1,2].

Generally, when assessing pesticide exposure in epidemiological studies, several approaches have to be distinguished [3]. The most basic assessment is the use of questionnaires with no temporal correlation with the actual application of the pesticides [1,2]. The application event and resulting exposure are mainly reconstructed via a written procedure or interviews. While procedurally easy to perform and cost effective, this approach, by method, bears the risk of recall bias, unless it is substantiated by further data. This can be data from geographic information systems (GISs) such as information from registries of residents living in close proximity to the fields treated, pesticide use registers or working accounts of operators and workers. Additionally, without additional serum measurements or further substantiation with biomarkers, these approaches are inherently limited as they rely on indirect exposure assessments only. This poses a significant challenge for any further verification and interpretation of the results which, in turn, severely impacts their potential scientific, toxicological or regulatory use.

A promising route towards improvement in the quality, reliability and practical relevance of epidemiological studies is to combine the collection of health outcomes with human biomonitoring (HBM) data for the assessment of real exposure. HBM methods comprise the analytical measurement of active compounds or specific metabolites. Typically, measurements will be conducted in serum or urine [4–6], yet potential matrices range from breast milk through hair and nails to dents. Apart from exposure markers, biological effect markers are also frequently measured. Examples comprise the inhibition of plasma or red blood cell acetylcholinesterase, DNA adduct formation or chromosome anomalies [7]. Depending on the marker, the corresponding measurements allow conclusions regarding exposure to single substances or classes of pesticides. For example, inhibition of plasma or red blood cell acetylcholinesterase will indicate exposure to organophosphorous insecticides as a class rather than a single substance. Crucially, the aim of HBM studies is, however, the quantitation of systemic exposure and not the detection of single exposure events (which might well lie in the past). This so-called "back (or reverse) calculation" currently often relies on physiology-based pharmacological or toxicokinetic models (PBPK/PBTK). The reliability of such calculations increases greatly when kinetic data are available as it is the case, for example, for some pyrethroids [8,9]. The resulting exposure estimates can subsequently be compared to the corresponding pesticides' reference doses. Depending on the scenario, such data can help to retrospectively quantify exposure and evaluate health policies but might also be helpful when trying to substantiate individual case reports or etiologies [10–14]. The combination of HBM and epidemiological tools will certainly increase the reliability of results but is also clearly more demanding in terms of effort, time and costs [3]. Proper assessment of systemic exposure imminently requires that two study criteria be met. First, the biomonitoring sampling has to be timed to the pesticide application because many pesticides, in fact, have a half-life which actually is only in the range of hours [15,16]. Second, repeated sampling is necessary to warrant capturing the respective toxicokinetics [15].

In the present review, we therefore focused on biomonitoring studies which reliably assessed pesticide exposure by means of biomonitoring and, preferably, also quantified the systemic exposure. The latter requirement makes sure that the routes of exposure monitored comprised all three options, that is: (i) inhalative, (ii) oral and (iii) dermal uptake [17]. This is a major advantage as the extent of exposure by each route will obviously differ for residents as compared to operators or workers. Moreover, contrary to previous reviews by Dereumeaux et al. and Teysseire et al. [3,17,18], the focus of the present analysis is on biomonitoring not as an exploratory tool but as a regulatory tool. This includes addressing, for example, the question of observance of reference values.

Two studies were identified as preliminary examples for extensive biomonitoring of pesticide exposure, one being from the Netherlands [16], and the other from the UK [10,19]. Both studies used urine as the primary matrix for the biomonitoring. Apart from the sampling of urine being a non-invasive method, it has the advantage of the respective metabolite measurements being integrative across the various exposure pathways.

From a regulatory point of view, application techniques leading to a potentially high exposure to pesticides are of special concern. Such high (and thus conservative) exposure scenarios are expected particularly for overhead cultures with pesticide application often being performed either by machine-drawn air blast or as a hand-held overhead spray. Examples of such cultures are vine, tree-grown produce, such as apples, or hops. All of these are thus a good starting point for critically assessing pesticide exposure in the field and to draw conclusions for future studies. The further literature search hence was focused on vine, apples and hops with particular attention on studies that used urine or blood as integrative matrices for biomonitoring.

2. Materials and Methods

2.1. PRISMA Extension

Our scoping review was performed according to the PRISMA extension for scoping reviews (PRISMA-ScR) and the accompanying checklist as proposed by Tricco et al. [20]. In Table S1 of this review, we have added an overview providing information on which pages the various "reporting items" from the checklist have been addressed.

2.2. Search Strategy

The literature search was independently performed in the bibliographic databases Web of Science, Scopus and PubMed from 16 to 18 March 2021 by two authors (C.T.W. and J.P.). The construction of the search string followed our research objective and is a specialization of the string of Teysseire et al. [17]. The search string met the demands of our particular research question with its focus on biomonitoring studies with agricultural pesticide application in selected target cultures of potential high exposure (tree-grown produce such as apples, vine or hops). For exposure analysis, we included all groups possibly exposed to pesticides used in agricultural applications in order to assess the safety of pesticides in a regulatory and public health context. These comprised (i) operators, who apply pesticides to plants, (ii) workers, who reenter the crops after pesticide application, (iii) residents, who live nearby fields where pesticides are applied, and (iv) bystanders, who might be exposed accidentally. The inclusion of either urine or serum was mandatory. The search was restricted to articles in English. The construction of the respective search string followed the PICOS principles (participants, interventions, comparisons, outcomes and study design) [3] and read "(proximity or neighborhood) AND (field* OR crop* OR agriculture*) AND (agricultural* OR rural) AND (communit* OR area*) AND (oper* OR operator OR bystander OR appli* OR farm* OR work* OR worker OR workers OR residen* OR resident) AND (urin*OR blood*) AND ((pesticide* OR fungicide* OR herbicide* OR insecticide* OR agrochemical*) AND exposure*) AND (hop OR hops* OR fruit* OR wine* OR vine* OR "tree fruit" OR bushberr* OR orchard* OR pome* OR pomi* OR apple* OR cultivation)".

In this string, the participants are the operators, workers, residents and bystanders. Intervention is the application of pesticides to crops. It should be noted that this comparison has several dimensions: comparison of the same individuals before and after the spraying event, comparison between different individuals and groups and comparison to reference values. The outcome is the (measured or estimated) exposure to pesticides with the target cultures being part of the study design.

2.3. Data Collection and Storage

The search of all databases led to 86 non-duplicate articles. In total, 108 articles were obtained and regarded as potentially relevant. All articles were stored in an Endnote file which was shared amongst the authors together with an Excel file listing the respective bibliographies and abstracts of all studies.

2.4. Selection of Eligible Studies

The eligibility criteria for the selection of studies were based on analyses of existing extensive biomonitoring studies such as Figueiredo et al. [21], Vermeulen et al. [16] and Galea et al. [10,19,22,23]. In short, studies deficient in one or more of the following criteria were excluded:

- No biomonitoring for the assessment of pesticide exposure but sole reliance on a geographic information system, self-reporting or a questionnaire.
- Neither blood nor urine used as a matrix for biomonitoring.
- Non-human studies or methodological studies.
- Reviews.
- Non-observational studies.
- No relationship to pesticide exposure.

- Target cultures other than fruit growing, hops or vine ("wrong target culture").
- Focus on pesticides banned in Germany for more than six years (since 1 January 2015), e.g., chlorpyrifos.
- Measurement of unspecific biomarkers which only mark the presence of a group of pesticides such as dimethylthiophosphate (DMTP) or dialkylphosphate (DAP) for organophosphates. Such unspecific biomarkers were only considered in those few cases where the study design allowed for clear exposure correlations: for example, if only one pesticide was applied on the target crops during the study.

We did not exclude studies which investigated only one group of operators, workers, residents or bystanders. These groups often overlap, and we supposed that information gained out of the investigation of either one of these groups might be relevant for the others.

Assessment was independently performed by four authors (C.T.W., J.P., S.D. and L.N.) by screening titles and abstracts and applying the aforementioned exclusion criteria. Individual assessments were then subjected to within-group discussion. Subsequently, C.T.W. and J.P. conducted a full-text analysis of the selected studies and extracted detailed information on study design, characteristics and results.

3. Results

3.1. Basic Information

The literature search was performed as depicted in Figure 1. A systematic screen of PubMed, Web of Science and Scopus yielded a total of 155 articles, 86 of which were non-duplicates. Another 22 articles of potential interest were identified by screening their reference lists or full texts and, therefore, were considered to have been obtained "from other sources".

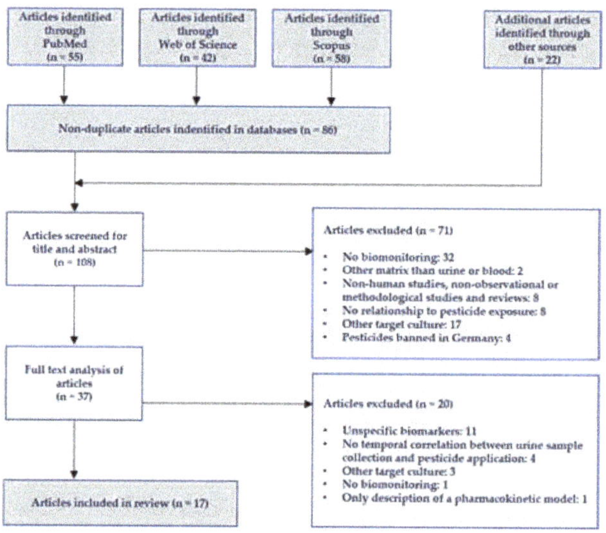

Figure 1. Flowchart of search strategy and study selection with exclusion criteria (based on [20]).

In a stepwise approach, subsequent analysis of titles/abstracts and full texts filtered out a total of 11 studies published in 17 articles which were eventually included in the review (see Tables 1–4). It should be noted that articles referring to the same study cohort (as in participants) are considered as belonging to the same study. This is, for example, the case for Kennedy et al. [24] and Fustinoni et al. [25].

3.1.1. Number of Articles Arranged by Target Cultures

We identified nine articles and six studies with vine as the target culture (Table 2) [24–32]. For tree-grown produce, we found eight articles based on five studies (Table 1) [10,19,23,33–37], two of which referred to apples [35–37]. No reports were found for hops as the target culture. Study data were collected in three regions, namely, North America (three studies [32,35–37]), Europe (seven studies [10,19,22–31,34]) and China (one study [33]), with six of these studies confined to a single region only [24–27,32,33,36,37].

3.1.2. Participants and Sample Size

In most studies, the participants were allocated to group cohorts depending on their (professional) role. Generally, the studies distinguished between operators, reentry workers and residents. Operators [24–27,29–31,33,35] and reentry workers [24–28,32,36,37] were considered in six studies and residents in three [10,19,22,23,33,34]. The corresponding cohort sizes varied significantly, with n = 7–170, 1–22 and 54–403 for operators, reentry workers and residents, respectively. It should be noted that depending on the study design, there can be overlaps with regard to the various groups. For example, in Mercadante et al. [27], operators also performed reentry work. Additionally, the definition of residents varied. All studies considered residents to be persons living in a certain radius of the treated fields. However, depending on the study, the size of the respective radius would range from 50 m to 250 m. Only one study explicitly included bystanders (n = 7) [26], and only two studies had control groups [28,33].

3.1.3. Pesticides Used

As might be expected, pesticide application varied depending on the culture investigated. For tree-grown produce (Table 1), the corresponding active substances covered a range of protective uses and comprised azinphos-methyl (organophosphate insecticide) [36,37], captan (fungicide) [10,19,35], chlormequat (growth regulator) [10,19], chlorpyrifos (insecticide) [10,19], cypermethrin (insecticide) [10,19], deltamethrin (insecticide) [10,19], diquat (herbicide) [10,19], iprodione (fungicide) [10,19], penconazole (fungicide) [10,19,34], pirimicarb (insecticide) [10,19], thiophanate-methyl (fungicide) [10,19,35], benomyl (fungicide) [35] and imidacloprid (insecticide) [33]. Contrastingly, in vine (Table 2), application mainly included fungicides such tebuconazole [24,25], mancozeb [26,28–31] and penconazole [27]. Lopez-Galvez et al. [32] also investigated the insecticides imidacloprid, clothianidin, thiamethoxam, acetamiprid and thiacloprid.

3.2. Sampling Strategies

The study duration ranged from 12 days [28,29] to up to 2 years [10,23,26,27], with all studies relying on urine as the preferred matrix and only two studies additionally sampling serum [19,28]. Measures for urinary metabolites are usually given as concentrations (µg/L, µmol/L, µg/g creatinine or µmol/mol creatinine), with only a few studies also referring to rates (excretion in µg/h [29]). In all cases, care was taken to carefully time sampling to actual substance usage with the participating farmers calling in prior to application. Concomitant measurements of background exposure were performed in seven studies looking at tree-grown produce (n = 4, please see Table 1) [10,19,23,33–35,37] or vine (n = 3, please see Table 2) [24,25,27,30]. Except for Simcox et al. [37], all these studies additionally employed a questionnaire asking the study participants about living conditions and/or possible confounders. Of the remaining studies, three relied on questionnaires only for their background information [26,28,32].

Table 1. Study characteristics of studies on fruit trees as target cultures.

Study	Participants, Sample Size, Location, Year	Culture	Pesticides (Type), Analyzed Metabolites/Pesticides Application Type	Biomonitoring Matrices and Other Samples	Biomonitoring Strategy	Systemic Exposure
Fenske et al., 2003 [36]	**Reentry workers** (apple thinners) ($n = 20$) Three orchards in Washington, DC, USA Year: 1994	Apples (fruit trees)	**Pesticide:** azinphos-methyl (insecticide) **Analyzed metabolites:** dimethylphosphate (DMP), dimethylthiophosphate (DMTP), dimethyldithiophosphate (DMDTP) (specific due to study design) **Application type:** spray	Urine	**Sampling timed to pesticide application:** yes **Background and repeated measurements:** no **24 h urine analyzed:** no **Strategy for urine measurements:** urine spot sample was assumed to represent steady state **Conversion factors determined:** no **Dermal exposure assessed:** no **Environmental measurements:** yes (collection of foliar samples of leaves, see Simcox et al., 1999 [37]) **Study duration:** six-week thinning season **Control group:** no **Personal protective equipment considered:** yes (Simcox et al., 1999 [37]) **Questionnaire/Confounders/Medical pre-examinations considered:** no	**Systemic exposure assessed:** yes, based on known conversion factors and assumptions of urine volume **Conversion factors used:** yes **Correlation between urine and dermal measurements:** no
Simcox et al., 1999 [37]	See Fenske et al., 2003 [36]	See Fenske et al., 2003 [36]	See Fenske et al., 2003 [36]	Urine, serum	**Sampling timed to pesticide application:** yes **Background and repeated measurements:** yes **24 h urine analyzed:** no **Strategy for urine measurements:** spot urine samples were collected daily at the end of the shift for the duration of the whole study, and collection continued one week after end of thinning season (reentry work) **Conversion factors determined:** no **Dermal exposure assessed:** no **Environmental measurements:** no **Study duration:** six-week thinning season **Control group:** no **Personal protective equipment considered:** no (long pants, long-sleeved shirts, cap, work boots or tennis shoes) **Questionnaire/Confounders:** no	Assessed in Fenske et al., 2003 [36]

Table 1. *Cont.*

Study	Participants, Sample Size, Location, Year	Culture	Pesticides (Type), Analyzed Metabolites/Pesticides Application Type	Biomonitoring Matrices and Other Samples	Biomonitoring Strategy	Systemic Exposure
Galea et al., 2011 [23]	**Residents** living within 100 m of the edge of a field ($n \geq$ 130 adults and 65 children) East Lothian, Kent and Norfolk (UK) Years: 2011 and 2012	Fruit trees, arable crop	**Pesticides:** captan (fungicide), chlormequat (growth regulator), chlorpyrifos (insecticide), cypermethrin (insecticide), deltamethrin (insecticide), diquat (herbicide), iprodione (fungicide), penconazole (fungicide), pirimicarb (insecticide), thiophanate-methyl (fungicide) **Analyzed metabolites/pesticides *:** not specified **Application type:** spray	Urine	**Sampling timed to pesticide application:** yes **Background and repeated measurements:** yes **24 h urine analyzed:** yes **Strategy for urine measurements:** urine samples collected within two days after spray event; background samples collected within and outside the spray season **Conversion factors determined:** no **Dermal exposure assessed:** no **Environmental measurements:** no **Study duration:** two years **Control group:** no **Personal protective equipment considered:** no **Questionnaire/Confounders/Medical pre-examinations considered:** yes	**Systemic exposure assessed:** no (for systemic exposure, see Galea et al., 2015 [10]) **Conversion factors used:** no **Correlation between urine and dermal measurements:** no
Galea et al., 2015 [19]	See Galea et al., 2011 [23]	Fruit trees, arable crop	**Pesticides:** captan (fungicide), chlormequat (growth regulator), chlorpyrifos (insecticide), cypermethrin (insecticide) **Analyzed metabolites/pesticides *:** cis-1,2,3,6-tetrahydrophthalimide (THPI), chlormequat *, 3,5,6-trichlorpyridinol (TCP), cis- and trans-2,2-dichlorovinyl-3,3-dimethylcyclopropane-1-carboxylic acid (DCVA) (specific) **Application type:** spray	Urine	See Galea et al., 2011 [23]	**Systemic exposure assessed:** no (for systemic exposure, see Galea et al., 2015 [10]) **Conversion factors used:** no (for conversion of urinary concentrations, a pharmacokinetic model was applied) **Correlation between urine and dermal measurements:** no

Table 1. *Cont.*

Study	Participants, Sample Size, Location, Year	Culture	Pesticides (Type), Analyzed Metabolites/Pesticides Application Type	Biomonitoring Matrices and Other Samples	Biomonitoring Strategy	Systemic Exposure
Galea et al., 2015 [10]	See Galea et al., 2011 [23]	Fruit trees, arable crop	See Galea et al., 2015 [19]	Urine	See Galea et al., 2011 [23]	**Systemic exposure assessed:** yes, systemic exposure estimated via regulatory exposure models (Europoem or REA) due to spray information (amount of pesticide sprayed, etc.). **Conversion factors used:** no (for conversion of urinary concentrations, a pharmacokinetic model was applied) **Correlation between urine and dermal measurements:** no
Hines et al., 2008 [35]	Private **operators** (*n* = 74 (73 men, 1 woman)) Iowa, NC, USA Years: 2002 and 2003	Fruit trees (apples and/or peaches)	**Pesticides:** captan (fungicide), thiophanate-methyl (fungicide), benomyl (fungicide) **Analyzed metabolite:** cis-1,2,3,6-tetrahydrophthalimide (THPI) (specific) **Application type:** air blast, hand spray	Urine, dermal exposure, air	**Sampling timed to pesticide application:** yes **Background and repeated measurements:** yes **24 h urine analyzed:** yes **Strategy for urine measurements:** first morning urine one day before pesticide application and 24 h urine after application **Conversion factors determined:** no **Dermal exposure assessed:** yes, via patches at ten different spots on clothes or skin **Environmental measurements:** air within breathing zone **Study duration:** two days for each participant (at least seven days apart) **Control group:** no **Personal protective equipment considered:** yes (input for the AHS algorithm) **Questionnaire/Confounders considered:** yes	**Systemic exposure assessed:** no **Conversion factors used:** no **Correlation between urine and dermal measurements:** yes (significant)

Table 1. *Cont.*

Study	Participants, Sample Size, Location, Year	Culture	Pesticides (Type), Analyzed Metabolites/Pesticides Application Type	Biomonitoring Matrices and Other Samples	Biomonitoring Strategy	Systemic Exposure
Sams et al., 2016 [34]	**Residents** living within 100 m of the edge of a field (n = 48 adults and 6 children) Kent (UK) Years: 2011 and 2012	Fruit trees	**Pesticide:** penconazole (fungicide) **Analyzed metabolites:** penconazole-OH, penconazole-COOH (specific) **Application type:** spray	Urine	See Galea et al., 2011 [23] **Conversion factors determined:** yes, based on a volunteer study (single oral dose of penconazole at the ADI)	**Systemic exposure assessed:** no **Conversion factors used:** yes, determined within a volunteer study **Correlation between urine and dermal measurements:** no
Tao et al., 2019 [33]	**Residents/operators** (n = 119), operators' family members (n = 156), urban **control group** (n = 42) **Children** (rural children (n = 247), urban **control group** (n = 53)) Henan Province, China Year: 2017	Fruit trees	**Pesticide:** imidacloprid (IMI) (insecticide) **Analyzed metabolites/pesticides *:** 6-chloronicotinic acid (6-CNA) and imidacloprid * resulting in ∑IMI (specific due to study design) **Application type:** spray	Urine	**Sampling timed to pesticide application:** yes **Background and repeated measurements:** yes **24 h urine analyzed:** no **Strategy for urine measurements:** six spot urine samples per participant (first morning urine 1 d before spray event and 1 d, 2 d, 3 d, 5 d and 7 d after spray event); one first morning urine sample from each control **Conversion factors determined:** no **Dermal exposure assessed:** no **Environmental measurements:** no **Study duration:** March to June 2017 **Control group:** yes **Personal protective equipment considered:** yes **Questionnaire/Confounders considered:** yes	**Systemic exposure assessed:** no **Conversion factors used:** no **Correlation between urine and dermal measurements:** no

Table 2. Study characteristics of studies on vine as the target culture.

Study	Participants, Sample Size, Location, Year	Culture	Pesticides (type), Analyzed Metabolites/Pesticides Application Type	Biomonitoring Matrices and Other Samples	Biomonitoring Strategy	Systemic Exposure
Fustinoni et al., 2014 [25]	**Operators/reentry workers** ($n = 7$) Piedemont, Monferrato region, Italy Year: 2011	Vine	**Pesticide:** tebuconazole (TEB) (fungicide) **Analyzed metabolites:** TEB-OH, TEB-COOH (specific) **Application type:** tractor-mounted air blast or spraying upward with hand-held application equipment	Urine, dermal exposure	**Sampling timed to pesticide application:** yes **Background and repeated measurements:** yes **24 h urine analyzed:** yes **Strategy for urine measurements:** workers collected urine 24 h before application and mostly 48 h after last shift **Conversion factors determined:** no **Dermal exposure assessed:** yes, hand exposure was assessed by collecting hand washing liquids during work for 24 h after exposure **Environmental measurements:** no **Study duration:** 12 working days **Control group:** no **Personal protective equipment considered:** yes **Questionnaire/Confounders considered:** yes	**Systemic exposure assessed:** no (for systemic exposure, see Kennedy et al., 2015 [24]) **Conversion factors used:** no **Correlation between urine and dermal measurements:** yes (significant)
Kennedy et al., 2015 [24]	See Fustinoni et al., 2014 [25]	Vine	See Fustinoni et al., 2014 [25]	See Fustinoni et al., 2014 [25]	See Fustinoni et al., 2014 [25]	**Systemic exposure assessed:** yes, assessed based on measurements of actual dermal exposure in Fustinoni et al., 2014 [25] **Conversion factors used:** no **Correlation between urine and dermal measurements:** yes (significant) (see also results from Fustinoni et al., 2014 [25])

Table 2. *Cont.*

Study	Participants, Sample Size, Location, Year	Culture	Pesticides (type), Analyzed Metabolites/Pesticides Application Type	Biomonitoring Matrices and Other Samples	Biomonitoring Strategy	Systemic Exposure
Lopez-Galvez et al., 2020 [32]	Male migrant **reentry workers** (*n* = 20) (neither involved in applying nor mixing pesticides) Sonora, Mexico Year: 2016	Vine	**Pesticides:** imidacloprid (insecticide), clothianidin (insecticide), thiamethoxam (insecticide), acetamiprid (insecticide), thiacloprid (insecticide) **Analyzed metabolites/pesticides *:** 5-hydroxy-imidacloprid (5-OH-IMI), imidacloprid *, clothianidin *, acetamiprid-N-desmethyl, acetamiprid *, thiacloprid * (specific) **Application type:** drip irrigation	Urine, dermal exposure, air	**Sampling timed to pesticide application:** yes **Background and repeated measurements:** no **24 h urine analyzed:** no **Strategy for urine measurements:** one urine sample per participant (first morning urine five days after pesticide application) **Conversion factors determined:** no **Dermal exposure assessed:** yes, assessed using hand wipes **Environmental measurements:** yes, air measurements within breathing zone of workers **Study duration:** winter and summer seasons in 2016 **Control group:** no **Personal protective equipment considered:** yes **Questionnaire/Confounders considered:** yes	**Systemic exposure assessed:** no **Conversion factors used:** no **Correlation between urine and dermal measurements:** yes (significant)
Mandic-Rajcevic et al., 2018 [30]	Male, healthy right-handed **operators** (mixing, application and equipment maintenance work) (*n* = 29) Mantova and Pavia Provinces of the region of Lombardy (Northern Italy) Year: 2011	Vine	**Pesticide:** mancozeb (fungicide) **Analyzed metabolite:** ethylene-bis-thiourea (ETU) **Application type:** closed and filtered tractors (*n* = 29) vs. open tractors (*n* = 9)	Urine, dermal exposure	**Sampling timed to pesticide application:** yes **Background and repeated measurements:** yes **24 h urine analyzed:** yes **Strategy for urine sampling:** 24 h pre-exposure and 24 h post-exposure urine **Conversion factors determined:** no **Dermal exposure assessed:** yes, external pads on clothes for potential exposure measurements and internal pads on skin for actual exposure measurements (modified OECD patch methodology), collection of hand washing liquid (24 h post-exposure) **Environmental measurements:** no **Study duration:** 38 working days in April to July 2011 **Control group:** no **Personal protective equipment considered:** yes, assessment of effect of coveralls and gloves on exposure **Questionnaire/Confounders considered:** yes	**Systemic exposure assessed:** no (systemic exposure and comparison with reference values stated in Mandic-Rajcevic et al., 2019 [29]) **Conversion factors used:** no **Correlation between urine and dermal measurements:** yes (significant)

Table 2. *Cont.*

Study	Participants, Sample Size, Location, Year	Culture	Pesticides (type), Analyzed Metabolites/Pesticides Application Type	Biomonitoring Matrices and Other Samples	Biomonitoring Strategy	Systemic Exposure
Mandic-Rajcevic et al., 2019 [29]	See Mandic-Rajcevic et al., 2018 [30]	Vine	See Mandic-Rajcevic et al., 2018 [30]	See Mandic-Rajcevic et al., 2018 [30]	See Mandic-Rajcevic et al., 2018 [30]	**Systemic exposure assessed:** Yes, systemic exposure assessed via patch measurements on clothes (potential exposure) and skin (actual exposure). **New method for data analysis accounting for duration of exposure compared to establish fixed fractional approach which assumes a standard working day of 8 h.** **Conversion factors used:** no **Correlation between urine and dermal measurements:** yes (significant) (see also results from Mandic-Rajcevic et al., 2018 [30])
Mandic-Rajcevic et al., 2020 [31]	Male **operators** (n = 16) (subgroup from Mandic-Rajcevic et al., 2018 [30]) Mantova and Pavia Provinces of the region of Lombardy (Northern Italy) Year: 2011	Vine	See Mandic-Rajcevic et al., 2018 [30] However, smaller sample size: closed and filtered tractors (n = 11), open tractors (n = 5)	See Mandic-Rajcevic et al., 2018 [30]	See Mandic-Rajcevic et al., 2018 [30]	**Systemic exposure assessed:** yes, systemic exposure assessed via patch measurements on clothes (potential exposure) and skin (actual exposure) **Conversion factors used:** no **Correlation between urine and dermal measurements:** yes (significant) (see also results from Mandic-Rajcevic et al., 2018 [30])

Table 2. *Cont.*

Study	Participants, Sample Size, Location, Year	Culture	Pesticides (type), Analyzed Metabolites/Pesticides Application Type	Biomonitoring Matrices and Other Samples	Biomonitoring Strategy	Systemic Exposure
Medda et al., 2017 [28]	Chianti (iodine-deficient growing area): Male **workers** (n = 29) Male **controls** (n = 24) Bolzano: Male **workers** (n = 148) Male **controls** (n = 40) n = 170 workers involved in mixture and application, n = 7 workers in reentry work vineyards in Chianti and Bolzano areas, Italy Year: 2017	Vine	**Pesticides:** mancozeb (fungicide) **Analyzed metabolite:** ethylene-bis-thiourea (ETU) (specific for EBDC fungicides and due to study design) **Application type:** no	Urine (ETU and urinary iodine concentration (UIC)), serum (iodine biomarkers for health assessment of thyroid (tyroglobulin (Tg), (free) tri-iodothyoidine (T3,FT3), free thyroxine (T4, FT4)), thyroid volume assessed by ultrasonography	**Sampling timed to pesticide application:** yes **Background and repeated measurements:** no **24 h urine analyzed:** no **Strategy for urine measurements:** spot urine samples collected the day after the treatment for operators/mixers and one day after the reentry in culture for reentry workers **Conversion factors determined:** no **Dermal exposure assessed:** no **Environmental measurements:** no **Study duration:** June to September 2017 **Control group:** yes **Personal protective equipment considered:** yes **Questionnaire/Confounders/Medical pre-examinations considered:** yes **Other:** blood samples for serum analysis for thyroid health effects collected after six weeks from the last treatment in October (time of grape harvest)	**Systemic exposure assessed:** no **Conversion factors used:** no **Correlation between urine and dermal measurements:** no
Mercadante et al., 2019 [27]	**Operators/reentry workers** (n = 22), different regions Lombardy, Italy Year: 2012	Vine	**Pesticide:** penconazole (fungicide) **Analyzed metabolites:** PEN-OH, PEN-COOH (specific) **Application type:** sideways spraying tractor-mounted air blast or spraying upwards with hand-held application equipment	Urine, dermal exposure	**Sampling timed to pesticide application:** yes **Background and repeated measurements:** yes **24 h urine analyzed:** yes **Strategy for urine measurements:** 24 h before and 48 h after last shift, 42 mixing and applications and 12 reentries monitored **Conversion factors determined:** no **Dermal exposure assessed:** yes, assessment of potential and actual body exposure via pads **Environmental measurements:** no **Study duration:** May to July 2012 **Control group:** no **Personal protective equipment considered:** yes **Questionnaire/Confounders considered:** yes	**Systemic exposure assessed:** no **Conversion factors used:** no **Correlation between urine and dermal measurements:** yes (significant)

Table 2. *Cont.*

Study	Participants, Sample Size, Location, Year	Culture	Pesticides (type), Analyzed Metabolites/Pesticides Application Type	Biomonitoring Matrices and Other Samples	Biomonitoring Strategy	Systemic Exposure
Sleeuwenhoek et al., 2007 [26]	**Operators** (n = 8), **reentry workers** (n = 1), **bystanders** (n = 7) UK Year: 2004	vine (potatoes)	**Pesticide:** mancozeb (fungicide), cypermethrin (for potatoes)[+] (insecticide) **Analyzed metabolite:** ethylenethiourea (ETU) (specific for EBDTC fungicides) **Application type:** hand-held, air-assisted, boom	Urine	**Sampling timed to pesticide application:** yes **Background and repeated measurements:** no **24 h urine analyzed:** no **Strategy for urine measurements:** one urine sample per participant (first morning urine after last exposure of the week) **Conversion factors determined:** no **Dermal exposure assessed:** yes, estimated based on EUROPOEM database and the regulatory risk assessment process **Environmental measurements:** no **Study duration:** one year **Control group:** no **Personal protective equipment considered:** yes **Questionnaire/Confounders/Medical pre-examinations considered:** yes **Other:** inhalative and oral exposure routes were assumed to be negligible	**Systemic exposure assessed:** yes, systemic exposure estimated via regulatory exposure models (EUROPOEM and REA) due to spray information (amount of pesticide sprayed, etc.) **Conversion factors used:** no **Correlation between urine and dermal measurements:** no (but estimates for dermal exposure used to predict urinary concentrations)

Table 3. Results of studies on fruit trees as target cultures.

Study	Exposure Measurements	Conclusion/Results	Critique
Fenske et al., 2003 [36]	**Units of urine measurements:** μmol/L **Comparison to reference values:** reference values mostly observed; 2.4% of doses exceeded California EPA reference value for reentry > 14 days (geometric mean of 19 μg/kg and day), but 27% of doses exceeded reference value for reentry < 14 days (geometric mean 42 μg/kg and day). **Results of personal protective equipment:** PPE not considered	**Results of exposure:** Exposure after application higher than before when measured in μg/L urinary excretion and dermal exposure. TEB-OH metabolite peaked within 24 h after application. None of the doses exceeded US EPA guidance value of 560 μg/kg and day, but in 2.4 (reentry > 14 days) to 27% (reentry < 14 days) of all cases, there was transgression for reference value of CAL US EPA of 76 μg/kg and day. Comparison of reentry periods of more or less than 14 days. Bioconversion factors for azinphos-methyl applied for estimation of systemic exposure. Exposure was found to depend on reentry timing. Exposure was lower for reentry after more than 14 days. **Health-related outcomes:** not assessed	No determination of background concentrations prior to pesticide application (important parameter for the study design). No repeated measurements subsequent to exposure. Investigation of only a single sample. Systemic exposure was assessed by assuming the excreted urine volume since the beginning of the pesticide exposure. No residents and operators considered (only reentry workers). No control group. Only azinphos-methyl investigated.

Table 3. *Cont.*

Study	Exposure Measurements	Conclusion/Results	Critique
Simcox et al., 1999 [37]	See Fenske et al., 2003 [36]	See Fenske et al., 2003 [36] **Health-related outcomes:** cholinesterase (ChE) activity monitored, no adverse outcomes noticed	See Fenske et al., 2003 [36]
Galea et al., 2011 [23]	Results presented in Galea et al., 2015a [19], Galea et al., 2015b [10] and Sams et al., 2016 [34]	Description of study protocol. Results are presented in Galea et al., 2015a [19], Galea et al., 2015b [10], Galea et al., 2017 [22] and Sams et al., 2016 [34]. **Health-related outcomes:** not assessed	See Galea et al., 2015 [19] and Galea et al., 2015 [10]
Galea et al., 2015 [19]	**Units of urine measurements:** μg/L and μg/g creatinine **Comparison to reference values:** no **Results of personal protective equipment:** PPE not considered	Very low biomarker concentrations for captan and cypermethrin were found (approximately 90% of the samples < LOD). **Results of exposure: For chlorpyrifos and chlormequat, no significant in-crease in urinary biomarker concentrations upon spray event were observed.** Urinary biomarker concentrations for chlormequat were found to be higher within compared to outside the spray season. **Health-related outcomes:** not assessed	Systemic exposure not assessed. Only residents considered, no operators or reentry workers. No control group. No comparison to reference values.
Galea et al., 2015 [10]	**Units of urine measurements:** μg/L and μg/g creatinine **Comparison to reference values:** yes, AOEL was not reached for any of the pesticides **Results of personal protective equipment:** PPE not considered Blindly estimated urinary excretion of metabolites was compared to actual measurements from Galea et al., 2015a [19] and Sams et al., 2016 [34]	**Results of exposure: Very low effect of spray event on the overall pesticide exposure of residents.** Predictive values based on the REA-PK model were found to be sufficiently conservative. **Health-related outcomes:** not assessed	Systemic exposure only indirectly assessed via exposure models. No assessment of the systemic exposure on the basis of urine measurements and conversion factors. Only residents considered, no operators or reentry workers. No control group.
Hines et al., 2008 [35]	**Units of urine measurements:** μg/L and μg/g creatinine, excretion rates in μg/h **Comparison to reference values:** no **Results of personal protective equipment:** reduction factor for PPE as input for the AHS algorithm	**Results of exposure:** Generally, captan and THPI were most frequently detected when pesticides were applied via air blast. Highest concentrations of THPI in urine were found on the morning after pesticide application. Significant correlations between internal (urine) and external (air, hand rinse, dermal patches) exposure measurements were found (**strongest correlation between captan concentrations in hand rinse samples and THPI concentrations in urine**). Highest dermal exposure was found on hands, forearms and thighs. AHS algorithm was significantly and marginally predictive of thigh and forearm exposures, but did not predict air, hand rinse or urinary THPI exposures. **Health-related outcomes:** not assessed	**Systemic exposure not assessed.** Since the private application of pesticides was investigated, a proper separation between operators and residents is not possible. No control group. No comparison to reference values.

Table 3. *Cont.*

Study	Exposure Measurements	Conclusion/Results	Critique
Sams et al., 2016 [34]	**Units of urine measurements:** µmol/mol creatinine **Comparison to reference values:** yes, highest urinary concentration was approximately 100 x lower than the peak excretion after an oral dose at the ADI **Results of personal protective equipment:** PPE not considered	Penconazole-OH and penconazole-COOH suitable as urinary biomarkers to assess systemic exposure of penconazole. **Results of exposure:** Very low biomarker concentrations for penconazole were found (>80% of the samples < LOD). Results used to estimate systemic exposure of penconazole within a study from Fustinoni et al., 2015 [38]. **Health-related outcomes:** not assessed	**Systemic exposure not assessed.** Only residents living in close proximity to target fields were considered. No operators and/or reentry workers considered. No control group. Human volunteer studies of this type in order to determine conversion factors are not allowed in Germany.
Tao et al., 2019 [33]	**Units of urine measurements:** ng/mL and µg/g creatinine **Comparison to reference values:** no **Results of personal protective equipment:** personal protective measures of operators were found to be insufficient	**Results of exposure:** Target metabolites were found in 100% of the urine samples. Increase in IMI exposure in rural residents after pesticide application, with significantly higher concentrations for operators. **Highest IMI concentrations in rural residents were found 2 d after spray event.** Significant impacts of diet, sex, age and region on exposure to IMI observed. **Health-related outcomes:** not assessed	Systemic exposure not assessed. No comparison to reference values. Not clear whether apple cultivation was considered (only "orchards" in general mentioned).

Table 4. Results of studies on vine as the target culture.

Study	Exposure Measurements	Conclusion/Results	Critique
Fustinoni et al., 2014 [25]	**Units of urine measurements:** µg/L and µg/g creatinine, excretion rates in µg/h **Comparison to reference values:** reference values assessed in Kennedy et al., 2015 [24] **Results of personal protective equipment:** Non-uniform PPE and different efficacy might be two of the factors for the observed range of excretion rates for TEB-OH and TEB-COOH. Median protection factor of 98% provided by wearing overalls.	**Results of exposure:** Exposure after application higher than before when measured in µg/L urinary excretion and dermal exposure. TEB-OH metabolite peaked within 24 h after application. Correlation of total dermal exposure measurements and post-application 24 h urinary biomarker measurements of TEB- OH and TEB-COOH was significant, $r = 0.756$ and $r = 0.577$. Possibly suitable approach for future studies. Very small number of participants ($n = 7$) noted as major limit. Hands accounted for 17 to 86% of actual skin exposure. **Health-related outcomes:** not assessed	Systemic exposure not stated (see Kennedy et al., 2015 [24]). In Kennedy et al., 2015 [24], systemic exposure only determined via dermal exposure and model calculations, but not directly on the basis of urine measurements. No use of conversion factors. Very small sample size ($n = 7$). No residents. No control group.

Table 4. *Cont.*

Study	Exposure Measurements	Conclusion/Results	Critique
Kennedy et al., 2015 [24]	**Units of urine measurements:** see Fustinoni et al., 2014 [25] **Comparison to reference values:** Reference values assessed by dermal exposure and model predictions of ACROPOLIS model. Reference values observed in both cases. Applied model gives good prediction of exposure based on urinary measurements. **Results of personal protective equipment:** higher actual dermal exposure predicted if pesticide is applied without gloves	**Results of exposure:** Used study data from Fustinoni et al., 2014 [25]. Urinary measurements were compared with prediction from exposure model for all sources of exposure (dietary and non-dietary). Systemic exposure was measured based on dermal exposure measurements. Correlation with urinary measurements was assessed. Model predictions of systemic exposure seemed to be reliable, but very small (n = 7) sample size. **Health-related outcomes:** not assessed	Systemic exposure only determined via dermal exposure and model calculations, but not directly on the basis of urine measurements. No use of conversion factors. Very small sample size (n = 7). No residents. No control group.
Lopez-Galvez et al., 2020 [32]	**Units of urine measurements:** µg/L and µg/g creatinine **Comparison to reference values:** no **Results of personal protective equipment:** training on PPE usage was found to significantly reduce IMI concentrations in hand wipes	**Results of exposure:** Imidacloprid was most frequently detected among all neonicotinoid biomarkers (in 95% of the urine samples). **Strong correlation between imidacloprid concentrations measured in hand wipes and urinary 5-OH-IMI.** Hand wipes stated as a possible alternative to urinary biomonitoring. Concentrations of imidacloprid in air < LOD. Concentrations of urinary 5-OH-IMI significantly higher in summer. **Health-related outcomes:** not assessed	No background or repeated measurements. Systemic exposure not assessed. No control group. No comparison to reference values.
Mandic-Rajcevic et al., 2018 [30]	**Units of urine measurements:** µg/L and µg/g creatinine **Comparison to reference values:** for reference values, see Mandic-Rajcevic et al., 2019 [29] **Results of personal protective equipment:** Coveralls reduced skin exposure by 4 times (open tractors) and 10 times (closed tractors). Gloves led to 10 times lower hand exposure during application with open tractors. Gloves led to an increase in exposure when closed tractors were used, due to suspected transport of contaminated gloves in tractor cabins.	**Results of exposure:** Comparison of individual levels pre- and post-exposure dependent on if the tractor was open or closed during application. ETU level post-exposure significantly higher than pre-exposure ($p < 0.001$). Absolute levels higher in most individual comparisons. Statistically significant positive correlation between total skin exposure and ETU levels ($r = 0.55$, $p < 0.001$). Dermal exposure contributed more than 90% of total skin dose. ETU is a suitable biomarker for occupational exposure to mancozeb, but urinary measurement values cannot be assessed directly because of a lack of biological exposure limits. **Further studies concerning use of correlation between total skin exposure and urine biomonitoring advocated.** **Health-related outcomes:** not assessed	Systemic exposure determined via dermal measurements. In a second step, comparison/correlation with urine measurements. No direct assessment of the systemic exposure on the basis of urine measurements and conversion factors. Small sample size (n = 29), only operators (only men). No residents. No control group.

Table 4. *Cont.*

Study	Exposure Measurements	Conclusion/Results	Critique
Mandic-Rajcevic et al., 2019 [29]	**Units of urine measurements:** µg/L and µg/g creatinine **Comparison to reference values:** yes, even less conservative methods give estimate of several hundred times below AOEL reference value of 0.02 mg/kg bw **Results of personal protective equipment:** see Mandic-Rajcevic et al., 2018 [30]	**Results of exposure:** Median pre-exposure 24 h ETU urine 0.93 and 0.51 µg/g creatine for open and closed tractors. Median post-exposure 24 h ETU urine 1.83 and 1.22 µg/g creatine for open and closed tractors. Use of new calculation method for calculation of systemic exposure accounting for duration of exposure yields a reduced dose of 50%, 81% and 80% for body, hands and total absorbed dose when compared to established fixed fractional approach. Systemic exposure assessment dependent on chosen approach. New model yielded better correlation of dermal pad methodology for total body dose and urine measurements post-exposure. It was noted that hand exposure contributed 97 % to total skin exposure but the correlation with post-exposure urine measurements of free ETU for dermal hand exposure (ρ = 0.41) (time-adjusted) was lower than for the dermal body exposure (ρ = 0.58) (fixed time) and total dose (ρ = 0.51) (fixed time). Correlation of hand exposure and free ETU urine levels improved by duration-adjusted method. **Health protection recommendations:** regular hand washing might considerably lower the absorbed dose via skin	See Mandic-Rajcevic et al., 2018 [30].
Mandic-Rajcevic et al., 2020 [31]	**Units of urine measurements:** µg/L **Comparison to reference values:** yes, exposure of highest exposed operator was 1000 times lower compared to the AOEL of 0.035 mg/kg for mancozeb **Results of personal protective equipment:** see Mandic-Rajcevic et al., 2018 [30]	**Results of exposure:** Mean absorbed dose (dermal exposure) was 0.9 ng/kg (from body exposure: 0.1 ng/kg; from hand exposure: 0.6 ng/kg). Estimation of EBEL (equivalent biologic exposure value) for mancozeb by combining the results for dermal exposure (see Mandic-Rajceciv et al., 2018 [30]) and consideration of the AOEL for mancozeb as guidance value. Approach resulted in an EBEL of 0.15 mg (free ETU in urine) and 0.7 mg (total ETU in urine) for mancozeb. Exposure of the highest exposed operator was at the level of 20% of the EBEL; the values for all the other operators were less than or equal to 3% of the EBEL. In future, the concept of EBEL might be applied as a screening method in order to estimate the risk of pesticide operators. **Health-related outcomes:** not assessed	See Mandic-Rajcevic et al., 2018 [30]

Table 4. *Cont.*

Study	Exposure Measurements	Conclusion/Results	Critique
Medda et al., 2017 [28]	**Units of urine measurements:** µg/L **Comparison to reference values:** no **Results of personal protective equipment:** frequency of use of personal protective equipment in Bolzano workers higher (97.2%) than in Chianti workers (85.2%) (significant, $p = 0.01$)	Mild thyroid-disrupting effect due to mancozeb exposure. Higher thyroid health effects in workers in areas with an iodine deficit (Chianti) than in areas (Bolzano) with an established program for iodine supplementation. **Health-related outcomes:** Lower mean FT4 iodine serum levels in exposed workers in iodine-deficient Chianti area. Increased iodine urinary excretion of >250 µg/L more frequently in higher exposed workers (>20 µg/L ETU) than in less exposed workers. This effect was stronger in Chianti (iodine deficit) than in Bolzano (iodine sufficient) area. Workers in an area with iodine deficit had lower thyroid volumes (≤6 mL) than the respective control groups.	No background or repeated measurements. Systemic exposure not assessed. No comparison to reference values.
Mercadante et al., 2019 [27]	**Units of urine measurements:** µg/L **Comparison to reference values:** no **Results of personal protective equipment:** potential body exposure without clothing and actual body exposure compared, clothing provided good protection (1/100 to 1/1000 lower than potential exposure)	**Results of exposure:** Measurements of exposure ranging from 15.6 to 27.6 µg/L for PEN-OH and 2.5 to 10.2 µg/L for PEN-COOH. Excretion rate of PEN-OH had a peak within 24 h post-exposure. PEN-OH could possibly be used for biomonitoring on a regular basis. Hand exposure of reentry workers was found to be a major factor for overall exposure. Correlation found between total dermal exposure and urinary excretion of metabolites. Concentration of PEN-OH 24 h and 24 to 48 h after work shift correlated with actual body and total dermal exposure ($0.279 \leq r \leq 0.562$). Described a method for assessing if participants provided all urine samples based on creatine levels. **Health-related outcomes:** not assessed	Systemic exposure not assessed. Very small sample size ($n = 22$). No residents. No control group. No comparison to reference values.

Table 4. *Cont.*

Study	Exposure Measurements	Conclusion/Results	Critique
Sleeuwenhoek et al., 2007 [26]	**Units of urine measurements:** µg/L and µg/g creatinine **Comparison to reference values:** yes, measured urine biomarker concentrations indicate that the ADI is not reached **Results of personal protective equipment:** no protective equipment assumed for bystanders in modeling	Urinary ETU concentrations were highest for sprayers. Median ETU concentrations for post-application workers and bystanders were < LOD. **Overall, predicted urinary concentrations were higher than the observed ones. Estimated values based on regulatory risk assessment were found to be sufficiently conservative.** Pre-exposures of mancozeb have to be taken into account due to its long half-life of 100 h. Results possibly not representative due to small sample size. **Results of exposure:** Measurements of exposure ranging from 15.6 to 27.6 µg/L for PEN-OH and 2.5 to 10.2 µg/L for PEN-COOH. Excretion rate of PEN-OH had a peak within 24 h post-exposure. The results for cypermethrin are not discussed here, since different target cultures were covered. **Health-related outcomes:** not assessed	No background or repeated measurements. Systemic exposure only estimated by using a model calculation including the amount of pesticides. No direct assessment of the systemic exposure on the basis of urine measurements. No use of conversion factors. Small sample size: 7 bystanders, 8 sprayers, 1 worker. No control group.

The studies measuring background exposure showed considerable variation with regard to the respective sampling schemes. While Hines et al. [35] and Tao et al. [33] collected the first morning urine on the day before active substance application, Simcox et al. [37] relied on baseline measurements over a period of at least two weeks prior to the first application. In addition, the latter also collected reference samples from the general, non-agricultural population. Contrastingly, Galea et al. and Sams et al. collected up to three samples during the spraying season which were not correlated with an application event, and up to three additional samples outside the spraying season [10,19,23,34].

For the subsequent sampling of actual exposure, the studies either relied on spot sampling or continuous sampling. Simcox et al. [37] collected daily spot urine samples at the end of each work shift. Sampling was commenced for another week after application and occupational exposure in the thinning season. Other studies used less systematic sampling regimes such as Galea et al. and Sams et al., who collected urine samples within two days after the spray event [10,19,23,34], or Tao et al. [33], who relied on spot urine of the first, second, third, fifth and seventh days after the substance application. Meanwhile, Lopez-Galvez et al. [32] collected morning urine on the first five days after application. Comparable sampling strategies were followed by Medda et al. [28], who collected spot urine samples from operators and mixers the day after substance treatment and from reentry workers the day after starting reentry work. Likewise, Sleeuwenhoek et al. [26] collected one sample per participant following the last exposure of the week. In contrast, several other studies [24,25,27,30,35,37] collected the complete 24 h urine after substance application, with some studies extending the sampling period to include the 24 h before [25,27,30] and 48 h thereafter [27,38]. Notably, in the three latter studies, the amount of urinary metabolites correlated best with the respective dermal exposure assessments.

3.3. *Exposure*

3.3.1. Routes of Exposure

There are three major routes of exposure. These are: (i) the inhalative route, (ii) the dermal route and (iii) the oral route. The individual contributions of these routes to the overall exposure will obviously vary for the different cohorts monitored. For example, an operator inhaling vapors will experience a different exposure than a worker coming into contact with contaminated foliage or somebody ingesting dust particles. While urine measurements are integrative, other matrices or environmental measurements might therefore prove to be more appropriate when it comes to further dissecting the particular contributions to overall exposure. Furthermore, absorption and metabolism will strongly depend on the route of uptake, explaining some of the observed differences in metabolite patterns and kinetics [16,35].

3.3.2. Inhalative Exposure

Inhalative exposure can be measured in a non-invasive way by analysis of (breathing) filters or measurements of ambient air in the respective breathing zones [25,35,39,40]. The latter approach is often the method of choice due to ease of use and was thus used in the few studies that looked at the respective contribution of inhalation to overall exposure. For imidacloprid, the inhalative route accounted for less than 1% of the total exposure, with air concentrations being below the detection limit in the breathing zone of the workers [41,42]. Contrastingly, captan was detected in more than 60% and 32.8% of personal breathing zone samples during application by air blast or hand spray, respectively [35]. This is in line with previous studies which also found inhalation to be a major route of exposure for captan [43–46].

3.3.3. Dermal Exposure

Skin exposure was assessed experimentally in several studies, most of them on vine [24,25]. One study also modeled exposure based on EUROPOEM [26]. For apples, Hines et al. used patches and hand rinses to assess potential substance exposure for operators [35]. In this study, rinse sampling was restricted to only one hand per person to minimize uncertainties due to error propagation.

Other studies used different approaches. For example, Fustinoni et al. relied on the collection of hand washing liquids in order to distinguish between potential and actual exposure to tebuconazole in vineyards 24 h after application [25]. This sets the amount detected in clothes against what was washed off the skin. Exposure of the head was recorded separately using non-woven fabric head covers. Likewise, Mandic-Rajcevic et al. used a modified OECD patch methodology and the collection of hand washing liquids over 24 h post-exposure for determining actual and potential body skin exposure to mancozeb in vineyards [30], while Mercadante et al. similarly looked at penconazole [27]. Meanwhile, Lopez-Galvez et al. analyzed hand wipes of reentry workers for imidacloprid [32].

Several of the aforementioned studies identified hand exposure as a major contributor to the overall dermal substance load as well as to total exposure (Tables 3 and 4). Other sites of contact were the forearms and thighs [35]. Depending on the substance, study design and group looked at, hands accounted for 17 to 97% of skin exposure [29,30]. This contribution was particularly large for reentry workers for whom the data also showed the largest variability [27]. Such variability could pose a problem with regard to data evaluation. With the respective range spanning an order of magnitude of 10^4, the most likely explanation is a varying use of gloves in this group.

3.3.4. Other Factors Influencing Exposure—Repeated Exposure, Substance Kinetics and Means of Application

Many of the less systematic studies focused on actual substance application as the major exposure event. The logic behind this is that, for groups such as operators and bystanders, application will constitute a good part of the overall load experienced. Indeed, for captan and cypermethrin, residential biomarker concentrations clearly correlated with substance application, with approximately 90% of the samples being below the limit of detection (LOD) outside the season [19]. However, repeated contact can also have a significant impact. This is most clearly seen for azinphos-methyl, where total worker exposure strongly depended on the number of days between application and reentry (Table 3) [36]. When the reentry time was longer than 14 days, only 2.7% of the specimens exceeded the reference value of 76 µg/kg bw and day for CAL EPA, whereas a reentry time of less than 14 days led to 27% of all specimens being above the reference value. For reentry, the respective dose–response relationship therefore strongly depends on the start of the work. Befittingly, Simcox et al. reported generally low urinary baselines before reentry, except for the one cohort that had started work prior to the start of the study [37].

In addition, studies need to accommodate individual substance kinetics to obtain reliable measurements. For imidacloprid, there was an 8.5–15-fold increase in metabolite levels in rural residents and operators one day after the spray event, with the residential levels peaking on day two [33]. Moreover, exposure to imidacloprid was found to be ubiquitous for the respective rural population, with metabolites found in 100% of the sampled cohort. Notably, other substances will follow different kinetics. This is exemplified by penconazole, where urinary metabolite excretion peaked 24 h after exposure at work [27].

Finally, the application technique (air blast or hand-held spray) will, of course, influence the exposure route of a particular active substance and should thus be taken into account [35].

3.3.5. Use of Personal and Other Protective Equipment

The use of personal protective equipment (PPE) and other protective equipment obviously has a huge impact on potential exposure. It is hence a key parameter when assessing potential health impacts for workers and operators alike. Unsurprisingly, several studies focused on the impact of PPE on substance exposure. For apples, two studies investigated the influence of PPE [35]. Both studies saw clear effects, recommending the inclusion of a reduction factor for pesticide exposure in the applied exposure model (AHS model) as a result. For vine, the varying use of PPE was suspected to be partly responsible for the observed range of the metabolite excretion rates [25]. In particular, the use of work suits and gloves was found to have a significant impact, with the former reducing dermal exposure by up to 98% [25]. Training on the correct use of PPE was found to be another significant factor for exposure minimization [32].

For operators, PPE proved to be similarly effective. In tractors, work suits reduced skin exposure by 4 to 10 times, depending on whether it was an open (OT) or closed and filtered system (CFT) [30]. Interestingly, while the use of gloves led to a 10 times lower exposure during application in OTs, exposure in CFTs increased. The most likely reason for this initial paradoxical result is operators taking their used gloves into the cabins. In terms of CFTs, these are semi-closed systems with any contaminated PPE, hence leading to an increased inhalative load. Likewise, Mercadante et al. found that clothing provided good protection since actual exposure was 100 or even 1000 times lower than the potential exposure [27]. Nevertheless, it should be noted that the use of PPE can vary significantly depending on availability, awareness or work culture. This is exemplified by the results of Medda et al., which found percentages of PPE use amongst workers to differ from 85.2% to 97.2% [28]. Additionally, broad use of PPE is obviously not an option for bystanders.

3.4. Assessment of Systemic Exposure

Biomarker measurements allow conclusions on the overall substance load experienced. However, any further assessment needs information on the systemic exposure as the potentially effective dose. To calculate this, more information is needed such as substance-specific conversion factors and toxicokinetics. Systemic exposure was assessed in several of the reviewed studies [36].

Most authors relied on conversion factors to calculate the absorbed dose from the respective (urinary) metabolite measurements. Fenske et al. [36] and Sams et al. [34] used oral conversion factors to assess systemic exposure for azinphos-methyl and penconazole, respectively. In the case of azinphos-methyl, the exposure remained well below the US EPA guidance value of 560 μg/kg bw and day but would exceed the more conservative CAL EPA reference value in up to 27% of cases when reentry started prior to a 14-day period [36]. Meanwhile, residential exposure to penconazole was determined to be extremely low throughout [34]. The maximum measured value (1.8 μmol/mol creatinine) was 100 times lower than what was excreted in the human dosing study used for establishing the conversion factor. Several of the studies looking at the fungicide exposure of workers in vineyards derived the systemic exposure from the total actual dermal exposure, as the inhalative route is negligible for many of the examined fungicides [25,27]. According to several studies, this also applies to operators for whom inhalative exposure is reported to be as low as 1.1% [39]. The resulting systemic exposure did not exceed the respective reference values in any of the cases investigated. For the cohort of Fustinoni et al. [25], the mean daily exposure was 1.73 ($\pm$1.31) μg active substance/kg bw and, as such, clearly below the ADI and AOEL of 0.03 mg/kg bw per day [24,47]. Likewise, in the case of mancozeb, worst-case assumptions resulted in an estimate several hundred times below the AOEL of 0.02 mg/kg bw for the corresponding metabolite (median saturation of 0.01%) [29]. Even the highest individual exposure of workers was still a thousand times below the daily limit, encumbering the reference value of only 1% [31].

Statistically significant correlations between dermal and urinary measurements of metabolites were found in five studies including apples [35] and vine and were observed for operators as well as for workers [24,25,27,29–32] (Tables 3 and 4). Reliable knowledge on the correlation of dermal and urinary measurements facilitates the determination of systemic exposure tremendously as it renders the elaborate and costly assessment of dermal patches obsolete, or any other measurements for that matter [25]. Yet, although dermal exposure often constitutes a major route of uptake, other potential routes such as inhalation should always be accounted for. However, the extent to which inhalation contributes to the respective urinary biomarker profile strongly depends on the respective substance and application scenario [35].

A few studies applied exposure models to assess systemic exposure and to predict urinary biomarker concentrations based on information about the type and amount of pesticide applied and culture treated [10,24,26,35].

Notably, for most cases, modeling systemic exposure resulted in overestimates when compared to the actual measured values. This applied for the ACROPOLIS model (www.acropolis-eu.com, accessed on 28 February 2022) [24], the UK's residential exposure assessment (REA) model [26] and the EUROPOEM model [10,26]. With regard to health protection, this suggests that these models were sufficiently conservative.

Predictions of urinary metabolites for captan and penconazole by the REA model, for example, exceeded the actual measured amounts in 98% and 97% of cases. Additionally, for chlorpyrifos and chlormequat, the urinary biomarker concentrations were still higher than predicted, but not significantly different from the background measurements [9].

Similarly, the ACROPOLIS model demonstrated a good predictive performance for the dermal uptake of tebuconazole, albeit in a small cohort ($n = 7$) [24]. Here, the model predicted a mean daily exposure of 1.77 µg active substance/kg bw (±1.96) as opposed to the 1.73 µg active substance/kg bw (±1.31) measured.

Other models such as the AHS pesticide exposure intensity algorithm performed less well [48]. When applied for the assessment of captan in apples and peaches, the algorithm yielded a sufficiently predictive performance for dermal exposure of the thighs and forearms but proved less reliable with regard to overall urinary metabolites or estimates regarding the breathing zone or hand rinse. Hines et al. therefore suggested a revision of the respective exposure weights depending on the method used for application (air blast vs. hand spray) [35].

3.5. Health-Related Outcomes

Health-related aspects were explicitly assessed in only one study which looked at the acute effects of mancozeb on the thyroid (Tables 3 and 4) [28,37]. The study compared two vine-growing areas in Italy, one being iodine-deficient (Chianti), and the other iodine-sufficient (Bolzano). Exposed workers of the first region had lower serum means of free thyroxine (fT4) and showed higher iodine urinary excretion (>250 µg/L ETU as opposed to >20 µg/L ETU). Moreover, workers exposed to mancozeb in the iodine-deficient area tended to have lower thyroid volumes (≤6 mL). These findings fit well with experimental evidence that the thyroid was a major toxicological target of mancozeb as found both in vitro by Lori et al. (2021) and in laboratory animals as summarized by the European Food Safety Authority (EFSA) in a recent evaluation in 2019 [49,50].

4. Discussion

This review extends the existing analyses on the exposure of operators, workers, residents and bystanders to pesticides [3,17,18]. It also highlights the need for more systematic studies on this issue. In striking contrast to the numerous claims made in the literature on pesticide exposure, there are surprisingly few studies that looked at this systematically and only in a limited range of target cultures. Focusing on tree-grown produce, vine and hops as typical "overheads" representing a worst case, we found only 11 studies to meet the minimal requirements for any further in-depth analysis. Notably, this was a fraction of the 108 articles initially identified and restricted mainly to apples and vine. Altogether, the respective studies covered a period of 22 years, although the majority were performed in 2010 or thereafter. This helps in so much as one goal of this review was a compilation of the lessons learnt together with an outlook on what could be improved in future studies.

Key issues to consider are the selection of the substance and culture as well as what parameters to (bio)monitor, how to do so and how frequently. One of the lessons learnt in this context is that questionnaires should only serve as means to support the collection of "hard" data, not replace them. Likewise, sole measurement of urinary metabolites without any further data on toxicokinetics or conversion factors will only allow for relative comparisons and thus severely limit any further analysis regarding, for example, the absorbed dose or systemic exposure. This aspect should be considered from the start as both of the latter factors are a precondition for the assessment of any potential health effects, even if only speculative. Conveniently, for some pesticides, the inhalative and oral routes apparently play a negligible role when it comes to total exposure, at least for certain cohorts. In these cases where the dermal route is the main or sole route of exposure, actual skin exposure can be used as a good approximation for systemic exposure.

Sampling of urinary specimens is often the method of choice for its ease of sampling and non-invasive and integrative nature. Yet, the suitability of the matrices selected for monitoring depends on the cohorts, routes of exposure to be assessed, the type of pesticide, the application technique and any potential personal protective equipment used. For the aforementioned groups, the main routes of exposure are usually dermal or inhalative, with the oral route being mainly important in the case of intake of food or ingestion of particles (dust), or as a proxy for inhalation [16]. Therefore, concomitant dermal or inhalative sampling should be considered as appropriate. Both routes are experimentally accessible in a non-invasive manner using pad dosimetry and rinse collection or respiratory filters and air sampling, respectively. Literature searches or experimental pre-screens will also provide data on which routes to consider and by what means [39,40,51]. The detection of particular metabolites shows the occurrence of exposure but fails to inform on the timing, amount or frequency. Without this additional information, further health assessments are not possible. The abundance of pesticide metabolites in urine might hence sufficiently fuel public concern about possible health impacts, but toxicologically, it will inherently remain of limited relevance. State-of-the-art biomonitoring should thus aim for repeated and well-timed sampling to establish levels of systemic exposure, absorbed doses and the respective biokinetics. Only then will it be possible to compare the data to the corresponding reference values and to draw informed conclusions on the likelihood and plausibility of potential adverse health effects.

Depending on the route of uptake, metabolism and the excretion rate, the amount of detectable metabolites can differ widely [6]. This is due to the varying potential influence of the first pass effect. In the case of oral exposure, any xenobiotic will primarily be subjected to hepatic metabolism, but not in cases of dermal or inhalative uptake [16]. This obviously also affects options for back-modeling, for example, when substances with predominantly dermal exposure feature a significant correlation between external measurements and the excreted amount of metabolites.

Kinetic measurements or data pending such cases allow for the calculation of systemic exposure or an absorbed dose merely based on the urinary metabolite concentrations [15,29].

Alternatively, one can rely on conversion factors which likewise have to be determined in advance [25,29–31].

Finally, biomonitoring constitutes an important pillar in the validation and refinement of toxicokinetic exposure models such as the ACROPOLIS model, the EUROPOEM model or the UK's REA model. Such models have already shown a high degree of accordance and accuracy in studies where they were used for metabolite and exposure prediction [10,24,26,29,35]. Refined further, they will hence undoubtedly play an increasingly important role in future studies.

5. Strengths and Limitations

Modern biomonitoring of pesticide exposure to humans is a topic of high complexity warranting in-depth evaluation of the respective studies and data. This is particularly important when the results are intended to be used for toxicological evaluation or risk assessment.

The structure and presentation of this scoping review followed the extension of the PRISMA guidelines by Tricco et al. [20]. Comparable methodological approaches for reviews of pesticide exposures were applied in existing reviews [3,17,18]. This includes the use of three different databases and of at least two independent assessments regarding study selection and data extraction [17]. Important features of the respective study design are presented in tabular form, as are the results of the exposure assessment. A further strength of this review is the assessment of the quantification of the respective systemic exposures. Unlike in other reviews which focused on residents only, this review also included groups with potentially higher exposures such as operators and workers [3,17,18].

On the other hand, our review was limited to some overhead cultures where a high degree of exposure to pesticides would be expected. In future, additional overhead cultures such as olive trees should also be considered, in addition to apples, vine and hops. A further limitation of our review might be that it was confined to substances which are currently allowed in Germany and the EU. Although it is true that biomonitoring studies are highly specific to the cultures and substances under investigation, and thus the exclusion of substances which are no longer in use such as DDT might be justified, the inclusion of additional substances would have been certainly of interest, at least from a methodological point of view.

While sampling of at least urine or serum was a mandatory requirement for the inclusion of articles in this review, analysis of the data clearly highlighted the importance of dermal sampling and inhalative dosimetry. Establishing clear correlations with urinary measurements for both these parameters remains a challenge though, as does establishing clear and unequivocal links to adverse health outcomes. With regard to the latter, one has to differentiate between acute and long-term effects. As shown by the example of mancozeb, it is acute health effects, in particular, that are accessible by biomonitoring if the study is based on a confined cohort and planned accordingly. The situation is more difficult for long-term health effects or chronic endpoints, particularly when relying on large cohort studies such as the Agricultural Health Study (AHS) (https://aghealth.nih.gov, accessed on 28 February 2022). Here, confounding becomes a larger issue, as does exposure monitoring. Long-term exposures can, for example, be monitored by hair measurements, measurement of house dust or environmental parameters such as residues on foliage [3]. However, establishing sufficiently robust dose–response relationships from these data remains scientifically challenging.

6. Conclusions

This review is the first to systematically analyze the exposure of operators, workers, residents or bystanders in cultures with potentially high exposure. In these cultures, exposure is predominantly driven by the actual substance application as well as by reentry. As might be expected, this leads to higher exposure for operators and workers than for residents or bystanders. In nearly all cases, comparison of the relevant toxicological reference values to the respective estimated systemic exposure showed the latter to be well below, indicating no reason for concern. This analysis shows the potential of biomonitoring, empowering a hitherto predominantly observational tool to become a toxicologically informative asset for public health protection. As a first step, future studies in this field should therefore aim to systematically establish specific active substance and culture correlations for dermal or inhalative exposure and urinary biomarkers. A framework of established correlations would allow future studies to be based solely on the detection and quantification of metabolites in urine, but performed regularly, such studies could be combined with the monitoring of short- and long-term health effects.

Supplementary Materials: The following supporting information can be downloaded at: https://www.mdpi.com/article/10.3390/toxics10040170/s1, Table S1: Checklist of preferred reporting items for systematic reviews and meta-analyses extension for Scoping Reviews (PRISMA-ScR)—where these items are addressed in the review article by Willenbockel et al. (2022).

Author Contributions: Conceptualization: C.T.W., P.M.-S., C.W. and L.N.; methodology and conduct of literature search: C.T.W. and J.P.; epidemiological advice: C.W. and S.D.; selection of studies to be included: C.T.W., J.P., S.D. and L.N.; full-text analysis of elected studies: C.T.W. and J.P.; writing—original draft preparation: C.T.W. and J.P.; writing—review and editing: T.T., C.W., P.M.-S. and L.N.; supervision and project administration: P.M.-S. and L.N. All authors have read and agreed to the published version of the manuscript.

Funding: This research received no external funding.

Institutional Review Board Statement: Not applicable.

Informed Consent Statement: Not applicable.

Data Availability Statement: Not applicable.

Acknowledgments: We are grateful to Barbara Freytag and Sandra Sieke for their support and assistance in the literature search and data management.

Conflicts of Interest: The authors declare no conflict of interest.

References

1. Ntzani, E.E.; Ntritsos, G.C.M.; Evangelou, E.; Tzoulaki, I. *Literature Review on Epidemiological Studies Linking Exposure to Pesticides and Health Effects*; EFSA Supporting Publications, 2013; Volume 10, p. 497E. [CrossRef]
2. Ockleford, C.; Adriaanse, P.; Berny, P.; Brock, T.; Duquesne, S.; Grilli, S.; Hougaard, S.; Klein, M.; Kuhl, T.; Laskowski, R.; et al. Scientific Opinion of the PPR Panel on the follow-up of the findings of the External Scientific Report 'Literature review of epidemiological studies linking exposure to pesticides and health effects'. *EFSA J.* **2017**, *15*, e05007. [PubMed]
3. Dereumeaux, C.; Fillol, C.; Quenel, P.; Denys, S. Pesticide exposures for residents living close to agricultural lands: A review. *Environ. Int.* **2020**, *134*, 105210. [CrossRef]
4. Heudorf, U.; Butte, W.; Schulz, C.; Angerer, J. Reference values for metabolites of pyrethroid and organophosphorous insecticides in urine for human biomonitoring in environmental medicine. *Int. J. Hyg. Environ. Health* **2006**, *209*, 293–299. [CrossRef]
5. Simaremare, S.R.S.; Hung, C.-C.; Hsieh, C.-J.; Yiin, L.-M. Relationship between organophosphate and pyrethroid insecticides in blood and their metabolites in urine: A pilot study. *Int. J. Environ. Res. Public Health* **2019**, *17*, 34. [CrossRef] [PubMed]
6. Glorennec, P.; Serrano, T.; Fravallo, M.; Warembourg, C.; Monfort, C.; Cordier, S.; Viel, J.F.; Le Gléau, F.; Le Bot, B.; Chevrier, C. Determinants of children's exposure to pyrethroid insecticides in western France. *Environ. Int.* **2017**, *104*, 76–82. [CrossRef]
7. Roussel, C.; Witt, K.L.; Shaw, P.B.; Connor, T.H. Meta-analysis of chromosomal aberrations as a biomarker of exposure in healthcare workers occupationally exposed to antineoplastic drugs. *Mutat. Res. Rev. Mutat. Res.* **2019**, *781*, 207–217. [CrossRef] [PubMed]
8. Sams, C.; Jones, K. Biological monitoring for exposure to deltamethrin: A human oral dosing study and background levels in the UK general population. *Toxicol. Lett.* **2012**, *213*, 35–38. [CrossRef] [PubMed]
9. Eadsforth, C.V.; Bragt, P.C.; Van Sittert, N.J. Human dose-excretion studies with pyrethroid insecticides cypermethrin and alphacypermethrin: Relevance for biological monitoring. *Xenobiotica* **1988**, *18*, 603–614. [CrossRef] [PubMed]
10. Galea, K.S.; MacCalman, L.; Jones, K.; Cocker, J.; Teedon, P.; Cherrie, J.W.; van Tongeren, M. Comparison of residents' pesticide exposure with predictions obtained using the UK regulatory exposure assessment approach. *Regul. Toxicol. Pharmacol.* **2015**, *73*, 634–643. [CrossRef] [PubMed]
11. Hays, S.M.; Aylward, L.L.; Gagné, M.; Krishnan, K. Derivation of Biomonitoring Equivalents for cyfluthrin. *Regul. Toxicol. Pharmacol.* **2009**, *55*, 268–275. [CrossRef]
12. Quindroit, P.; Beaudouin, R.; Brochot, C. Estimating the cumulative human exposures to pyrethroids by combined multi-route PBPK models: Application to the French population. *Toxicol. Lett.* **2019**, *312*, 125–138. [CrossRef] [PubMed]
13. Côté, J.; Bouchard, M. Dose reconstruction in workers exposed to two major pyrethroid pesticides and determination of biological reference values using a toxicokinetic model. *J. Expo. Sci. Environ. Epidemiol.* **2018**, *28*, 599–614. [CrossRef]
14. Aylward, L.L.; Irwin, K.; St-Amand, A.; Nong, A.; Hays, S.M. Screening-level Biomonitoring Equivalents for tiered interpretation of urinary 3-phenoxybenzoic acid (3-PBA) in a risk assessment context. *Regul. Toxicol. Pharmacol.* **2018**, *92*, 29–38. [CrossRef]
15. Zoller, O.; Rhyn, P.; Zarn, J.A.; Dudler, V. Urine glyphosate level as a quantitative biomarker of oral exposure. *Int. J. Hyg. Environ. Health* **2020**, *228*, 113526. [CrossRef]
16. Vermeulen, R.C.H.; Gooijer, I.Y.M.; Hoftijser, D.G.W.; Lageschaar, I.L.C.C.; Oerlemans, D.A.; Scheepers, D.i.P.T.J.; Kivits, I.C.M.; Duyzer, D.J.; Gerritsen-Ebben, D.M.G.; Figueiredo, I.D.M.; et al. *Research on Exposure of Residents to Pesticides in The Netherlands OBO Flower Bulbs*; Rijksinstituut voor Volksgezondheid en Milieu RIVM: Utrecht, The Netherlands, 2019. Available online: https://www.rivm.nl/sites/default/files/2019-04/Onderzoeksrapport%20OBO.pdf (accessed on 28 February 2022).
17. Teysseire, R.; Manangama, G.; Baldi, I.; Carles, C.; Brochard, P.; Bedos, C.; Delva, F. Assessment of residential exposures to agricultural pesticides: A scoping review. *PLoS ONE* **2020**, *15*, e0232258. [CrossRef] [PubMed]
18. Teysseire, R.; Manangama, G.; Baldi, I.; Carles, C.; Brochard, P.; Bedos, C.; Delva, F. Determinants of non-dietary exposure to agricultural pesticides in populations living close to fields: A systematic review. *Sci. Total Environ.* **2021**, *761*, 143294. [CrossRef]
19. Galea, K.S.; MacCalman, L.; Jones, K.; Cocker, J.; Teedon, P.; Cherrie, J.W.; van Tongeren, M. Urinary biomarker concentrations of captan, chlormequat, chlorpyrifos and cypermethrin in UK adults and children living near agricultural land. *J. Expo. Sci. Environ. Epidemiol.* **2015**, *25*, 623–631. [CrossRef]

20. Tricco, A.C.; Lillie, E.; Zarin, W.; O'Brien, K.K.; Colquhoun, H.; Levac, D.; Moher, D.; Peters, M.D.J.; Horsley, T.; Weeks, L.; et al. PRISMA extension for scoping reviews (PRISMA-ScR): Checklist and explanation. *Ann. Intern. Med.* **2018**, *169*, 467–473. [CrossRef] [PubMed]
21. Figueiredo, D.A.-O.; Krop, E.A.-O.; Duyzer, J.A.-O.; Gerritsen-Ebben, R.A.-O.; Gooijer, Y.A.-O.; Holterman, H.A.-O.; Huss, A.A.-O.; Jacobs, C.A.-O.; Kivits, C.A.-O.; Kruijne, R.A.-O.X.; et al. Pesticide Exposure of residents living close to agricultural fields in The Netherlands: Protocol for an observational study. *JMIR Res. Protoc.* **2021**, *10*, e27883. [CrossRef] [PubMed]
22. Galea, K.S.; MacCalman, L.; Jones, K.; Cocker, J.; Teedon, P.; Cherrie, J.W.; van Tongeren, M. Biological monitoring of pesticides exposure in residents living near agricultural land. *Outlooks Pest Manag.* **2017**, *28*, 52–54. [CrossRef]
23. Galea, K.S.; MacCalman, L.; Jones, K.; Cocker, J.; Teedon, P.; Sleeuwenhoek, A.J.; Cherrie, J.W.; van Tongeren, M. Biological monitoring of pesticide exposures in residents living near agricultural land. *BMC Public Health* **2011**, *11*, 856. Available online: https://bmcpublichealth.biomedcentral.com/articles/10.1186/1471-2458-11-856 (accessed on 28 February 2022). [CrossRef]
24. Kennedy, M.C.; Glass, C.R.; Fustinoni, S.; Moretto, A.; Mandic-Rajcevic, S.; Riso, P.; Turrini, A.; van der Voet, H.; Hetmanski, M.T.; Fussell, R.J.; et al. Testing a cumulative and aggregate exposure model using biomonitoring studies and dietary records for Italian vineyard spray operators. *Food. Chem. Toxicol.* **2015**, *79*, 45–53. [CrossRef]
25. Fustinoni, S.; Mercadante, R.; Polledri, E.; Rubino, F.M.; Mandic-Rajcevic, S.; Vianello, G.; Colosio, C.; Moretto, A. Biological monitoring of exposure to tebuconazole in winegrowers. *J. Expo. Sci. Environ. Epidemiol.* **2014**, *24*, 643–649. [CrossRef]
26. Sleeuwenhoek, A.; Cocker, J.; Jones, K.; Cherrie, J.W. Biological monitoring of pesticide exposures. In *IOM Research Report TM/07/02*; 2007.
27. Mercadante, R.; Polledri, E.; Rubino, F.M.; Mandic-Rajcevic, S.; Vaiani, A.; Colosio, C.; Moretto, A.; Fustinoni, S. Assessment of penconazole exposure in winegrowers using urinary biomarkers. *Environ. Res.* **2019**, *168*, 54–61. [CrossRef]
28. Medda, E.; Santini, F.; De Angelis, S.; Franzellin, F.; Fiumalbi, C.; Perico, A.; Gilardi, E.; Mechi, M.T.; Marsili, A.; Citroni, A.; et al. Iodine nutritional status and thyroid effects of exposure to ethylenebisdithiocarbamates. *Environ. Res.* **2017**, *154*, 152–159. [CrossRef]
29. Mandic-Rajcevic, S.; Rubino, F.M.; Ariano, E.; Cottica, D.; Negri, S.; Colosio, C. Exposure duration and absorbed dose assessment in pesticide-exposed agricultural workers: Implications for risk assessment and modeling. *Int. J. Hyg. Environ. Health* **2019**, *222*, 494–502. [CrossRef] [PubMed]
30. Mandic-Rajcevic, S.; Rubino, F.M.; Ariano, E.; Cottica, D.; Neri, S.; Colosio, C. Environmental and biological monitoring for the identification of main exposure determinants in vineyard mancozeb applicators. *J. Expo. Sci. Environ. Epidemiol.* **2018**, *28*, 289–296. [CrossRef]
31. Mandić-Rajčević, S.; Rubino, F.M.; Colosio, C. Establishing health-based biological exposure limits for pesticides: A proof of principle study using mancozeb. *Regul. Toxicol. Pharmacol.* **2020**, *115*, 104689. [CrossRef] [PubMed]
32. Lopez-Galvez, N.; Wagoner, R.; Canales, R.A.; de Zapien, J.; Calafat, A.M.; Ospina, M.; Rosales, C.; Beamer, P. Evaluating imidacloprid exposure among grape field male workers using biological and environmental assessment tools: An exploratory study. *Int. J. Hyg. Environ. Health* **2020**, *230*, 113625. [CrossRef] [PubMed]
33. Tao, Y.; Dong, F.; Xu, J.; Phung, D.; Liu, Q.; Li, R.; Liu, X.; Wu, X.; He, M.; Zheng, Y. Characteristics of neonicotinoid imidacloprid in urine following exposure of humans to orchards in China. *Environ. Int.* **2019**, *132*, 105079. [CrossRef]
34. Sams, C.; Jones, K.A.-O.; Galea, K.S.; MacCalman, L.; Cocker, J.; Teedon, P.; Cherrie, J.W.; van Tongeren, M. Development of a biomarker for Penconazole: A human oral dosing study and a survey of UK residents' exposure. *Toxics* **2016**, *4*, 10. [CrossRef]
35. Hines, C.J.; Deddens, J.A.; Jaycox, L.B.; Andrews, R.N.; Striley, C.A.; Alavanja, M.C. Captan exposure and evaluation of a pesticide exposure algorithm among orchard pesticide applicators in the Agricultural Health Study. *Ann. Occup. Hyg.* **2008**, *52*, 153–166.
36. Fenske, R.A.; Curl, C.L.; Kissel, J.C. The effect of the 14-day agricultural restricted entry interval on azinphosmethyl exposures in a group of apple thinners in Washington state. *Regul. Toxicol. Pharm.* **2003**, *38*, 91–97. [CrossRef]
37. Simcox, N.J.; Camp, J.; Kalman, D.; Stebbins, A.; Bellamy, G.; Lee, I.-C.; Fenske, R. Farmworker exposure to organophosphorus pesticide residues during apple thinning in central Washington State. *Am. Ind. Hyg. Assoc. J.* **1999**, *60*, 752–761. [CrossRef] [PubMed]
38. Fustinoni, S. Biomonitoring of exposure to penconazole in agriculture. In Proceedings of the 31st International Congress on Occupational Health, Seoul, Korea, 31 May–5 June 2015.
39. Aprea, C.; Terenzoni, B.; De Angelis, V.; Sciarra, G.; Lunghini, L.; Borzacchi, G.; Vasconi, D.; Fani, D.; Quercia, A.; Salvan, A.; et al. Evaluation of skin and respiratory doses and urinary excretion of alkylphosphates in workers exposed to dimethoate during treatment of olive trees. *Arch. Environ. Contam. Toxicol.* **2004**, *48*, 127–134. [CrossRef] [PubMed]
40. Baldi, I.; Lebailly, P.; Jean, S.; Rougetet, L.; Dulaurent, S.; Marquet, P. Pesticide contamination of workers in vineyards in France. *J. Expo. Sci. Environ. Epidemiol.* **2006**, *16*, 115–124. [CrossRef] [PubMed]
41. Cao, L.; Chen, B.; Zheng, L.; Wang, D.; Liu, F.; Huang, Q. Assessment of potential dermal and inhalation exposure of workers to the insecticide imidacloprid using whole-body dosimetry in China. *J. Environ. Sci.* **2015**, *27*, 139–146. [CrossRef] [PubMed]
42. Cao, L.; Zhang, H.; Li, F.; Zhou, Z.; Wang, W.; Ma, D.; Yang, L.; Zhou, P.; Huang, Q. Potential dermal and inhalation exposure to imidacloprid and risk assessment among applicators during treatment in cotton field in China. *Sci. Total Environ.* **2018**, *624*, 1195–1201. [CrossRef] [PubMed]
43. Oudbier, A.J.; Bloomer, A.W.; Price, H.A.; Welch, R.L. Respiratory route of pesticide exposure as a potential health hazard. *Bull Env. Contam Toxicol* **1974**, *12*, 1–9. [CrossRef]

44. Hansen, J.D.; Schneider, B.A.; Olive, B.M.; Bates, J.J. Personnel safety and foliage residue in an orchard spray program using azinphosmethyl and captan. *Arch. Environ. Contam. Toxicol.* **1978**, *7*, 63–71. [CrossRef] [PubMed]
45. McJllton, C.E.; Berckman, G.E.; Deer, H.M. Captan exposure in apple orchards. *Am. Ind. Hyg. Assoc. J.* **1983**, *44*, 209–210. [CrossRef] [PubMed]
46. de Cock, J.; Heederik, D.; Boleij, J.S.; Kromhout, H.; Hoek, F.; Wegh, H.; Ny, E.T. Exposure to captan in fruit growing. *Am. Ind. Hyg. Assoc. J.* **1998**, *59*, 158–165. [CrossRef] [PubMed]
47. EFSA. Conclusion on the peer review of the pesticide risk assessment of the active substance tebuconazole. *EFSA J.* **2014**, *12*, 98. [CrossRef]
48. Dosemeci, M.; Alavanja, M.C.; Rowland, A.S.; Mage, D.; Zahm, S.H.; Rothman, N.; Lubin, J.H.; Hoppin, J.A.; Sandler, D.P.; Blair, A. A quantitative approach for estimating exposure to pesticides in the Agricultural Health Study. *Ann. Occup. Hyg.* **2002**, *46*, 245–260. [PubMed]
49. Lori, G.; Tassinari, R.; Narciso, L.; Udroiu, I.; Sgura, A.; Maranghi, F.; Tait, S. Toxicological comparison of mancozeb and zoxamide fungicides at environmentally relevant concentrations by an in vitro approach. *Int. J Environ. Res. Public Health* **2021**, *18*, 8591. [CrossRef] [PubMed]
50. European Food Safety Authority (EFSA); Abdourahime, H.; Anastassiadou, M.; Arena, M.; Auteri, D.; Barmaz, S.; Brancato, A.; Bura, L.; Cabrera, L.C.; Chaideftou, E.; et al. Conclusion on the peer review of the pesticide risk assessment of the active substance mancozeb. *EFSA J.* **2019**, *18*, 28. [CrossRef]
51. Baldi, I.; Lebailly, P.; Bouvier, G.; Rondeau, V.; Kientz-Bouchart, V.; Canal-Raffin, M.; Garrigou, A. Levels and determinants of pesticide exposure in re-entry workers in vineyards: Results of the PESTEXPO study. *Environ. Res.* **2014**, *132*, 360–369. [CrossRef]

toxics

Systematic Review

Towards Reference Values for Malondialdehyde on Exhaled Breath Condensate: A Systematic Literature Review and Meta-Analysis

Veronica Turcu [1], Pascal Wild [1], Maud Hemmendinger [1], Jean-Jacques Sauvain [1], Enrico Bergamaschi [2], Nancy B. Hopf [1,†] and Irina Guseva Canu [1,*,†]

[1] Center for Primary Care and Public Health (Unisanté), University of Lausanne, Route de la Corniche 2, 1066 Epalinges, Switzerland; veronica.turcu@unisante.ch (V.T.); pascal.wild@unisante.ch (P.W.); maud.hemmindinger@unisante.ch (M.H.); jean-jacques.sauvain@unisante.ch (J.-J.S.); nancy.hopf@unisante.ch (N.B.H.)

[2] Department of Public Health and Pediatrics, University of Turin, Via Zuretti 29, 10125 Turin, Italy; enrico.bergamaschi@unito.it

* Correspondence: irina.guseva-canu@unisante.ch

† These authors contributed equally to this work.

Citation: Turcu, V.; Wild, P.; Hemmendinger, M.; Sauvain, J.-J.; Bergamaschi, E.; Hopf, N.B.; Canu, I.G. Towards Reference Values for Malondialdehyde on Exhaled Breath Condensate: A Systematic Literature Review and Meta-Analysis. *Toxics* **2022**, *10*, 258. https://doi.org/10.3390/toxics10050258

Academic Editor: Christophe Rousselle

Received: 30 March 2022
Accepted: 6 May 2022
Published: 18 May 2022

Publisher's Note: MDPI stays neutral with regard to jurisdictional claims in published maps and institutional affiliations.

Copyright: © 2022 by the authors. Licensee MDPI, Basel, Switzerland. This article is an open access article distributed under the terms and conditions of the Creative Commons Attribution (CC BY) license (https://creativecommons.org/licenses/by/4.0/).

Abstract: Many pathological conditions and certain airway exposures are associated with oxidative stress (OS). Malondialdehyde (MDA) is an end-product of the oxidation of lipids in our cells and is present in all biological matrices including exhaled breath condensate (EBC). To use MDA as a biomarker of OS in EBC, a reference interval should be defined. Thus, we sought to summarize reference values reported in healthy adult populations by performing a systematic review and meta-analysis using a standardized protocol registered in PROSPERO (CRD42020146623). Articles were retrieved from four major databases and 25 studies with 28 subgroups were included. Defining the distribution of MDA measured in reference populations with a detection combined with a separation technique still represents a challenge due to the low number of studies available, different analytical methods used, and questionable methodological qualities of many studies. The most salient methodological drawbacks have been in data collection and reporting of methods and study results by the researchers. The lack of compliance with the recommendations of the European Respiratory Society and American Thoracic Society was the major limitation in the current research involving EBC. Consequently, we were unable to establish a reference interval for MDA in EBC.

Keywords: malondialdehyde; exhaled breath condensate; meta-analysis; reference values

1. Introduction

Oxidative stress is an imbalance between production and elimination of reactive oxygen species (ROS), and represents a common pathological pathway for many inflammatory diseases, including chronic and acute lung conditions, such as chronic obstructive disease and asthma [1]. The biological effects of excess ROS production are their interactions with membrane lipids, a process known as lipid peroxidation, which yields a series of secondary molecules capable of exacerbating the oxidative damage [2]. Malondialdehyde (MDA) is one of the end-products of lipid peroxidation and has been largely investigated as a biomarker of oxidative stress. MDA correlates with different pathological states, such as cardiovascular disease and atherosclerosis [3,4], diabetic disease [5], Alzheimer's, DNA damage, and ageing [6]. MDA has also been extensively investigated as a biomarker of pulmonary oxidative stress [7] and in evaluating the impact of air pollution, smoking [8], or respiratory exposure in occupational environments such as exposure to heavy metals [9] and particles [10].

MDA has attracted attention and interest as a biomarker of oxidative stress, because it can be detected in a variety of biological matrices, i.e., blood, urine, and exhaled breath condensate (EBC), and has shown good correlations with other biomarkers of lipid peroxidation [11].

EBC, the biological matrix considered in this meta-analysis, can be obtained by cooling an individual's exhaled air using a noninvasive and safe procedure. This condensate contains low-volatile compounds and aerosolized airway lining fluid [12] which have been diluted by water [13]. According to current recommendations from the joint task force of American Thoracic Society and European Respiratory Society, EBC collection can be performed by asking an individual to breathe normally into a tube connected to a cooling apparatus while wearing a nose clip for a period of time until a reasonable amount of condensate is collected [14]. The current ERS technical standards also include specific recommendations as follows: (a) define and report EBC volume, time of collection, collection temperature, breathing pattern, prevention of salivary and environmental contamination, ambient conditions; (b) keep the time between collection of EBC and analysis as short as possible; (c) report results in terms of concentration per 100l of EBC or per minute, in order to reduce the confounding factor of the dilution of substances in the EBC; (d) report data concerning the intra- and inter-assay variability, limit of detection, sensitivity of detection methods, and more generally describe the methods and techniques used to collect and analyze the EBC; (e) note and report the time between food intake and EBC collection; (f) schedule collection for the same time of day.

The same task force, in 2005, also released a first series of methodological recommendations which were more generic than the 2017 recommendations, and the use of a nose clip and a saliva trap have been formulated since the earlier version. The biomarkers in EBC originate from different regions of the respiratory tract, from the lower airways and lungs up to the oropharynx and nasopharynx [15]. Hence, the use of a nose clip is recommended to prevent collection of mediators formed in the nose and the sinuses and to exclusively collect from the lower airways, as well as to ensure that no sample is lost through nasal respiration [16]. Similarly, saliva contamination needs to be avoided since it may contain the same mediators of interests but derived from other anatomical sites than the lower respiratory zone. It has been proposed that a saliva trap is sufficient, since there is a need for a sensitive test for salivary alpha-amylase.

Other recommendation in the 2005 version included performing the analysis as soon as possible after collection, noting and reporting temperature collection, and detailing the collection devices if custom materials are used especially for the contact surface. Although a collection time of 10 min has been recommended, further research on the effect of collection time and temperature versus expiratory volume is needed [17]. Concerning the collection and analysis methods, the 2005 ATS/ERS recommendations were to provide sufficient details to ensure the reproducibility of technique.

Furthermore, reporting participants' details, such as smoking habits, race, age, sex, as well as eating and drinking behaviors have also been recommended since the first version of the task force [14]. We found it crucial to detail these two series of recommendations from the joint task force of the American Thoracic Society and European Respiratory Society, as this could have affected the different study protocols in the studies reviewed for this meta-analysis.

MDA concentrations in EBC are usually quantified after collection and storage at low temperatures for up to 6 months, by using different analytical and separation methods: high performance liquid chromatography (HPLC) coupled with mass spectrometry (MS) or UV photometry detection. MDA is derivatized with either thiobarbitoric acid (TBA) or 2,4-dinitrophenylhydrazine (DNPH), with TBA being less specific for MDA [18].

Different levels of MDA in the EBC have been detected in some physiological or pathological conditions characterized by oxidative stress, but the absence of established or recognized reference intervals has limited the interpretation of the findings of the MDA as a biomarker in different settings. Thus, the current research has as a primary objective to investigate the possibility of defining reference intervals for MDA in EBC.

2. Materials and Methods

The present study focuses on MDA measured in EBC and is part of a series of systematic reviews and meta-analyses for several oxidative stress biomarkers in two different matrices: urine and EBC [19].

Our protocol followed recommendations from the Preferred Reporting Items for Systematic Reviews and Meta-Analysis (PRISMA) [20] and is registered with the International Prospective Register of Systematic Reviews (PROSPERO, registration number CRD 42020146623) [21].

2.1. Literature Research

Research for scientific literature was conducted on four different databases (The Cochrane Central Register, EMBASE, PubMed, and Web of Science), from journal apparition up to October 2020. We performed a quick search for the period covering the end of 2020 to end of 2021 to confirm that no new data had been published that could change our outcome. The research string was constructed in collaboration with the library of Unisanté and their documentarist using a combination of Medical Subject Headings (MeSH) terms and key words. The complete research string has been described elsewhere (www.doi.org/10.16909/dataset/17, accessed on 15 February 2022) and is available at the Unisanté data repository.

2.2. Study Selection

We included only original research studies written in English that reported MDA concentrations in EBC among human healthy adults. The flowchart in Figure 1 represents the selection process of the studies that were ultimately included in our meta-analysis. First, we identified and excluded duplicates, and second, we used the Rayyan software [22] for title and abstract screening.

The exclusion criteria were: MDA measured in any other matrix than EBC (e.g., blood/plasma and sperm) in vitro or non-human studies (animal or plant), incorrect MDA abbreviation (e.g., mass drug administration and methylenedioxyamphetamine), and inappropriate types of publications such as reviews or conference papers. In accordance with our registered protocol, our research string was built to cover an entire series of OS biomarkers, and was not limited to MDA. Consequently, our initial pool of articles contained articles that did not refer to MDA. Another factor that made this pool large was that authors often chose keywords that were very generic such as "oxidative stress" and "biomarkers of oxidative stress".

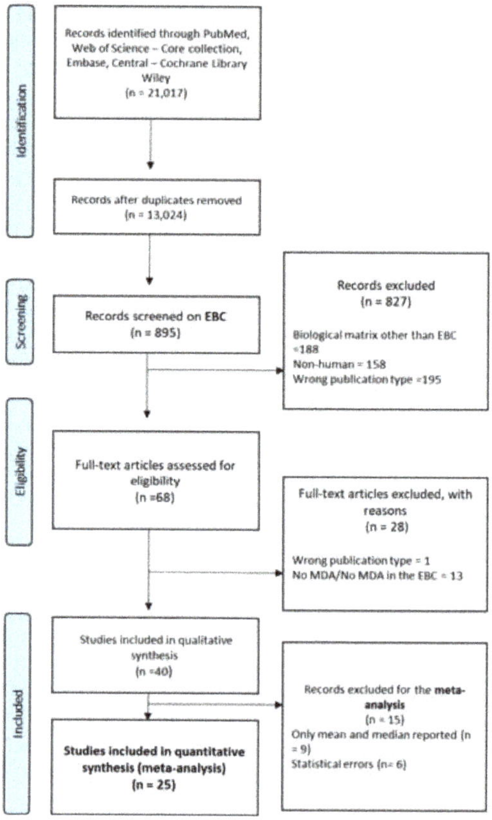

Figure 1. Flowchart of the article selection process.

2.3. Data Extraction

We extracted data according to a previously published protocol [19], including a list of parameters regarding technical aspects of the analytical methods used to measure MDA in EBC and lifestyle factors such as body mass index (BMI) and vitamin supplement intake. We only extracted data on healthy subjects who were free of any disease, especially inflammatory conditions, and who were free of known inhalation exposure, or, where possible, who were healthy subjects before an exposure, i.e., baseline data and, if possible, all subgroup-specific data.

2.4. Quality Assessment

Articles that underwent data extraction were also evaluated using a standardized quality assessment checklist, previously used in other studies [22]. This allowed us to appraise four different domains: study design and risk of bias, technical and analytical methods, study sample, and data interpretation. We assessed different criteria for each domain and graded them with scores from 1 to 3. The output was an overall score ranging from 0 to 27. In this range, two cut-offs and three quality categories were established: "low" (scores less than or equal to $\leq$9), "moderate" (scores less than or equal to $\leq$18), and "high" (scores greater than 19). The checklist used to perform the quality assessment is detailed in Supplementary Table S1 and the overall scores of articles elected for data extraction are listed in Table 1.

Table 1. Qualitative synthesis of articles that underwent data extraction.

Author, Year, Country	Study Objectives	Population Studied and Number of Participants	Control Population	Method of EBC Collection and Analysis	Main Findings	Quality Score
Aksu, 2012, Turkey [23]	To investigate lower airway inflammation using malondialdehyde and total protein measurement in exhaled breath condensate	12 Mild asthma, 53 persistent rhinitis, and 16 concomitant asthma and rhinitis	13 Control subjects (5 males and 8 females)	EBC was collected with a homemade apparatus for max 15 min at 10.00 am, no nose clip, immediately stored at −80 °C	MDA and total protein levels in EBC did not differ among the groups	14 Moderate
Andreoli, 2003, Italy [24]	Method development	None	2 Non-smoking and 13 asymptomatic smoking control subjects	EBC collected on EcoScreen, 15 min with nose clip, and then quickly stored at −80 °C	MDA concentrations were consistent with those determined in EBC using an independent HPLC-FLD method, mass spectrometry to EBC analysis could play a fundamental role in the validation of EBC as a new accessible biological matrix and in the definition of standardized EBC collection	12 Moderate
Antus, 2015, Hungary [25]	To investigate oxidative stress in cystic fibrosis (CF) patients	40 Patients with CF	25 Control subjects (13 males and 12 females)	EBC collection with an EcoScreen condenser using MDA reagent kit, MDA concentrations in EBC were measured with an isocratic HPLC system		19 High
Antus, 2013, Hungary [26]	To investigate MDA levels in patients with acute COPD exacerbation	55 Patients with COPD	20 Healthy controls, ex-smokers (10 females and 10 males)	EBC collected for 10 min EcoScreen condenser, samples were stored frozen at −80 °C before analysis, isocratic HPLC system using MDA reagent kit		12 Moderate
Antczak, 1997, Poland [27]	To determine whether asthmatic patients exhale more H_2O_2 and TBARS than healthy subjects	21 Asthmatic patients	10 Healthy subjects (6 males and 4 females)	EBC collected in a tube installed in a polystyrene foam container filled with ice and salt, for 20 min, and aliquotes were transferred in Eppendorf tubes stored at −80 °C for no longer than 7 days until measurement, TBAR species in expired breath condensate was determined according to the method of YAGI	H_2O_2 was 26 times higher in asthmatic EBC as compared with TBARS, these findings were statistically significant	15 Moderate
Araneda, 2005, Chile [28]	To investigate lung oxidative stress in subjects exercising in moderate and high altitude	None	6 Healthy males	EBC collected with self-constructed device for 20 min, condensated at −5 °C, and stored in liquid nitrogen, MDA was determined with HPLC according to Larstad	Lung OS may constitute a pathogenic factor in acute mountain sickness	11 Moderate

Table 1. *Cont.*

Author, Year, Country	Study Objectives	Population Studied and Number of Participants	Control Population	Method of EBC Collection and Analysis	Main Findings	Quality Score
Araneda, 2012, Chile [29]	To investigate the impact of an endurance race on pulmonary pro-oxidative formation and lipoperoxidation	41 Healthy recreational runners divided into 3 groups according to the races they decided to run (10, 21.1, and 41.2 km) non-smokers	Two groups (10 and 42.2 km) with MDA in EBC, pre-race data was available just for these subjects	EBC collected with self-constructed device for 10 to 20 min, with nose clip, condensated at −5 °C, and stored in liquid nitrogen, MDA was determined with HPLC according to Larstad	Intense prolonged exercise favors an increase in pulmonary pro-oxidative levels, with no modifications on lipoperoxidation; Running time relates to the magnitude of acute post exercise pro-oxidative formation	12 Moderate
Araneda, 2013, Chile [30]	To investigate pro-oxidative, lipid peroxidation, and inflammation in EBC, in subjects running 10 km	None	10 Physically active healthy subjects (9 males and 1 female), EBC measured 20 min before the race	EBC collected with self-constructed device for 10 to 15 min, with nose clip, condensated at −5 °C, and stored in liquid nitrogen, MDA was determined with HPLC according to Larstad	Unlike the previous results obtained in amateur runners, in physically active subjects, 10 km produces an increase in oxygen- and nitrogen-derived pro-oxidative species, without early peroxidation	7 Low
Barregard, 2007, Sweden [31]	To examine whether short-term exposure to wood smoke in healthy subjects affects markers of pulmonary inflammation and oxidative stress		13 Healty subjects (6 males and 7 females), EBC measured prior to exposure	EBC was collected with an Rtube, with nose clip, until 80 L were collected, samples were frozen at −20 °C, MDA was determined with HPLC according to Larstad	Wood smoke exposure affects the respiratory tract, especially the lower airways	10 Moderate
Bartoli, 2011, Italy [32]	To assess the usefulness of MDA in EBC in different groups of pulmonary diseases	64 Subjects with asthma, 19 with bronchiectasies, 73 with COPD, 38 with idiopathic pulmonary fibrosis	14 Healthy non-smoking subjects	EBC was collected on an EcoScreen device, during 15 min, samples were immediately stored at −80 °C and analyzed within 6 months, with HPLC according to Larstad	Subjects with chronic airway disorders have increased levels of MDA in EBC, MDA concentrations in EBC are related to FEV1 and neutrophilic inflammation, particularly in COPD patients.	15 Moderate
Brand, 2013, Germany [33]	To investigate if short-term exposure to welding fumes results in changes in lung function and early stages of inflammatory reactions		12 Healthy non-smoking males	EBC was collecte on an ECOSCREEN device wearing a nose clip, for 20 min, samples stored at −80 °C until analysis, MDA was derivatized with DNPH, and then separated by HPLC and determined by mass spectrometry	In healthy, young subjects, neither changes in spirometry nor changes in inflammatory markers measured in exhaled breath condensate could be detected after short-term exposure	12 Moderate
Caglieri, 2006, Italy [34]	To investigate chromium levels in EBC of workers exposed to Cr(VI) and to assess their relationship with biochemical changes in the airways by analyzing EBC biomarkers of oxidative stress	24 Chrome-plating workers	25 Control subjects (13 males and 12 females, 5 ex-smokers and 20 non-smokers)	EBC was collected on TURBO DECCS, condensation temperature of −5 °C, during 15 min; MDA-EBC was measured by tandem liquid chromatography–mass spectrometry	Cr-EBC levels correlated with those of H2O2-EBC and MDA-EBC, as well as with urinary Cr levels.	14 Moderate
Casimirri, 2015, Italy [35]	Casimirri, 2015, Italy	Casimirri, 2015, Italy	Casimirri, 2015, Italy	Casimirri, 2015, Italy	Professional exposure to chlorinated agents increases EBC biomarkers of oxidativestress and inflammation	15 Moderate

Table 1. *Cont.*

Author, Year, Country	Study Objectives	Population Studied and Number of Participants	Control Population	Method of EBC Collection and Analysis	Main Findings	Quality Score
Corradi, 2003, Italy [36]	To evaluate if aldehydes could be measured in EBC, to assess the influence of sampling procedures, to compare levels of different pulmonary disease groups with those of a control group	20 Patients with stable COPD (18 males and 2 females)	12 Smoking (9 males and 3 females) and 20 non-smoking (17 males and 3 females) control subjects	EBC was collected on a Tygon Tube immersed in thawing ice, and then frozen to −80 °C, no nose clip, MDA measured with LC-MS tandem	Aldehydes were identified in EBC, all, but were lower in control groups	16 Moderate
Corradi, 2002, Italy [37]	To investigate if short-term exposure to ozone (O3) induces changes in biomarkers of lung inflammation and oxidative stress in EBC		22 Non-smoking healthy control subjects (12 males and females)	EBC was collected on an ECO SCREEN device, during 15 min, frozen at −80 °C, MDA was measured as TBARS according to Nowak	A single 2-hour exposure to 0.1 ppm of O3 induces changes in biomarkers of inflammation and oxidative stress in those susceptible	11 Moderate
Corradi, 2004, Italy [38]	To compare aldehyde levels resulting from lipid peroxidation in EBC and induced sputum (IS) supernatant of subjects with asthma and chronic obstructive pulmonary disease	21 Subjects with COPD, 10 asthmatics	9 Healthy non-smoking control subjects (8 females and 1 male)	EBC was collected on a two glass chamber device from Incofar, no nose clip for 20 min, samples were stored at −80 °C, and MDA measured according to Larstad	Aldehydes can be detected in both exhaled breath condensate and supernatant of induced sputum, but their relative concentrations are different and not correlated with each other	12 Moderate
Doruk, 2011, Turkey [39]	To investigate oxidative stress in the lungs associated with tobacco smoke and to evaluate the effect of this stress with pulmonary function tests		69 Healthy subjects divided into 3 groups according to their exposure to tobacco smoke: 26 current smokers (23 males and 3 females), 21 subjects (15 males and 6 females) who did not smoke within the last year but had second-hand smoking and 22 (100 males and 12 females) with no tobacco smoke exposure	EBC was collected on an ECOSCREEN device for 15 min, wearing a nose clip, samples stored at −70 °C, MDA measured as TBARS	The levels of MDA, 8-OHdG, SOD, and GSH-Px were higher in smokers; NO levels gradually increased from Group I to Group III; MDA levels were lower in Group III than Group II	14 Moderate
Goldoni, 2004, Italy [40]	To investigate whether EBC can be used as a suitable matrix to assess target tissue dose and effects of inhaled cobalt and tungsten, using EBC malondialdehyde (MDA) as a biomarker of pulmonary oxidative stress	33 Workers exposed to Co and W in workshops producing either diamond tools or hard-metal mechanical parts	16 Control subjects (11 males and 5 females)	EBC was collected on a homemade apparatus, during 10 min, samples were transported in ice to the lab, and then stored at −80 °C, MDA was measured by liquid chromatography–tandem mass spectrometry (LC-MS/MS)	MDA levels were increased depending on cobalt concentration and were enhanced by coexposure to tungsten.	11 Moderate

Table 1. *Cont.*

Author, Year, Country	Study Objectives	Population Studied and Number of Participants	Control Population	Method of EBC Collection and Analysis	Main Findings	Quality Score
Goldoni, 2013, Italy [41]	To compare the concentration of several biomarkers in whole (W-EBC) and fractionated EBC (A-EBC)		45 Healthy control subjects (10 males and 35 females, 6 smokers and 39 non-smokers)	EBC was collected on a homemade apparatus for the fractioned EBC and a TURBODECCS for the whole EBC for 15 min without a nose clip, EBC was centrifuged, and then stored at −80 °C, MDA was measured by liquid chromatography–tandem mass spectrometry	H2O2, 8-isoprostane, malondialdehyde, and 4-hydroxy-2-nonhenal were all higher in W-EBC, suggesting a contribution from the upper airways to oxidative stress biomarkers in apparently healthy subjects	10 Moderate
Goldoni, 2005, Italy [42]	To test the effect of condensation temperature on the parameters of exhaled breath condensate and the levels of selected biomarkers		24 Healthy control subjects (13 males and 11 females, 3 ex-smokers and 21 non-smokers)	EBC was collected on a TURBODECCS, during 10 min, at different temperatures, samples were centrifuged, and then stored at −80 °C, MDA was measured by liquid chromatography–tandem mass spectrometry (LC-MS/MS) within 2 weeks	Cooling temperature of exhaled breath condensate collection influenced selected biomarkers	13 Moderate
Gong, 2014, USA/China [43]	To compare ultrafine particles (UFPs) and fine particles (PM2.5) with respect to their associations with biomarkers reflecting multiple pathophysiological pathways linking exposure and cardiorespiratory events		125 Healthy non-smoking individuals (initially 64 males and 64 females) working on a campus near Peking's University Health Sciences Centre	EBC was collected on an ECOSCREEN device, collected for 20 min, with nose clip, aliquotes stored at −70 °C, MDA was measured as MDA-TBA in HPLC	Associations of certain biomarkers with UFPs had different lag patterns as compared with those with PM2.5, suggesting that the ultrafine size fraction and the fine size fraction of PM2.5 are likely to affect PM-induced pathophysiological pathways independently	15 Moderate
Graczyk, 2016, Switzerland [44]	To investigate time course changes of particle-associated oxidative stress in exposed tungsten inert gas welders		20 Non-smoking healthy volunteers, with less than 1 year of apprentice in welding	EBC was collected on an Rtube, during 10 min, while wearing a nose clip, EBC samples were stored at −70 °C, MDA was measured by HPLC separation and fluorescence detection TBARS	A 60-minute exposure to TIG welding fume in a controlled, well-ventilated setting induced acute oxidative stress at 3 h post exposure in healthy, non-smoking apprentice welders not chronically exposed to welding fumes	10 Moderate
Gube, 2010, Germany [45]	To investigate the effect of welding as well as the impact of smoking and protection measures on biological markers in EBC	45 Male welders	24 Healthy males, non-exposed	EBC was collected on an ECOSCREEN device, wearing a nose clip, as long as 200 L of exhaled breath was collected, then stored at −80 °C, MDA was measued with a previously described but slightly changed method, i.e., MDA derivatized with DNPH and separated by HPLC, and then tandem mass spectrometry	Welders showed significantly increased concentrations of all these parameters at baseline as compared with non-exposed controls	14 Moderate

Table 1. *Cont.*

Author, Year, Country	Study Objectives	Population Studied and Number of Participants	Control Population	Method of EBC Collection and Analysis	Main Findings	Quality Score
Inonu, 2012, Turkey [46]	To evaluate the differences in the burden of oxidative stress in patients with COPD, smokers, and non-smokers by measuring hydrogen peroxide (H_2O_2), malondialdehyde (MDA), and 8-isoprostane levels in EBC	25 Male COPD smokers	26 Smokers and 29 non-smokers, males, healthy	EBC collected on an EcoScreen device, 15 min with nose clip, and then quickly stored at −70 °C for 6 months, MDA was measured using a commercial kit in a fluorescence detector in HPLC	Even if respiratory function tests were within normal limits, oxidant burden in the lungs of smokers was equivalent to that in COPD patients, 8-isoprostane could be useful in assessing symptom severity and health status of COPD patients	16 Moderate
Larstad, 2001, Sweden [47]	To develop a method of MDA quantification	29 Patients with asthma, 7 of which with wheezing	15 control subjects without asthma	EBC was collected on an ECOSCREEN d evice, during 4 min, with a nose clip, stored at −20 °C until analysis, MDA was measured using TBS at 95 °C, and then separated by HPLC and MDA-TBA measured by fluorescence	No statistically significant difference between patients with asthma and patients without asthma; however, among females, subjects with asthma had higher MDA levels as compared with females without asthma; the use of the method when studying airway inflammation has to be further evaluated	12 Moderate
Laumbach, 2014, USA [48]	To determine if exposure to traffic-related pollutant particles (TRAPs) during commuting causes acute oxidative stress in the respiratory tract or changes in heart rate variability		21 Young volunteers (15 males and 6 females)	EBC was collected with an ECOSCREEN device, for 20 min, samples were frozen to −80 °C, MDA analysiswas performed by using a mixture of EBC, phosphoric acid, and thiobarbituric acid, which was heated and injected into an HPLC fluorescence system	Increases in markers of oxidative stress in EBC may represent early biological responses to widespread exposures to TRAPs particles that affect passengers in vehicles on heavily trafficked roadways	9 Low
Lee, 2015, Korea [49]	To investigate the actual health effects of multi wall carbon nanotubes in manufacturing workers	9 Male carbon nanotube manufacturing workers, 5 smokers	4 Office workers (3 males and 1 female), 2 smokers	EBC was collected on an Rtube, during 10 min, with a noseclip, stored in dry ice for 2 weeks. MDA was measured using fluorescence HPLC	Analyzed biomarkers in the MWCNT manufacturing workers were significantly higher than those in the office workers; MDA, n-hexanal, and molybdenum could be useful biomarkers of MWCNT exposure	10 Moderate
Majewska, 2004, Poland [50]	To determine whether concentrations of H_2O_2 and TBARs in EBC are elevated and correlate with systemic response to pneumonia during 10 days of hospital treatment	43 Inpatients with community acquired pneumonia (12 females and 31 males)	20 Healthy non-smoking control subjects (6 females and 14 males)	EBC was collected on a device from Jaeger, cooled with ethanol at −9 °C, during 20 min, wearing a nose clip, stored on ice until measurement was performed, MDA was measured as TBARS	Pneumonia is accompanied by oxidative stress in airways that moderately correlates with intensity of systemic inflammatory response, determination of H_2O_2 in EBC may be helpful for noninvasive monitoring of oxidant production during lower respiratory tract infection	13 Moderate

Table 1. *Cont.*

Author, Year, Country	Study Objectives	Population Studied and Number of Participants	Control Population	Method of EBC Collection and Analysis	Main Findings	Quality Score
Nowak, 2001, Poland [51]	To investigate the concentration of H_2O_2 and TBARs in EBC and influencing factors		58 Healthy volunteers (18 smokers and 40 non-smokers, 31 males and 27 females)	EBC was collected on a device from Jaeger Toennies, during 20 min, wearing a noseclip, MDA was measured as TBARS spectrofotometrically	Neither moderate exercise nor one puff of salbutamol nor ipratropium significantly influenced the concentration of H_2O_2 and TBARs in EBC, only 4 of 120 EBC specimens from non-smoker subjects revealed detectable levels of TBARs, cigarette smokers exhaled more TBARs	11 Moderate
Pelclova, 2018, Czech Republic [52]	To investigate lung oxidative stress in workers handling nanocomposites	19 Nanocomposite-synthesizing and processing researchers (14 males and 5 females), all non-smokers	19 Control subjects (13 males and 6 females, all non-smokers)	EBC was collected on an ECOSCREEN device, until a minimum of 120 L of exhaled breath, wearing a nose clip, immediately stored at −80 °C, MDA was measured	Significant associations were found between working in nanocomposite synthesis and EBC biomarkers; more research is needed to understand the contribution of nanoparticles from nanocomposite processing in inducing oxidative stress, relative to other co-exposures generated during welding, smelting, and secondary oxidation processes	12 Moderate
Pelclova, 2014, Czech Republic [53]	To search for optimal markers in the exhaled breath condensate (EBC), plasma, and urine that would reflect the activity/severity of occupational asthma (OA) after withdrawal from the exposure to the allergen	43 Subjects with previously diagnosed immunological OA (18 males and 25 females)	20 Control subjects, working as office or healthcare employees and having no symptoms of asthma (10 males and 10 females)	EBC was collected for 15–20 min with EcoScreen, wearing a nose clip, samples were stored at −80 °C for a maximum of 2 months, MDA was measured with LC-ESI-MS/MS	Improvement in OA is very slow and objective impairments persist years after removal from the exposure; cysteinyl LTs and 8-ISO in EBC and 8-ISO in plasma might enrich the spectrum of useful objective tests for the follow-up of OA	12 Moderate
Pelclova, 2016, Czech Republic [54]	To investigate the utility of oxidative stress biomarkers in EBC in iron oxide pigment production workers	14 male workers, 43 ± 7 years, 43% smokers	14 Males, 39 ± 4 years, 50% smokers, non-exposed	EBC was collected for 15 min with an EcoScreen device, wearing a nose clip, samples were stored at −80 °C, MDA was measured with LC-ESI-MS/MS	Almost all markers of lipid, nucleic acid, and protein oxidation were elevated in the EBC of workers as compared with control subjects; markers in urine were not elevated	10 Moderate
Ricelli, 2018, Italy [55]	To investigate the cromium and nickel content of EBC of stainless steel (SS) tungsten inert gas (TIG) welders, and relate their concentrations with oxidative stress and inflammatory biomarkers		100 SS welding workers, aged 18–65 years, smokers/non-smokers/ex-smokers 33/36/31, values were considered pre-shift	EBC was collected on an ECOSCREEN device, during 15 min, and stored at −80 °C, MDA was determined by tandem LC-MS/MS	Given the weak relationship between the biomarkers and effects of exposure, we speculate that other substances generated during SS TIG welding also play a role in generating lung oxidative stress	11 Moderate
Rundell, 2008, USA [56]	To investigate PM1 inhalation during exercise on lung function, exhaled nitric oxide (eNO), total nitrate (NO3), S-nitrosoglutathione (GSNO), and malondialdehyde (MDA) in EBC		Twelve physically fit, non-asthmatic, non-smoking males	EBC was collected on an ECOSCREEN device, during 15 min, wearing a nose clip and stored at −80 °C, MDA was measured after derivatization with TBA by HPLC fluorescence	MDA increased 40% after low PM exercise, high PM1 inhalation during exercise caused a reduced alveolar contribution to eNO, NO3 and eNO variables were decreased and were related to impaired lung function	9 Low

Table 1. *Cont.*

Author, Year, Country	Study Objectives	Population Studied and Number of Participants	Control Population	Method of EBC Collection and Analysis	Main Findings	Quality Score
Sakhvidi, 2015, Iran [57]	To measure exhaled breath malondialdehyde (EBC-MDA) in workers exposed to dust containing silica as compared with that of a non-exposed control group	25 Male workers in a ceramic factory	25 Male control subjects from administrative departments of the same factory	EBC was collected with a homemade apparatus, during 5 min, wearing a nose clip, samples were stored in a freezer at −20 °C until examination, MDA was analyzed using an HPLC equipped with a fluorescent detector	Significant correlation between respirable dust exposure intensity and the level of EBC-MDA of the exposed subjects, but no significant correlation with lung functions	13 Moderate
Sauvain, 2014, Switzerland [58]	To evaluate the feasibility of using exhaled breath condensate (EBC) from healthy volunteers for (1) assessing the lung deposited dose of combustion nanoparticles and (2) determining the resulting oxidative stress		15 Healthy non-smoking control subjects	EBC collected on a Rtube, during 10 min, wearing a nose clip, a aliquote was stored at −78 °C, MDA was derivatized with TBA, and then measured with HPLC fluorescence	The results suggest two phases of oxidation markers in EBC: first, the initial deposition of particles and gases in the lung lining liquid, and later the start of oxidative stress with associated cell membrane damage	11 Moderate
Stockfelt, 2012, Sweden [59]	To examine airway effects of two kinds of wood smoke in a chamber study		16 Healthy non-smoking control subjects (8 males and 8 females), 3 excluded for respiratory problems before the experiment	EBC collected according to Barregard, MDA measured always according to Barregard	Relatively low levels of wood smoke exposure induce effects on airways	8 Low
Syslova, 2009, Czech Republic [60]	To develop a sensitive method for a parallel, rapid and precise determination of the most prominent oxidative stress biomarkers for patients with silica or asbestos disease	20 Subjects with previous exposure to silica or asbestos and related disease	10 Control subjects	EBC was collected on an ECOSCREEN device, for 5–10 min, wearing a nose clip, frozen at −80 °C, and stored for a maximum of 1 month, MDA was measured with optimized LC–ESI-MS/MS	The differences in concentration levels of biomarkers between the two groups was perceptible in all the body fluids (the difference observed in an exhaled breath condensatewas statistically most significant)	12 Moderate
Szkudlarek, 2003, Poland [61]	To test whether exhalation of H_2O_2 and TBARs by healthy subjects depends on reactive oxygen species generation from blood phagocytes		41 Healthy, non-smoking control subjects, mean age 20.770.8 years (18 males and 23 females)	EBC was collected on a homemade mouth piece connected to a glass tube cooler, wearing a nose clip, during 20 min, MDA was measured as TBARS according to Nowak	No association between exhaled TBARs and blood phagocytes activity was found	10 Moderate
Pelclova, 2020, Czech Republic [62]	To analyze biomarkers in EBC of nanocomposite workers and understand the health effects	20 Researchers handling nanocomposites (15 males and 5 females), 19 non-smokers, mean age of 41.8 years	21 Control subjects (15 males and 6 females), office emplyees, 19 non-smokers, mean age of 42.7 years	EBC was collected on ECOSCREEN and stored at −80 °C, MDA was measured with LC-ESI- MS/MS	among inflammation markers, LT4 and tumor necrosis factor were the most useful	10 moderate

2.5. Statistical Analysis

We computed geometric means (GMs) and geometric standard deviations (GSDs) as the basis of the meta-analysis. For value conversion and unit harmonization, we used different formulae according to types of data available and as previously described in [63].

Regarding the meta-analysis itself, the first step involved estimating the geometric means (GMs), and then representing the data with forest plots since they illustrated the heterogeneity and the overall combined results from individual studies. Between-study heterogeneity was evaluated with a Q test and displayed for each category in the forest plots. Finally, we modeled the Log-transformed GMs using linear models. We reported the results in ng/mL of MDA in EBC.

3. Results

Among the 40 full-text articles from which we extracted the data, several studies had to be excluded due to statistical errors such as lack of variability estimates or impossible reported ranges for the studied groups. We also excluded from the meta-analysis several study groups that demonstrated a coefficient of variation (CV), either <20% or >200%. The former corresponds to the analytical variability alone, and implies no between-subject variability. The latter implies that the study is uninformative. One study was excluded as the results had been reported using inappropriate units. The whole selection process is illustrated in Figure 1.

Since different factors were indicated as potential confounding factors in the literature, we wanted to assess their influence. Therefore, the results were stratified by gender, age, smoking status, EBC collection device, duration of EBC collection, use of nose clip, and analytical methods combined with their detection methods. We created a separate group identified by the abbreviation NA, meaning "not available", which was assigned when the information was not given in the article. This resulted in the description of 28 study subgroups. Figure 2 shows the boxplots concerning the collection device (Figure 2a), duration of collection (Figure 2b), and gender groups (Figure 2c).

Among the included articles, we identified the following analytical and separation methods combined with a derivatizing agent: high-performance liquid chromatography/thiobarbituric acid HPLC/TBA, spectrophotometry/TBA, HPLC—mass spectroscopy, HPLC-MS/initrophenylhydrazine (DNPH), and HPLC without mentioning the derivatizing agent and detection technique. A mixed effect regression analysis was performed to verify differences among the MDA GMs measured by different analytical methods. The same was done concerning the use of a nose clip versus no use of a nose clip, age categories, and among smokers and non-smokers. For each category, a factor-specific Wald test within the model was equally performed. Table 2 shows the results of the mixed-effects regression analysis. As compared with MDA GMs measured in EBC by HPLC/TBA, which was considered to be the reference method, the MDA GMs measured by MS/DNPH were significantly lower. The GMs of EBC MDA collected in uncertain conditions were higher as compared with the GMs of EBC collected while wearing a nose clip, which was the reference group. GMs of MDA in EBC higher than the reference were also found in smokers and in subjects 50+ years old as compared with in subjects younger than 30 years old.

These results were equally presented as separate forest plots, which better illustrate the heterogeneity for the age categories (Figure 3) and smoking habits (Figure 4).

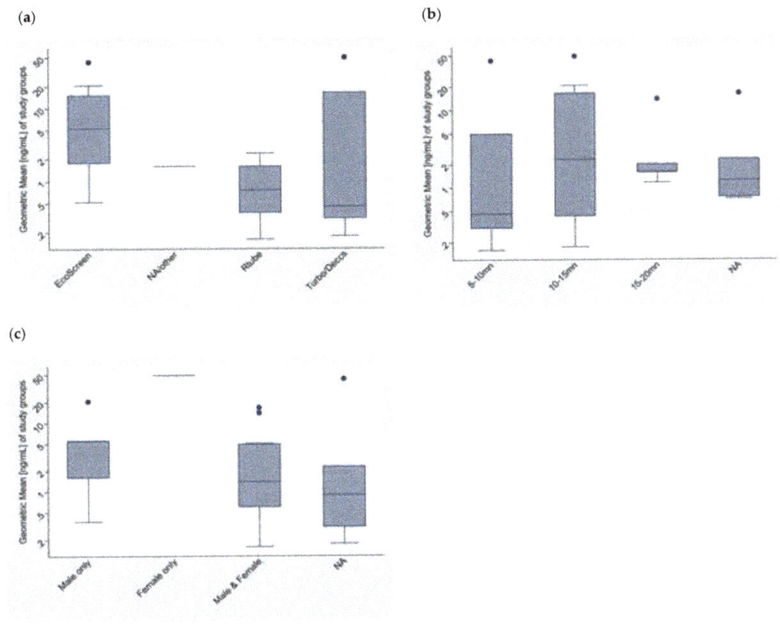

Figure 2. MDA concentration in EBC stratified according to: (**a**) Collection device; (**b**) duration of collection (minutes); (**c**) gender group (no papers with females only). All presented results exclude the outliers and units of the GMs are ng/mL.

Table 2. Mixed effects regression analysis according to analytical method, use of nose clip, smoking status, and age category.

Parameters	β Coefficient	[95% Conf. Interval]	P > |t|
Analytical method			
HPLCfluo/TBA (n = 9)	reference		
Spectro/TBA (n = 2)	0.30	[−0.902; 1.49]	0.59
MS/DNPH (n = 9)	−1.03	[−2.02; 0.032]	0.04
HPLCfluo/NA (n = 2)	1.26	[−0.079; 2.6]	0.06
Factor-specific Wald test			0.04
Use of nose clip			
Yes (n = 9)	reference		
No (n = 8)	0.64	[−0.52; 1.81]	0.25
NA/other (n = 5)	1.14	[0.32; 1.96]	0.01
Factor-specific Wald test			0.04
Smoking habit			
Non-smokers (n = 11)	reference		
Mixed population (n = 3)	−0.71	[−1.70; 0.28]	0.14
Smokers (n = 5)	0.8	[0.08; 1.51]	0.03
NA (n = 3)	0.37	[−0.57; 1.31]	0.4
Factor-specific Wald test			0.04
Mean age (years)			
<30 (n = 4)	reference		
30–<40 (n = 8)	0.88	[−0.014; 1.77]	0.05
40–<50 (n = 3)	0.82	[−0.51; 2.16]	0.2
≥50 (n = 7)	1.21	[0.09; 2.33]	0.04
Factor-specific Wald test			0.15
Intercept	−1.15	[−2.07; −0.22]	0.02

HPLCfluo/TBA—high-performance liquid chromatography fluorescence/thiobarbituric acid; spectro/TBA—spectrophotometry/thiobarbituric acid; MS—mass spectroscopy; NA—information not available.

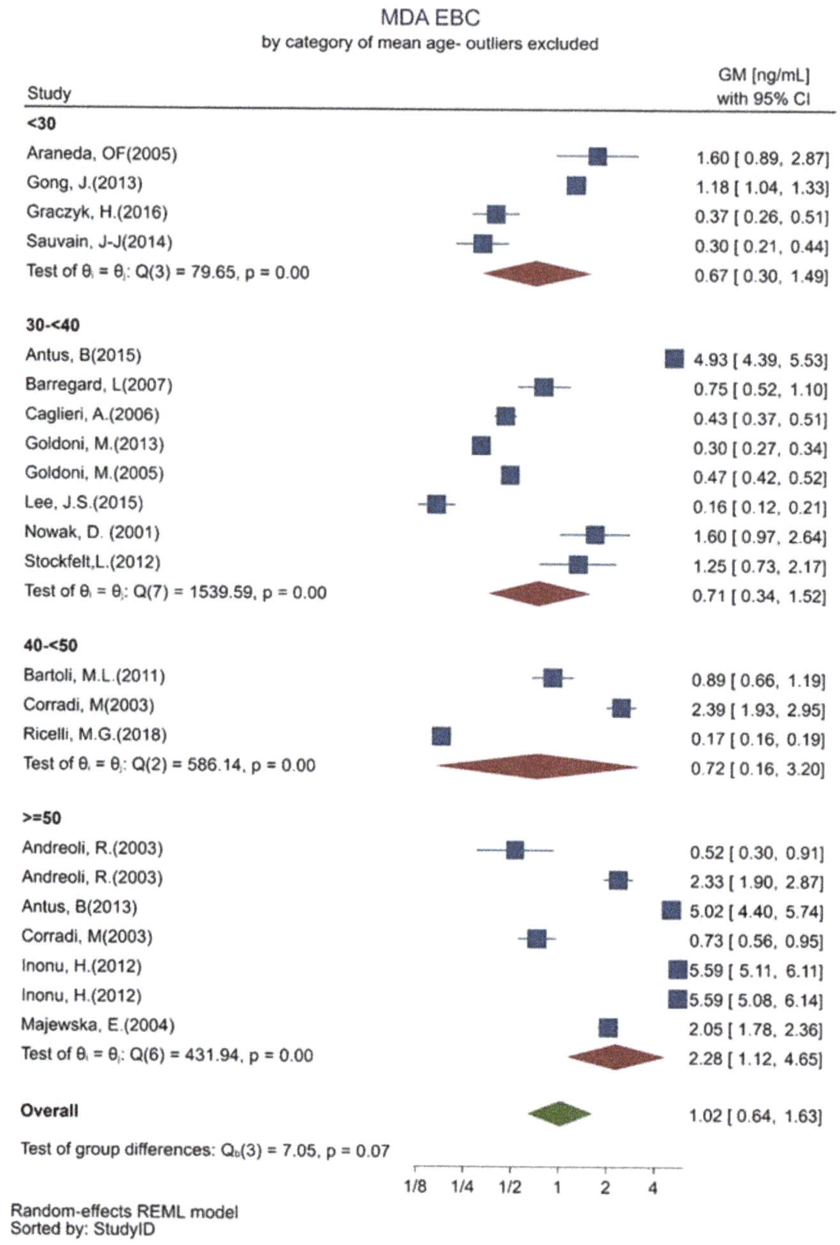

Figure 3. Forest plots of MDA in EBC concentration quantified in healthy adult participants by age group: population with a mean age less than 30 years [28,43,46,58]; between 30 and 40 years [25,31,34,41,42,49,51,59], between 40 and 50 years [32,36,55] and above 50 years [24,26,36,46,50].

The predicted GMs in ng/mL for MDA measured in EBC is provided in Table 3, as well as the 95% confidence interval for each parameter considered. We observe that most of our predicted values fall in the interval of 0.18–2.87 ng/mL.

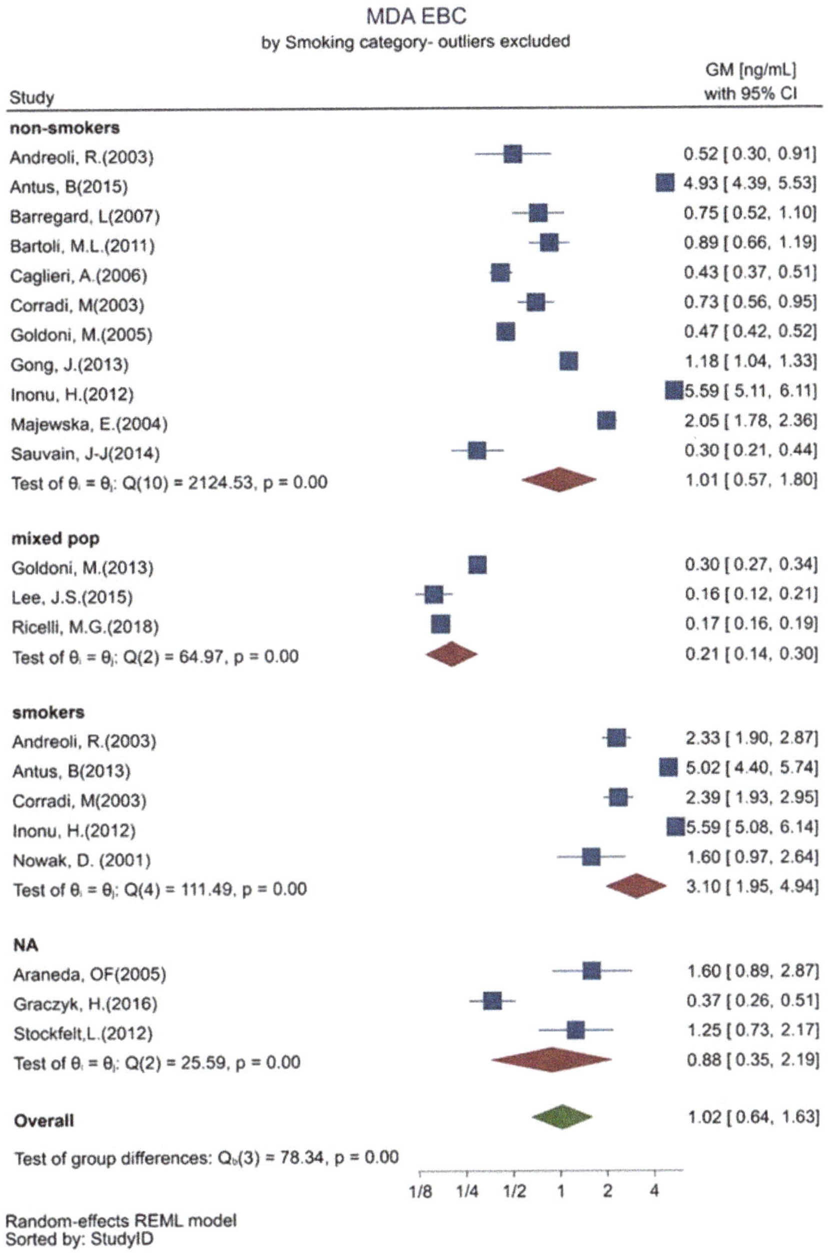

Figure 4. Forest plots of MDA in EBC concentration quantified in healthy adult participants considering their smoking habits: non-smokers exclusively [24,25,31,32,34,36,42,43,46,50,58]; mixed population of smokers and non-smokers [41,49,55], smokers exclusively [24,26,36,46,51] and unknown smoking status [28,44,59].

Table 3. Predicted concentrations of MDA (ng/mL) in EBC using the linear model.

Parameters	GM	95% Confidence Interval
Analytical method		
HPLCfluo/TBA (n = 9)	1.8	[0.795; 2.814]
Spectro/TBA (n = 2)	2.43	[0.142; 4.710]
MS/DNPH (n = 9)	0.65	[0.279; 1.015]
HPLCfluo/NA (n = 2)	6.35	[−0.252; 12.954]
Use of nose Clip		
NA/other (n = 5)	3.72	[1.150; 6.292]
No (n = 8)	2.27	[0.141; 4.403]
Yes (n = 9)	1.19	[0.656; 1.727]
Smoking habits		
Non-smokers (n = 11)	1.2	[0.751; 1.665]
Mixed population (n = 3)	0.6	[0.074; 1.111]
Smokers (n = 5)	2.69	[1.297; 4.065]
NA (n = 3)	1.75	[0.294; 3.207]
Mean Age (years)		
<30 (n = 4)	0.68	[0.186; 1.178]
30–<40 (n = 8)	1.64	[0.767; 2.520]
40–<50 (n = 3)	1.56	[0.238; 2.875]
≥50 (n = 7)	2.28	[1.150; 3.425]

4. Discussion

This meta-analysis showed that subjects older than 50 years had different MDA in EBC values from those younger than 30 years of age. Hence, for healthy subjects aged 30–50 years, we observed MDA in EBC values ranging from 0.18 to 2.87 ng/mL. Another observation we made was that smokers had higher values, almost double, of MDA in EBC as compared with non-smokers.

These findings are in line with recent studies that have investigated the link between aging and oxidative stress and have demonstrated the age dependence of oxidative stress in humans, although on different matrices. When measuring several biomarkers of oxidative stress, Pinchuk et al. revealed that MDA increased with age, and especially in post-menopausal women [63], and they attributed this to changes in estrogen levels. A meta-analysis regarding urinary MDA has made similar findings, that is, a trend of MDA values increasing with age, probably through the pathway of proteasome dysfunction. In addition, the association between smoking and increased oxidative stress has been well established, i.e., smoking induces the production of reactive oxygen species (ROS) that may activate a cascade and generate further molecular damage [64]. Although limited, there is evidence that MDA also increases in the EBC of smokers and not just in biological matrices such as the blood [65].

Our meta-analysis revealed significant variations in MDA levels according to the use of a nose clip during the EBC collection. The interpretation of such a finding is flawed by the inclusion of studies that did not specify if they used the nose clip or not. The finding itself would be coherent with the present recommendations to use a nose clip.

Nevertheless, we do not suggest the above-mentioned value intervals as reference intervals, due to several factors, which are discussed in the following paragraphs, such as the high heterogeneity, the low number of studies, or the use of proximity samples in the majority of the selected studies.

4.1. Heterogeneity

Our meta-analysis had a high level of heterogeneity, which we failed to reduce in subgroup analysis. This may be due to several factors, such as the heterogeneity in several steps: sample collection, chemical analytical analysis, data collection, and data reporting.

Missing information regarding these steps also led to mitigated quality scores, mostly being moderate to low (Table 1).

4.1.1. Heterogeneity Related to the Collection of EBC

We checked the compliance with current methodological recommendations for EBC collection according to the date of article publication, and found three studies performed after 2017; two of these studies respected the guidelines concerning the EBC collection. The majority of the 26 studies included in the qualitative analysis was published between 2005 and 2017. Even though these studies should have adhered to the ERS/ATS 2005 recommendations, ten studies stated that they did not use a nose clip or did not provide the information regarding the use of a nose clip. The reasons for omitting this information or for not providing a nose clip to the participants as recommended by the guidelines remain to be unclear. Similar observations were made concerning the EBC devices, collection duration, and time of day the EBC collection was performed. Further, for about a fourth of the studies, we did not have enough information from the published article to allow us to understand if saliva contamination had been excluded.

4.1.2. Heterogeneity Related to the Analytical Method

Studies in our meta-analysis reported the use of different MDA derivation methods. The authors presented improved or modified already existing methodologies, which probably contributed to the overall heterogeneity. Moreover, data concerning the limit of detection and/or the limit of quantification, exclusion of potential contamination, and parameters of transportation or storage were partially revealed in the articles.

4.1.3. Heterogeneity Related to Data Collection and Reporting

Concerning the data collection and reporting, almost 20% of the studies did not specify the distribution of males and females among their control participants, 25% of the studies did not clearly report the smoking status, and 60% of the studies did not provide information on the BMI. Ideally, in addition to these basic data, more information concerning lifestyle factors and dietary information, such as vitamin intake, medical regime, occasional drug use, or living in an area with traffic-related heavy air pollution, should be taken into consideration, to better study their influence on MDA levels [66]. As these factors might influence the level of MDA in a reference sample, MDA reference intervals should be stratified accordingly, avoiding any misleading estimation.

5. Study Strengths and Limitations

The strengths of our meta-analysis are our rigorous research protocol, the exhaustive literature search by a documentarist using four databases, and our effort to harmonize units and generate background values as geometric means. Given the time lapse between literature research and data extraction, we also checked for studies published after October 2020, but this did not change our dataset.

Nevertheless, in this meta-analysis, the reduced number of performed studies and their quality represented major limitations. Ultimately, this also led to the high overall heterogeneity, which we did not manage to reduce even by performing a subgroup analysis.

6. Recommendations

Good quality and complete datasets are required to derive a reference interval of MDA in the EBC; however, the heterogeneity among studies observed here, prompts us to strongly recommend that researchers collect and report complete data information in future studies. This pertains especially to:

(a) Collection and reporting of demographics and health status, namely gender, age, smoking status, BMI, diet, living area, sports, and respiratory functions to insure that the study sample does really correspond to the reference sample, and that it is representative of the general healthy population and all its subgroups.

(b) Collection and reporting of the time of EBC collection and volume of collected EBC as this is one ERS/ATS recommendation that has not yet been fulfilled. We strongly recommend avoiding the use of non-validated EBC collection devices since it has been shown that the inner surface of the collection equipment can interfere with the determination of biomarkers [67], presumably, also with MDA determination. Therefore, EBC collections need to be standardized. A greater number of studies respecting the current guidelines would help to get closer to establishing a standardized EBC collection method.

(c) Reporting of analytical methods that may affect MDA quantification and other factors influencing or supposed to influence the results, related to the analysis itself or collection device and material.

7. Conclusions

Our systematic review and meta-analysis could not provide any reference interval for MDA in EBC, due to the large heterogeneity among studies and the methodological limitations encountered. Further research is needed to, first, understand the influence of demographic and collection parameters on the MDA values in EBC, and secondly, to harmonize analytical methods. This and a sufficient number of high-quality studies would help to define MDA reference values in EBC.

Supplementary Materials: The following supporting information can be downloaded at https://www.mdpi.com/article/10.3390/toxics10050258/s1, Supplementary Table S1: Quality assessment criteria. References [22–61] are cited in the supplementary materials.

Author Contributions: The manuscript was written by V.T. and revised by all authors; N.B.H. and I.G.C. conceptualized and designed the study; M.H. initiated the research protocol; V.T. extracted and screened the studies; J.-J.S. served as an advisor for the manuscript and technic information; E.B. served as an advisor for the manuscript drafts; P.W. performed all statistical analyses. All authors have read and agreed to the published version of the manuscript.

Funding: This study was conducted within the framework of the EU Life Project "NanoExplore" (grant no. LIFE17ENV/GR/000285) and the Franco-Swiss project "ROBoCoP" (Swiss National Science Foundation grant no IZCOZ0_177067).

Institutional Review Board Statement: Not applicable.

Informed Consent Statement: Not applicable.

Data Availability Statement: Not applicable.

Conflicts of Interest: The authors declare no conflict of interest. The funders had no role in the design of the study; in the collection, analyses, or interpretation of data; in the writing of the manuscript, or in the decision to publish the results.

References

1. Borrill, Z.L.; Roy, K.; Singh, D. Exhaled breath condensate biomarkers in COPD. *Eur. Respir. J.* **2008**, *32*, 472–486. [CrossRef] [PubMed]
2. Repetto, M.; Semprine, J.; Boveris, A. Lipid Peroxidation: Chemical Mechanism, Biological Implications and Analytical Determination. In *Lipid Peroxidation*; Catala, A., Ed.; InTech: London, UK, 2012; Available online: http://www.intechopen.com/books/lipid-peroxidation/lipid-peroxidation-chemical-mechanism-biological-implications-and-analytical-determination (accessed on 19 October 2021).
3. Ito, F.; Sono, Y.; Ito, T. Measurement and Clinical Significance of Lipid Peroxidation as a Biomarker of Oxidative Stress: Oxidative Stress in Diabetes, Atherosclerosis, and Chronic Inflammation. *Antioxidants* **2019**, *8*, 72. [CrossRef] [PubMed]
4. Cherubini, A.; Ruggiero, C.; Polidori, M.C.; Mecocci, P. Potential markers of oxidative stress in stroke. *Free. Radic. Biol. Med.* **2005**, *39*, 841–852. [CrossRef]
5. Cvetković, T.; Mitić, B.; Lazarević, G.; Vlahović, P.; Antić, S.; Stefanović, V. Oxidative stress parameters as possible urine markers in patients with diabetic nephropathy. *J. Diabetes Its Complicat.* **2009**, *23*, 337–342. [CrossRef]
6. Jové, M.; Mota-Martorell, N.; Pradas, I.; Martín-Gari, M.; Ayala, V.; Pamplona, R. The Advanced Lipoxidation End-Product Malondialdehyde-Lysine in Aging and Longevity. *Antioxidants* **2020**, *9*, 1132. [CrossRef] [PubMed]

7. Del Rio, D.; Stewart, A.J.; Pellegrini, N. A review of recent studies on malondialdehyde as toxic molecule and biological marker of oxidative stress. *Nutr. Metab. Cardiovasc. Dis.* **2005**, *15*, 316–328. [CrossRef]
8. Kostikas, K.; Minas, M.; Nikolaou, E.; Papaioannou, A.I.; Liakos, P.; Gougoura, S.; Gourgoulianis, K.I.; Dinas, P.C.; Metsios, G.S.; Jamurtas, A.Z.; et al. Secondhand smoke exposure induces acutely airway acidification and oxidative stress. *Respir. Med.* **2013**, *107*, 172–179. [CrossRef]
9. Denise, G.; Santa, M.L.; Juliana, V.; Clóvis, P.; Cristina, S.G.G.S.; Pomblum; Juarez, V.; Rocha, J.B.T.; Marcelo, F. Importance of the lipid peroxidation biomarkers and methodological aspects FOR malondialdehyde quantification. *Quim. Nova* **2009**, *32*, 169–174.
10. Sauvain, J.J.; Edmé, J.L.; Wild, P.; Suarez, G.; Bezerra, O.M.P.A.; Talvani, A.; Algranti, E.; Carneiro, A.P.S.; Chérot-Kornobis, N.; Sobaszek, A.; et al. Does exposure to inflammatory particles modify the pattern of anion in exhaled breath condensate? *J. Breath Res.* **2020**, *14*, 026005. [CrossRef]
11. Gaggini, M.; Sabatino, L.; Vassalle, C. Conventional and innovative methods to assess oxidative stress biomarkers in the clinical cardiovascular setting. *Biotechniques* **2020**, *68*, 223–231. [CrossRef]
12. Hunt, J. Exhaled breath condensate: An evolving tool for noninvasive evaluation of lung disease. *J. Allergy Clin. Immunol.* **2002**, *110*, 28–34. [CrossRef] [PubMed]
13. Davis, M.D.; Montpetit, A.; Hunt, J. Exhaled Breath Condensate: An overview. *Immunol. Allergy Clin. N. Am.* **2012**, *32*, 363–375. [CrossRef] [PubMed]
14. Horváth, I.; Barnes, P.J.; Loukides, S.; Sterk, P.J.; Högman, M.; Olin, A.C.; Amann, A.; Antus, B.; Baraldi, E.; Bikov, A. A European Respiratory Society technical standard: Exhaled biomarkers in lung disease. *Eur. Respir. J.* **2017**, *49*, 1600965. [CrossRef] [PubMed]
15. Davis, M.D.; Montpetit, A.J. Exhaled Breath Condensate: An update. *Immunol. Allergy Clin. N. Am.* **2018**, *38*, 667–678. [CrossRef] [PubMed]
16. Latzin, P.; Beck, J.; Bartenstein, A.; Griese, M. Comparison of exhaled breath condensate from nasal and oral collection. *Eur. J. Med Res.* **2003**, *8*, 505–510. [PubMed]
17. Horváth, I.; Hunt, J.; Barnes, P.J. Exhaled breath condensate: Methodological recommendations and unresolved questions. *Eur. Respir. J.* **2005**, *26*, 523–548. [CrossRef] [PubMed]
18. Forman, H.J.; Augusto, O.; Brigelius-Flohe, R.; Dennery, P.A.; Kalyanaraman, B.; Ischiropoulos, H.; Mann, G.E.; Radi, R.; Roberts, L.J.; Vina, J. Even free radicals should follow some rules: A Guide to free radical research terminology and methodology. *Free. Radic. Biol. Med.* **2015**, *78*, 233–235. [CrossRef]
19. Hemmendinger, M.; Wild, P.; Shoman, Y.; Graille, M.; Bergamaschi, E.; Hopf, N.; Guseva Canu, I. Reference ranges of oxidative stress biomarkers selected for non-invasive biological surveillance of nanotechnology workers: Study protocol and meta-analysis results for 8-OHdG in exhaled breath condensate. *Toxicol. Lett.* **2020**, *327*, 41–47. [CrossRef]
20. Liberati, A.; Altman, D.G.; Tetzlaff, J.; Mulrow, C.; Gøtzsche, P.C.; Ioannidis, J.P.A.; Clarke, M.; Devereaux, P.J.; Kleijnen, J.; Moher, D. The PRISMA Statement for Reporting Systematic Reviews and Meta-Analyses of Studies That Evaluate Health Care Interventions: Explanation and Elaboration. *PLoS Med.* **2009**, *6*, e1000100. [CrossRef]
21. Shoman, Y.; Wild, P.; Hemmendinger, M.; Graille, M.; Sauvain, J.J.; Hopf, N.B.; Guseva Canu, I. Reference Ranges of 8-Isoprostane Concentrations in Exhaled Breath Condensate (EBC): A Systematic Review and Meta-Analysis. *Int. J. Mol. Sci.* **2020**, *21*, 3822. [CrossRef]
22. Ouzzani, M.; Hammady, H.; Fedorowicz, Z.; Elmagarmid, A. Rayyan—A web and mobile app for systematic reviews. *Syst Rev. Déc* **2016**, *5*, 210.
23. Aksu, K.; Kurt, H.; Gündüz, E.; Değirmenci, I.; Kurt, E. Inflammatory markers in exhaled breath condensate in patients with asthma and rhinitis. *Tuberk Toraks* **2012**, *60*, 321–326.
24. Andreoli, R.; Manini, P.; Corradi, M.; Mutti, A.; Niessen, W.M.A. Determination of patterns of biologically relevant aldehydes in exhaled breath condensate of healthy subjects by liquid chromatography/atmospheric chemical ionization tandem mass spectrometry. *Rapid Commun. Mass Spectrom.* **2003**, *17*, 637–645. [CrossRef] [PubMed]
25. Antus, B.; Drozdovszky, O.; Barta, I.; Kelemen, K. Comparison of Airway and Systemic Malondialdehyde Levels for Assessment of Oxidative Stress in Cystic Fibrosis. *Lung* **2015**, *193*, 597–604. [CrossRef]
26. Antus, B.; Harnasi, G.; Drozdovszky, O.; Barta, I. Monitoring oxidative stress during chronic obstructive pulmonary disease exacerbations using malondialdehyde: MDA and COPD exacerbations. *Respirology* **2014**, *19*, 74–79. [CrossRef] [PubMed]
27. Antczak, A.; Nowak, D.; Shariati, B.; Król, M.; Piasecka, G.; Kurmanowska, Z. Increased hydrogen peroxide and thiobarbituric acid-reactive products in expired breath condensate of asthmatic patients. *Eur. Respir. J.* **1997**, *10*, 1235–1241. [CrossRef] [PubMed]
28. Araneda, O.F.; García, C.; Lagos, N.; Quiroga, G.B.; Cajigal, J.Y.; Salazar, M.P.; Behn, C. Lung oxidative stress as related to exercise and altitude. Lipid peroxidation evidence in exhaled breath condensate: A possible predictor of acute mountain sickness. *Eur. J. Appl. Physiol.* **2005**, *95*, 383–390. [CrossRef]
29. Araneda, O.; Guevara, A.; Contreras, C.; Lagos, N.; Berral, F. Exhaled Breath Condensate Analysis after Long Distance Races. *Int. J. Sports Med.* **2012**, *33*, 955–961. [CrossRef]
30. Araneda, O.F.; Urbina-Stagno, R.; Tuesta, M.; Haichelis, D.; Alvear, M.; Salazar, M.P.; García, C. Increase of pro-oxidants with no evidence of lipid peroxidation in exhaled breath condensate after a 10-km race in non-athletes. *J. Physiol. Biochem.* **2014**, *70*, 107–115. [CrossRef]
31. Barregard, L.; Sällsten, G.; Andersson, L.; Almstrand, A.C.; Gustafson, P.; Andersson, M.; Olin, A.C. Experimental exposure to wood smoke: Effects on airway inflammation and oxidative stress. *Occup. Environ. Med.* **2008**, *65*, 319–324. [CrossRef]

32. Bartoli, M.L.; Novelli, F.; Costa, F.; Malagrinò, L.; Melosini, L.; Bacci, E.; Cianchetti, S.; Dente, F.L.; Di Franco, A.; Vagaggini, B.; et al. Malondialdehyde in Exhaled Breath Condensate as a Marker of Oxidative Stress in Different Pulmonary Diseases. *Mediat. Inflamm.* **2011**, *2011*, 891752. [CrossRef]
33. Brand, P.; Bischof, K.; Siry, L.; Bertram, J.; Schettgen, T.; Reisgen, U.; Kraus, T.; Gube, M. Exposure of healthy subjects with emissions from a gas metal arc welding process: Part 3—biological effect markers and lung function. *Int. Arch. Occup. Environ. Health* **2013**, *86*, 39–45. [CrossRef] [PubMed]
34. Caglieri, A.; Goldoni, M.; Acampa, O.; Andreoli, R.; Vettori, M.V.; Corradi, M.; Apostoli, P.; Mutti, A. The Effect of Inhaled Chromium on Different Exhaled Breath Condensate Biomarkers among Chrome-Plating Workers. *Environ. Health Perspect.* **2006**, *114*, 542–546. [CrossRef] [PubMed]
35. Casimirri, E.; Stendardo, M.; Bonci, M.; Andreoli, R.; Bottazzi, B.; Leone, R.; Schito, M.; Vaccari, A.; Papi, A.; Contoli, M.; et al. Biomarkers of oxidative-stress and inflammation in exhaled breath condensate from hospital cleaners. *Biomarkers* **2016**, *21*, 115–122. [CrossRef] [PubMed]
36. Corradi, M.; Rubinstein, I.; Andreoli, R.; Manini, P.; Caglieri, A.; Poli, D.; Alinovi, R.; Mutti, A. Aldehydes in Exhaled Breath Condensate of Patients with Chronic Obstructive Pulmonary Disease. *Am. J. Respir. Crit. Care Med.* **2003**, *167*, 1380–1386. [CrossRef]
37. Corradi, M.; Alinovi, R.; Goldoni, M.; Vettori, M.; Folesani, G.; Mozzoni, P.; Cavazzini, S.; Bergamaschi, E.; Rossi, L.; Mutti, A. Biomarkers of oxidative stress after controlled human exposure to ozone. *Toxicol. Lett.* **2002**, *134*, 219–225. [CrossRef]
38. Corradi, M. Comparison between exhaled and sputum oxidative stress biomarkers in chronic airway inflammation. *Eur. Respir. J.* **2004**, *24*, 1011–1017. [CrossRef]
39. Doruk, S.; Ozyurt, H.; Inonu, H.; Erkorkmaz, U.; Saylan, O.; Seyfikli, Z. Oxidative status in the lungs associated with tobacco smoke exposure. *Clin. Chem. Lab. Med.* **2011**, *49*, 2007–2012. Available online: https://www.degruyter.com/document/doi/10.1515/CCLM.2011.698/html (accessed on 30 March 2022). [CrossRef]
40. Goldoni, M.; Catalani, S.; De Palma, G.; Manini, P.; Acampa, O.; Corradi, M.; Bergonzi, R.; Apostoli, P.; Mutti, A. Breath Condensate as a Suitable Matrix to Assess Lung Dose and Effects in Workers Exposed to Cobalt and Tungsten. *Environ. Health Perspect.* **2004**, *112*, 1293–1298. [CrossRef]
41. Goldoni, M.; Corradi, M.; Mozzoni, P.; Folesani, G.; Alinovi, R.; Pinelli, S.; Andreoli, R.; Pigini, D.; Tillo, R.; Filetti, A.; et al. Concentration of exhaled breath condensate biomarkers after fractionated collection based on exhaled CO_2 signal. *J. Breath Res.* **2013**, *7*, 017101. [CrossRef]
42. Goldoni, M.; Caglieri, A.; Andreoli, R.; Poli, D.; Manini, P.; Vettori, M.V.; Corradi, M.; Mutti, A. Influence of condensation temperature on selected exhaled breath parameters. *BMC Pulm. Med.* **2005**, *5*, 10. [CrossRef]
43. Gong, J.; Zhu, T.; Kipen, H.; Wang, G.; Hu, M.; Ohman-Strickland, P.; Lu, S.E.; Zhang, L.; Wang, Y.; Zhu, P.; et al. Malondialdehyde in exhaled breath condensate and urine as a biomarker of air pollution induced oxidative stress. *J. Expo. Sci. Environ. Epidemiol.* **2013**, *23*, 322–327. [CrossRef]
44. Graczyk, H.; Lewinski, N.; Zhao, J.; Sauvain, J.J.; Suarez, G.; Wild, P.; Danuser, B.; Riediker, M. Increase in oxidative stress levels following welding fume inhalation: A controlled human exposure study. *Part. Fibre Toxicol.* **2015**, *13*, 31. [CrossRef] [PubMed]
45. Gube, M.; Ebel, J.; Brand, P.; Göen, T.; Holzinger, K.; Reisgen, U.; Kraus, T. Biological effect markers in exhaled breath condensate and biomonitoring in welders: Impact of smoking and protection equipment. *Int. Arch. Occup. Environ. Health* **2010**, *83*, 803–811. [CrossRef] [PubMed]
46. Inonu, H.; Doruk, S.; Sahin, S.; Erkorkmaz, U.; Celik, D.; Celikel, S.; Seyfikli, Z. Oxidative Stress Levels in Exhaled Breath Condensate Associated With COPD and Smoking. *Respir. Care* **2012**, *57*, 413–419. [CrossRef] [PubMed]
47. Lärstad, M.; Ljungkvist, G.; Olin, A.-C.; Torén, K. Determination of malondialdehyde in breath condensate by high-performance liquid chromatography with fluorescence detection. *J. Chromatogr. B* **2002**, *766*, 107–114. [CrossRef]
48. Laumbach, R.J.; Kipen, H.M.; Ko, S.; Kelly-McNeil, K.; Cepeda, C.; Pettit, A.; Ohman-Strickland, P.; Zhang, L.; Zhang, J.; Gong, J.; et al. A controlled trial of acute effects of human exposure to traffic particles on pulmonary oxidative stress and heart rate variability. *Part. Fibre Toxicol.* **2014**, *11*, 45. [CrossRef]
49. Lee, J.S.; Choi, Y.C.; Shin, J.H.; Lee, J.H.; Lee, Y.; Park, S.Y.; Baek, J.E.; Park, J.D.; Ahn, K.; Yu, I.J. Health surveillance study of workers who manufacture multi-walled carbon nanotubes. *Nanotoxicology* **2015**, *9*, 802–811. [CrossRef]
50. Majewska, E.; Kasielski, M.; Luczynski, R.; Bartosz, G.; Bialasiewicz, P.; Nowak, D. Elevated exhalation of hydrogen peroxide and thiobarbituric acid reactive substances in patients with community acquired pneumonia. *Respir. Med.* **2004**, *98*, 669–676. [CrossRef]
51. Nowak, D.; Kałucka, S.; Białasiewicz, P.; Król, M. Exhalation of H_2O_2 and thiobarbituric acid reactive substances (TBARs) by healthy subjects. *Free Radic. Biol. Med.* **2001**, *30*, 178–186. [CrossRef]
52. Pelclova, D.; Zdimal, V.; Schwarz, J.; Dvorackova, S.; Komarc, M.; Ondracek, J.; Kostejn, M.; Kacer, P.; Vlckova, S.; Fenclova, Z.; et al. Markers of Oxidative Stress in the Exhaled Breath Condensate of Workers Handling Nanocomposites. *Nanomaterials* **2018**, *8*, 611. [CrossRef]
53. Pelclová, D.; Fenclová, Z.; Vlčková, S.; Klusáčková, P.; Lebedová, J.; Syslová, K.; Běláček, J.; Kuzma, M.; Navrátil, T.; Zakharov, S. Occupational asthma follow-up—which markers are elevated in exhaled breath condensate and plasma? *Int. J. Occup. Med. Environ. Health* **2014**, *27*, 206–215. Available online: http://ijomeh.eu/Occupational-asthma-follow-up-Which-markers-are-elevated-in-exhaled-breath-condensate-and-plasma-,2044,0,2.html (accessed on 30 March 2022). [CrossRef] [PubMed]

54. Pelclova, D.; Zdimal, V.; Kacer, P.; Fenclova, Z.; Vlckova, S.; Syslova, K.; Navratil, T.; Schwarz, J.; Zikova, N.; Barosova, H. Oxidative stress markers are elevated in exhaled breath condensate of workers exposed to nanoparticles during iron oxide pigment production. *J. Breath Res.* **2016**, *10*, 016004. [CrossRef]
55. Riccelli, M.G.; Goldoni, M.; Andreoli, R.; Mozzoni, P.; Pinelli, S.; Alinovi, R.; Selis, L.; Mutti, A.; Corradi, M. Biomarkers of exposure to stainless steel tungsten inert gas welding fumes and the effect of exposure on exhaled breath condensate. *Toxicol. Lett.* **2018**, *292*, 108–114. [CrossRef] [PubMed]
56. Rundell, K.W.; Slee, J.B.; Caviston, R.; Hollenbach, A.M. Decreased Lung Function After Inhalation of Ultrafine and Fine Particulate Matter During Exercise is Related to Decreased Total Nitrate in Exhaled Breath Condensate. *Inhal. Toxicol.* **2008**, *20*, 1–9. [CrossRef] [PubMed]
57. Sakhvidi, M.J.; Biabani Ardekani, J.; Firoozichahak, A.; Zavarreza, J.; Hajaghazade, M.; Mostaghaci, M.; Mehrparvar, A.; Barkhordari, A. Exhaled breath malondialdehyde, spirometric results and dust exposure assessment in ceramics production workers. *Int. J. Occup. Med. Environ. Health* **2015**, *28*, 81–89. [CrossRef] [PubMed]
58. Sauvain, J.-J.; Hohl, M.S.S.; Wild, P.; Pralong, J.A.; Riediker, M. Exhaled Breath Condensate as a Matrix for Combustion-Based Nanoparticle Exposure and Health Effect Evaluation. *J. Aerosol Med. Pulm. Drug Deliv.* **2014**, *27*, 449–458. [CrossRef] [PubMed]
59. Stockfelt, L.; Sallsten, G.; Olin, A.C.; Almerud, P.; Samuelsson, L.; Johannesson, S.; Molnar, P.; Strandberg, B.; Almstrand, A.C.; Bergemalm-Rynell, K. Effects on airways of short-term exposure to two kinds of wood smoke in a chamber study of healthy humans. *Inhal. Toxicol.* **2012**, *24*, 47–59. [CrossRef]
60. Syslová, K.; Kacer, P.; Kuzma, M.; Najmanová, V.; Fenclová, Z.; Vlcková, S.; Lebedová, J.; Pelclová, D. Rapid and easy method for monitoring oxidative stress markers in body fluids of patients with asbestos or silica-induced lung diseases. *J. Chromatogr. B* **2009**, *877*, 2477–2486. [CrossRef]
61. Szkudlarek, U.; Maria, L.; Kasielski, M.; Kaucka, S.; Nowak, D. Exhaled hydrogen peroxide correlates with the release of reactive oxygen species by blood phagocytes in healthy subjects. *Respir. Med.* **2003**, *97*, 718–725. [CrossRef]
62. Tanger, S.R.O. (Ed.) Proceedings of the 10th Anniversary International Conference on Nanomaterials—Research and Application (NANOCON 2018), Brno, Czech Republic, 17–19 October 2018; Curran Associates: Red Hook, NY, USA, 2019; p. 741.
63. Pinchuk, I.; Weber, D.; Kochlik, B.; Stuetz, W.; Toussaint, O.; Debacq-Chainiaux, F.; Dollé, M.E.T.; Jansen, E.H.J.M.; Gonos, E.S.; Sikora, E. Gender- and age-dependencies of oxidative stress, as detected based on the steady state concentrations of different biomarkers in the MARK-AGE study. *Redox Biol.* **2019**, *24*, 101204. [CrossRef]
64. Caliri, A.W.; Tommasi, S.; Besaratinia, A. Relationships among smoking, oxidative stress, inflammation, macromolecular damage, and cancer. *Mutat. Res./Rev. Mutat. Res.* **2021**, *787*, 108365. [CrossRef] [PubMed]
65. Lykkesfeldt, J. Malondialdehyde as biomarker of oxidative damage to lipids caused by smoking. *Clin. Chim. Acta* **2007**, *380*, 50–58. [CrossRef] [PubMed]
66. Carraro, E.; Schilirò, T.; Biorci, F.; Romanazzi, V.; Degan, R.; Buonocore, D.; Verri, M.; Dossena, M.; Bonetta, S.; Gilli, G. Physical Activity, Lifestyle Factors and Oxidative Stress in Middle Age Healthy Subjects. *Int. J. Environ. Res. Public Health* **2018**, *15*, 1152. [CrossRef]
67. Rosias, P. Methodological aspects of exhaled breath condensate collection and analysis. *J. Breath Res.* **2012**, *6*, 027102. [CrossRef]

toxics MDPI

Systematic Review

Urinary Malondialdehyde (MDA) Concentrations in the General Population—A Systematic Literature Review and Meta-Analysis

Antonio Toto [1], Pascal Wild [1], Mélanie Graille [1], Veronica Turcu [1], Camille Crézé [1], Maud Hemmendinger [1], Jean-Jacques Sauvain [1], Enrico Bergamaschi [2], Irina Guseva Canu [1,†] and Nancy B. Hopf [1,*,†]

1 Center for Primary Care and Public Health (Unisanté), University of Lausanne, Route de la Corniche 2, 1066 Lausanne, Switzerland; antonio.toto@unisante.ch (A.T.); pascal.wild@unisante.ch (P.W.); melanie.graille@unisante.ch (M.G.); veronica.turcu@unisante.ch (V.T.); crezecamille@gmail.com (C.C.); maud.hemmendinger@unisante.ch (M.H.); jean-jacques.sauvain@unisante.ch (J.-J.S.); irina.guseva-canu@unisante.ch (I.G.C.)

2 Laboratory of Toxicology and Industrial Epidemiology, Department of Public Health and Pediatrics, University of Turin, Via Zuretti 29, 10125 Turin, Italy; enrico.bergamaschi@unito.it

* Correspondence: nancy.hopf@unisante.ch

† These authors contributed equally to the work.

Citation: Toto, A.; Wild, P.; Graille, M.; Turcu, V.; Crézé, C.; Hemmendinger, M.; Sauvain, J.-J.; Bergamaschi, E.; Guseva Canu, I.; Hopf, N.B. Urinary Malondialdehyde (MDA) Concentrations in the General Population—A Systematic Literature Review and Meta-Analysis. *Toxics* **2022**, *10*, 160. https://doi.org/10.3390/toxics10040160

Academic Editor: Christophe Rousselle

Received: 21 February 2022
Accepted: 26 March 2022
Published: 29 March 2022

Publisher's Note: MDPI stays neutral with regard to jurisdictional claims in published maps and institutional affiliations.

Copyright: © 2022 by the authors. Licensee MDPI, Basel, Switzerland. This article is an open access article distributed under the terms and conditions of the Creative Commons Attribution (CC BY) license (https://creativecommons.org/licenses/by/4.0/).

Abstract: Oxidative stress has been associated with various inflammation-related human diseases. It is defined as an imbalance between the production and elimination of reactive oxygen species (ROS). ROS can oxidize proteins, lipids, and DNA, and some of these oxidized products are excreted in urine, such as malondialdehyde (MDA), which is considered a biomarker for oxidative damage of lipids. To interpret changes of this biomarker as a measure of oxidative species overproduction in humans, a background range for urinary MDA concentration in the general population is needed. We sought to establish urinary MDA concentration ranges for healthy adult populations based on reported values in the available scientific literature. We conducted a systematic review and meta-analysis using the standardized protocol registered in PROSPERO (CRD42020146623). EMBASE, PubMed, Web of Science, and Cochrane library databases were searched from journal inception up to October 2020. We included 35 studies (divided into 47 subgroups for the quantitative analysis). Only studies that measured creatinine-corrected urinary MDA with high-performance liquid chromatography (HPLC) with mass spectrometry (MS), fluorescence detection, or UV photometry were included. The geometric mean (GM) of urinary MDA concentration was 0.10 mg/g creatinine and 95% percentile confidence interval (CI) 0.07–0.12. Age, geographical location but not sex, and smoking status had a significant effect on urinary MDA concentrations. There was a significant increasing trend of urinary MDA concentrations with age. These urinary MDA values should be considered preliminary, as they are based on mostly moderate to some low-quality evidence studies. Although urinary MDA can reliably reflect excessive oxidative stress in a population, the influence of physiological parameters that affect its meaning needs to be addressed as well as harmonizing the chemical analytical methods.

Keywords: oxidative stress; MDA; systematic review; meta-analysis; urinary biomarker; reference range; general population

1. Introduction

Oxidative stress is defined as an imbalance between the production and elimination of reactive oxygen species (ROS) in the body [1]. ROS targets many biological entities but mainly lipids and polyunsaturated fatty acids [2]. When these biomolecules undergo peroxidation, the chain reaction evolves in three steps: initiation, propagation, and termination, with various reactive products generated at each step [3]. Malondialdehyde (MDA) (chemical structure shown in Figure 1), together with other aldehydes, is one of polyunsaturated fatty acids' peroxidation best-studied end-products [4].

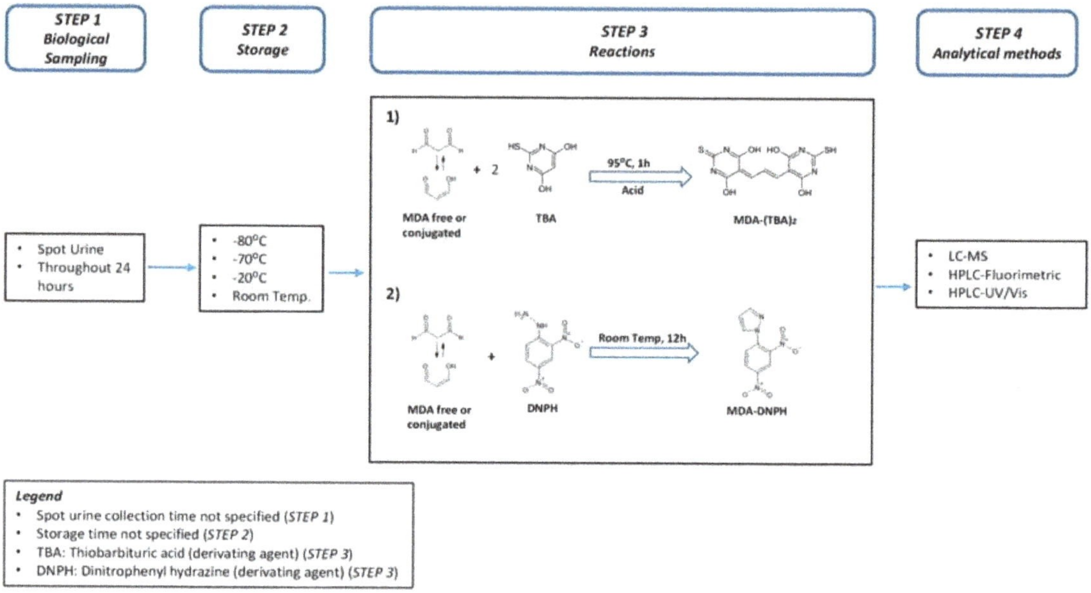

Figure 1. Simplified schematic process for the determination of MDA in biological samples (Urine).

MDA can be generated both through an enzymatic pathway, identical to thromboxane A2 and prostaglandins, as well as through a non-enzymatic process [5]. MDA is not just a biomarker of oxidative stress but also a biologically active compound having several biological roles [5,6]. Owing to its several biological functions, MDA can also be regarded as a biomarker of peroxidation of cell membrane fatty acids when produced via the enzymatic process [6]. MDA can act as a signaling messenger in insulin secretion [7] and as an inducer of collagen-gene expression in hepatic cells [8]. MDA generated via a non-enzymatic process would, however, interact with other biomolecules, such as proteins, amino groups, and DNA [9], to generate a multitude of adducts, ultimately resulting in a genotoxic effect. MDA has been indicated as putatively being the most mutagenic molecule among ROS end-products [4].

Circulating MDA can be detectable either free (unconjugated) or conjugated [10], and the sum of the two forms is labeled total MDA. MDA forms adducts with many biological molecules, and the majority of MDA produced is found in the conjugated form. MDA adducts are highly immunogenic, i.e., able to trigger an immune response. They have been found to be associated with autoimmune diseases, such as lupus erythematosus and nephritis [11], while others have shown a correlation with the development and progression of atherosclerosis [12] and longevity [5].

MDA concentrations in different biological samples collected from several sub-populations have been investigated. Plasma MDA levels tend to be higher in smokers than in non-smokers and in populations exposed to high compared to low air pollution [13]. Plasma MDA levels are consistently higher in patients with acute stroke [14], diabetes, and chronic inflammation [10], such as chronic obstructive disease (COPD) [15] and asthma [13] compared with healthy people. Urinary MDA levels are mostly evaluated as a biomarker of systemic oxidative stress and found to be elevated in conditions such as urinary infections [16], diabetic nephropathy [17], but also after air pollution exposure [18]. Body mass index (BMI), sex, and age have been mentioned as potential confounding factors in a sizable part of the literature.

Urinary MDA concentrations are quantified either by immunochemical assays or chemical analytical methods. Figure 1 outlines the different steps generally needed for urinary MDA quantification, i.e., urine sample collection (step 1), storage and transportation (step 2), and derivatization of urinary MDA with a complexing agent (step 3). This step is needed to increase sensitivity and reduce the limit of detection. Both thiobarbituric acid (TBA) and dinitrophenyl hydrazine (DNPH) are used as derivatization reagents. TBA is less specific compared to DNPH [19]. The last step is the quantification of MDA with a chemical analytical method (step 4) [6,20].

Collecting urine samples is easy, convenient, and non-invasive. These are reasons why urinary MDA is an interesting quantifiable biomarker of systemic oxidative stress. To use this biomarker tool to measure oxidative stress in selected populations, we need a general population reference range. As of yet, a general population reference range has not been generated. In addition, urinary MDA concentration ranges reported for the general population in the scientific literature vary greatly and are inconsistent [21]. We, therefore, sought to close this data gap by performing a systematic review and meta-analysis [22] according to our registered Prospero protocol ((https://www.ncbi.nlm.nih.gov/pmc/articles/PMC7311981/pdf/ijms-21-03822.pdf) Section 4.1), and providing reference ranges for urinary MDA concentrations in a healthy adult population. Our aim was two-fold:

Aim 1: Provide the geometric mean (GM) and standard deviation (GSD) for urinary MDA concentrations,

Aim 2: Assess the influence of age, smoking status, geographic locations (countries), and sex on urinary MDA concentrations.

2. Materials and Methods

We registered our study protocol in the International Prospective Register of Systematic Reviews (PROSPERO; registration number CRD42020146623). We report our results here following the recommendations from the Preferred Reporting Items for Systematic Reviews and Meta-Analysis (PRISMA) [23].

2.1. Literature Search

We searched the published scientific literature from journal inception and up to October 2020 in the following bibliographic electronic databases: EMBASE, PubMed, Web of Science, and Cochrane library. The full search strategy, including the search string used, can be found in (https://www.doi.org/10.16909/dataset/17, accessed on 20 February 2022). Only original research studies written in either English or French were included. Two researchers (CC and AT) conducted two rounds of selection, i.e., abstract screening and full-text reading. We excluded studies without quantitative data for MDA, non-human studies, reviews, correspondence, conference papers, expert opinions, and editorials, as well as abstracts without full text. The reviewers (CC and AT) independently performed a first screening of titles and abstracts retrieved during the searches, using Rayyan software [24], a systematic review web application for title and abstract screening [25].

2.2. Study Selection

We only included original research studies conducted on healthy adult human participants (aged > 18 years, no known disease), measuring urinary MDA. The flow chart outlining the study selection is presented in Figure 2. An initial 21,017 records were retrieved and exported to the Rayyan software. We excluded 3579 articles after abstract screening. Studies showing non-creatinine-adjusted data and values with suspected unit mistakes were excluded during the second round of selection. Overall, 135 studies were deemed eligible, and the corresponding papers were downloaded into the EndNote software. We conducted a standardized quality assessment, which was developed as part of our study protocol [26]. In a third selection round, we selected studies purposely for the quantitative (meta-)analysis. We thus excluded studies reporting no standard deviations (SD) or confidence intervals (CI) with their mean and median values, suspiciously low coefficient of variation (CV < 20%), missing chemical analytical method entirely or descriptions partially missing, such as separation techniques and detection method, and aberrant units.

A total of 35 articles remained comprised of 47 different exposure groups, which we carried forward to the meta-analysis.

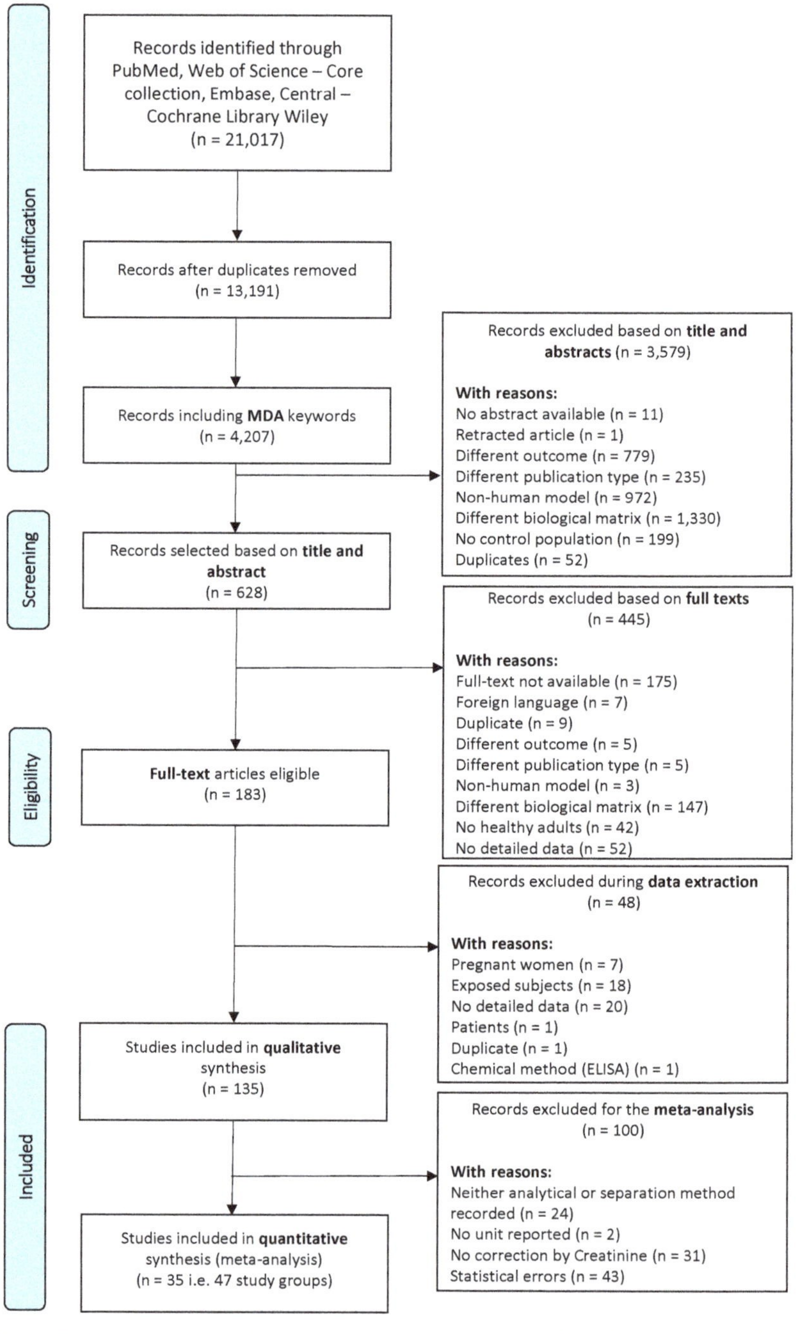

Figure 2. Flow chart describing the selection process of the 135 studies included in the quality assessment and the subset of 35 studies included in this meta-analysis.

Figure 2 flow chart describing the selection process of the 135 studies included in the quality assessment and the subset of 35 studies included in this meta-analysis.

2.3. Data Extraction

We used the standardized data extraction form developed as part of our study protocol [26]. In addition to populations with known diseases, groups with known occupational and/or environmental chemical exposures were also excluded. When data on several subgroups were available in the published article, we extracted all subgroup-specific data. Only baseline data were extracted whenever data from several time points were available. We also recorded possible covariates that affect the oxidative stress concentrations such as study design, sample collection methods (spot urine samples or 24 h urine samples), sample storage, pre-analytical methods, and vitamin supplements, as well as statistical analysis. A statistician (PW) cross-checked all data extracted for the meta-analysis.

2.4. Quality Assessment

Whether a qualitative or a quantitative approach is most appropriate depends on the nature and state of the existing literature, the research questions, and theoretical and empirical issues. We used a standardized quality assessment checklist previously used in other studies [26,27]. Briefly, the quality checklist covered four domains: (I) study sample, (II) study design and risk of bias, (III) technical and analytical methods, (IV) data processing, analysis, and result reporting. We assessed each domain based on a number of objective criteria (Supplementary Table S1) by grading these criteria with sub-scores from 1 to 3. The resulting sub-scores were first summarized in a quality score for each of the four domains, then into an overall study quality score as described in the GRADE guidelines [28]. The total quality scores ranged between 9 and 27 (Supplementary Table S2) and were considered "high" when scores were equal or higher than 20, "moderate" between 14 and 19, and "low" for equal or lower than 13 [28]. The quality assessments of the included studies were performed by one (AT) and reviewed by two independent reviewers (NBH, IGC).

2.5. Statistical Analysis

Values of urinary biomarkers are generally log-normally distributed; we, therefore, computed geometric means (GM) and geometric standard deviations (GSD) as the basis for the meta-analysis, i.e., equivalently *mu*L (log geometric means (GM)) = ln(GM) and *sdL* (log geometric standard deviations (GSD)) = ln(GSD). Details of the computations are given in Graille et al. 2020 [27]. All study-specific results were then converted to mg/g creatinine using the molar weights of MDA and creatinine when necessary.

The chemical analytical methods used were gas chromatography (GC) and high-performance liquid chromatography (HPLC) separation with either mass spectrometry (MS), fluorescence, and spectrophotometry (UV) detection (we did not include immunohistochemistry analyses, e.g., ELISA, but focused on chemical analytical quantification). Forest plots were used to display GMs and 95% confidence intervals (in mg/g creatinine) of the different study groups both graphically (the squares represent the GM and the lines around the CI) with the study groups re-grouped by age and smoking categories, respectively. The diamonds represent the summary GM of the categories. Between-study heterogeneity was assessed using the Q test for homogeneity within each category and displayed on the forest plots [29]. If the between-study heterogeneity is larger than the between-subject heterogeneity, then any attempt of obtaining a summary value for individual participants will not be valid. We further modeled the study group-specific log-transformed GMs using a linear mixed model with the study ID as a random effect, without further considering the within-study heterogeneity. Such an analysis is warranted when the between-study heterogeneity dominates the within-study heterogeneity. The study ID as a random effect was included in order to account for between-study heterogeneity when assessing the effect of other parameters. We used STATA, version 16 software for data management and statistical analysis.

3. Results

We included 135 studies in the qualitative analysis, with a subset of 35 studies in the quantitative analysis (Figure 2). We described the subgroups by sex (Figure 3a), age (Figure 3b), smoking status (Figure 3c), geographical location (Figure 3d), chemical analytical methods (Figure 3e), and body mass index (BMI) (Figure 3f) subgroups, as far as these results were available, which resulted in the description of 47 study subgroups.

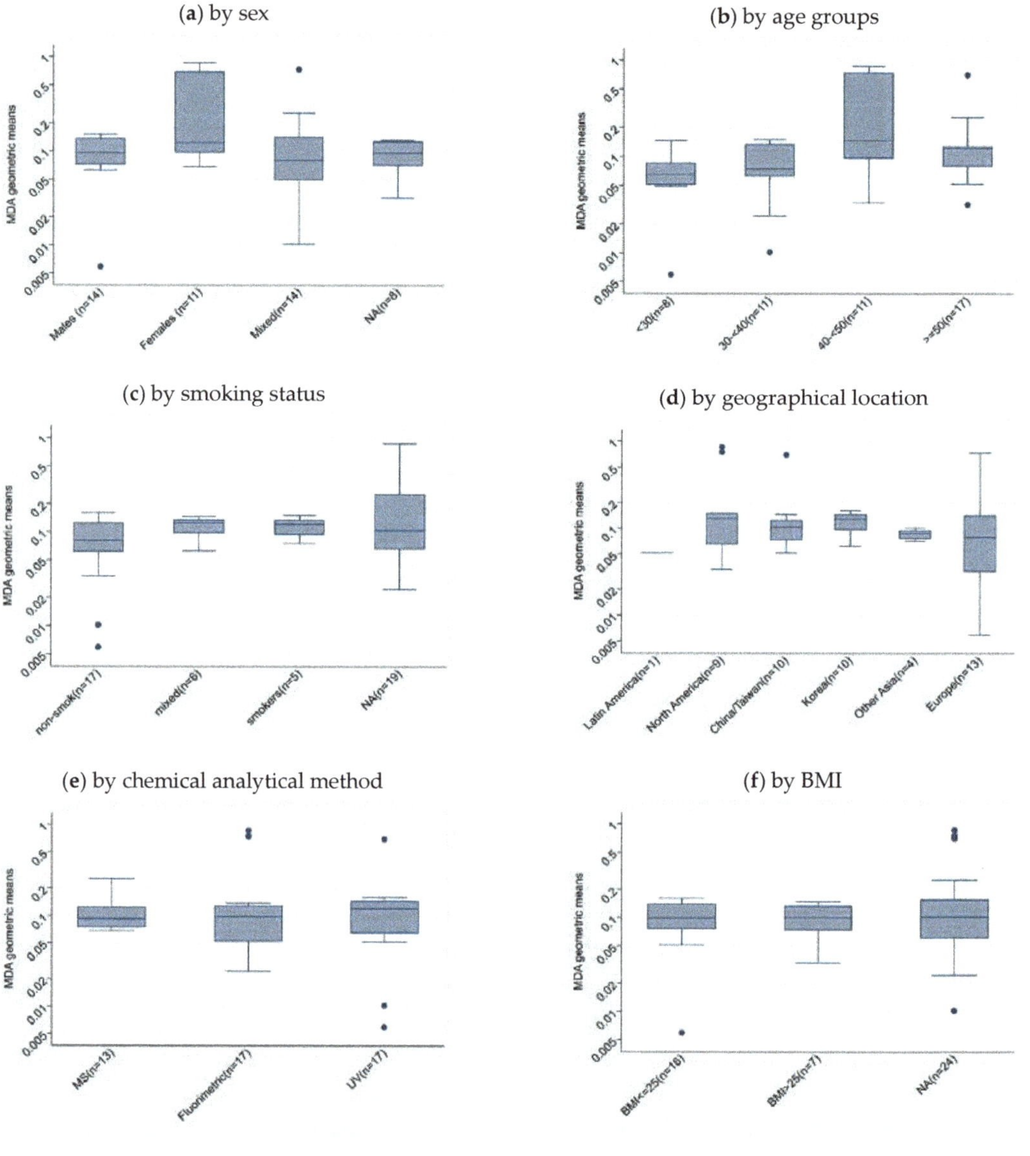

Figure 3. Boxplots of log-transformed urinary MDA concentration (mg/g creatinine) GM (y-axis) by subgroups (x-axis): (**a**) sex, (**b**) age, (**c**) smoking status, (**d**) geographical location, (**e**) chemical analytical method, and (**f**) BMI.

We combined the chemical separation methods (LC and GC) as they did not show significantly different results (Figure 3e). We merged UV/VIS and photometry detection and labeled this "UV/VIS". Consequently, the chemical analytical methods are represented with their detection method in Figure 3e. We were unable to include chemical analytical variables such as derivatization method, clean-up procedures, or instrumental parameters in our analysis as these were not reported in most studies.

The chemical analytical methods were comparable (Figure 3e); thus, we did not include this variable in the statistical analysis. This was also true for BMI (Figure 3f); thus, we did not include this variable either. Smoking status, age group, and geographical location were variables included in the statistical models.

The overall GM and 95% CI for urinary MDA are provided in Table 1, as well as the GM and 95% CI for each age group. The overall between-study 95% reference range of the study-specific GMs 0.01–0.65 (data not shown) is, of course, much wider.

Table 1. Model-based estimates of geometric mean of urinary MDA concentrations (mg/g creatinine) by age group.

	GM	95% CI	Age Group	GM	95% CI
Overall	0.10	0.07–0.12	<30	0.05	0.03–0.10
			30–40	0.09	0.06–0.13
			40–50	0.13	0.08–0.17
			>50	0.12	0.09–0.18

Table 2 gives the results from the mixed-effect regression analysis. The majority of the studies were conducted in Europe (n = 23) with North America, China/Taiwan, and Korea with half as many studies. Urinary MDA concentrations were seldom reported for Africa and Latin America. Geographical location has an impact on healthy populations in Asian countries (China, Korea, and Taiwan), having higher urinary MDA levels than in the European studies. The test for trend with age group with higher urinary MDA concentrations in older participants was significant (p = 0.041). Smoking status had a significant effect, even though this effect was due to differences with the included studies that did not stratify participants by smoking status or report smoking status. Since age and smoking status were the most relevant factors as seen from the statistical analysis, we present the data as forest plots according to these two factors.

Table 2. Results of the mixed-effect regression analyses according to geographical location, smoking status, and mean age of the population.

logGM	Coef.	Std. Err.	P > |z|	(95% Conf. Interval)	
CatCountry					
Latin America (n = 1)	−0.324	0.339	0.340	−0.988	0.341
North America (n = 9)	0.283	0.156	0.070	−0.024	0.590
China/Taiwan (n = 10)	0.490	0.164	0.003	0.169	0.812
Korea (n = 10)	0.421	0.199	0.034	0.032	0.811
Other Asia (n = 4)	0.384	0.257	0.134	−0.119	0.890
Europe (n = 13)	0	(base)			
SmokCat					
non-smok (n = 17)	0	(base)			
Mixed (n = 6)	0.008	0.084	0.922	−0.157	0.174
Smokers (n = 5)	0.102	0.061	0.096	−0.018	0.221
Not reported (n = 19)	0.389	0.141	0.006	0.113	0.665
MeanAgeCat					
<30 (n = 8)	0	(base)			
30–<40 (n = 11)	0.240	0.175	0.170	−0.103	0.584
40–<50 (n = 11)	0.369	0.181	0.042	0.014	0.723
>=50 (n = 17)	0.403	0.177	0.023	0.056	0.750
_cons	−1.75	0.181	0.000	−2.107	−1.40

We present the results in separate forest plots of urinary MDA concentrations (mg/g creatinine) in Figure 4 by age groups and Figure 5 by smoking/non-smoking groups. It is apparent from these forest plots that the between-study groups heterogeneity was much larger than the within-study heterogeneity uncertainty to the point that the within-study confidence intervals are barely distinguishable. This is confirmed by the highly significant Q tests within each all the smoking and age categories.

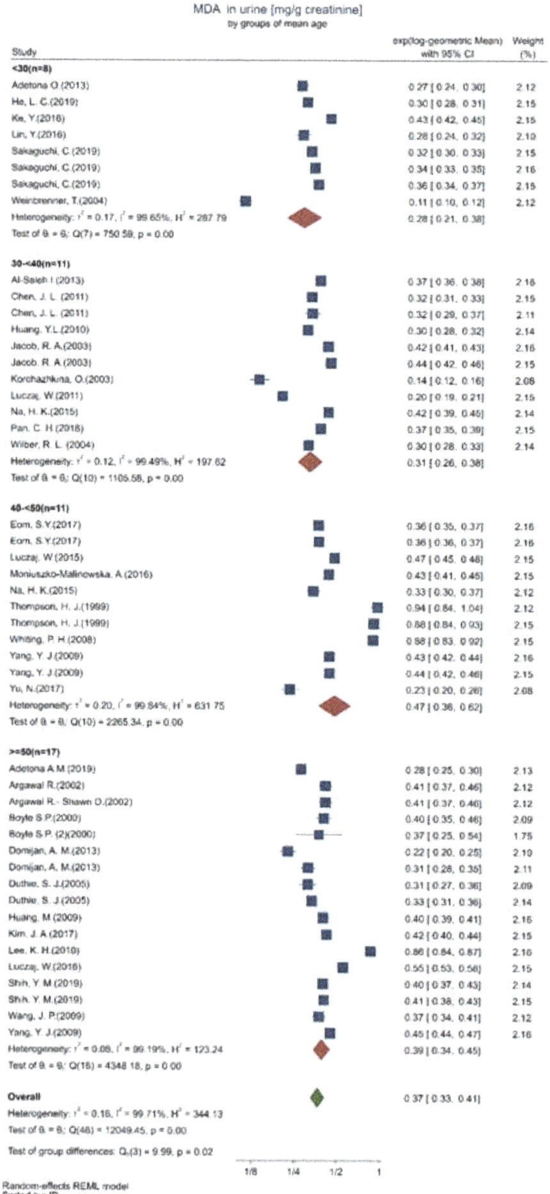

Figure 4. Forest plot of urinary MDA concentrations quantified in healthy adults (18+ years) participants by age groups (<30: [30–35], 30–<40: [36–44], 40–<50: [42,45–51], >=50: [7,46,50,52–62]).

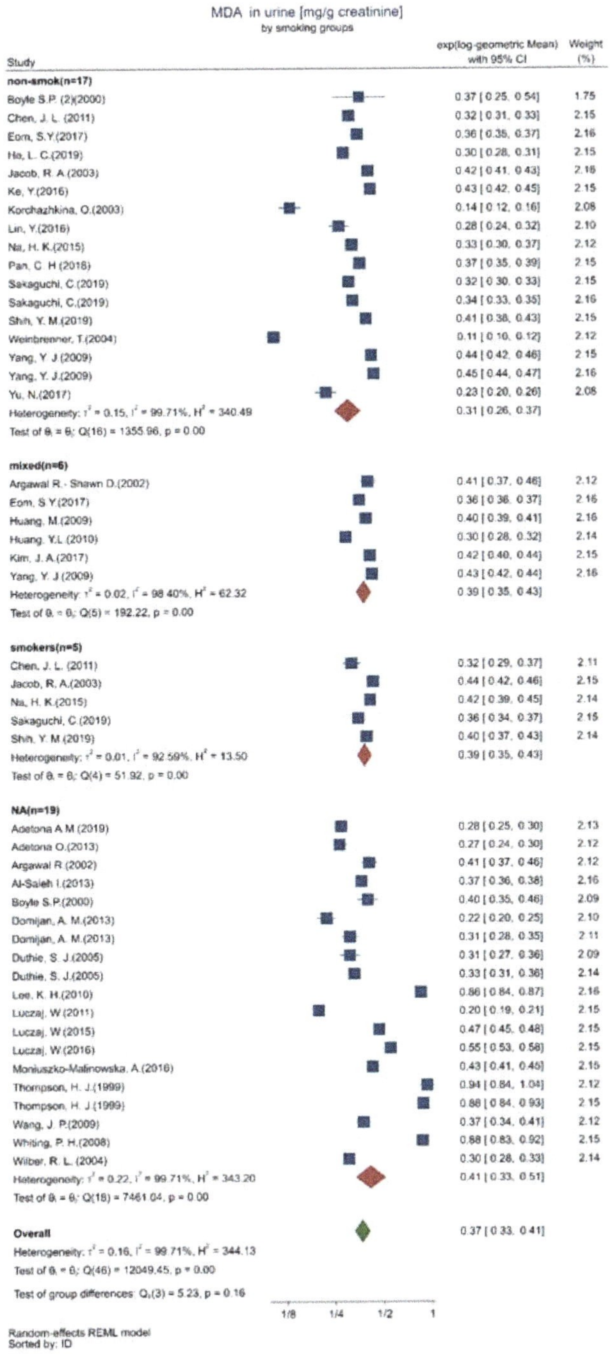

Figure 5. Forest plot of urinary MDA concentration (mg/g creatinine) in healthy adult (18+ years) by smoking/non-smoking groups (non-smok: [31–35,37,39,40,42,43,45,50,51,56,62], mixed: [38,45,50,54,59,60], smokers: [34,37,39,42,62], NA: [7,30,36,41,44,46–49,52,53,55,57,58,61,63].

4. Discussion

4.1. Interpretation of Findings

From the analysis of the literature, we can extrapolate and suggest a mean value of 0.10 mg/g creatinine and an overall reference range of 0.01–0.65 mg/g creatinine for study-population GMs of free urinary MDA concentrations in healthy adults. These values are valid for chemical analytical methods and not necessarily for colorimetric assays (e.g., ELISA).

The urinary MDA concentrations increase with age, which corroborates previous findings [64]. We were able to analyze the age factor in our meta-analysis as most of the included studies (77%) had recorded the age of their subjects. This age-related trend for increasing urinary MDA concentrations might reflect an increase in oxidative stress, which is expected with aging. This has previously been thoroughly discussed [65] in a joint effort from several researchers studying oxidative stress and health-related outcomes as well as the underlying biochemical mechanisms. Aging has been found to be related to the dysfunction of proteasome-mediated degradation of oxidized proteins, a critical player for protein homeostasis maintenance. Yet, proteasome up-regulation has been shown to successfully decelerate the aging progression by enhancing resistance to oxidative stress in genetically modified animals. Thus, increases in MDA levels with age might reflect increased oxidative stress through the progressive dysfunction of the protective proteasome pathway [66].

4.2. Heterogeneity

We found an overall high heterogeneity in studies included in this systematic review. A number of unknown factors, e.g., vitamin supplements, biological variability, air pollution, also contribute to the modification of oxidative stress levels that have yet to be characterized. These unknowns probably contributed to the great variability in our meta-analysis [67,68]. The lack of information on these covariates led to quality scores in moderate (65% of the studies were scored as moderate) and low (35%) levels. The heterogeneity could be related to study designs, sample collection methods (spot urine samples or 24 h urine samples), sample storage, pre-analytical methods, and statistical analysis. We included only 26% (35 out of 135) of the selected studies in the quantitative synthesis. Indeed, we discarded 32% due to statistical errors; i.e., presence of extreme values (outliers); or undescribed data distribution (GSD, IQR, CI were missing, coefficient of variations either too high >300% or too low <20%), 23% of the studies did not have their data creatinine-corrected, 1% had no units, and 18% did not have either a complete chemical analytical or separation method description.

4.2.1. Heterogeneity in Data Collection of Demographics

We found greater urinary MDA concentrations in the Asian populations. In spite of no apparent reasons, environmental factors and dietary habits may account for differences between Asian and Western countries. Air pollution measured as particular matter with an aerodynamic diameter of 2.5 μm or smaller (PM2.5) has been associated with greater urinary MDA concentrations [69], but so has other geographically linked parameters such as vitamin intake [67,68]. It is therefore difficult at this stage to assess what role geographical location plays with regard to other factors.

An increased urinary MDA value with increasing BMI has been reported [70]. However, we could not assess this effect in our meta-analysis, as BMI was missing in 54% of the included studies. Sex differences in MDA concentrations have been demonstrated in healthy adults [71]. For instance, urinary MDA concentrations were reported to be higher in healthy young men compared to age-matched women [72]. Three-quarters of the studies reported separate values for men and women; however, we could not detect a sex difference in urinary MDA levels.

Smoking is considered a source of oxidants leading to lipid peroxidation and a factor depleting antioxidants. Even though urinary and plasma MDA concentrations are associated, and plasma MDA concentrations have been shown to be different in smokers and non-smokers [73,74], we did not find a clear difference in smoking status, albeit the median urinary MDA concentration values seemed slightly higher among smokers. Other studies [34,75] have shown urinary MDA concentrations to be significantly greater in smokers compared to non-smokers. The lack of observed difference between these groups in our review might be related to the number of cigarettes smoked, as suggested in a previous review [74], which found a dose-dependent relationship between cigarette smoke exposure and plasma MDA concentration. Furthermore, urinary MDA might not be a sensitive biomarker for detecting the increase in oxidative stress from tobacco smoking [76], as studies show divergent results. This has been suggested by other authors [77]. We believe one reason might be that the use of thiobarbituric acid (TBA) as a derivatization agent in the analysis is not sufficiently sensitive. Although no difference could be detected between the groups of smokers, the groups of non-smokers, and the mixed smokers/nonsmoker groups in our review, the MDA concentration was higher in nearly one-half of the study groups for which the smoking status was not reported, leading to overall statistical significance. Consequently, we cannot rule out that smoking has an effect on urinary MDA levels.

4.2.2. Heterogeneity in Collection of Biological Samples

Most of the studies used spot urine samples (69 study groups; some of them without any indication of collection time) rather than 24 h urine collection (4 study groups). MDA levels fluctuate [78] during the day depending on activity, and urine concentrations represent MDA excretions from the last urine void until the next. The first urine void provides a measure of cumulative MDA concentrations (representing excretion overnight) and correlates well with 24 h urinary collections [27]. We included studies reporting any spot or 24 h urine samples. This difference in urine collection time probably contributed to the heterogeneity of the results. We found varying storage temperatures, and information on storage time and conditions were often omitted (56 creatinine-corrected studies). Storage conditions might have an effect on the measured concentration of MDA, knowing that urinary MDA concentrations need to be analyzed within 24 h of collection and stored in an airtight container at 0 °C [40]. In fact, one study [40] has shown a 43% ± 15% reduction in MDA concentration when the urine sample was left at −20 °C for more than 3 weeks. This decay needs to be confirmed, and standardized storage methods need to be developed, as most population studies cannot analyze the urine sample immediately after collection.

4.2.3. Heterogeneity in MDA Analysis

The included studies reported using different derivation methods for MDA analysis, which probably contributes to the overall heterogeneity. The most common, e.g., 96% of the included studies used the TBARS derivatization method, which quantifies TBARS formed as a byproduct of lipid peroxidation. This method of assay requires high temperatures (80–100 °C) for an extended incubation time under strong acidic conditions. These harsh conditions can lead to reactions with several other materials such as non-lipid-related materials and fatty peroxide-derived decomposition products. Consequently, urinary MDA is overestimated using TBA by a factor of almost 10 compared to another derivatization agent 2,4 dinitrophenylhyldrazine (DNPH) [40]. DNPH requires lower temperatures (37 °C) and slightly acidic pH, which should lead to results that are more specific. The DNPH derivatization method was used only in 4% of cases of all selected studies. Therefore, we could not compare the results from different derivatization methods. We recommend that future studies use specific derivatizing agents such as DNPH.

We found that authors reported urinary MDA concentrations either as free, conjugated or total, but this was rarely specified in the studies. This can contribute to the observed heterogeneity as well. If this specification was reported, then authors more often reported free MDA over total MDA as a biomarker of oxidative stress. Total urinary MDA would better estimate the total body burden, but the ease and the convenience of only quantifying free MDA, skipping the hydrolysis step will not just shorten the analysis time but also the cost. Urinary concentrations of free MDA and total MDA are reported to be significantly correlated [13]; thus, the use of free or total MDA may reflect similar oxidative stress levels.

One limitation of our study is that we did not systematically record results from studies using ELISA. Chemical analytical methods and colorimetric assays, and ELISA have not been compared for urinary MDA analysis. Analytical methods have been shown to provide different values for other oxidative stress biomarkers such as 8-isoprostane [27]. We cannot compare urinary MDA concentrations between ELISA and chemical analytical methods, as we did not include ELISA in our meta-analysis.

Another limitation is that we were unable to read articles that were not in English or French. We, therefore, do not know how many articles with relevant information we missed.

4.3. Recommendations

We believe that the between-studies heterogeneity can be reduced and controlled if future studies will address the effects of additional factors of interest, such as biological mechanisms and physiological variables. Studies should clearly state whether they have quantified total MDA or free MDA in urine. Harmonizing the unit metrics (mg/g creatinine) for reporting urinary MDA would also be helpful. The variation in urine flow rate, body mass, and workload could certainly affect urinary MDA values, as is common for other effect biomarkers. For this purpose, urinary creatinine should be used to normalize MDA concentrations [21,79]. Creatinine normalization is appropriate whenever spot urine samples are collected, while 24 h urine samples do not need adjustments [49]. Reporting efforts also include the descriptive statistics provided (GM and GSD): for this, we suggest reporting the median and the first and third quartile for a better interpretation of GSDs. In terms of analytical methods, we recommend using the DNPH derivatization agent over TBA and reporting both storage time and temperature.

5. Conclusions

Our systematic review and meta-analysis indicate a general population concentration range of 0.07–0.12 mg/g creatinine for GMs of urinary MDA in healthy adults. These GMs increase with the mean age of the study populations and cannot be used for comparison with individual results. There were several challenges encountered when analyzing the published data. The lack of homogeneity in data collection and storage conditions likely affected our meta-analysis. Consequently, the values determined in this study should be considered preliminary as they are based on moderate to low-quality studies. Further research efforts regarding the use of urinary MDA as a biomarker for oxidative stress need to address the following: standardize the reporting, understand the ideal urine collection time, elucidate optimal sample storage temperature, and the best derivatization agent as well as harmonize the chemical analytical methods.

Supplementary Materials: The following supporting information can be downloaded at: https://www.mdpi.com/article/10.3390/toxics10040160/s1, Table S1: Quality assessment Criteria; Table S2: Quality appraisal.

Author Contributions: The manuscript was written by A.T., P.W., and N.B.H. and revised with all authors. N.B.H. and I.G.C. conceptualized and designed the study. M.G. and M.H. initiated the research protocol. C.C. extracted and screened the studies. A.T. read and organized all the studies and wrote the initial manuscript. J.-J.S. and M.H. contributed to the technical discussions and V.T. for medical discussions. E.B. served as an advisor for the manuscript drafts. P.W. performed all statistical analyses. N.B.H. contributed to the interpretation of the results. All authors have read and agreed to the published version of the manuscript.

Funding: This study was conducted within the framework of the EU Life Project"NanoExplore" (grant N°LIFE17 ENV/GR/000285) and the Swiss National Science Foundation (grant N IZCOZ0_177067) for the Respiratory disease Occupational Biomonitoring Collaborative Project (ROBoCop).

Institutional Review Board Statement: Not applicable.

Informed Consent Statement: Not applicable.

Data Availability Statement: Not applicable.

Conflicts of Interest: The authors declare no conflict of interest. The funders had no role in the design of the study, in the collection, analyses, or interpretation of data, in the writing of the manuscript, or in the decision to publish the results.

Abbreviations

MDA	Malondialdehyde
TBARS	Thiobarbituric acid reactive substances
ROS	Reactive oxygen species
GM	Geometric mean
GSD	Geometric standard deviation
BMI	Body mass index
SD	Standard deviation
SEM	Standard error of the mean
CV	Coefficient of variation
IQR	Interquartile range

References

1. Lushchak, V.I. Free radicals, reactive oxygen species, oxidative stress and its classification. *Chem. Biol. Interact.* **2014**, *224*, 164–175. [CrossRef]
2. Del Rio, D.; Stewart, A.J.; Pellegrini, N. A review of recent studies on malondialdehyde as toxic molecule and biological marker of oxidative stress. *Nutr. Metab. Cardiovasc. Dis.* **2005**, *15*, 316–328. [CrossRef]
3. Repetto, M.; Semprine, J.; Boveris, A. Lipid Peroxidation: Chemical Mechanism, Biological Implications and Analytical Determination. In *Lipid Peroxidation*; Catala, A., Ed.; IntechOpen: London, UK, 2012.
4. Ayala, A.; Muñoz, M.F.; Argüelles, S. Lipid peroxidation: Production, metabolism, and signaling mechanisms of malondialdehyde and 4-hydroxy-2-nonenal. *Oxidative Med. Cell Longev.* **2014**, *2014*, 360438. [CrossRef]
5. Jové, M.; Mota-Martorell, N.; Pradas, I.; Martín-Gari, M.; Ayala, V.; Pamplona, R. The Advanced Lipoxidation End-Product Malondialdehyde-Lysine in Aging and Longevity. *Antioxidants* **2020**, *9*, 1132. [CrossRef]
6. Tsikas, D. Assessment of lipid peroxidation by measuring malondialdehyde (MDA) and relatives in biological samples: Analytical and biological challenges. *Anal. Biochem.* **2017**, *524*, 13–30. [CrossRef]
7. Wang, J.P.; Maddalena, R.; Zheng, B.; Zai, C.; Liu, F.; Ng, J.C. Arsenicosis status and urinary malondialdehyde (MDA) in people exposed to arsenic contaminated-coal in China. *Environ. Int.* **2009**, *35*, 502–506. [CrossRef]
8. García-Ruiz, I.; de la Torre, P.; Díaz, T.; Esteban, E.; Fernández, I.; Muñoz-Yagüe, T.; Solís-Herruzo, J.A. Sp1 and Sp3 Transcription Factors Mediate Malondialdehyde-induced Collagen α1(I) Gene Expression in Cultured Hepatic Stellate Cells. *J. Biol. Chem.* **2002**, *277*, 30551–30558. [CrossRef]
9. Onyango, A.N.; Baba, N. New hypotheses on the pathways of formation of malondialdehyde and isofurans. *Free Radic. Biol. Med.* **2010**, *49*, 1594–1600. [CrossRef]
10. Ito, K.; Watanabe, C.; Nakamura, A.; Oikawa-Tada, S.; Murata, M. Reduced Coenzyme Q10 Decreases Urinary 8-Oxo-7,8-Dihydro-2′-Deoxyguanosine Concentrations in Healthy Young Female Subjects. *J. Med. Food* **2015**, *18*, 835–840. [CrossRef]
11. Hardt, U.; Larsson, A.; Gunnarsson, I.; Clancy, R.M.; Petri, M.; Buyon, J.P.; Silverman, G.J.; Svenungsson, E.; Grönwall, C. Autoimmune reactivity to malondialdehyde adducts in systemic lupus erythematosus is associated with disease activity and nephritis. *Arthritis Res. Ther.* **2018**, *20*, 36. [CrossRef]
12. Duryee, M.J.; Klassen, L.W.; Schaffert, C.S.; Tuma, D.J.; Hunter, C.D.; Garvin, R.P.; Anderson, D.R.; Thiele, G.M. Malondialdehyde–acetaldehyde adduct is the dominant epitope after MDA modification of proteins in atherosclerosis. *Free Radic. Biol. Med.* **2010**, *49*, 1480–1486. [CrossRef]
13. Cui, X.; Gong, J.; Han, H.; He, L.; Teng, Y.; Tetley, T.; Sinharay, R.; Chung, K.F.; Islam, T.; Gilliland, F.; et al. Relationship between free and total malondialdehyde, a well-established marker of oxidative stress, in various types of human biospecimens. *J. Thorac. Dis.* **2018**, *10*, 3088–3197. [CrossRef]
14. Cherubini, A.; Ruggiero, C.; Polidori, M.C.; Mecocci, P. Potential markers of oxidative stress in stroke. *Free Radic. Biol. Med.* **2005**, *39*, 841–852. [CrossRef]
15. Paliogiannis, P.; Fois, A.G.; Sotgia, S.; Mangoni, A.A.; Zinellu, E.; Pirina, P.; Carru, C.; Zinellu, A. Circulating malondialdehyde concentrations in patients with stable chronic obstructive pulmonary disease: A systematic review and meta-analysis. *Biomark. Med.* **2018**, *12*, 771–781. [CrossRef]

16. Kurutas, E.B.; Gumusalan, Y.; Cetinkaya, A.; Dogan, E. Evaluation of method performance for oxidative stress biomarkers in urine and biological variations in urine of patients with type 2 diabetes mellitus and diabetic nephropathy. *Biol. Proced. Online* **2015**, *17*, 3. [CrossRef]
17. Cvetković, T.; Mitić, B.; Lazarević, G.; Vlahović, P.; Antić, S.; Stefanović, V. Oxidative stress parameters as possible urine markers in patients with diabetic nephropathy. *J. Diabetes Complicat.* **2009**, *23*, 337–342. [CrossRef]
18. Gong, J.; Zhu, T.; Kipen, H.; Wang, G.; Hu, M.; Ohman-Strickland, P.; Lu, S.-E.; Zhang, L.; Wang, Y.; Zhu, P.; et al. Malondialdehyde in exhaled breath condensate and urine as a biomarker of air pollution induced oxidative stress. *J. Expo. Sci. Environ. Epidemiol.* **2013**, *23*, 322–327. [CrossRef]
19. Mendonça, R.; Gning, O.; Di Cesaré, C.; Lachat, L.; Bennett, N.C.; Helfenstein, F.; Glauser, G. Sensitive and selective quantification of free and total malondialdehyde in plasma using UHPLC-HRMS. *J. Lipid Res.* **2017**, *58*, 1924–1931. [CrossRef]
20. Hemmendinger, M.; Sauvain, J.-J.; Hopf, N.B.; Wild, P.; Suárez, G.; Canu, I.G. Method Validation and Characterization of the Associated Uncertainty for Malondialdehyde Quantification in Exhaled Breath Condensate. *Antioxidants* **2021**, *10*, 1661. [CrossRef]
21. Martinez-Moral, M.-P.; Kannan, K. How stable is oxidative stress level? An observational study of intra- and inter-individual variability in urinary oxidative stress biomarkers of DNA, proteins, and lipids in healthy individuals. *Environ. Int.* **2019**, *123*, 382–389. [CrossRef]
22. Yuan, Y.; Hunt, R.H. Systematic Reviews: The Good, the Bad and the Ugly. *Am. J. Gastroenterol.* **2009**, *104*, 1086–1092. [CrossRef]
23. Liberati, A.; Altman, D.G.; Tetzlaff, J.; Mulrow, C.; Gøtzsche, P.C.; Ioannidis, J.P.A.; Clarke, M.; Devereaux, P.J.; Kleijnen, J.; Moher, D. The PRISMA Statement for Reporting Systematic Reviews and Meta-Analyses of Studies That Evaluate Health Care Interventions: Explanation and Elaboration. *PLoS Med.* **2009**, *6*, e1000100. [CrossRef]
24. Ouzzani, M.; Hammady, H.; Fedorowicz, Z.; Elmagarmid, A. Rayyan—A web and mobile app for systematic reviews. *Syst. Rev.* **2016**, *5*, 210. [CrossRef]
25. Siddaway, A.P.; Wood, A.M.; Hedges, L.V. How to Do a Systematic Review: A Best Practice Guide for Conducting and Reporting Narrative Reviews, Meta-Analyses, and Meta-Syntheses. *Annu. Rev. Psychol.* **2019**, *70*, 747–770. [CrossRef]
26. Hemmendinger, M.; Wild, P.; Shoman, Y.; Graille, M.; Bergamaschi, E.; Hopf, N.; Canu, I.G. Reference ranges of oxidative stress biomarkers selected for non-invasive biological surveillance of nanotechnology workers: Study protocol and meta-analysis results for 8-OHdG in exhaled breath condensate. *Toxicol. Lett.* **2020**, *327*, 41–47. [CrossRef]
27. Graille, M.; Wild, P.; Sauvain, J.-J.; Hemmendinger, M.; Canu, I.G.; Hopf, N. Urinary 8-isoprostane as a biomarker for oxidative stress. A systematic review and meta-analysis. *Toxicol. Lett.* **2020**, *328*, 19–27. [CrossRef]
28. Guyatt, G.H.; Oxman, A.D.; Vist, G.E.; Kunz, R.; Falck-Ytter, Y.; Alonso-Coello, P.; Schünemann, H.J. GRADE: An emerging consensus on rating quality of evidence and strength of recommendations. *BMJ* **2008**, *336*, 924–926. [CrossRef]
29. Higgins, J.P.T.; Thompson, S.G. Quantifying heterogeneity in a meta-analysis. *Stat. Med.* **2002**, *21*, 1539–1558. [CrossRef]
30. Adetona, O.; Zhang, J.; Hall, D.B.; Wang, J.-S.; Vena, J.E.; Naeher, L.P. Occupational exposure to woodsmoke and oxidative stress in wildland firefighters. *Sci. Total Environ.* **2013**, *449*, 269–275. [CrossRef]
31. He, L.; Cui, X.; Xia, Q.; Li, F.; Mo, J.; Gong, J.; Zhang, Y.; Zhang, J. Effects of personal air pollutant exposure on oxidative stress: Potential confounding by natural variation in melatonin levels. *Int. J. Hyg. Environ. Health* **2020**, *223*, 116–123. [CrossRef]
32. Ke, Y.; Huang, L.; Xia, J.; Xu, X.; Liu, H.; Li, Y.R. Comparative study of oxidative stress biomarkers in urine of cooks exposed to three types of cooking-related particles. *Toxicol. Lett.* **2016**, *255*, 36–42. [CrossRef]
33. Lin, Y.; Qiu, X.; Yu, N.; Yang, Q.; Araujo, J.A.; Zhu, Y. Urinary Metabolites of Polycyclic Aromatic Hydrocarbons and the Association with Lipid Peroxidation: A Biomarker-Based Study between Los Angeles and Beijing. *Environ. Sci. Technol.* **2016**, *50*, 3738–3745. [CrossRef]
34. Sakaguchi, C.; Miura, N.; Ohara, H.; Nagata, Y. Effects of reduced exposure to cigarette smoking on changes in biomarkers of potential harm in adult smokers: Results of combined analysis of two clinical studies. *Biomarkers* **2019**, *24*, 457–468. [CrossRef]
35. Weinbrenner, T.; Fitó, M.; de la Torre, R.; Sáez, G.; Rijken, P.; Tormos, C.; Coolen, S.; Albaladejo, M.F.; Abanades, S.; Schroder, H.; et al. Olive Oils High in Phenolic Compounds Modulate Oxidative/Antioxidative Status in Men. *J. Nutr.* **2004**, *134*, 2314–2321. [CrossRef]
36. Al-Saleh, I.; Abduljabbar, M.; Al-Rouqi, R.; Elkhatib, R.; Alshabbaheen, A.; Shinwari, N. Mercury (Hg) Exposure in Breast-Fed Infants and Their Mothers and the Evidence of Oxidative Stress. *Biol. Trace Elem. Res.* **2013**, *153*, 145–154. [CrossRef]
37. Chen, J.-L.; Huang, Y.-J.; Pan, C.-H.; Hu, C.-W.; Chao, M.-R. Determination of urinary malondialdehyde by isotope dilution LC-MS/MS with automated solid-phase extraction: A cautionary note on derivatization optimization. *Free Radic. Biol. Med.* **2011**, *51*, 1823–1829. [CrossRef]
38. Huang, Y.-L. Lipid Peroxidation in Workers Exposed to Hexavalent Chromium. *J. Toxicol. Environ. Health Part A* **1999**, *56*, 235–247. [CrossRef]
39. Jacob, R.A.; Aiello, G.M.; Stephensen, C.B.; Blumberg, J.B.; Milbury, P.E.; Wallock, L.M.; Ames, B.N. Moderate Antioxidant Supplementation Has No Effect on Biomarkers of Oxidant Damage in Healthy Men with Low Fruit and Vegetable Intakes. *J. Nutr.* **2003**, *133*, 740–743. [CrossRef]
40. Korchazhkina, O.; Exley, C.; Spencer, S.A. Measurement by reversed-phase high-performance liquid chromatography of malondialdehyde in normal human urine following derivatisation with 2,4-dinitrophenylhydrazine. *J. Chromatogr. B* **2003**, *794*, 353–362. [CrossRef]

41. Łuczaj, W.; Moniuszko, A.; Rusak, M.; Pancewicz, S.; Zajkowska, J.; Skrzydlewska, E. Lipid peroxidation products as potential bioindicators of Lyme arthritis. *Eur. J. Clin. Microbiol.* **2010**, *30*, 415–422. [CrossRef]
42. Na, H.-K.; Kim, M.; Chang, S.-S.; Kim, S.-Y.; Park, J.Y.; Chung, M.W.; Yang, M. Tobacco smoking-response genes in blood and buccal cells. *Toxicol. Lett.* **2015**, *232*, 429–437. [CrossRef]
43. Pan, C.-H.; Jeng, H.A.; Lai, C.-H. Biomarkers of oxidative stress in electroplating workers exposed to hexavalent chromium. *J. Expo. Sci. Environ. Epidemiol.* **2018**, *28*, 76–83. [CrossRef]
44. Wilber, R.L.; Holm, P.L.; Morris, D.M.; Dallam, G.M.; Subudhi, A.W.; Murray, D.M.; Callan, S.D. Effect of FIO2 on Oxidative Stress during Interval Training at Moderate Altitude. *Med. Sci. Sports Exerc.* **2004**, *36*, 1888–1894. [CrossRef]
45. Eom, S.-Y.; Seo, M.-N.; Lee, Y.-S.; Park, K.-S.; Hong, Y.-S.; Sohn, S.-J.; Kim, Y.-D.; Choi, B.-S.; Lim, J.-A.; Kwon, H.-J.; et al. Low-Level Environmental Cadmium Exposure Induces Kidney Tubule Damage in the General Population of Korean Adults. *Arch. Environ. Contam. Toxicol.* **2017**, *73*, 401–409. [CrossRef]
46. Łuczaj, W.; Moniuszko, A.; Jarocka-Karpowicz, I.; Pancewicz, S.; Andrisic, L.; Zarkovic, N.; Skrzydlewska, E. Tick-borne encephalitis–lipid peroxidation and its consequences. *Scand. J. Clin. Lab. Investig.* **2016**, *76*, 1–9. [CrossRef]
47. Moniuszko-Malinowska, A.; Łuczaj, W.; Jarocka-Karpowicz, I.; Pancewicz, S.; Zajkowska, J.; Andrisic, L.; Zarkovic, N.; Skrzydlewska, E. Lipid peroxidation in the pathogenesis of neuroborreliosis. *Free Radic. Biol. Med.* **2016**, *96*, 255–263. [CrossRef]
48. Thompson, H.J.; Heimendinger, J.; Haegele, A.; Sedlacek, S.M.; Gillette, C.; O'Neill, C.; Wolfe, P.; Conry, C. Effect of increased vegetable and fruit consumption on markers of oxidative cellular damage. *Carcinogenesis* **1999**, *20*, 2261–2266. [CrossRef]
49. Whiting, P.; Kalansooriya, A.; Holbrook, I.; Haddad, F.; Jennings, P. The relationship between chronic glycaemic control and oxidative stress in type 2 diabetes mellitus. *Br. J. Biomed. Sci.* **2008**, *65*, 71–74. [CrossRef]
50. Yang, Y.J.; Hong, Y.-C.; Oh, S.-Y.; Park, M.-S.; Kim, H.; Leem, J.-H.; Ha, E.-H. Bisphenol A exposure is associated with oxidative stress and inflammation in postmenopausal women. *Environ. Res.* **2009**, *109*, 797–801. [CrossRef]
51. Yu, N.; Shu, S.; Lin, Y.; She, J.; Ip, H.S.S.; Qiu, X.; Zhu, Y. High efficiency cabin air filter in vehicles reduces drivers' roadway particulate matter exposures and associated lipid peroxidation. *PLoS ONE* **2017**, *12*, e0188498. [CrossRef]
52. Adetona, A.; Martin, W.K.; Warren, S.H.; Hanley, N.M.; Adetona, O.; Zhang, J.; Simpson, C.; Paulsen, M.H.; Rathbun, S.L.; Wang, J.-S.; et al. Urinary mutagenicity and other biomarkers of occupational smoke exposure of wildland firefighters and oxidative stress. *Inhal. Toxicol.* **2019**, *31*, 73–87. [CrossRef] [PubMed]
53. Agarwal, R. Proinflammatory effects of oxidative stress in chronic kidney disease: Role of additional angiotensin II blockade. *Am. J. Physiol. Renal Physiol.* **2003**, *284*, F863–F869. [CrossRef] [PubMed]
54. Agarwal, R.; Chase, S.D. Rapid, fluorimetric–liquid chromatographic determination of malondialdehyde in biological samples. *J. Chromatogr. B* **2002**, *775*, 121–126. [CrossRef]
55. Boyle, S.P.; Dobson, V.L.; Duthie, S.; Hinselwood, D.C.; Kyle, J.; Collins, A. Bioavailability and efficiency of rutin as an antioxidant: A human supplementation study. *Eur. J. Clin. Nutr.* **2000**, *54*, 774–782. [CrossRef] [PubMed]
56. Boyle, S.P.; Dobson, V.L.; Duthie, S.; Kyle, J.A.M.; Collins, A. Absorption and DNA protective effects of flavonoid glycosides from an onion meal. *Eur. J. Nutr.* **2000**, *39*, 213–223. [CrossRef]
57. Domijan, A.M.; Miletić-Medved, M.; Peraica, M.; Loft, S. Malondialdehyde and 8-oxo-7.8-dihydro-2′deoxyguanosine in the urine of residents from Balkan endemic nephropathy area in Croatia—A pilot study. *Coll. Antropol.* **2013**, *37*, 1195–1198.
58. Duthie, S.J.; Jenkinson, A.M.E.; Crozier, A.; Mullen, W.; Pirie, L.; Kyle, J.; Yap, L.S.; Christen, P.; Duthie, G.G. The effects of cranberry juice consumption on antioxidant status and biomarkers relating to heart disease and cancer in healthy human volunteers. *Eur. J. Nutr.* **2005**, *45*, 113–122. [CrossRef]
59. Huang, M.; Choi, S.-J.; Kim, D.-W.; Kim, N.-Y.; Park, C.-H.; Yu, S.-D.; Kim, D.-S.; Park, K.-S.; Song, J.-S.; Kim, H.; et al. Risk Assessment of Low-Level Cadmium and Arsenic on the Kidney. *J. Toxicol. Environ. Health Part A* **2009**, *72*, 1493–1498. [CrossRef]
60. Kim, J.-A.; Noh, S.R.; Cheong, H.-K.; Ha, M.; Eom, S.-Y.; Kim, H.; Park, M.-S.; Chu, Y.; Lee, S.-H.; Choi, K. Urinary oxidative stress biomarkers among local residents measured 6 years after the Hebei Spirit oil spill. *Sci. Total Environ.* **2017**, *580*, 946–952. [CrossRef]
61. Lee, K.-H.; Shu, X.-O.; Gao, Y.-T.; Ji, B.-T.; Yang, G.; Blair, A.; Rothman, N.; Zheng, W.; Chow, W.-H.; Kang, D. Breast Cancer and Urinary Biomarkers of Polycyclic Aromatic Hydrocarbon and Oxidative Stress in the Shanghai Women's Health Study. *Cancer Epidemiol. Biomark. Prev.* **2010**, *19*, 877–883. [CrossRef]
62. Shih, Y.-M.; Cooke, M.; Pan, C.-H.; Chao, M.-R.; Hu, C.-W. Clinical relevance of guanine-derived urinary biomarkers of oxidative stress, determined by LC-MS/MS. *Redox Biol.* **2018**, *20*, 556–565. [CrossRef] [PubMed]
63. Łuczaj, W.; Gindzienska-Sieskiewicz, E.; Jarocka-Karpowicz, I.; Andrisic, L.; Sierakowski, S.; Zarkovic, N.; Waeg, G.; Skrzydlewska, E. The onset of lipid peroxidation in rheumatoid arthritis: Consequences and monitoring. *Free Radic. Res.* **2016**, *50*, 304–313. [CrossRef] [PubMed]
64. Bokov, A.; Chaudhuri, A.; Richardson, A. The role of oxidative damage and stress in aging. *Mech. Ageing Dev.* **2004**, *125*, 811–826. [CrossRef] [PubMed]
65. Egea, J.; Fabregat, I.; Frapart, Y.M.; Ghezzi, P.; Görlach, A.; Kietzmann, T.; Kubaichuk, K.; Knaus, U.G.; Lopez, M.G.; Olaso-Gonzalez, G.; et al. European contribution to the study of ROS: A summary of the findings and prospects for the future from the COST action BM1203 (EU-ROS). *Redox Biol.* **2017**, *13*, 94–162. [CrossRef]
66. Ruano, D. Proteostasis Dysfunction in Aged Mammalian Cells. The Stressful Role of Inflammation. *Front. Mol. Biosci.* **2021**, *8*, 658742. [CrossRef]

67. Bergin, P.; Leggett, A.; Cardwell, C.R.; Woodside, J.V.; Thakkinstian, A.; Maxwell, A.P.; McKay, G.J. The effects of vitamin E supplementation on malondialdehyde as a biomarker of oxidative stress in haemodialysis patients: A systematic review and meta-analysis. *BMC Nephrol.* **2021**, *22*, 126. [CrossRef]
68. Huang, H.-Y.; Appel, L.J.; Croft, K.; Miller, E.R.; Mori, T.A.; Puddey, I.B. Effects of vitamin C and vitamin E on in vivo lipid peroxidation: Results of a randomized controlled trial. *Am. J. Clin. Nutr.* **2002**, *76*, 549–555. [CrossRef]
69. Hu, W.; Wang, Y.; Wang, T.; Ji, Q.; Jia, Q.; Meng, T.; Ma, S.; Zhang, Z.; Li, Y.; Chen, R.; et al. Ambient particulate matter compositions and increased oxidative stress: Exposure-response analysis among high-level exposed population. *Environ. Int.* **2021**, *147*, 106341. [CrossRef]
70. Jia, X.-J.; Liu, L.-X.; Tian, Y.-M.; Wang, R.; Lu, Q. The correlation between oxidative stress level and intra-abdominal fat in obese males. *Medicine* **2019**, *98*, e14469. [CrossRef]
71. Zhu, M.; Liu, Z.; Guo, Y.; Sultana, M.S.; Wu, K.; Lang, X.; Lv, Q.; Huang, X.; Yi, Z.; Li, Z. Sex difference in the interrelationship between TNF-α and oxidative stress status in first-episode drug-naïve schizophrenia. *J. Neuroinflamm.* **2021**, *18*, 202. [CrossRef]
72. Ide, T.; Tsutsui, H.; Ohashi, N.; Hayashidani, S.; Suematsu, N.; Tsuchihashi, M.; Tamai, H.; Takeshita, A. Greater Oxidative Stress in Healthy Young Men Compared With Premenopausal Women. *Arter. Thromb. Vasc. Biol.* **2002**, *22*, 438–442. [CrossRef]
73. Lykkesfeldt, J.; Viscovich, M.; Poulsen, H.E. Plasma malondialdehyde is induced by smoking: A study with balanced antioxidant profiles. *Br. J. Nutr.* **2004**, *92*, 203–206. [CrossRef]
74. Lykkesfeldt, J. Malondialdehyde as biomarker of oxidative damage to lipids caused by smoking. *Clin. Chim. Acta* **2007**, *380*, 50–58. [CrossRef]
75. Li, N.; Jia, X.; Chen, C.-Y.O.; Blumberg, J.B.; Song, Y.; Zhang, W.; Zhang, X.; Ma, G.; Chen, J. Almond Consumption Reduces Oxidative DNA Damage and Lipid Peroxidation in Male Smokers. *J. Nutr.* **2007**, *137*, 2717–2722. [CrossRef]
76. Megson, I.L.; Haw, S.J.; Newby, D.E.; Pell, J.P. Association between Exposure to Environmental Tobacco Smoke and Biomarkers of Oxidative Stress Among Patients Hospitalised with Acute Myocardial Infarction. *PLoS ONE* **2013**, *8*, e81209. [CrossRef]
77. Campos, C.; Guzmán, R.; López-Fernández, E.; Casado, Á. Urinary biomarkers of oxidative/nitrosative stress in healthy smokers. *Inhal. Toxicol.* **2011**, *23*, 148–156. [CrossRef]
78. Kanabrocki, E.L.; Murray, D.; Hermida, R.C.; Scott, G.S.; Bremner, W.F.; Ryan, M.D.; Ayala, D.E.; Third, J.L.; Shirazi, P.; Nemchausky, B.A.; et al. Circadian variation in oxidative stress markers in healthy and type II diabetic men. *Chronobiol. Int.* **2002**, *19*, 423–439. [CrossRef]
79. Stiegel, M.A.; Pleil, J.D.; Sobus, J.R.; Angrish, M.M.; Morgan, M.K. Kidney injury biomarkers and urinary creatinine variability in nominally healthy adults. *Biomarkers* **2015**, *20*, 436–452. [CrossRef]

MDPI
St. Alban-Anlage 66
4052 Basel
Switzerland
Tel. +41 61 683 77 34
Fax +41 61 302 89 18
www.mdpi.com

Toxics Editorial Office
E-mail: toxics@mdpi.com
www.mdpi.com/journal/toxics

www.ingramcontent.com/pod-product-compliance
Lightning Source LLC
Chambersburg PA
CBHW061824210926
51850CB00006B/795

* 9 7 8 3 0 3 6 5 7 0 0 4 4 *